# POWER SEMICONDUCTOR DEVICES

# POWER SEMICONDUCTOR DEVICES
## THEORY AND APPLICATIONS

**Vítězslav Benda**
*Czech Technical University, Prague, Czech Republic*
**John Gowar**
**Duncan A. Grant**
*University of Bristol, UK*

JOHN WILEY & SONS
Chichester · New York · Weinheim · Brisbane · Singapore · Toronto

A revised and updated translation of the original Czech edition published by
Vydavatelství ČVUT, Zikova 4, 16635 Praha 6, Czech Republic, under the
title 'Výkonové polovodičové součástky a integrované struktury'
© 1994 Vítězslav Benda.

Copyright © 1999 by John Wiley & Sons Ltd,
Baffins Lane, Chichester,
West Sussex PO19 1UD, England

National  01243 779777
International  (+44) 1243 779777

e-mail (for orders and customer service enquiries): cs-books@wiley.co.uk
Visit our Home Page on http://www.wiley.co.uk or http://www.wiley.com

Reprinted July 1999

All Rights Reserved. No part of this publication may be reproduced, stored in a
retrieval system, or transmitted, in any form or by any means, electronic,
mechanical, photocopying, recording, scanning or otherwise, except under the
terms of the Copyright, Designs and Patents Act 1988 or under the terms of a
licence issued by the Copyright Licensing Agency, 90 Tottenham Court Road,
London W1P 9HE, UK, without the permission in writing of the Publisher.

V. Benda, J. Gowar and D. A. Grant have asserted their rights under the Copyright,
Designs and Patents Act 1988 to be identified as the authors of this work.

*Other Wiley Editorial Offices*

John Wiley & Sons, Inc., 605 Third Avenue,
New York, NY 10158-0012, USA

Weinheim • Brisbane • Singapore • Toronto

*Library of Congress Cataloging-in-Publication Data*

Benda, Vítězslav.
 [Výkonové polovodičové součástky a integrované struktury.
 English]
 Power semiconductor devices: theory and applications/Vítězslav
Benda, John Gowar, Duncan A. Grant.
  p.   cm.
 Includes bibliographical references and index.
 ISBN 0-471-97644-X
 1. Power semiconductors.  I. Gowar, John, 1938– .  II. Grant,
Duncan A. (Duncan Andrew)  III. Title.
TK7871.85.B38313  1999                             98-22030
621.3815 — dc21                                     CIP

*British Library Cataloguing in Publication Data*

A catalogue record for this book is available from the British Library

ISBN 0 471 97644 X

Typeset in 10/12pt Times from the authors' disks by Dobbie Typesetting Limited.
Printed and bound in Great Britain by Bookcraft (Bath) Ltd.
This book is printed on acid-free paper responsibly manufactured from sustainable
forestry in which at least two trees are planted for each one used for paper production.

# CONTENTS

**Preface** xi

**1 Properties of Semiconductors** 1
   1.1   Charge Carriers and Semiconductor Band Structure    1
   1.2   The Concentration of Free Carriers    3
   1.3   The Conductivity of Semiconductors    9
       1.3.1   Factors Determining Conductivity    9
       1.3.2   Factors Determining the Carrier Mobilities in Silicon    11
   1.4   The Generation and Recombination of Excess Charge Carriers    16
       1.4.1   Equilibrium and Non-Equilibrium Conditions    16
       1.4.2   Recombination via Local Trapping Centres    18
       1.4.3   Other Recombination Processes    24
   1.5   Diffusion and Drift of Charge Carriers    26
   1.6   Effects of Non-Uniform Doping    29
   Summary    32
   References    32

**2 Elementary Semiconductor Structures** 33
   2.1   The p–n Junction and its Basic Properties    33
       2.1.1   Current–Voltage Characteristics    33
       2.1.2   Reverse Bias    41
       2.1.3   Electrical Breakdown    45
       2.1.4   Thermal Breakdown and Thermal Runaway    49
       2.1.5   The Influence of Illumination on p–n Junction Characteristics    51
       2.1.6   Transient Behaviour of the p–n Junction    53
   2.2   $n$–$n^+$ and $p$–$p^+$ Junctions    57
   2.3   Surface Effects and MOS Structures    59
   2.4   Metal–Semiconductor Contacts    62
       2.4.1   Rectifying (Schottky) Contacts    62
       2.4.2   Ohmic Contacts    65
   Summary    66
   References    67

**3 Devices, Fabrication and Modelling** 69
   3.1   The Range of Power Semiconductor Devices    69
       3.1.1   Power Diodes    72

|  |  |  |  |
|---|---|---|---|
|  | 3.1.2 | Conventional Power Bipolar Transistors | 75 |
|  | 3.1.3 | Thyristor Structures | 77 |
|  | 3.1.4 | Junction Field-Effect (Static Induction) Devices | 80 |
|  | 3.1.5 | Power MOS Structures 1: The Power MOSFET | 83 |
|  | 3.1.6 | Power MOS Structures 2: The Insulated Gate Bipolar Transistor (IGBT) | 85 |
|  | 3.1.7 | Power MOS Structures 3: The MOS-Controlled Thyristor (MCT) | 87 |
|  | 3.1.8 | A Summary of Device Characteristics | 88 |
| 3.2 | Fabrication Processes | | 93 |
|  | 3.2.1 | The Preparation of High Purity Single-Crystal Silicon | 94 |
|  | 3.2.2 | Wafer Preparation | 97 |
|  | 3.2.3 | Epitaxial Growth | 99 |
|  | 3.2.4 | Thermal Oxidation | 100 |
|  | 3.2.5 | Photolithography | 101 |
|  | 3.2.6 | Etching Processes | 103 |
|  | 3.2.7 | The Introduction and Redistribution of Impurities 1: Diffusion | 104 |
|  | 3.2.8 | The Introduction and Redistribution of Impurities 2: Ion Implantation | 109 |
|  | 3.2.9 | Chemical Vapour Deposition Techniques | 110 |
|  | 3.2.10 | The Preparation of Contacts | 112 |
| 3.3 | The Control of Carrier Lifetime | | 114 |
|  | 3.3.1 | Techniques for Obtaining Long Carrier Lifetimes | 114 |
|  | 3.3.2 | Techniques for Reducing Carrier Lifetimes | 115 |
| 3.4 | High Voltage Structures | | 118 |
| 3.5 | Computer Modelling and Simulation Techniques | | 124 |
| Summary | | | 125 |
| References | | | 126 |

## 4  Power Semiconductor Device Applications — 127

| | | |
|---|---|---|
| 4.1 | Uncontrolled Rectification | 127 |
| 4.2 | Controlled Rectification | 129 |
| 4.3 | Conversion of ac to ac | 130 |
| 4.4 | Inverters | 132 |
| 4.5 | Non-Isolated dc to dc Converters | 138 |
| 4.6 | Transformer-Isolated dc to dc Converters | 140 |
| 4.7 | Power Factor Correction | 142 |
| 4.8 | Resonant Circuits | 143 |
| Summary | | 145 |
| References | | 145 |

## 5  Power Diodes — 147

|  |  |  |  |
|---|---|---|---|
| 5.1 | The Forward-Biased Diode | | 147 |
| 5.2 | Reverse Characteristics of Power Diodes | | 157 |
| 5.3 | Transient Processes in Power Diodes | | 160 |
|  | 5.3.1 | The Transition from Reverse Bias to Forward Bias | 160 |
|  | 5.3.2 | The Transition from Forward Bias to Reverse Bias | 162 |
| 5.4 | Power Schottky Diodes | | 171 |
| 5.5 | Silicon Power Diode Applications | | 174 |
|  | 5.5.1 | Power Frequency Applications | 174 |
|  | 5.5.2 | Fast-Switching Applications | 175 |

|   |   | Summary | 175 |
|---|---|---|---|
|   |   | References | 176 |

## 6 Bipolar Junction Power Transistors — 177

- 6.1 Basic Characteristics of the Bipolar Junction Transistor Structure — 177
- 6.2 Basic Characteristics of Power Transistors — 183
  - 6.2.1 High Voltage Considerations — 183
  - 6.2.2 Transistor Operating Regions — 185
  - 6.2.3 High Current Considerations: Collector Conductivity Modulation and Base Widening Effects — 186
  - 6.2.4 Other High Current Density and Temperature Effects — 192
- 6.3 The Dynamic Behaviour of Power Transistors — 195
  - 6.3.1 Turn-On — 195
  - 6.3.2 Turn-Off — 198
- 6.4 Safe Operating Area — 202
- 6.5 The Power Darlington Configuration — 205
- 6.6 Power Transistor Applications — 207

Summary — 208
References — 208

## 7 Thyristors: Basic Operating Principles — 209

- 7.1 Steady-State Operation — 209
  - 7.1.1 The Reverse Blocking State — 209
  - 7.1.2 The Forward Blocking State — 211
  - 7.1.3 Surface Profiles for High Breakdown Voltages — 212
  - 7.1.4 The Forward Conducting State — 214
- 7.2 The Two-Transistor Model for Thyristor Switching — 216
- 7.3 Transient Processes during Turn-On — 220
  - 7.3.1 Gate Turn-On — 222
  - 7.3.2 Critical $di/dt$ — 226
  - 7.3.3 Critical $dV/dt$ — 230
- 7.4 Transient Processes during Turn-Off — 232
  - 7.4.1 Turn-Off using Circuit Commutation — 233
  - 7.4.2 Turn-Off by a Decrease of Forward Current — 235
  - 7.4.3 Gate Turn-Off — 236

Summary — 244
References — 245

## 8 Thyristor Types and Applications — 247

- 8.1 Phase-Control Thyristors — 247
- 8.2 Thyristors for High Speed Applications — 249
  - 8.2.1 The Asymmetric Thyristor (ASCR) — 250
  - 8.2.2 The Reverse Conducting Thyristor (RCT) — 253
  - 8.2.3 The Gate-Assisted Turn-Off Thyristor (GATT) — 254
- 8.3 The Gate Turn-Off Thyristor (GTO) — 257
- 8.4 The Triac — 264
- 8.5 Light-Triggered Thyristors (LTTs) — 266
- 8.6 Breakover Diodes (BODs) — 269

Summary — 271
References — 271

## 9 Static Induction Power Devices — 273
- 9.1 The Static Induction Transistor — 273
- 9.2 The Field-Controlled Diode or Static Induction Thyristor — 280
  - 9.2.1 Gated Turn-Off — 282
  - 9.2.2 The $dV/dt$ Capability — 283
  - 9.2.3 The Turn-On Process — 284
- Summary — 284
- References — 285

## 10 Power Metal–Oxide–Semiconductor Field-Effect Transistors — 287
- 10.1 Principles of MOS Transistor Operation — 287
  - 10.1.1 The On-State — 288
  - 10.1.2 The Saturation Condition — 293
- 10.2 Vertical Power MOSFET Designs — 293
  - 10.2.1 Static Characteristics — 294
  - 10.2.2 The Frequency Dependence of Parameters — 303
- 10.3 The Switching of Power MOSFETs — 305
  - 10.3.1 The Turn-On Process — 306
  - 10.3.2 The Turn-Off Process — 309
  - 10.3.3 The Frequency Dependence of the Switching Parameters — 311
- 10.4 Safe Operating Area (SOA) — 312
- Summary — 315
- References — 315

## 11 Power Bipolar–MOS Devices — 317
- 11.1 The Insulated Gate Bipolar Transistor (IGBT) — 318
  - 11.1.1 Static Parameters — 319
  - 11.1.2 Switching Characteristics — 323
  - 11.1.3 The Frequency Dependence of Parameters and the SOA — 329
  - 11.1.4 Punch-Through (PT) and Non-Punch-Through (NPT) IGBTs — 331
  - 11.1.5 IGBT Developments — 333
- 11.2 The MOS-Controlled Thyristor (MCT) — 333
- 11.3 Other Bipolar–MOS Structures — 337
  - 11.3.1 The Lateral IGBT — 337
  - 11.3.2 The IBT — 338
  - 11.3.3 Monolithic Integration of Parallel-Connected Devices — 338
- Summary — 340
- References — 341

## 12 Power Modules and Integrated Structures — 343
- 12.1 Power Modules — 344
- 12.2 Power Integrated Circuits — 349
- 12.3 Smart Power: Intelligent Power Devices and Integrated Circuits — 353
- Summary — 356
- References — 357

## 13 Conditions for Reliable Operation — 359
- 13.1 The Cooling of Power Semiconductor Devices — 359
  - 13.1.1 Thermal Resistance and Transient Thermal Impedance — 360
  - 13.1.2 The Encapsulation of Power Semiconductor Devices — 364

|  | 13.1.3 Heat Sinks | 371 |
|---|---|---|
|  | 13.1.4 The Thermal Resistance of the Device–Heat Sink Interface | 378 |
| 13.2 | The Parallel and Series Connection of Power Semiconductor Devices | 379 |
|  | 13.2.1 Devices in Parallel | 379 |
|  | 13.2.2 Devices in Series | 382 |
| 13.3 | Overcurrent and Overvoltage Protection of Power Semiconductor Devices | 385 |
|  | 13.3.1 Overvoltage Protection | 385 |
|  | 13.3.2 Overcurrent Protection | 390 |
| 13.4 | The Operating Reliability of Power Semiconductor Devices | 391 |
| Summary | | 394 |
| References | | 395 |

## 14 Future Materials and Devices — 397

| 14.1 | Materials other than Silicon | 397 |
|---|---|---|
| 14.2 | Gallium Arsenide Devices | 400 |
| 14.3 | Silicon Carbide Devices | 401 |
| 14.4 | Diamond Devices | 401 |
| References | | 402 |

## Appendix  The Diffusion Equation — 405

| A.1 | Basic Concepts | 405 |
|---|---|---|
| A.2 | The Effect of Recombination | 406 |
| A.3 | Methods of Solution | 406 |
|  | A.3.1 The Laplace Transform | 406 |
|  | A.3.2 Separation of the Variables | 408 |
| A.4 | The Constant Current Condition | 408 |
| Reference | | 409 |

## Index — 411

# PREFACE

It is not always appreciated that the developments in power semiconductor devices in recent years have been as many and varied as those in integrated circuits. Indeed there are many examples of integrated circuit technology influencing the development of new types of power device. All three authors have felt for some time that there has been a need for a comprehensive text that linked the design, physical processes and applications performance of these devices. And in 1994 that gap was filled by the publication in Prague of Dr Benda's Czech-language textbook *Výkonové polovodičové součástky a integrované struktury* (*Power Semiconductor Devices and Integrated Structures*). The present book is a revised and expanded English version of that work.

The book is intended as a reference source for practising engineers on the structure, operation and application of the full range of power semiconductor devices that are now available. In addition, it is designed to be used as a supporting text for college courses at senior undergraduate and masters levels in power electronics. The authors have long experience in studying the basic physical processes that determine the operational behaviour of power semiconductor devices and in the development of new applications to exploit them.

In a review article written more than 20 years ago, J. Te Winkel of Philips Research Laboratories, distinguishing the respective needs of solid-state physicists, device technologists and circuit engineers, wrote that device models should serve as communication channels between members of these three groups and should ideally 'be securely founded in solid-state physics, incorporate the main technological parameters, and provide the data necessary for efficient circuit design.' Our aim has been to set out the available models for power semiconductor devices and to assess the extent to which they fulfil these objectives.

Of course, the needs of circuit designers vary considerably. At one extreme it may be adequate to represent an active device as no more than an ideal switch: one that presents zero impedance in the on-state and infinite impedance in the off-state, and switches between the two conditions in zero time and without dissipation. Other designers may only require a knowledge of switching-time delays. In other situations it may be important to know about the on- and off-states in greater detail, to be able to estimate the time involved and the dissipation that arises in changes between them. And so a hierarchy of models of increasing sophistication is needed, but each level should be well founded in the device physics and reflect the detailed compromises of the device technology. Our objective has been to set out a foundation for such models for each of the main types of power electronic device.

In the preface to the original Czech text, Dr Benda expressed his aim as one of providing the power semiconductor user with a good understanding of the structure, function, characteristics and features of the important discrete and integrated power devices. The physical models were developed without overcomplicated mathematics and the limitations imposed by technology were emphasised. It is our hope that these intentions will be carried to a wider audience in this expanded edition.

The main sources of information on current research activity on power semiconductor devices and their applications are the several IEEE journals that deal with these subjects, in particular *Transactions on Electron Devices* (T-ED), *Transactions on Power Electronics* (T-PE) and *Transactions on Industry Applications* (T-IA). There are also the proceedings of the main international conferences, in particular: the IEEE Annual International Symposium on Power Semiconductor Devices and ICs (ISPSD), the Annual European Conference on Power Electronics and Applications (EPE) and the IEEE Annual International Electron Devices Meeting (IEDM). Readers who wish to keep up to date with the many activities in this challenging field are recommended to look at these publications in the first instance.

All three authors are grateful to many colleagues for advice and help on various matters during the preparation of the text, and to their families for their support and understanding over a long period.

# 1

# PROPERTIES OF SEMICONDUCTORS

In order to understand the behaviour of semiconductor devices and the dependence of their operating characteristics on the working conditions, a knowledge of the basic theory of such elementary semiconductor structures as the p–n junction, the metal–semiconductor contact and the metal–insulator–semiconductor (MIS) system is required. Many books discuss the properties of semiconductor materials in detail [1.1 to 1.4], and in this first chapter we present no more than a short survey of some of the more important properties of bulk semiconductors. In Chapter 2 we continue with a brief account of the essential characteristics of p–n junctions, metal–semiconductor junctions and metal–oxide–semiconductor (MOS) structures. We give particular attention to those features that are important in power devices, such as the characteristics of very lightly doped material, the conductivity of material in the presence of very high carrier concentrations and the design of structures that support high voltages.

## 1.1 Charge Carriers and Semiconductor Band Structure

Electrons in a semiconductor may exist in one of two conditions: they may either be free, i.e. in an interstitial position, or they may be bound, as in a covalent bond. These two possible states are separated by an energy gap. The set of all allowed energy levels in the covalent bond make up the so-called *valence* band. The set of all allowed energy levels of the free electrons make up the *conduction* band. The two sets of states are separated by an energy barrier $W_g$ between the highest level in the valence band $W_v$ and the lowest level in the conduction band $W_c$. This is normally called the bandgap.

When impurity atoms, vacancies or other imperfections are present in the single crystal structure of a semiconductor, the local disturbance caused to the covalent bond may lead to a localised energy level within the bandgap. For instance, if a pentavalent impurity atom such as phosphorus (P), arsenic (As) or antimony (Sb) is substituted for one of the normal tetravalent atoms of a semiconductor such as silicon (Si), four of its valence electrons are fixed strongly in the covalent bond. However, the fifth electron is relatively weakly bound by electrostatic attraction and can quite easily be made free if it obtains an energy $W_D$ which is less than $W_g$. An electron liberated in this way leaves the impurity centre positively charged. It is then said to be ionised. Such impurities are called *donors*.

As well as the donors, there are other impurities and imperfections that easily capture electrons, thereby becoming negatively charged and creating unoccupied levels, or *holes*, in the covalent bond. These impurities are called *acceptors*. An acceptor is also associated with a local energy level in the bandgap, this time lying at an energy $W_A < W_g$ above the top of the valence band. An example of one way by which an acceptor can be created in a tetravalent semiconductor such as silicon is by the substitution of a trivalent atom such as boron (B), aluminium (Al), gallium (Ga) or indium (In). The two types of energy level are illustrated in Figure 1.1(a) and (b).

When one of the electrons forming the covalent bond acquires an energy exceeding $W_g$, it is possible for it to become free to move around inside the crystal. Effectively, it is excited across the bandgap and it leaves behind a vacancy in the covalent bond. Such vacancies (holes) can be filled by electrons bound to adjacent atoms. They too, therefore, appear to move freely within the crystal. Band-to-band excitation creates both a free electron and a free hole.

Very often acceptor and donor impurities are present together in a semiconductor. Semiconductors containing impurities of both types are said to be *compensated*. In such material, some electrons from donor levels occupy acceptor states and the number of active levels decreases, as shown in Figure 1.1(c). If the donor concentration in a semiconductor is $N_D$ and the acceptor concentration is $N_A$, when $N_D > N_A$, $(N_D - N_A)$ free electrons may be generated from the donor levels and when $N_A > N_D$, $(N_A - N_D)$ free holes may be generated from the acceptor levels.

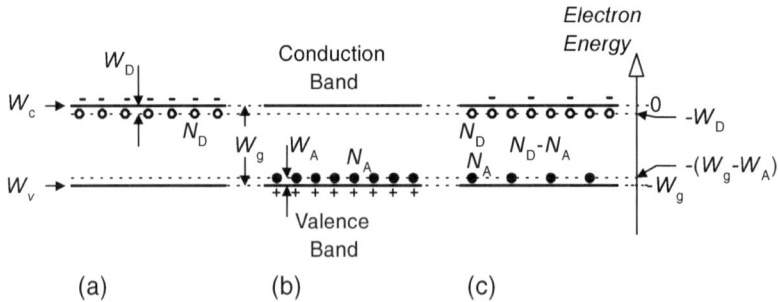

○     empty donor levels: neutral when occupied

●     filled acceptor levels: neutral when unoccupied

**Figure 1.1** Energy level (band) diagrams for doped semiconductors: (a) n-type material, doped with donor atoms that cause localised energy levels at an energy, $W_D$, below the bottom of the conduction band; (b) p-type material, doped with acceptor atoms that generate localised energy levels at an energy, $W_A$, above the top of the valence band; (c) compensated n-type material, doped with both donors and acceptors

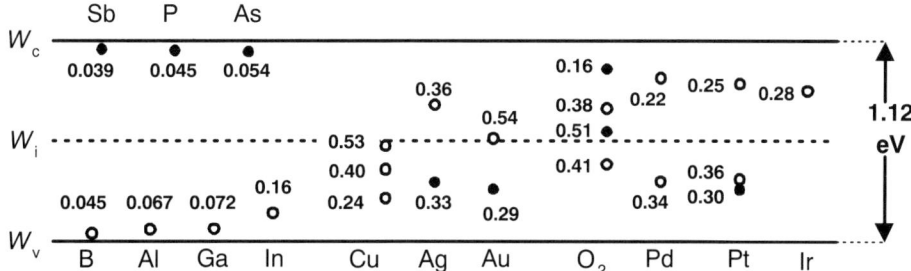

**Figure 1.2** Localised energy levels, within the bandgap, created by doping silicon with different impurities. The energy of each level, either above the top of the valence band or below the bottom of the conduction band, is given in [eV]; donor levels ● are shown filled, acceptor levels ○ are shown empty

Semiconductors with a higher concentration of free electrons than holes are called *n-type* semiconductors. Those with a higher concentration of holes than electrons are called *p-type*. Figure 1.2 shows where the energy levels created by some of the impurities that may be introduced during the fabrication of power semiconductor devices lie within the bandgap of silicon. The energy $W_i$ shown in the figure, corresponds to the middle of the bandgap.

## 1.2 The Concentration of Free Carriers

The behaviour of the free electrons in a semiconductor may be modelled using an ideal electron gas approximation. At an absolute temperature $T$, the probability $F_0(W)$ that an electron occupies an energy level of energy $W$ is given by the Fermi–Dirac distribution function

$$F_0(W) = \frac{1}{1 + \exp[(W - W_F)/kT]} \tag{1.1}$$

where $W_F$ is the Fermi energy (otherwise known as the electrochemical potential) and $k = 1.38 \times 10^{-23}$ J/K is Boltzmann's constant.

If the density of energy states in the conduction band is g($W$), the total concentration of electrons in the conduction band is

$$n = \int_{W_c}^{\infty} F_0(W) g(W) \, dW \tag{1.2}$$

where $W_c$ is the lowest energy in the conduction band and it is assumed that the upper limit of the integral may be taken to be infinity. It is often convenient to set $W_c = 0$, so that $W > 0$ represents the kinetic energy of the free electrons, and any energy level at an energy $W < 0$ represents a bound electron state. When $W_F < -3kT$, i.e. the Fermi level lies sufficiently deep in the bandgap, the semiconductor is said to be non-degenerate. Then, for the free electrons, the Fermi–Dirac distribution (1.1) may be approximated by the Maxwell–Boltzmann distribution

$$F_0(W) = \exp\left\{\frac{(W_F - W)}{kT}\right\} \quad (1.3)$$

If we put $W_c=0$, then by substituting (1.3) into (1.2) and integrating, it is possible to express the free electron concentration as

$$n = N_c \exp\left(\frac{W_F}{kT}\right) \quad (1.4)$$

where $N_c$ is an effective state density in the conduction band that depends on the effective mass of the free electrons $m_e^*$ and the absolute temperature $T$:

$$N_c = 2\left(\frac{2\pi m_e^* kT}{h^2}\right)^{3/2} \quad (1.4a)$$

In equation (1.4a), $h$ is Planck's constant: $h = 6.626 \times 10^{-34}$ J s.

Using a similar argument it is possible to express the concentration of holes as

$$p = N_v \exp\left[\frac{(-W_g - W_F)}{kT}\right] \quad (1.5)$$

where

$$N_v = 2\left(\frac{2\pi m_h^* kT}{h^2}\right)^{3/2} \quad (1.5a)$$

is the effective density of states in the valence band; it depends on the effective mass of the holes $m_h^*$ and on the temperature.

For silicon, $m_e^* = 1.18\, m_{e0}$ and $m_h^* = 0.5\, m_{e0}$, where $m_{e0} = 9.11 \times 10^{-31}$ kg is the free electron rest mass. Thus, $N_c = 4.83 \times 10^{21} T^{3/2}\,\text{m}^{-3}$ and $N_v = 1.71 \times 10^{21} T^{3/2}\,\text{m}^{-3}$. Values for the material properties of silicon have been obtained from the literature [1.4, 1.5].

From (1.4) and (1.5) it follows that the product of the electron concentration and the hole concentration is independent of the position of the Fermi level and depends only on the temperature and the band structure:

$$np = N_c N_v \exp\left(\frac{-W_g}{kT}\right) = n_i^2 \quad (1.6)$$

where $n_i$ is known as the *intrinsic carrier concentration*, i.e. the carrier concentration in a semiconductor with no impurities or defects.

The bandgap energy decreases with increasing temperature. For silicon it has been measured at room temperature ($T = 293$ K) as $W_g = 1.126$ eV. The variation with temperature can be fitted to $W_g = (1.1785 - 9.025 \times 10^{-5} T - 3.05 \times 10^{-7} T^2)$ eV. Thus, the intrinsic carrier concentration in silicon can be expressed as

## 1.2 THE CONCENTRATION OF FREE CARRIERS

$$n_i = 3.86 \times 10^{23} T^{3/2} \exp\left(\frac{T}{565} - \frac{6838}{T}\right) \text{ m}^{-3} \quad (1.6a)$$

It also follows from (1.4) and (1.5) that the individual concentrations of the free electrons and the holes do depend on the position of the Fermi energy. They are a function of the concentrations of donor and acceptor impurity atoms, the semiconductor band structure and the temperature. They can be determined from the condition for electrical neutrality which, for a doped semiconductor, is

$$n + N_A^- = p + N_D^+ \quad (1.7)$$

where $N_A^-$ is the concentration of ionised (i.e. filled) acceptors and $N_D^+$ is the concentration of ionised (i.e. empty) donors.

The temperature dependence can be divided into three regions. At very low temperatures the probability of excitation across the bandgap is negligible and free carriers can be generated only from shallow donor or acceptor levels. In this temperature region the Fermi energy lies between the impurity level and the energy of the nearest band edge. Then, in n-type material, the free electron concentration is given by

$$n = \left(\frac{1}{2} N_c N_D\right)^{1/2} \exp\left(\frac{-W_D}{2kT}\right) \quad (1.8)$$

In p-type material at low temperatures, the hole concentration is given by

$$p = (4 N_v N_A)^{1/2} \exp\left(\frac{-W_A}{2kT}\right) \quad (1.9)$$

These temperature dependencies are valid up to a temperature $T_s$ at which all the donor or acceptor levels are ionised. For $T > T_s$ and provided that the probability of direct excitation across the bandgap is still negligible, the majority carrier concentration does not vary with the temperature. Then in n-type material

$$n = N_D \quad (1.10a)$$

whereas in p-type material

$$p = N_A \quad (1.10b)$$

In thermodynamic equilibrium the concentration of minority carriers is related to the concentration of majority carriers by (1.6), so that over this range of temperature the concentration of minority holes in n-type semiconductor rises rapidly with temperature according to the relation

$$p_n = \frac{n_i^2}{n_D} \quad (1.11)$$

Similarly, the concentration of minority electrons in a p-type semiconductor is given by

$$n_p = \frac{n_i^2}{N_A} \qquad (1.12)$$

With partially compensated n-type material, $N_D$ in (1.8), (1.10a) and (1.11) should be interpreted as the *net* donor concentration ($N_D - N_A$). With partially compensated p-type material, $N_A$ in (1.9), (1.10b) and (1.12) should be interpreted as the *net* acceptor concentration ($N_A - N_D$).

With increasing temperature there is a rapidly increasing probability of creating electron–hole pairs by direct excitation from the valence band to the conduction band, i.e. by breaking the covalent bond. If the temperature is high enough ($T > T_i$), the intrinsic carrier concentration $n_i$ becomes much greater than the net impurity concentration, i.e. $n_i \gg (N_D - N_A)$ or $n_i \gg (N_A - N_D)$, and becomes characteristic of the intrinsic semiconductor. Effectively, all the free carriers are generated by direct excitation from the valence band to the conduction band, and

$$n = p = n_i = \sqrt{N_c N_v} \exp\left(\frac{-W_g}{2kT}\right) \qquad (1.13)$$

Figure 1.3 shows the variation of the position of the Fermi energy within the bandgap as a function of temperature for n-type and p-type materials. Figure 1.4 shows the variation of the free carrier concentration with the reciprocal of the absolute temperature. Both figures clearly identify the three temperature regions defined by $T_s$ and $T_i$. A more detailed analysis of these matters can be found in the literature [1.1, Section 4.3].

Most semiconductor devices make use of the temperature range over which the majority carrier concentration is independent of temperature and is determined by

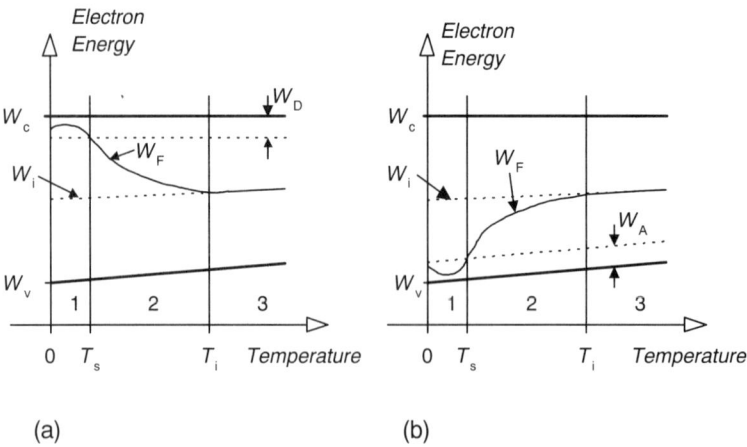

**Figure 1.3** The effect of temperature on the position of the Fermi level within the bandgap: (a) in n-type material; (b) in p-type material. In Region (1), the temperature is too low for carriers to be excited thermally from the impurity levels. In Region (2) the impurities are fully excited and the carrier concentration is approximately equal to the net impurity concentration. In Region (3), thermal band-to-band excitation, characteristic of the intrinsic semiconductor, dominates the effect of the impurities

## 1.2 THE CONCENTRATION OF FREE CARRIERS

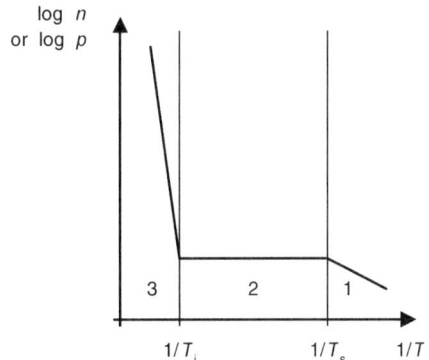

**Figure 1.4** The effect of temperature on the concentration of carriers in doped semiconductor material. Note that the variation is plotted against $1/T$

the concentration of impurities, i.e. from $T_s$ to $T_i$. It is here that the normal rectifying property of a p–n junction is observed. The values of both $T_s$ and $T_i$ increase with the doping concentration. For silicon, $T_s$ typically lies between $-100\,°\text{C}$ and $-200\,°\text{C}$, while $T_i$ is normally in the range $300\,°\text{C}$ to $700\,°\text{C}$. Other considerations limit normal device operation to the range $-80\,°\text{C}$ to $+200\,°\text{C}$, and often well within these temperatures.

As the concentrations of donor and acceptor impurities increase into the range $10^{22}$ to $10^{23}\,\text{m}^{-3}$ ($10^{16}$ to $10^{17}\,\text{cm}^{-3}$), several effects complicate the simple theory for the band structure and the carrier concentrations that we have given. First, the impurities interact with each other and the impurity levels themselves broaden into bands. Next, because the impurity atoms are randomly distributed, the band edges form tails that alter the distribution of energy levels in the valence and conduction bands and reduce the width of the bandgap. Finally, the minority carriers are effectively electrostatically screened by the high concentration of majority carriers, with the result that the thermal energy needed to create an electron–hole pair is reduced and the bandgap is narrowed. The overall effect is that the impurity bands merge with the conduction and valence bands, as illustrated in Figure 1.5. The bandgap energy is reduced and the effective density of states $N_v$ is affected. Accurate modelling is difficult [1.4, Section 2.4] but one important consequence is the resultant increase, at any given temperature, in the intrinsic carrier concentration and hence the concentration of minority carriers.

We may define an effective bandgap narrowing energy $\Delta W_g$ such that the intrinsic carrier concentration $n_i$ is given by

$$n_i^2(T,N) = n_{i0}^2(T)\exp\left[\frac{\Delta W_g(N)}{kT}\right] \tag{1.14}$$

where $n_{i0}$ is the intrinsic carrier concentration at low doping levels and $N$ is the dopant (acceptor or donor) concentration. For silicon, at temperatures close to room temperature, the variation of $\Delta W_g$ with $N$ is shown in Figure 1.6.

The position of the Fermi energy in the bandgap depends on the net effective impurity concentration $N_D - N_A$. This is shown in Figure 1.7, where the bandgap

# 1 PROPERTIES OF SEMICONDUCTORS

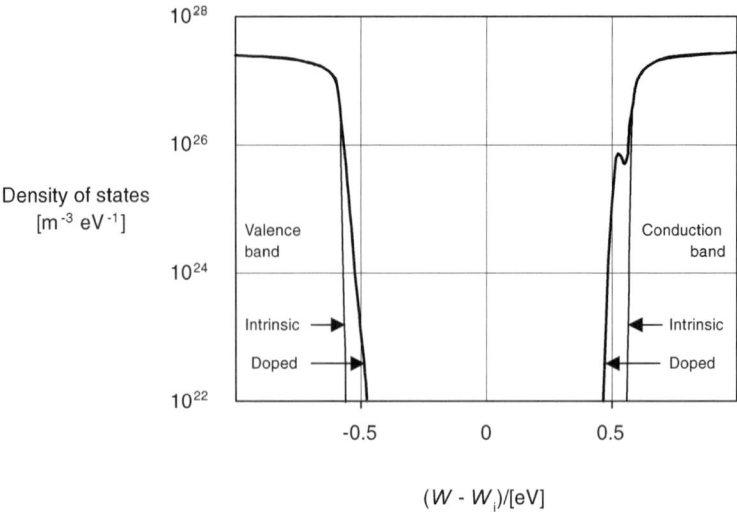

**Figure 1.5** An illustration of the effect of a donor concentration of $10^{24}$ m$^3$ on the densities of states at the band edges

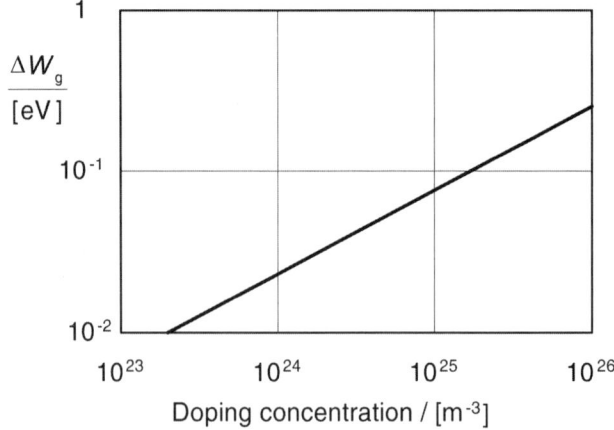

**Figure 1.6** The effective bandgap narrowing as a function of the doping concentration

narrowing that occurs at high doping levels is illustrated schematically. It can also be seen that in very heavily doped n-type semiconductors, with donor concentrations exceeding about $10^{24}$ m$^{-3}$, the Fermi energy lies within the conduction band. In similarly heavily doped p-type semiconductors it lies within the valence band. In each case the material is said to be *degenerate*. Degenerate semiconductors have conduction properties similar to those of metals.

The following symbols are often used to indicate the dopant levels and hence the position of the Fermi level in different layers of semiconductor structures:

n$^+$ heavily doped (degenerate) n-type semiconductor
n normally doped (non-degenerate) n-type semiconductor

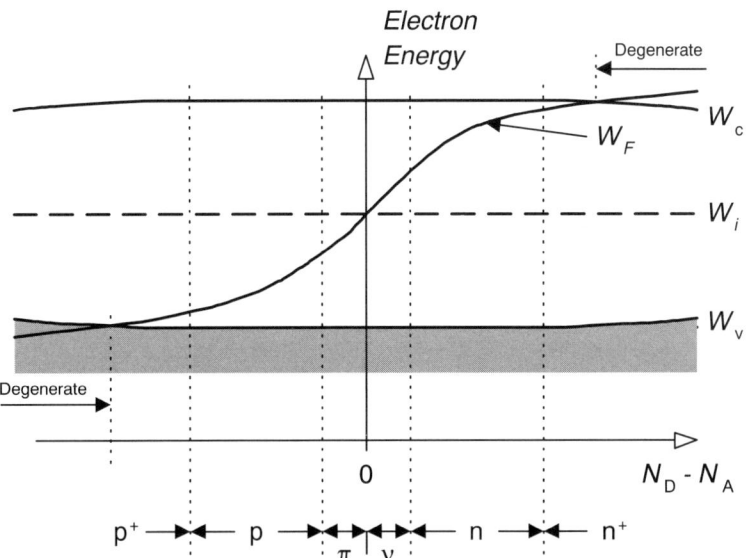

**Figure 1.7** The variation of the Fermi energy with doping concentration. The narrowing of the band gap that occurs at very high doping concentrations, when the impurity levels merge with the band edges, is shown schematically

ν    lightly doped n-type semiconductor (sometimes shown as n⁻)
i    intrinsic (fully compensated) semiconductor
π    lightly doped p-type semiconductor (sometimes shown as p⁻)
p    normally doped (non-degenerate) p-type semiconductor
p⁺    heavily doped (degenerate) p-type semiconductor

In this section and throughout the book, it is important to distinguish between quantities we have defined as energies or energy levels, such as $W$, $W_F$, $W_c$, $W_v$, and those we have defined as energy differences such, as $W_g$, $W_D$ and $W_A$. Energy levels may take on either positive or negative values, depending on the reference energy chosen to be zero. In general, we take this to be $W_c$. Energy differences are always positive quantities. Notice that, with these definitions, $W_v = -W_g$.

Another energy level that it is often convenient to use for reference is the level corresponding to the Fermi energy in intrinsic material. This is very near to the middle of the bandgap and is the correct way to define the energy level $W_i$ shown in Figures 1.2, 1.3 and 1.7.

## 1.3 The Conductivity of Semiconductors

### 1.3.1 Factors Determining Conductivity

In the absence of an electric field, the motion of the free charge carriers (electrons and holes) is chaotic. For each carrier with a velocity **v** there exists a carrier with a velocity −**v**, so that the vector sum of the velocities of all the carriers is zero. There is therefore no *net* transport of charge through the crystal. However, the carrier velocities may be high. At a temperature $T$ the average carrier kinetic energy is given by

$$W_{kin} = \frac{3}{2}kT = \frac{1}{2}m^*v_{th}^2 \qquad (1.15)$$

where $v_{th}$ is the thermal carrier velocity and $m^*$ is the effective carrier mass.

In silicon the thermal velocity of the electrons $v_{thn}$ is approximately $1.32 \times 10^4 \, T^{1/2}$, and the thermal velocity of the holes $v_{thp}$ is about $1.08 \times 10^4 \, T^{1/2}$. At room temperature these velocities are in the region of $2 \times 10^5$ m/s.

In the presence of an electric field, $\mathbf{E}$, a force $\mathbf{F} = -e\mathbf{E}$ is exerted on the electrons and a force, $\mathbf{F} = e\mathbf{E}$ on the holes. The carriers are initially accelerated by these forces, but are scattered by interactions with crystal lattice vibrations and with imperfections in the crystal. The consequence of these effects is that small average drift velocities $\mathbf{v}_{dn}$ and $\mathbf{v}_{dp}$ are superimposed on the random motion of the electrons and holes. The current density that results from the transport of the electrons is $\mathbf{J}_n = -en\mathbf{v}_{dn}$, and that resulting from the transport of the holes is $\mathbf{J}_p = ep\mathbf{v}_{dp}$.

When the electric field is relatively weak, so that the average drift velocities are much smaller than the thermal velocities, they are directly proportional to the electric field and may be expressed as

$$\mathbf{v}_{dn} = -\mu_n \mathbf{E} \qquad (1.16a)$$

$$\mathbf{v}_{dp} = \mu_p \mathbf{E} \qquad (1.16b)$$

where $\mu_n$ and $\mu_p$ are the mobilities of the free electrons and the holes, respectively. The mobility may be defined as the drift velocity caused by unit electric field.

The total current density $\mathbf{J}$ is the sum of the current densities of electrons and holes:

$$\mathbf{J} = \mathbf{J}_n + \mathbf{J}_p = e(n\mu_n + p\mu_p)\mathbf{E} = \sigma\mathbf{E} = \mathbf{E}/\rho \qquad (1.17)$$

where $\sigma$ is the electrical conductivity and $\rho$ the resistivity of the semiconductor material. Thus

$$\frac{1}{\rho} = \sigma = e(n\mu_n + p\mu_p) \qquad (1.18)$$

The electrical conductivity of a piece of semiconductor depends on the concentrations of carriers and their mobilities, hence it depends on the concentration of impurities and the temperature. The effects are summarised in Figure 1.8. At very low temperatures the conductivity increases with temperature as a result of the increasing concentration of carriers ionised from impurity levels. At such low temperatures the mobility increases with temperature too. At higher temperatures the majority carrier concentration is determined by the impurity concentrations (in n-type material $n = N_D - N_A$; in p-type material $p = N_A - N_D$) and the conductivity varies with temperature only because the carrier mobilities vary. In general, they decrease with increasing temperature over this range of temperatures, as discussed in detail in the next section, and the conductivity decreases too. This continues until the

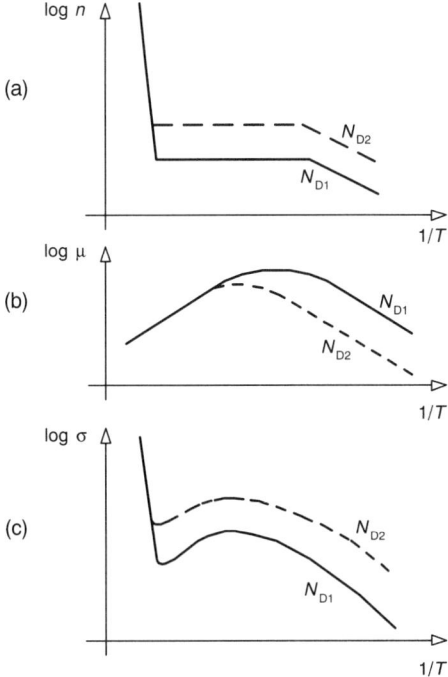

**Figure 1.8** The variation with temperature of the properties of n-type silicon: (a) carrier concentration; (b) electron mobility; (c) electrical conductivity. In each case, two levels of net doping concentration. $N_{D1}$ and $N_{D2}$, are shown with $N_{D2} > N_{D1}$

direct excitation of carriers across the bandgap becomes significant. Then the conductivity rises once again with further rises in temperature, the increase in the carrier concentration offsetting any reduction in mobility. At temperatures higher than $T_i$ the material becomes intrinsic, $n = p = n_i$, and the conductivity may be expressed as

$$\sigma = en_i(\mu_n + \mu_p) = KT^{1.5-x}\exp\left(\frac{-W_g}{2kT}\right) \qquad (1.19)$$

where $K$ and $x$ are empirical constants. This assumes that the mobility varies with temperature as $T^{-x}$. In order to simplify analysis, the approximation $x = 1.5$ is often assumed. The measured room temperature resistivity of silicon, either doped n-type with phosphorus or p-type with boron, is shown in Figure 1.9, plotted against the impurity concentration.

### 1.3.2 Factors Determining the Carrier Mobilities in Silicon

In this section we examine in more detail the various factors that influence the mobilities of electrons and holes in silicon and their variation with temperature and material parameters. Both the absolute and the relative values of the carrier mobilities have important effects on the characteristics and limitations of devices. They are, however, difficult either to derive from first principles or to measure independently and there is much discussion and disagreement about precise values.

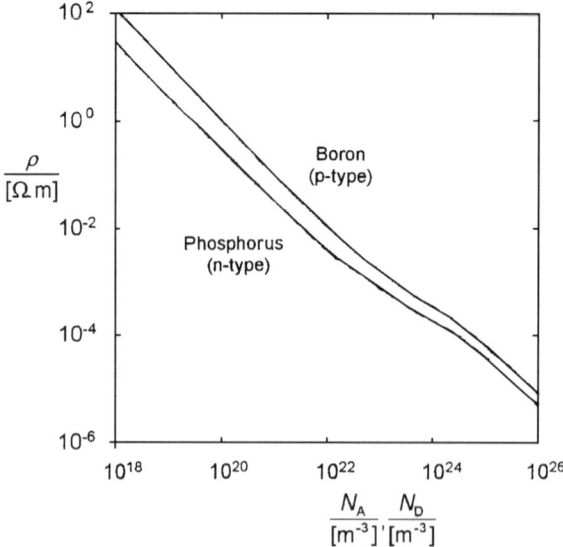

**Figure 1.9** The variation of the resistivity of silicon with donor (P) and acceptor (B) impurity concentration

When $T > 200\,\text{K}$ the dominant carrier scattering mechanism is caused by the vibrations of the crystal lattice. As the vibrations are quantised, this is often called phonon scattering. The scattering probability increases as the temperature rises and the mobility is therefore a decreasing function of temperature. For pure silicon, theory predicts that $\mu \propto T^{-5/2}$.

At any given temperature, increasing impurity concentration also causes the carrier mobilities to decrease, as shown in Figure 1.10. This is a consequence of the

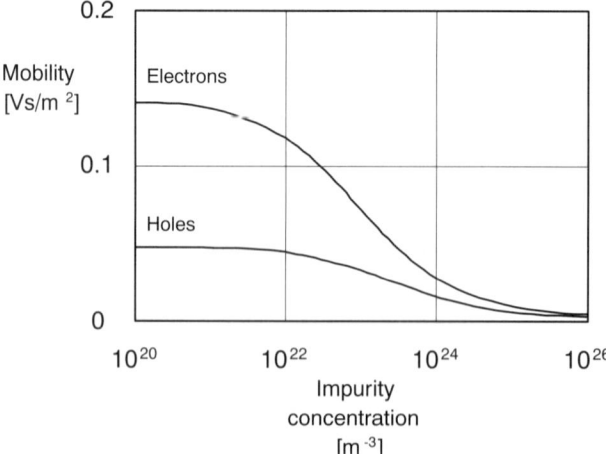

**Figure 1.10** The variation of the electron and hole mobilities in silicon with impurity concentration at 300 K

## 1.3 THE CONDUCTIVITY OF SEMICONDUCTORS

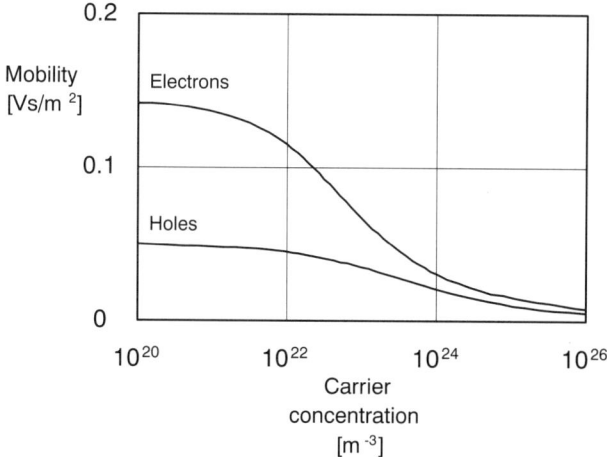

**Figure 1.11** The variation of the electron and hole mobilities in silicon with the concentration of free carriers

Coulomb scattering that occurs at the charged lattice sites of the ionised impurity levels. Another effect is to reduce the rates at which the carrier mobilities vary with temperature.

When devices are operated at high current densities, high concentrations of free carriers may be present in certain regions. Then carrier–carrier scattering, i.e. the scattering of electrons by holes and the scattering of holes by electrons, becomes significant. This results in a further reduction of the carrier mobility, as shown in Figure 1.11.

When several scattering processes are operating independently of one another, it is to be expected that the total scattering probability is the sum of the scattering probabilities of each. The overall mobility is then expected to follow Mathiessen's law

$$\frac{1}{\mu} = \sum_i \frac{1}{\mu_i} \tag{1.20}$$

where $\mu$ is the overall mobility and the $\mu_i$ are mobilities resulting from each individual scattering process acting alone. However, phonon scattering, ionised impurity scattering and carrier–carrier scattering are not truly independent processes. For example, high concentrations of carriers screen the ionised impurities and reduce their scattering probabilities. Thus, Mathiessen's law has to be applied with caution to the carrier mobilities in silicon and the various semi-empirical equations that have been proposed to represent them are complex and difficult to use.

In any experimental measurement of carrier mobilities, it is not easy to isolate the individual scattering mechanisms. In a model that is widely used for device simulation [1.6], an attempt is made to derive the dependence of each of the scattering processes on the concentrations of acceptor and donor impurities, the free

**Table 1.1** Examples of majority carrier mobilities predicted by equations (1.21) to (1.24)[a]

| $T$ | $N_D$ or $N_A$ or $2(pn)^{0.5}$ (m$^{-3}$) | $\mu_n$ (m$^2$/V s) | $\mu_p$ (m$^2$/V s) | $b$ | $\delta$ |
|---|---|---|---|---|---|
| $-55\,°C$ | 0 | 0.296 | 0.095 | 3.12 | 0.514 |
|  | $10^{24}$ | 0.045 | 0.023 | 1.935 | 0.318 |
| 25 °C | 0 | 0.144 | 0.048 | 3.023 | 0.503 |
|  | $10^{24}$ | 0.030 | 0.016 | 1.910 | 0.312 |
| 150 °C | 0 | 0.064 | 0.022 | 2.920 | 0.490 |
|  | $10^{24}$ | 0.019 | 0.010 | 1.902 | 0.311 |

[a] The parameters $b = \mu_n/\mu_p$ and $\delta = (\mu_n - \mu_p)/(\mu_n + \mu_p)$ are introduced in Section 5.1.

carrier concentration and the temperature. The data are then reassembled using the Mathiessen law and based on semi-empirical formulae that reflect the underlying physical processes governing each type of scattering. The equations are inevitably quite complex and, for our purposes, an earlier, simpler, empirical representation of the data of Figure 1.10 can be extended to include the effects of temperature and carrier–carrier scattering:

$$\mu = \mu_{\min} + \frac{(\mu_{\max} - \mu_{\min})}{[1 + (N/N_0)^a + ((pn)^{1/2}/2.0N_0)^a]} \quad (1.21)$$

where $N = N_A + N_D$ and the constants $\mu_{\max}$, $\mu_{\min}$, $N_0$ and $a$ are all functions of temperature. For $T > 200$ K and measured in kelvins,

$$a = 0.7\left(\frac{T}{300}\right)^{0.065} \quad (1.22)$$

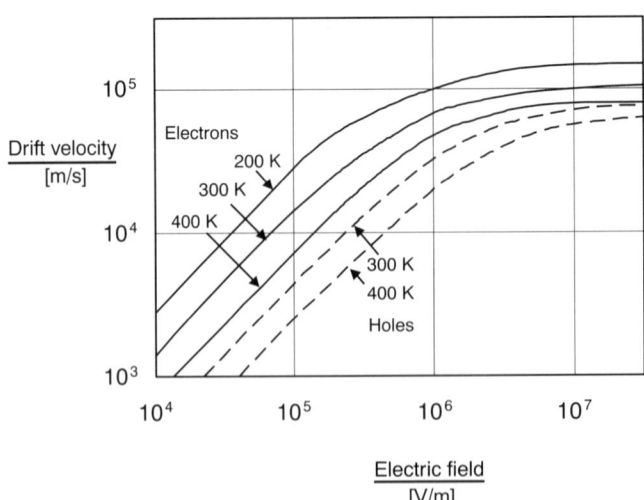

**Figure 1.12** The drift velocity of electrons and holes in bulk silicon showing the variation with the electric field at different temperatures

## 1.3 THE CONDUCTIVITY OF SEMICONDUCTORS

In n-type silicon,

$$\mu_{\text{max}} = 0.142 \left(\frac{T}{300}\right)^{-2.3} \quad (\text{m}^2/V\,s) \tag{1.23a}$$

$$\mu_{\text{min}} = 0.007 \left(\frac{T}{300}\right)^{-0.45} \quad (\text{m}^2/V\,s) \tag{1.23b}$$

and

$$N_0 = 1.0 \times 10^{23} \left(\frac{T}{300}\right)^{1.5} \quad (\text{m}^{-3}) \tag{1.23c}$$

In p-type silicon,

$$\mu_{\text{max}} = 0.047 \left(\frac{T}{300}\right)^{-2.2} \quad (\text{m}^2/V\,s) \tag{1.24a}$$

$$\mu_{\text{min}} = 0.0045 \left(\frac{T}{300}\right)^{-0.45} \quad (\text{m}^2/V\,s) \tag{1.24b}$$

and

$$N_0 = 2.2 \times 10^{23} \left(\frac{T}{300}\right)^{1.5} \quad (\text{m}^{-3}) \tag{1.24c}$$

Table 1.1 shows some values for the carrier mobilities in silicon predicted by these equations.

At higher electric field strengths, the drift velocity $v_d$ may no longer be negligible in comparison with the thermal velocity $v_{\text{th}}$, and new scattering interactions between the carriers and the crystal lattice become possible. This decreases the carrier mobility as the field strength $E$ increases. At high fields the drift velocity becomes independent of the field strength and reaches a maximum value $v_{d\,\text{max}}$, known as the saturation drift velocity. An illustration of the variation of the electron and hole drift velocities in silicon with the electric field strength is given in Figure 1.12. The saturation drift velocity decreases slowly with increasing temperature but is largely unaffected by the doping concentration.

The dependence of the mobility on the electric field strength can be described analytically for electrons as

$$\mu_n(E) = \frac{v_{\text{dn}}}{E} = \frac{\mu_{n0}}{[1 + (\mu_{n0}E/v_{\text{dn max}})^2]^{1/2}} \tag{1.25}$$

and for holes as

$$\mu_p(E) = \frac{v_{\text{dp}}}{E} = \frac{\mu_{p0}}{[1 + (\mu_{p0}E/v_{\text{dp max}})]} \tag{1.26}$$

where $v_{\text{dn max}} = 1.05 \times 10^5$ m/s and $v_{\text{dp max}} = 0.90 \times 10^5$ m/s at room temperature. The variation of the saturation drift velocities with temperature $T$ (K), can be fitted to

$$v_{\text{dn max}} = 1.05 \times 10^5 \left(\frac{T}{300}\right)^{-0.87} \tag{1.27}$$

for the electrons, and to

$$v_{\text{dp max}} = 0.9 \times 10^5 \left(\frac{T}{300}\right)^{-0.52} \tag{1.28}$$

for the holes.

Detailed theoretical and experimental discussions of the dependence of the electron and hole mobilities on temperature, impurity concentration, free carrier concentration and electric field can be found in the literature [1.4, 1.5].

Note that our discussion so far has dealt with carrier mobility in bulk semiconductor material. Conduction in very thin layers near the surface can be affected by the presence of additional scattering centres associated with the surface. As a result, values of mobility are generally lower and are a function of the direction of the flow with respect to the crystal lattice. This is important in many of the MOS devices discussed in Chapters 10 and 11, in which the current is usually controlled as it flows through a thin surface layer.

## 1.4 The Generation and Recombination of Excess Charge Carriers

### 1.4.1 Equilibrium and Non-Equilibrium Conditions

As a result of the interactions with the crystal lattice, carriers are generated thermally in a semiconductor at the rate $G$ per unit volume per unit time. Simultaneously, in the converse process, the free electrons reoccupy unoccupied bound energy levels in the valence band at the rate $R$, also per unit volume and per unit time. At thermodynamic equilibrium $G = R$, so the average concentration of free carriers is constant. However, external influences such as photon or particle irradiation or the presence of a high electric field strength can cause the covalent bonds to be broken, creating additional electron–hole pairs.

For example, in an interaction between a photon of energy $h\nu$ and the crystal lattice, provided that $h\nu > W_g$, a number $\beta$ of pairs of free electrons and holes can be generated ($\beta \geqslant 1$). Then the carrier generation as a function of the depth $x$ below the illuminated semiconductor surface can be expressed as

$$G = \left(\frac{d\Delta n}{dt}\right)_{\text{gen}} = \left(\frac{d\Delta p}{dt}\right)_{\text{gen}} = \alpha\beta\Phi_\lambda \exp(-\alpha x) \tag{1.29}$$

where $\Phi_\lambda$ is the spectral power density of the incident radiation at wavelength $\lambda$, entering the semiconductor and $\alpha$ is the absorption coefficient for light of this wavelength.

## 1.4 EXCESS CHARGE CARRIERS

The carrier concentration may be increased above the equilibrium level in other ways, perhaps as a result of the injection of carriers from other regions, perhaps by avalanche ionisation (Chapter 2).

At thermodynamic equilibrium, the concentrations of free electrons $n_0$ and holes $p_0$ are given by (1.4) and (1.5). Under external influence, the increased concentrations of carriers may be expressed as

$$n = n_0 + \Delta n \tag{1.30a}$$

$$p = p_0 + \Delta p \tag{1.30b}$$

The increased concentration, $\Delta n$ or $\Delta p$, caused by external processes is called the excess carrier concentration. Usually $\Delta n = \Delta p$ and their magnitude should be compared to the equilibrium concentration of the majority carriers, either $n_0$ or $p_0$. *Low injection conditions* describe the situation when $\Delta n \ll n_0$, or $\Delta p \ll p_0$, depending on which is the majority carrier. If $\Delta n$ and $\Delta p$ are much higher than the majority carrier concentration, $n_0$ or $p_0$, *high injection conditions* are said to apply. Situations often arise in power semiconductor devices when $\Delta n$ or $\Delta p$ are of comparable magnitude to the concentration of majority carriers, then special treatment is required.

The presence of excess carriers implies that the thermodynamic equilibrium is disturbed, and this means the Fermi level strictly cannot be defined. The carrier concentrations are not correctly given by (1.4) and (1.5). However, they can be represented by similar expressions that take the form

$$n = n_0 + \Delta n = N_c \exp\left(\frac{W_{Fn}}{kT}\right) \tag{1.31}$$

and

$$p = p_0 + \Delta p = N_v \exp\left[\frac{-(W_{Fp} + W_g)}{kT}\right] \tag{1.32}$$

The energy levels $W_{Fn}$ and $W_{Fp}$ are known as *quasi Fermi levels* and they differ from the true Fermi level whenever $(\Delta n/n_0)$ and $(\Delta p/p_0)$ are significant. Note that

$$np = N_c N_v \exp\left(\frac{-W_g}{kT}\right) \exp\left[\frac{(W_{Fn} - W_{Fp})}{kT}\right] = n_i^2 \exp\left[\frac{(W_{Fn} - W_{Fp})}{kT}\right] \tag{1.33}$$

When an external influence that has caused an excess of carriers to be generated in a particular region is discontinued, the excess carrier concentration decreases because the rate of recombination exceeds the rate of generation. The rate of recombination is given by

$$R = \left(\frac{d\Delta n}{dt}\right)_{rec} = \left(\frac{d\Delta p}{dt}\right)_{rec} = -\frac{\Delta n}{\tau} \tag{1.34}$$

where $\tau$ is known as the excess carrier lifetime. The recombination of the excess carriers is a relatively complicated process that can occur in several different ways. One is the direct band-to-band recombination of a free electron with a hole, i.e. the

# 1 PROPERTIES OF SEMICONDUCTORS

electron falls from a free energy state in the conduction band to a bound state in the valence band. Another involves impurity energy levels in the bandgap, in which a carrier is first captured by one of these so-called *trapping* levels and then, in a second interaction, it recombines with a carrier of opposite polarity. This is called trap recombination or recombination through local centres.

The energy of recombination can be lost in three different ways. The first is by the generation of photons in the process known as *radiative recombination*. The second involves the generation of a quantised lattice vibration known as *phonon generation*. In the third process, the recombination energy increases the kinetic energy of other free carriers and the process is known as *Auger recombination*. We consider each of them in turn.

## 1.4.2 Recombination via Local Trapping Centres

In the case of silicon, because of the nature of the band structure, the most common recombination process involves trapping levels whose energy states lie deep in the middle of the bandgap. This process is illustrated in Figure 1.13. An electron from the conduction band is captured by the trap, which lies at an energy $W_t$ below the bottom of the conduction band. The electron later interacts with and fills a vacant level in the valence band (i.e. it recombines with a hole), thus completing the recombination process. Phonon generation dominates the energy loss mechanism.

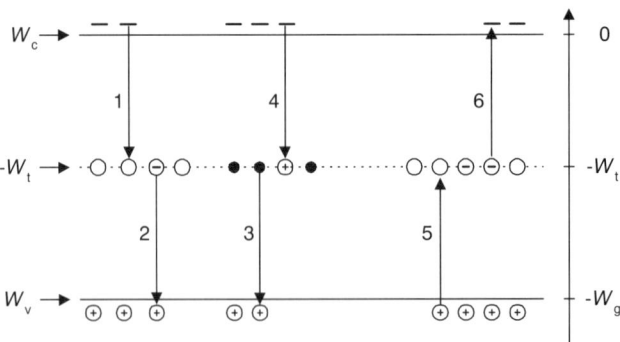

**Figure 1.13** A schematic energy level diagram illustrating the processes of generation and recombination via localised trapping centres whose energy levels lie near to the middle of the band gap: Processes (1) and (2) show first a free electron interacting with and filling an empty acceptor trap and then later, in a separate process, attracting and recombining with a hole. Processes (3) and (4) show first an electron in a filled donor level interacting and recombining with a hole and then later attracting and capturing a free electron to complete the recombination process. Processes (5) and (6) demonstrate the converse process of carrier generation via traps: first an electron is excited from the valence band into a trapping level, generating a hole, then it is later excited into the conduction band, generating a free electron

## 1.4 EXCESS CHARGE CARRIERS

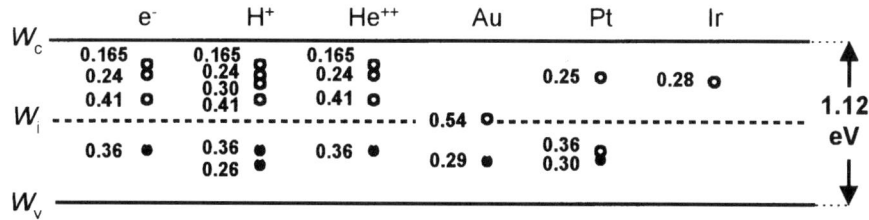

**Figure 1.14** Energy levels of some of the recombination centres used to control the carrier lifetime in silicon power devices. Donor levels ● are shown filled, acceptors ○ are shown empty

If the concentration of the recombination centres is $N_t$, the excess carrier lifetime $\tau_t$ that results from this mechanism can be expressed as

$$\tau_t = \frac{1}{C_t N_t} \tag{1.35}$$

where $C_t$ is a recombination coefficient.

Carrier lifetime usually increases with temperature. It also depends on the level of injection, the type of recombination centre and the relationship of the energy level of the recombination centre relative to the Fermi level. For instance, in the lightly doped n-type silicon frequently used in the fabrication of power semiconductor components, the energy of the recombination centres introduced by doping with gold lies below the Fermi level, as shown in Figure 1.14. Consequently, the carrier lifetime at low injection levels $\tau_L$ is shorter than that at high injection levels $\tau_H$. However, when this material is doped with platinum, the opposite is usually true: the Fermi energy remains below the energy of the recombination centres (Figure 1.14) and $\tau_H < \tau_L$.

It is important to recognise that the trapping centres also act as centres for the generation of free carriers. The theory of generation and recombination via local traps is usually known as the Shockley–Read–Hall theory and, because of its importance in the functioning of power semiconductor devices, we consider it here in some detail. As a working example, we take the particular case of an n-type semiconductor with a concentration of donor recombination centres $N_t$ lying at an energy level $W_t$ below $W_c$. In thermal equilibrium, designated by the subscript 0, some of the traps are ionised ($N_{t0}^+$) and the rest are neutral ($N_{t0}^*$):

$$N_t = N_{t0}^+ + N_{t0}^* \tag{1.36}$$

The ratio of $N_{t0}^+$ to $N_t$ is given by the Fermi function $F(-W_t)$ of equation (1.1). Thus

$$\frac{N_{t0}^+}{N_{t0}^*} = \exp\left(\frac{(-W_t - W_F)}{kT}\right) \tag{1.37}$$

In the case of steady-state non-equilibrium conditions, the concentrations of ionised and neutral centres are altered to values representative of a dynamic equilibrium:

$$N_t = N_t^+ + N_t^* \tag{1.38}$$

This dynamic equilibrium is a result of a balance between the electron capture rate $c_{rn}$ and the electron emission rate $c_{en}$ at the centres. The rate of capture of electrons from the conduction band on trapping centres is proportional to the concentration of ionised (empty) traps and to the concentration of free electrons, hence it can be written as

$$c_{rn} = v_{th,n} \sigma_n n N_t^+ \qquad (1.39)$$

where $v_{th,n}$ is the thermal velocity of electrons described by (1.15) and $\sigma_n$ is the capture cross-section of the trapping centre for conduction band electrons.

The rate of emission of electrons from the traps to the conduction band is proportional to the concentration of neutral (filled) trapping centres $N_t^*$, and to the concentration of empty energy levels in the conduction band, which may be represented by $N_c$, as defined in (1.4). Thus

$$c_{en} = \kappa_1 N_t^* N_c \qquad (1.40)$$

where $\kappa_1$ is a proportionality coefficient.

In thermal equilibrium $c_{en} = c_{rn}$ so, using (1.4), (1.37) and (1.39),

$$\kappa_1 = v_{th,n} \sigma_n \exp\left(\frac{-W_t}{kT}\right) \qquad (1.41)$$

Remember that we regard $W_t$ as a positive quantity.

By similar arguments, the rate of capture of holes can be written as

$$c_{rp} = v_{th,p} \sigma_p p N_t^* \qquad (1.42)$$

where $v_{th,p}$ is the average thermal velocity of the holes and $\sigma_p$ is the capture cross-section of the trapping centre for holes. Likewise, the rate of emission of holes into the valence band can be expressed as

$$c_{ep} = \kappa_2 N_t^+ N_v \qquad (1.43)$$

where the proportionality coefficient $\kappa_2$ is given by

$$\kappa_2 = v_{th,p} \sigma_p \exp\left(\frac{-(W_g - W_t)}{kT}\right) \qquad (1.44)$$

Note that $(W_g - W_t)$ is the energy difference between the trap energy level and the top of the valence band.

Assuming the coefficients $\kappa_1$ and $\kappa_2$ to be constants, the difference between the capture rate and the emission rate for electrons, i.e. the net recombination rate, is given by

$$R_n = -\frac{d\Delta n}{dt} = c_{rn} - c_{en} = v_{th,n} \sigma_n (n N_t^+ - n_1 N_t^*) \qquad (1.45)$$

where

$$n_1 = N_c \exp\left(\frac{-W_t}{kT}\right) \qquad (1.46)$$

Similarly, the difference between the capture rate and the emission rate for holes is

## 1.4 EXCESS CHARGE CARRIERS

$$R_p = -\frac{d\Delta p}{dt} = c_{rp} - c_{ep} = v_{th,p}\sigma_p(pN_t^* - p_1 N_t^+) \tag{1.47}$$

where

$$p_1 = N_v \exp\left(\frac{-(W_g - W_t)}{kT}\right) \tag{1.48}$$

Note that $n_1 p_1 = n_i^2$ and that $n_1$ and $p_1$ can also be expressed in terms of the position of the Fermi level $W_i$ in undoped (intrinsic) semiconductor:

$$n_1 = n_i \exp\left(\frac{-(W_i + W_t)}{kT}\right) \tag{1.49}$$

and

$$p_1 = n_i \exp\left(\frac{(W_i + W_t)}{kT}\right) \tag{1.50}$$

When the trapping level lies at the same energy as the Fermi level in intrinsic material, i.e. $W_t = -W_i$, then $n_1 = p_1 = n_i$.

In order to maintain the dynamic equilibrium, the rates of recombination of the excess electrons and the excess holes with the trapping centres must be the same. That is, $R_n = R_p = R = \Delta n/\tau_t$.

Equating (1.45) and (1.47) and defining $\tau_{p0} = 1/v_{th,p}\sigma_p N_t$ and $\tau_{t0} = 1/v_{th,n}\sigma_n N_t$ yields

$$nN_t^+ - n_1 N_t^* = \frac{\tau_{n0}}{\tau_{p0}}(pN_t^* - p_1 N_t^+) \tag{1.51}$$

Hence

$$\frac{N_t^+}{N_t^*} = \frac{p\tau_{n0} + n_1\tau_{p0}}{n\tau_{p0} + p_1\tau_{n0}} \tag{1.52}$$

$$\frac{N_t^*}{N_t} = \frac{n\tau_{p0} + p_1\tau_{n0}}{(n + n_1)\tau_{p0} + (p + p_1)\tau_{n0}} \tag{1.53}$$

and

$$\frac{N_t^+}{N_t} = \frac{p\tau_{n0} + n_1\tau_{p0}}{(n + n_1)\tau_{p0} + (p + p_1)\tau_{n0}} \tag{1.54}$$

Note that under conditions of high injection, when $n$ and $p$ become very large and approximately equal in value, the ratio of empty to filled trapping levels given by (1.52) approaches $(\tau_{n0}/\tau_{p0})$.

Substituting (1.53) and (1.54) back into (1.45) shows that the rate of recombination in dynamic equilibrium is given by

$$R = \frac{1}{\tau_{n0}N_t}(nN_t^+ - n_1 N_t^*) = \frac{1}{\tau_{n0}}\left[\frac{n(p\tau_{n0} + n_1\tau_{p0}) - n_1(n\tau_{p0} + p_1\tau_{n0})}{(n+n_1)\tau_{p0} + (p+p_1)\tau_{n0}}\right]$$

$$= \frac{(np - n_i^2)}{(n+n_1)\tau_{p0} + (p+p_1)\tau_{n0}} \tag{1.55}$$

As long as space charge neutrality is maintained, $\Delta n = \Delta p$ and

$$(np - n_i^2) = (n_0 + \Delta n)(p_0 + \Delta n) - n_0 p_0 = (n_0 + p_0 + \Delta n)\Delta n \tag{1.56}$$

The carrier lifetime is then given by

$$\tau_t = \frac{\Delta n}{R} = \tau_{p0}\frac{(n_0 + n_1 + \Delta n)}{(n_0 + p_0 + \Delta n)} + \tau_{n0}\frac{(p_0 + p_1 + \Delta n)}{(n_0 + p_0 + \Delta n)} \tag{1.57}$$

Several important results can be derived from equation (1.57).

First of all, $\tau_t$ is minimised when $n_1$ and $p_1$ *together* are as small as possible. Of course, they are not independent of one another, as can be seen in Figure 1.15, and it follows that $\tau_t$ is least when the trap level lies near to the middle of the bandgap. The optimum value depends on the relative magnitudes of $\tau_{n0}$ and $\tau_{p0}$, and also of $N_c$ and $N_v$. If the trapping level lies away from the midpoint, then either $n_1$ or $p_1$ is larger and, at any given concentration of trapping levels, $\tau_t$ is increased.

Next consider the dependence of the carrier lifetime on the excess carrier concentration, which we mentioned earlier in this section. Differentiating (1.57) with respect to $\Delta n$ yields

$$\frac{d\tau_t}{d\Delta n} = \frac{\tau_{p0}(p_0 - n_1) + \tau_{n0}(n_0 - p_1)}{(n_0 + p_0 + \Delta n)^2} \tag{1.58}$$

For any given semiconductor in which the recombination is dominated by one particular trapping level, we expect the rate of change of lifetime with carrier

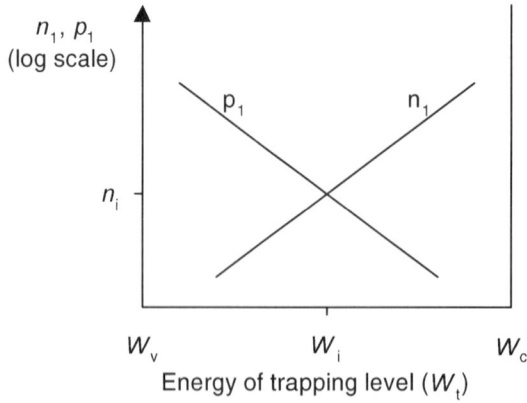

**Figure 1.15** A schematic illustration of the variation of the parameters $n_1$ and $p_1$ with the position of the trapping level within the band gap

## 1.4 EXCESS CHARGE CARRIERS

concentration to diminish as the excess carrier concentration increases. However, whether $\tau_t$ increases or decreases with $\Delta n$ depends on several factors. We can illustrate this for the case of n-type material for which the Fermi energy lies above the trap energy level. In the numerator of (1.58), the $p_0$ term is always negligible in comparison with the others, whereas the $n_0$ term dominates. Thus, $d\tau_t/d\Delta n$ is positive. If traps are introduced having an energy level that lies higher in the bandgap, the $n_1$ term is bigger, whereas the $p_1$ term becomes insignificant. If the trap level becomes higher than the Fermi level, we expect the $n_1$ term to become dominant and then $d\tau_t/d\Delta n$ is negative. Converse arguments apply in the case of p-type material. Thus, when a single trapping level dominates recombination, there is a monotonic variation of the carrier lifetime $\tau_t$ with the excess carrier concentration $\Delta n$. And whether it increases or decreases depends on the relative position of the trapping level and the Fermi level, as set out in Table 1.2. If $\tau_{p0}$ and $\tau_{n0}$ are not of the same order of magnitude as one another, the thresholds given in the table may be shifted by several $kT$.

At low injection levels, $\Delta n \ll n_0 + p_0$, the carrier lifetime $\tau_L$ is given by

$$\tau_L = \tau_{p0} \frac{(n_0 + n_1)}{(n_0 + p_0)} + \tau_{n0} \frac{(p_0 + p_1)}{(n_0 + p_0)} \quad (1.59)$$

In n-type semiconductor material in which the trapping centre lies near to the middle of the bandgap, $n_0 \gg p_0$, $n_0 \gg n_1$ and $n_0 \gg p_1$. Thus

$$\tau_L = \tau_{p0}\left(1 + \frac{n_1}{n_0}\right) \quad (1.60a)$$

In p-type semiconductor under similar conditions, the low injection carrier lifetime is given by

$$\tau_L = \tau_{n0}\left(1 + \frac{p_1}{p_0}\right) \quad (1.60b)$$

It follows from (1.60a) and (1.60b) that the low injection carrier lifetime depends on both the energy level $W_t$ of the trapping centre and its capture cross-sections $\sigma_n$ and $\sigma_p$ for electrons and holes. The lifetime is least when the energy level of the trapping centres lies near the centre of the bandgap. Otherwise, either $n_1$ or $p_1$ becomes large compared to the concentration of majority carriers and cannot be neglected in (1.59).

Deviations from (1.57), (1.59), (1.60) and (1.61) may occur [1.1, p. 276] when the concentration of the trapping levels approaches that of the shallow impurities. There is also some compensation of the donor or acceptor impurities.

**Table 1.2** Variation of carrier lifetime with excess carrier concentration

| Material type | Position of trapping level | Effect of $\Delta n$ on lifetime |
|---|---|---|
| n-type | $W_t < -W_F$ | $\tau$ decreases with $\Delta n$ |
| n-type | $W_t > -W_F$ | $\tau$ increases with $\Delta n$ |
| p-type | $W_t < W_g + W_F$ | $\tau$ decreases with $\Delta n$ |
| p-type | $W_t > W_g + W_F$ | $\tau$ increases with $\Delta n$ |

Under conditions of high injection, $\Delta n \gg n_0 + p_0$ and $\Delta n \gg N_t$, the carrier lifetime $\tau_H$ is given by

$$\tau_H = \tau_{n0} + \tau_{p0} \tag{1.61}$$

It is now independent of the energy level of the trapping centres and depends only on the trap concentration $N_t$ and its carrier capture cross-sections $\sigma_n$ and $\sigma_p$.

Another particular case that is important occurs in the depletion layer that forms in many devices, e.g. at a p–n junction, a heterojunction, a Schottky contact or a MOS structure (Chapter 2). Here we have a situation in which $np < n_i^2$. From (1.57) it follows there is now net generation of carriers, i.e. $d\Delta n/dt > 0$. If we assume that $n = p = 0$, the generation rate $G$ is given by

$$G = \frac{n_i^2}{(n_1 \tau_{p0} + p_1 \tau_{n0})} = \frac{n_i}{\tau_{sc}} \tag{1.62}$$

where

$$\tau_{sc} = \frac{(n_1 \tau_{p0} + p_1 \tau_{n0})}{n_i} = \tau_{p0} \exp\left(\frac{-(W_i + W_t)}{kT}\right) + \tau_{n0} \exp\left(\frac{(W_i + W_t)}{kT}\right) \tag{1.62a}$$

is the carrier lifetime in such a space charge region. It, too, depends on the position of the trapping energy level with respect to the position of the Fermi level in undoped semiconductor. If $\tau_{p0} = \tau_{n0}$ it is shortest when $W_t$ and $W_i$ coincide.

The carrier lifetime in silicon devices can be influenced during fabrication by the creation of recombination centres, either by the diffusion of impurities such as gold, platinum or iridium, or by creating crystal lattice defects using gamma radiation or high energy electrons, hydrogen ions (protons) or helium ions. The enhanced generation, caused by the presence of such recombination centres, increases the reverse current at a reverse-biased p–n junction (Section 2.1.2).

### 1.4.3 Other Recombination Processes

Direct band-to-band recombination, radiative and Auger, becomes significant at high levels of injection and in heavily doped material. Band-to-band radiative recombination results in the emission of a quantum of radiation of energy $h\nu \approx W_g$. In silicon this corresponds to a wavelength $\lambda = 1101$ nm. The probability of this process occurring is proportional to the free electron concentration and the hole concentration. At high injection levels, $\Delta n \gg n_0 + p_0$, we can express the net rate of carrier recombination as

$$-\frac{d(\Delta n)}{dt} = C_r(np - n_0 p_0) = C_r(n_0 + p_0 + \Delta n)\Delta n \approx C_r \Delta n^2 \tag{1.63a}$$

where $C_r$ is the radiation recombination coefficient.

In general, a carrier recombination time constant, resulting from a particular recombination process $i$, may be defined as $\tau_i = \Delta n/(-d(\Delta n)/dt)$, when that process alone is operating.

## 1.4 EXCESS CHARGE CARRIERS

In the case of radiative recombination, it follows from (1.63a) that the resulting recombination time constant $\tau_r$ is

$$\tau_r = \frac{1}{C_r \Delta n} \quad (1.63b)$$

The emission of recombination radiation allows us to measure the regions of high excess carrier concentration, but in silicon the overall carrier lifetime is not influenced significantly by this process, the value of $C_r$ being in the region of $2 \times 10^{-19}$ m$^3$/s ($2 \times 10^{-13}$ cm$^3$/s).

At very high levels of carrier concentration, the Auger recombination process becomes significant. It is a three-body process, in which the energy released in the recombination of a free electron with a hole is taken up by another free carrier, either an electron or a hole. The rate of recombination by this process depends on whether an electron or a hole is excited, so that two Auger recombination coefficients, $C_{An}$ and $C_{Ap}$, may be defined. The net rate of recombination of excess carriers by these processes can then be expressed as

$$-\frac{d(\Delta n)}{dt} = C_{An}(n^2 p - n_0^2 p_0) + C_{Ap}(np^2 - n_0 p_0^2) \quad (1.64)$$

In (1.64) we have assumed that $\Delta n = n - n_0 = \Delta p = p - p_0$. It is left as an exercise for the reader to show that Auger carrier lifetimes, defined as $\tau_A = \Delta n/(-d(\Delta n)/dt)$, take on the following forms:

In heavily doped, n-type material $\quad \tau_{An} = \dfrac{1}{C_{An} n_0^2} \quad (1.65a)$

In heavily doped, p-type material $\quad \tau_{Ap} = \dfrac{1}{C_{Ap} p_0^2} \quad (1.65b)$

And in situations where there is a very high concentration of excess carriers,

$$\tau_A = \frac{1}{(C_{An} + C_{Ap})\Delta n^2} \quad (1.65c)$$

The sum of the Auger recombination coefficients in silicon has been measured [1.7] to be in the region of $3 \times 10^{-43}$ to $5 \times 10^{-43}$ m$^6$/s ($3 \times 10^{-31}$ to $5 \times 10^{-31}$ cm$^6$/s) at room temperature. Electron excitation is the more likely process. Auger recombination may dominate the other recombination processes in devices carrying high current densities, particularly in heavily doped layers.

If several recombination mechanisms occur simultaneously, or several different types of recombination centre are present in the material, each has is own recombination probability $P_i$ and can be associated with a recombination time constant $\tau_i$, inversely proportional to $P_i$. The resultant total recombination probability is the sum of the probabilities of each of the partial mechanisms.

# 1 PROPERTIES OF SEMICONDUCTORS

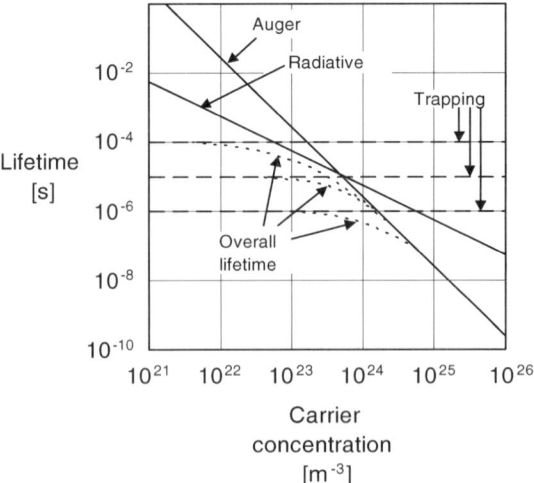

**Figure 1.16** A schematic illustration of the relative magnitudes of trapping, radiative and Auger recombination probabilities, as a function of carrier concentration

Thus, the resulting carrier lifetime is given by

$$\frac{1}{\tau} = \sum_i \frac{1}{\tau_i} \tag{1.66}$$

Figure 1.16 gives a schematic illustration of the relative magnitudes of the three principal recombination processes as a function of carrier concentration.

As well as the recombination that takes place in the bulk semiconductor regions, enhanced recombination of excess carriers may take place locally at particular interfaces within the material, and at its surface. Under some conditions, such surface recombination can be very important and can seriously affect the recombination rate in the surroundings of the interface. If the excess carrier concentration at the interface is $\Delta n$, the number of carriers recombining per unit area, per unit time is often expressed as $s\Delta n$, where $s$ is known as the surface recombination velocity. Note that the units of $s$ are indeed m/s. Its value varies very widely, according to the nature of the interface or the condition of the surface.

## 1.5 Diffusion and Drift of Charge Carriers

Whenever the carrier concentration varies with position, the resulting concentration gradient causes a flux of carriers $\Phi_{\text{diff}}$ by diffusion. According to Fick's first law of diffusion, this flux is directly proportional to the concentration gradient. For electrons

$$\Phi_{n,\text{diff}} = -D_n \, \mathbf{grad}\, n \tag{1.67a}$$

## 1.5 DIFFUSION AND DRIFT OF CHARGE CARRIERS

and for holes

$$\Phi_{p,\text{diff}} = -D_p \,\text{grad}\, p \tag{1.67b}$$

where $D_n$ and $D_p$ are the diffusion coefficients of the free electrons and holes, respectively.

Because each carrier carries a positive or negative electronic charge, any net transport of carriers results in the transport of charge. Thus, the diffusion fluxes of electrons and holes give rise to diffusion current densities

$$\mathbf{J}_{n,\text{diff}} = eD_n \,\text{grad}\, n \tag{1.68a}$$

$$\mathbf{J}_{p,\text{diff}} = -eD_p \,\text{grad}\, p \tag{1.68b}$$

As discussed in Section 1.3, transport of carriers also occurs when an electric field is applied, producing the so-called drift current. When an electric field and a concentration gradient are present simultaneously, the total current density is the sum of the diffusion and drift current densities and can be expressed for each carrier as

$$\mathbf{J}_n = e(n\mu_n \mathbf{E} + D_n \,\text{grad}\, n) \tag{1.69a}$$

$$\mathbf{J}_p = e(n\mu_p \mathbf{E} - D_p \,\text{grad}\, p) \tag{1.69b}$$

The carrier mobilities and the diffusion coefficients are both determined by the frequency of the collisions made by the carriers as a consequence of their random motion. As a result, they are directly related to one another by the Einstein equations

$$D_n = \frac{kT}{e}\mu_n \tag{1.70a}$$

and

$$D_p = \frac{kT}{e}\mu_p \tag{1.70b}$$

The total current density is the sum of the electron and hole current densities:

$$\mathbf{J} = \mathbf{J}_n + \mathbf{J}_p \tag{1.71}$$

If excess carriers are injected or created, they can move and give rise to currents at some distance from their point of origin. Furthermore, the concentration profiles may vary with time. The overall effect is determined by the processes of diffusion and drift, and by the generation and recombination of the excess carriers. It is described by Fick's second law:

$$\frac{\partial n}{\partial t} = G_n - \frac{\Delta n}{\tau_n} + \frac{1}{e}\,\text{div}\,\mathbf{J}_n \tag{1.72a}$$

$$\frac{\partial p}{\partial t} = G_p - \frac{\Delta p}{\tau_p} - \frac{1}{e}\text{div }\mathbf{J}_p \qquad (1.72b)$$

These equations are known as the *continuity equations* for electrons and holes.

Often the continuity equations are solved for conditions of electrical neutrality, when it is assumed that $\Delta n$ and $\Delta p$ are everywhere almost exactly equal. Then the evolution of the carrier concentration profiles can be simplified. We consider carrier flow in the x-direction, when there is no generation of excess carriers in the region under discussion, i.e. $G_n = G_p = 0$. We express the concentration profiles in terms of the excess concentration of free electrons. By substituting (1.69) into (1.72), it is straightforward to show that

$$\frac{\partial \Delta n}{\partial t} = \frac{\partial}{\partial x}\left(D\frac{\partial \Delta n}{\partial x}\right) + \frac{\partial}{\partial x}(n\mu_{\text{diff}}E_x) - \frac{\Delta n}{\tau} \qquad (1.73)$$

where

$$D = \frac{\sigma_p D_n + \sigma_n D_p}{\sigma_p + \sigma_n} \qquad (1.74)$$

$$\mu_{\text{diff}} = \frac{\sigma_p \mu_n - \sigma_n \mu_p}{\sigma_p + \sigma_n} \qquad (1.75)$$

and

$$\tau - \tau_p = \tau_n \qquad (1.76)$$

In (1.74) and (1.75), we have put $\sigma_n = ne\mu_n$ and $\sigma_p = pe\mu_p$.

The bipolar diffusion coefficient $D$ and the differential mobility $\mu_{\text{diff}}$ reflect the effect of the internal electric fields set up to maintain the equal flows of free electrons and holes. In calculating their values, it is necessary to take account of the effects on the carrier mobilities of carrier–carrier scattering at high carrier concentrations, of impurity scattering at high doping levels, and of temperature. Under conditions of high injection, we can assume that $n = p$ and then $D = D_a = 2D_n D_p/(D_n + D_p)$ and is known as the ambipolar diffusion coefficient and $\mu_{\text{diff}} = 0$.

In general, equation (1.73) is non-linear and not at all easily solved. When the electric field term is negligible, and provided that $D$ does not vary with the carrier concentration, equation (1.73) simplifies to

$$\frac{\partial \Delta n}{\partial t} = D\frac{\partial^2 \Delta n}{\partial x^2} - \frac{\Delta n}{\tau} \qquad (1.77)$$

This linear form of the diffusion equation often has to be solved in order to determine the time variation of the excess carrier concentration profile in the volume of a semiconductor structure following the generation or injection of excess carriers. With different coefficients and different initial and boundary conditions, it arises several times in future chapters in the discussions of various device structures. An overview is given in the appendix.

A further simplification that can be used is the so-called *stored charge* analysis, which leads on to the very important *charge-control model* for active devices. In essence, this model asserts that the terminal characteristics of a device are predictable functions of the charge contained within it and of the time derivatives of that charge.

In the absence of any further generation of carriers, equation (1.77) can be written as

$$\frac{\partial \Delta n}{\partial t} = \frac{1}{e} \text{div } \mathbf{J}_n - \frac{\Delta n}{\tau} \tag{1.78}$$

where $\mathbf{J}_n$ is the local current density of electrons. Multiplying through by the electronic charge $e$ and integrating over all or part of the device volume, equation (1.78) can be put in the form

$$\frac{dQ}{dt} = i(t) - \frac{Q}{\tau^*} \tag{1.79}$$

The integration is taken over the semiconductor volume of interest and then (1.79) expresses, in integral form, the time variation of the total excess charge $Q$ present in this region. We have applied Gauss's theorem and integrated the current density over the surface bounding the region. Part of this surface will comprise one or more metal–semiconductor contacts to electrodes that either feed the current $i(t)$ into the region or extract it. The remainder of the surface serves as a location for carrier recombination. The relaxation time constant $\tau^*$ in (1.79) represents an effective carrier lifetime that takes into account the effect of the recombination boundaries; it is different from the normal recombination lifetime $\tau$ that is characteristic of the bulk semiconductor material.

In the case of one-dimensional carrier flow across an area $A$, $Q/A = \int e \Delta n dx$ is the excess charge per unit cross-sectional area perpendicular to the direction of flow (the $x$-direction) and $i/A = \int \text{div } \mathbf{J}_n dx$ is the current density flowing in that direction.

The application of the charge-control model to the transient behaviour of power diodes and bipolar transistors is discussed in Sections 5.3.2 and 6.3, respectively. In terms of an equivalent circuit, the excess charge can usually be represented by a shunt capacitance, known as the diffusion capacitance $C_{\text{diff}}$, as described in Section 2.1.1.

## 1.6 Effects of Non-Uniform Doping

In semiconductors that are uniformly doped, the concentration of carriers (electrons and holes) does not depend on position, and at any point the condition for electrical neutrality (1.7) is valid. This situation is described by the energy band scheme shown in Figure 1.1. The concentration of free carriers in n-type semiconductor at a temperature $T$ is given by

$$n = N_c \exp\left(\frac{(W_F - W_c)}{kT}\right) \tag{1.80}$$

where $N_c$ is the effective state density in the conduction band, $k$ is Boltzmann's constant and $(W_F - W_c)$ is the distance of the Fermi energy from the bottom of the conduction band. In thermal equilibrium, $(W_F - W_c)$ is constant in the whole of the semiconductor crystal. In semiconductor doped with donors of concentration $N_D$, the electron concentration is impurity controlled over the temperature range $T_s$ to $T_i$. Then $n \approx N_D$ and the position of the Fermi level is given by

$$W_F - W_c = kT \ln\left(\frac{N_D}{N_c}\right) \tag{1.81}$$

When, on the other hand, the difference between the concentration of donors and the concentration of acceptors ($N_D \sim N_A$) is a function of position, the energy difference between the Fermi level and the bottom of the conduction band ($W_F - W_c$) is also a function of position. In thermal equilibrium the Fermi energy is uniform throughout the volume of the semiconductor, so the value of $W_c$, which represents the potential energy of the free electrons, now varies with position. This is illustrated schematically in Figure 1.17. The resulting potential gradient gives rise to an internal electric field, the so-called *built-in field*:

$$\mathbf{E} = -\operatorname{grad} V = \frac{-1}{e} \operatorname{grad}(W_F - W_c) = \frac{-kT}{e} \frac{1}{n} \operatorname{grad} n$$

$$\approx \frac{-kT}{e} \frac{1}{(N_D - N_A)} \operatorname{grad}(N_D - N_A) \tag{1.82}$$

By analogy, in p-type semiconductor the built-in field is given by

$$\mathbf{E} = \frac{kT}{e} \frac{1}{p} \operatorname{grad} p \approx \frac{kT}{e} \frac{1}{(N_A - N_D)} \operatorname{grad}(N_A - N_D) \tag{1.83}$$

These internal fields balance the tendency of the majority carriers to diffuse away from regions of high concentration. When the concentration gradients are large, electrical neutrality is not always maintained everywhere and regions of local space charge occur. The electric field is then related to the space charge density through Poisson's equation:

$$\operatorname{div}\mathbf{E} = -\operatorname{div}\operatorname{grad} V = \frac{e(p - n + N_D - N_A)}{\varepsilon_r \varepsilon_0} \tag{1.84}$$

Equation (1.84) is needed to solve problems of non-uniformly doped semiconductors whenever space charge regions occur. And if excess carriers are generated, the charge density, electric field and their variation with time can be determined by the

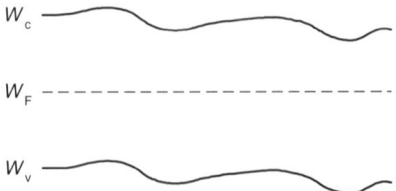

**Figure 1.17** A schematic illustration of the variation of the electron energy levels in a non-uniformly doped semiconductor. The constant Fermi level indicates thermodynamic equilibrium

simultaneous solution of Poisson's equation (1.84) and the equations of continuity (1.72). This set of equations is used for the accurate numerical modelling of transient processes in semiconductor devices.

With uniformly doped semiconductor, there is no built-in field and the application of an external electric field gives rise to electrical conduction according to Ohm's law. With non-uniform doping, the superposition of the internal and external electric fields may invalidate Ohm's law. When strong inhomogeneities are present, as at a p–n junction or a metal–semiconductor junction, they give rise to the highly non-linear effects such as rectification that are fundamental to the operation of most semiconductor devices. Some important examples of inhomogeneous doping in semiconductors and its consequences are discussed in more detail in Chapter 2.

It is clear from (1.82) and (1.83) that very significant electric fields are to be expected where the net doping concentration changes from n-type to p-type. In general, however, the electric fields set up as a result of a slight imbalance in the equilibrium charge concentrations are so strong that they are immediately shielded by the mobile carriers. As a result, they extend only over relatively short distances and for brief periods of time. The characteristic distance over which a disturbance in neutral semiconductor dies out is known as the Debye length $L_D = (\varepsilon_r \varepsilon_0 kT/e^2(n+p))^{1/2}$, and it decays over the dielectric relaxation time $\tau_R = \varepsilon_r \varepsilon_0 \rho$, where $\rho = 1/e(\mu_n n + \mu_p p)$ is the electrical resistivity of the material. This can be demonstrated by considering a piece of n-type material in which a flow of majority electrons suffers a local disturbance that momentarily increases the electron concentration $\Delta n$ above its equilibrium level. If any imbalance in the generation and recombination terms is negligible in comparison with the rate of change of current density, equation (1.72a) becomes

$$\frac{\partial n}{\partial t} = \frac{1}{e}\frac{\partial J_n}{\partial x} \tag{1.85}$$

The electron current density is given by (1.69a) as

$$J_n = en\mu_n E + eD_n \frac{\partial n}{\partial x} \tag{1.86}$$

Differentiating (1.86) and applying Poisson's equation

$$\frac{\partial E}{\partial x} = -\frac{e\Delta n}{\varepsilon_r \varepsilon_0} \tag{1.87}$$

yields

$$\frac{\partial n}{\partial t} + \frac{\Delta n}{\tau_R} - \frac{L_D^2}{\tau_R}\frac{\partial^2 n}{\partial x^2} = 0 \tag{1.88}$$

This is just another version of the linear diffusion equation (1.77). For the situation following a sudden change in $\Delta n$ to the new value $\Delta n(0)$ at $t=0$, in a volume of uniform material, it can be solved by the separation of variables. When the hole concentration is negligible, it gives solutions that decay in time as

$$\Delta n = \Delta n(0) \exp\left(-\frac{t}{\tau_R}\right) \tag{1.89}$$

In the situation where the excess carriers are generated only at $x=0$, but maintained there indefinitely, their steady-state concentration varies in space (see the appendix) as

$$\Delta n = \Delta n(0) \exp\left(-\frac{x}{L_\text{D}}\right) \qquad (1.90)$$

## Summary

The electrical characteristics of a semiconductor can be interpreted in terms of the electron energy level structure (band diagram) and the position of the Fermi energy within the bandgap.

The electrical conductivity of a semiconductor material is a function of the concentrations and mobilities of the carriers (free electrons and holes). These depend on the concentration of certain impurities and the temperature. The room temperature resistivity of silicon doped with either phosphorus or boron is shown in Figure 1.9.

Carrier generation and recombination are dynamic interactions involving several possible processes. In silicon, recombination via local trapping centres normally dominates, except at very high carrier concentrations, when Auger recombination may become important.

The flow of current in a semiconductor is the sum of the diffusion (in a concentration gradient) and drift (in an electric field) of both types of carrier. Under conditions of electrical neutrality, the flow is said to be bipolar. When the drift terms can be neglected, the charge control or stored charge model enables the device behaviour to be represented in terms of a simple equivalent circuit.

## References

[1.1] Smith, R. A. *Semiconductors* (Cambridge University Press, 2nd Edn, 1979).
[1.2] Benda, V. *Struktura a vlastnosti materiálů II – polovodiče* [*The Structure and Properties of Materials II – Semiconductors*] (ČVUT Praha, 1987).
[1.3] Sze, S. M. *Physics of Semiconductor Devices* (John Wiley & Sons, 2nd Edn, 1981).
[1.4] Selberherr, S. *Analysis and Simulation of Semiconductor Devices* (Springer-Verlag, 1984).
[1.5] INSPEC, *The Properties of Silicon*, EMIS Volume 4 (IEE, 2nd Edn, 1998).
[1.6] Klaassen, D. B. M. A Unified Mobility Model for Device Simulation – I. Model Equations and Concentration Dependence, *Solid-State Electronics*, **35**, 953–9 (1992); A Unified Mobility Model for Device Simulation – II. Temperature Dependence of Carrier Mobility and Lifetime, *Solid-State Electronics*, **35**, 961–7 (1992).
[1.7] Rosling, M., Bleichner, H., Lundquist, M. and Nordlander, E. A Novel Technique for the Simultaneous Measurement of Ambipolar Carrier Lifetime and Diffusion Coefficient in Silicon, *Solid-State Electronics*, **35**, 1223–7 (1992).

# 2

# ELEMENTARY SEMICONDUCTOR STRUCTURES

## 2.1 The p–n Junction and its Basic Properties

The p–n junction forms a very important part of most semiconductor devices. Its basic physics has been studied in great detail and is well described in many semiconductor device technology books [2.1, 2.2]. In Section 2.1 we discuss only the most important of the characteristics of the p–n junction.

### 2.1.1 Current–Voltage Characteristics

In order to fabricate p–n junctions, a number of different techniques are used to introduce different concentrations of donor and acceptor impurities into various regions of the semiconductor. Usually, a metallurgical boundary is formed between a region in which $N_A > N_D$ (the p-type region) and a region in which $N_D > N_A$ (the n-type region). At the junction, by definition, the material is exactly compensated and $N_A = N_D$.

The variation of the electron energy levels across an abrupt p–n junction in equilibrium is shown in Figure 2.1(a). In the n-type material, the Fermi level lies in the bandgap near to the bottom of the conduction band, whereas in the p-type material it lies near the top of the valence band. In thermal equilibrium the Fermi energy must be uniform throughout the semiconductor crystal. This condition is achieved by the transfer of electrons from the n-type side to the p-type side and by the movement of holes in the opposite direction, in each case by diffusion, as a result of the concentration gradients that occur.

A consequence of this charge transfer is the creation in the n-type region adjacent to the junction of a region of positive space charge. This results from the presence of ionised donors whose charge is not balanced by an equal concentration of free electrons. In the p-type region adjacent to the junction there is a region where the space charge of the ionised acceptors is not compensated by the presence of an equal

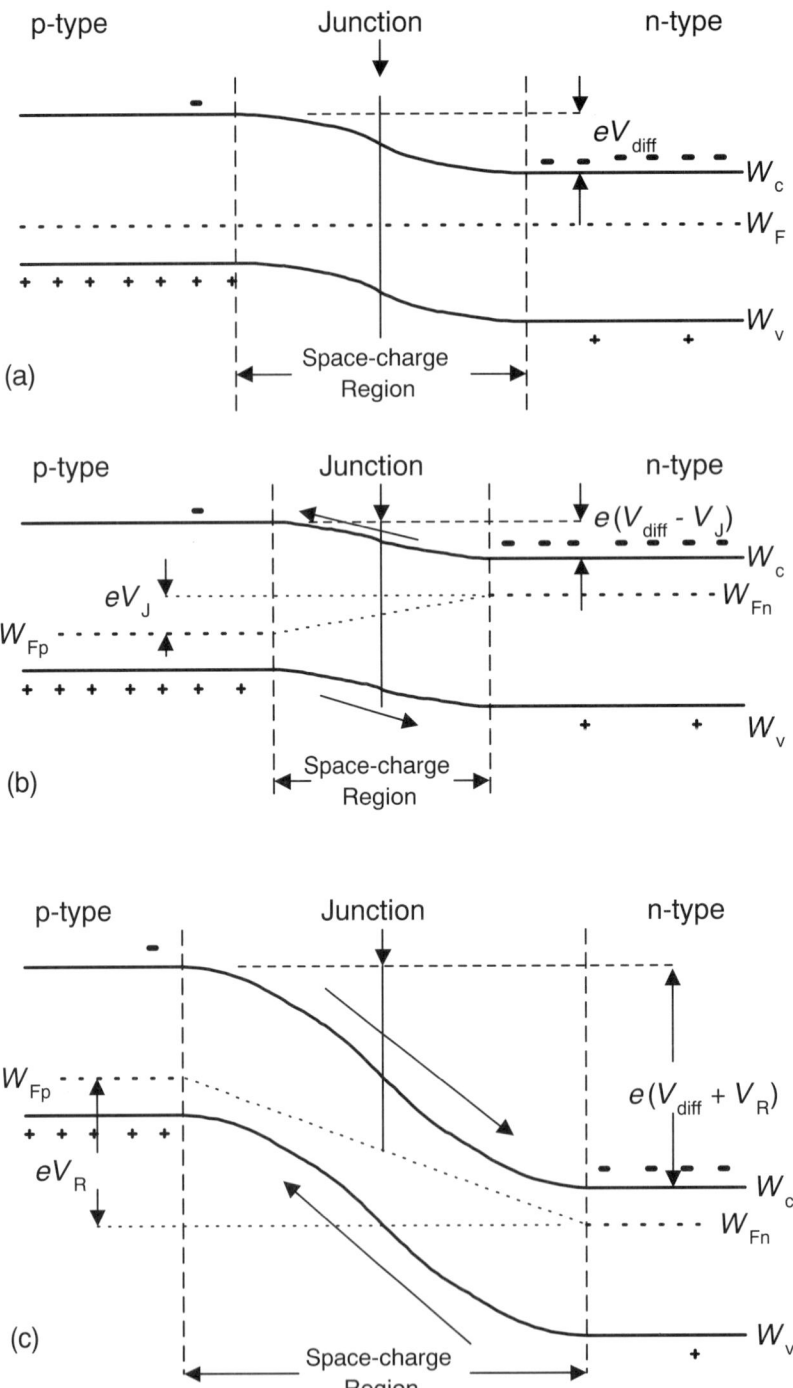

**Figure 2.1** Band diagrams, showing the variation of the electron energy levels across a p–n junction: (a) in thermal equilibrium; (b) under forward bias; (c) under reverse bias

concentration of holes. Within the space charge region, on either side of the junction, an electric field is established that can be described by (1.84). It is directed so as to counterbalance the tendency of the free carriers to diffuse.

Under conditions of thermal equilibrium, a potential energy barrier $eV_{\text{diff}}$ develops across the space charge layer. Using (1.82) and (1.83), it is clear that the diffusion potential $V_{\text{diff}}$, depends logarithmically on the doping concentrations on either side of the junction:

$$V_{\text{diff}} = \frac{kT}{e} \ln\left(\frac{n_{n0} p_{p0}}{n_i^2}\right) \tag{2.1}$$

where $n_{n0}$ is the equilibrium concentration of majority electrons in the n-type region and $p_{p0}$ is the equilibrium concentration of majority holes in the p-type region.

When an external voltage $V_J$ is applied across the p–n junction, the thermal equilibrium is disturbed and a current flows through the junction. When the p-type side is made positive with respect to the n-type side, the energy barrier at the junction is lowered to $e(V_{\text{diff}} - V_J)$ and the tendency of the electrons to diffuse into the p-type region exceeds their tendency to move in the opposite direction as a result of drift in the internal field. The p–n junction is said to be *forward biased*. The reduced height of the potential barrier also causes the thickness of the space charge region to be reduced.

The electron concentration at the edge of the space charge region in the p-type material is increased above its equilibrium value $n_{p0}$ by an amount $\Delta n(0)$, related to the reduction in the height of the potential barrier. In order to maintain space charge neutrality, there is an equal increase in the hole concentration, $\Delta p(0) = \Delta n(0)$. Provided the conditions are those of low injection, i.e. $\Delta n(0) \ll p_{p0}$, the free electrons remain in approximate thermodynamic equilibrium with the n-type region throughout the space charge layer. This implies that the Fermi level for electrons $W_{\text{Fn}}$ corresponds to that of the bulk semiconductor in the n-type region, whereas the Fermi level for holes $W_{\text{Fp}}$ corresponds to that of the bulk semiconductor in the p-type region. This is illustrated in Figure 2.1(b).

Thus, throughout the space charge region and, in particular, at its boundary in the p-type material ($x = 0$),

$$W_{\text{Fn}} - W_{\text{Fp}} = eV_J \tag{2.2}$$

Using (1.33) and putting $p_{p0}$ for the equilibrium hole concentration in the p-type region, we may therefore write

$$np = (n_{p0} + \Delta n(0))(p_{p0} + \Delta p(0)) = (n_{p0} + \Delta n(0))p_{p0} = n_i^2 \exp\left(\frac{eV_J}{kT}\right) \tag{2.3}$$

Thus

$$(n_{p0} + \Delta n(0)) = n_{p0} \exp\left(\frac{eV_J}{kT}\right) \tag{2.4}$$

hence

$$\Delta n(0) = n_{p0}\left[\exp\left(\frac{eV_J}{kT}\right) - 1\right] \quad (2.5)$$

Similarly, the hole concentration at the boundary of the space charge region on the n-type side is increased by an amount $\Delta p$ given by

$$\Delta p = p_{n0}\left[\exp\left(\frac{eV_J}{kT}\right) - 1\right] \quad (2.6)$$

The electrons injected into the p-type region diffuse away from the p–n junction and recombine either in the bulk semiconductor or at the contact. If the electric field outside the space charge region can be neglected, the carrier distribution is described by the continuity equation (1.77). Under conditions of low injection (so that $D_a = D_n$) and in the steady state (i.e. when $\partial n/\partial t = 0$), and provided that $n$ varies only in the direction perpendicular to the junction (the $x$-direction), equation (1.77) simplifies to

$$D_n \frac{d^2(\Delta n)}{dx^2} = \frac{\Delta n}{\tau_n} \quad (2.7)$$

where $\tau_n$ is the electron lifetime in the p-type semiconductor. If we set the origin of the $x$-coordinate at the edge of the space charge region in the p-type material and if the p-type region is assumed to be infinitely thick, the variation of the excess electron concentration is obtained by solving (2.7) to give

$$\Delta n(x) = \Delta n(0) \exp\left(\frac{-x}{L_n}\right) \quad (2.8)$$

where $L_n = \sqrt{D_n \tau_n}$ is known as the *electron diffusion length*.

A similar equation describes the distribution of the excess holes in the n-type region:

$$\Delta p(\xi) = \Delta p(0) \exp\left(\frac{-\xi}{L_p}\right) \quad (2.9)$$

where $\xi$ is the distance coordinate directed into the n-type region perpendicular to the junction, with its origin at the boundary of the space charge region on the n-type side of the junction. If $d$ is the thickness of the space charge region, $\xi = d - x$. By analogy with (2.8), $L_p = \sqrt{D_p \tau_p}$ is the hole diffusion length in the n-type region, where $\tau_p$ is the hole lifetime.

The electrons flow away from the boundary of the space charge region into the p-type semiconductor by diffusion, and the resulting current density is given by (1.68) as

$$J_n = eD_n \frac{d(\Delta n)}{dx} \quad (2.10)$$

Similarly, the current density carried by the holes entering the n-type region is

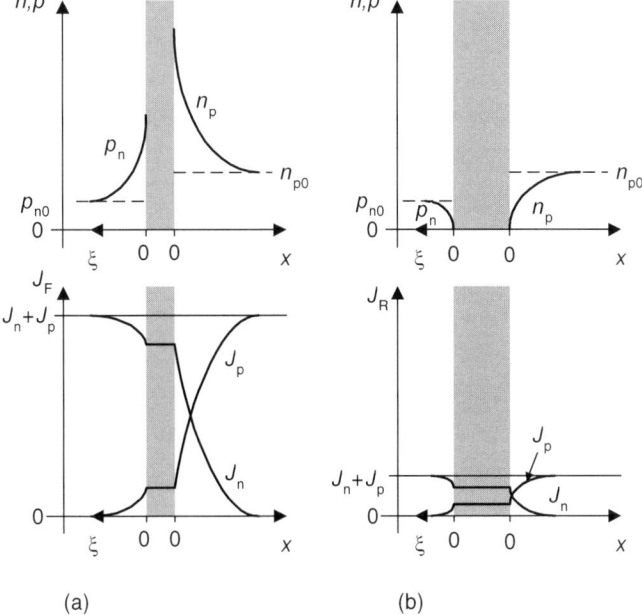

**Figure 2.2** Diagrams showing how the excess carrier concentrations and the proportion of the current density carried by electrons and holes vary with position across: (a) a forward biased p–n junction; (b) a reverse biased p–n junction. The space charge region is shaded

$$J_p = -eD_p \frac{d(\Delta p)}{d\xi} \tag{2.11}$$

The variation of the electron and hole concentrations and the current density carried by each type of carrier across a forward-biased p–n junction are shown in Figure 2.2(a).

Using (2.5), (2.6), (2.8), (2.9) and (2.11), the total current density flowing across the p–n junction is seen to be

$$J = J_n + J_p = e\left(\frac{D_n}{L_n} n_{p0} + \frac{D_p}{L_p} p_{n0}\right)\left[\exp\left(\frac{eV_J}{kT}\right) - 1\right] \tag{2.12}$$

Using (1.6), this can be rewritten as

$$J = n_i^2 e\left(\frac{D_n}{L_n}\frac{1}{p_{p0}} + \frac{D_p}{L_p}\frac{1}{n_{n0}}\right)\left[\exp\left(\frac{eV_J}{kT}\right) - 1\right] = J_0\left[\exp\left(\frac{eV_J}{kT}\right) - 1\right] \tag{2.13}$$

This ideal diode characteristic is illustrated in Figure 2.3(a) and (b). Note that

$$J_0 = n_i^2 e\left(\frac{D_n}{L_n}\frac{1}{p_{p0}} + \frac{D_p}{L_p}\frac{1}{n_{n0}}\right)$$

varies rapidly with temperature through the variation of $n_i^2$.

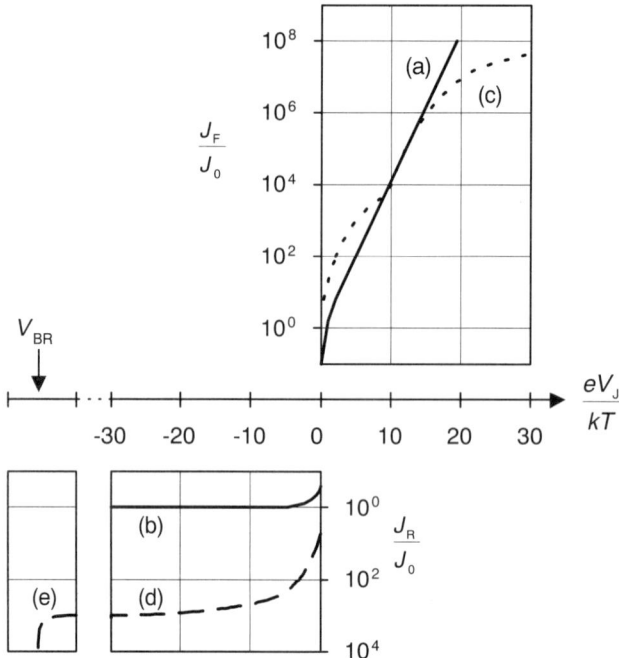

**Figure 2.3** Comparisons between real and ideal voltage–current characteristics of a p–n junction: (a) the ideal forward characteristic predicted by (2.11); (b) the ideal reverse characteristic; (c) a more realistic forward characteristic, taking account of generation–recombination current at low levels of current and the effects of high carrier injection at high currents; (d) a more realistic reverse characteristic, showing a higher leakage current that increases with the reverse bias voltage; (e) the rise of reverse current at the breakdown voltage

When the applied voltage $V_J$ is significantly greater than $+kT/e$ (at $T=300$ K, $kT/e=25.8$ mV), the exponential term in (2.13) is much greater than unity and the forward current density, $J=J_F$, increases exponentially with the applied voltage:

$$J_F = J_0 \exp\left(\frac{eV_J}{kT}\right) \tag{2.14}$$

Note that for a given current density, the forward voltage dropped across the junction varies linearly with temperature, an effect that can be used to determine the junction temperature. Substituting for $n_i$ using (1.6) and combining all the material constants into the single parameter $B$, we obtain

$$V_J = \frac{W_g}{e} - (B - \ln J_F)\frac{kT}{e} \tag{2.15}$$

The sensitivity to temperature decreases as $J_F$ increases.

The excess charge per unit area, $Q/A$, that has to be present in the neutral regions of the bulk semiconductor in order to support these diffusion currents can be found by integrating (2.8) and (2.9) as described at the end of Section 1.5. $Q/A$ varies

## 2.1 THE p–n JUNCTION

linearly with $\Delta p(0)$ and $\Delta n(0)$, hence it varies exponentially with $(eV_J/kT)$ under forward bias conditions.

The excess charge required to support a forward current $I = JA$ in a diode of junction area $A$ may be written as $Q = I\tau$, where $\tau$ is the average lifetime of the carriers. Since a variation of the current requires a variation of this stored charge, its effect is that of a capacitance across the junction. This is known as the diffusion capacitance and is given by

$$C_{\text{diff}} = \frac{dQ}{dV_J} = \frac{dQ}{dI}\frac{dI}{dV_J} = \tau\frac{eI}{kT} \qquad (2.16)$$

where the second term is obtained by differentiating (2.14) and multiplying by the junction area.

When the applied voltage $V_J$ has the opposite polarity, i.e. $V_J < 0$, the junction is said to be *reverse biased*. An argument that parallels our discussion of the forward-biased situation shows that (2.13) still applies. When $-V_J \gg kT/e$ the exponential term is much less than 1 and the current density is simply

$$J = -n_i^2 e\left(\frac{D_n}{L_n}\frac{1}{p_{p0}} + \frac{D_p}{L_p}\frac{1}{n_{n0}}\right) = -J_0 \qquad (2.17)$$

This reverse current results from the flow by diffusion into the space charge region of minority carriers that are thermally generated outside this region. When the holes that are thermally generated in the n-type region reach the edge of the space charge region, the built-in electric field transports them through the space charge region and into the p-type semiconductor. Similarly, electrons thermally generated in the p-type region are transported into the n-type region, as illustrated in Figures 2.1(c) and 2.2(b).

Equation (2.13) has been derived on the assumption that the space charge region is sufficiently thin for the thermal generation of carriers within it to be negligible. Under a reverse bias voltage, the potential barrier at the junction increases and the thickness of the space charge region also increases as a direct result. The thermal generation of electron–hole pairs in the space charge layer may then have to be taken into account. Carriers generated there drift in the built-in electric field. Electrons are transported into the n-type region and holes into the p-type region.

In silicon the main generation process occurs at local recombination centres, as discussed in Section 1.4.2, equation (1.62), where the rate of generation per unit volume per unit time was shown to be

$$G = \frac{n_i}{\tau_{\text{sc}}}$$

where $\tau_{\text{sc}}$ is the carrier lifetime in the space charge region. The additional current density that flows through the junction as a result of this process is known as the *generation–recombination* current. It depends on the magnitude of the reverse bias voltage through the width $d$ of the depletion layer, and it may be expressed as

$$J_{\text{gr}} = e\int_0^d G\,dx = \frac{en_i d}{\tau_{\text{sc}}} \qquad (2.18)$$

The total current density $J_R$ flowing across a reverse-biased junction is given by the sum of $J_0$ and $J_{gr}$:

$$J_R = n_i^2 e \left( \frac{D_n}{L_n} \frac{1}{p_{p0}} + \frac{D_p}{L_p} \frac{1}{n_{n0}} \right) + \frac{e n_i d}{\tau_{sc}^*} \tag{2.19}$$

In the case of silicon devices, the generation–recombination component of $J_R$ usually dominates the diffusion component, which normally becomes significant only at relatively high temperatures. It can be seen that $J_R$ depends on both the temperature and the applied reverse bias voltage. It is illustrated in Figure 2.3(d).

Carrier generation and recombination within the space charge region also influence the current–voltage characteristic of the forward-biased p–n junction, causing departures from (2.13). According to (2.3), throughout the space charge region, the electron and hole concentrations exceed their equilibrium values such that

$$np = n_i^2 \exp\left(\frac{eV_J}{kT}\right)$$

and there is net recombination. This causes an additional current of majority carriers to flow into the space charge layer. The rate of recombination is expected to be greatest at the point in the space charge layer where $n=p$. When the applied junction voltage $V_J$ exceeds $3kT/e$, the rate of recombination there is approximately $n/\tau_{sc}$ (virtually all the carriers are excess carriers). Averaging the recombination across the space charge region by defining an effective lifetime $\tau_{sc}^*$, the additional current density is $(en_i d/\tau_{sc}^*) \exp(eV_J/2kT)$. Taking this into account, the total forward current density may be expressed as

$$J_F = n_i^2 e \left( \frac{D_n}{L_n} \frac{1}{p_{p0}} + \frac{D_p}{L_p} \frac{1}{n_{n0}} \right) \exp\left(\frac{eV_J}{kT}\right) + \frac{en_i d}{\tau_{sc}^*} \exp\left(\frac{eV_J}{2kT}\right) \tag{2.20}$$

The second component (the generation–recombination current) is normally significant only at forward bias voltages less than 0.2 V. At higher forward voltages, the width of the space charge layer is reduced and the first component, the diffusion current, is dominant; then (2.20) becomes identical to (2.14).

Further increase of $V_J$ and hence of $J_F$ leads eventually to the onset of high injection conditions, when the concentration of the excess carriers, at the edge of the depletion layer on one side or the other, becomes comparable with the concentration of the majority carriers there. Dealing again with the edge of the space charge layer in the p-type region, when $\Delta n(0) \gg p_{p0} \approx N_A$, equation (2.3) becomes

$$(\Delta n(0))^2 = n_i^2 \exp\left(\frac{eV_J}{kT}\right)$$

and so

$$\Delta n(0) = n_i \exp\left(\frac{eV_J}{2kT}\right) \tag{2.21}$$

As before, we would expect the current density carried by each type of carrier to be roughly proportional to its concentration at the boundary of the space charge region where it is the minority carrier. Thus, $J_F$ tends once more to vary as $\exp(eV_J/2kT)$. However, at these higher current densities, the ohmic resistance of the bulk semiconductor and the contacts causes the junction voltage $V_J$ to be reduced below the forward voltage $V_F$ applied to the diode. All these effects are illustrated in Figure 2.3(c), which is based on an early paper by Moll [2.3].

At medium levels of injection, the excess carrier concentration at the edge of the space charge region is often expressed as being proportional to $\exp(eV_J/\alpha kT)$, where $\alpha$ takes on values between 1 and 2, increasing with the injection level. The current density follows a similar relationship with the applied voltage:

$$J_F = J_0' \exp\left(\frac{eV_J}{\alpha kT}\right) \quad (2.22)$$

where $J_0'$ is determined empirically and $\alpha$ increases with the current density.

The current flowing across the p–n junction consists of both electrons and holes. The term *injection efficiency* is used to express the fraction of the total current density transported by one or other of the carrier types. Thus, the electron injection efficiency is

$$\tilde{\gamma}_n = \frac{J_n}{J_n + J_p} \quad (2.23a)$$

and the hole injection efficiency is

$$\tilde{\gamma}_p = \frac{J_p}{J_n + J_p} \quad (2.23b)$$

An examination of (2.14) shows that when the p–n junction is forward biased and $n_{n0} \gg p_{p0}$, the current is carried mainly by the electrons injected into the p-type region ($\tilde{\gamma}_n \approx 1$, $\tilde{\gamma}_p \approx 0$). When $n_{n0} \ll p_{p0}$, $\tilde{\gamma}_n \approx 0$ and $\tilde{\gamma}_p \approx 1$.

The generation–recombination components of the current reduce the injection efficiency and therefore, at low current densities, it can be quite low. Another feature that affects the calculation of injection efficiency is the bandgap narrowing that occurs when one of the layers is very heavily doped, as described in Section 1.2. This causes the $np$ product, and hence the minority carrier concentration, to be increased in the heavily doped region by the factor $\exp(\Delta W_g/kT)$. Thus, at an n$^+$–p junction, the hole current density $J_p$ is increased by this factor and the electron injection efficiency predicted by (2.14) is too high. The converse applies at a p$^+$–n junction.

### 2.1.2 Reverse Bias

In Section 2.1.1 we were mainly concerned with the problem of the forward-biased p–n junction, where the most important effect is the minority carrier injection into the regions outside the space charge layer and their subsequent diffusion into the bulk semiconductor. The reverse-biased p–n junction is important because the relatively thick space charge region acts as an insulating layer that is essential for many device applications. In this section we examine it in more detail.

The dependence of the electric field and potential distributions on the donor and acceptor concentration profiles can be obtained from the solution of the Poisson equation (1.84). This is less complicated for the one-dimensional case, when the impurity concentration varies only in the $x$-direction. If the concentrations of free carriers can be neglected throughout the space charge region ($n=p=0$), equation (1.84) becomes

$$\frac{d^2 V}{dx^2} = -\frac{dE}{dx} = -\frac{e}{\varepsilon_r \varepsilon_0}[N_D(x) - N_A(x)] \quad (2.24)$$

To simplify the analysis of the space charge region, we reset the origin of the $x$-coordinate to be at the junction, i.e. at the point where $N_A = N_D$. Outside the space charge region, we assume electrical neutrality, i.e. $E=0$, and this requires that

$$\int_{-d_1}^{d_2} [N_A(x) - N_D(x)]dx = 0 \quad (2.25a)$$

when the space charge region extends from $-d_1$ in the p-type region to $d_2$ on the n-type side. More specifically

$$\int_{-d_1}^{0} [N_A(x) - N_D(x)]dx = \int_{0}^{d_2} [N_D(x) - N_A(x)]dx \quad (2.25b)$$

Thus, the space charge region extends further into the less heavily doped material.

The electric field and potential distributions depend on the distribution of the active impurities, $N_D(x) - N_A(x)$, which in turn depends on the technology used to fabricate the junction. In the case of power devices, this is usually by the diffusion of donor or acceptor impurities into material that is more or less uniformly doped with the opposite type of impurity, as described in Section 3.2. When, for example, acceptors are diffused into a layer of n-type semiconductor that is uniformly doped with a donor concentration $N_D$, the net dopant profile is

$$N_D(x) - N_A(x) = N_D - N_{A0}\,\text{erfc}\left(\frac{x + x_j}{2\sqrt{D_A t_{\text{diff}}}}\right) \quad (2.26)$$

where $N_{A0}$ is the acceptor concentration at the surface, $D_A$ is the diffusion coefficient for acceptor atoms at the temperature at which the diffusion is carried out, $t_{\text{diff}}$ is the diffusion time at this temperature and $x_j$ is the depth of the junction below the surface of the wafer. The profile is shown schematically in Figure 2.4(a). Note that $N_{A0}$ and $x_j$ are related by $N_D/N_{A0} = \text{erfc}(x_j/2\sqrt{D_A t_{\text{diff}}})$.

It is possible to obtain exact solutions of the electric field and potential distributions using numerical modelling. Approximate analytical solutions of the Poisson equation can be found if some simplifying assumptions are made. The first is that the junction is abrupt, with the dopant concentrations uniform on either side, as shown in Figure 2.4(b). This approximation is often used for modelling the p–n junction at high reverse bias voltages. In this case the assumed charge distribution is shown in Figure 2.5(a). In the p-type region ($-d_1 < x < 0$), $N_D = 0$ and the solution of Poisson's equation is

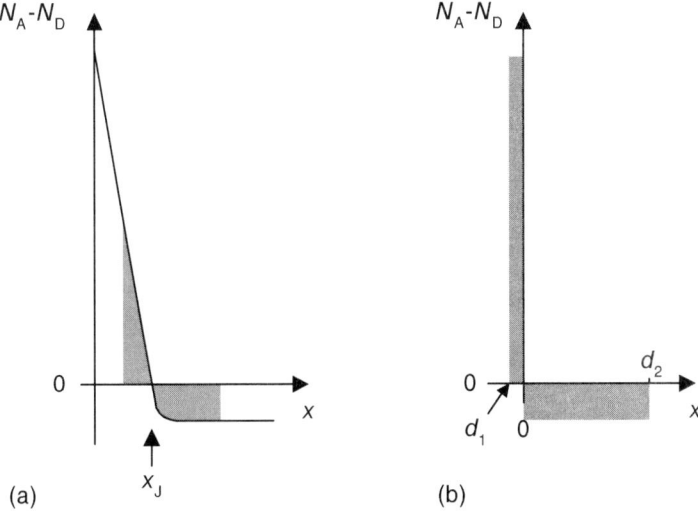

**Figure 2.4** A comparison of the doping profiles across: (a) a typical diffused p–n junction (b) an ideal abrupt junction

$$E(x) = \frac{e}{\varepsilon_r \varepsilon_0} \int_{-d_1}^{x} N_A dx' = -\frac{eN_A}{\varepsilon_r \varepsilon_0}(x + d_1) \quad (2.27)$$

where $x'$ is a dummy x-coordinate. We have assumed that $E$ falls to zero at the edge of the space charge region and remains zero throughout the bulk semiconductor. Thus, $E(-d_1) = 0$.

In the n-type portion of the space charge layer $(0 < x < d_2)$ we assume that $N_A = 0$, thus

$$E(x) = \frac{-e}{\varepsilon_r \varepsilon_0} \int_{x}^{d_2} N_D dx' = -\frac{eN_D}{\varepsilon_r \varepsilon_0}(d_2 - x) \quad (2.28)$$

assuming also that $E(d_2) = 0$.

The electric field distribution described by (2.27) and (2.28) is shown in Figure 2.5(b). The field is maximum at the junction $(x = 0)$, where it reaches the value

$$-E_{max} = \frac{eN_D d_2}{\varepsilon_r \varepsilon_0} = \frac{eN_A d_1}{\varepsilon_r \varepsilon_0} \quad (2.29)$$

The voltage distribution across the space charge region is obtained by integrating (2.24) and (2.25), as shown in Figure 2.5(c). If the applied reverse bias voltage is $V_{app} = -V_R$, the total voltage dropped across the space charge region is

$$V_{diff} - V_{app} = V_{diff} + V_R = \int_{-d_1}^{d_2} E(x) dx = \frac{e}{2\varepsilon_r \varepsilon_0}(N_A d_1^2 + N_D d_2^2) \quad (2.30)$$

Using the result of (2.25), namely $N_A d_1 = N_D d_2$, this can be re-expressed as

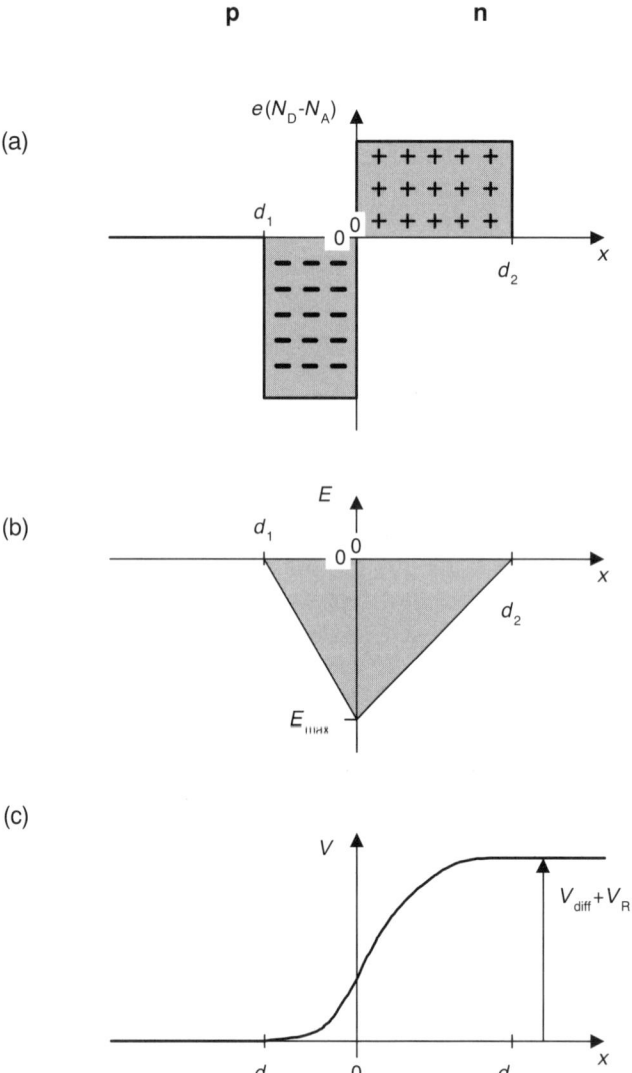

**Figure 2.5** Curves for an abrupt p–n junction showing the variation with position of: (a) the impurity concentration (and hence the charge distribution); (b) the electric field; (c) the potential

$$V_{\text{diff}} + V_R = \frac{ed^2}{2\varepsilon_r\varepsilon_0} \frac{N_A N_D}{(N_A + N_D)} \qquad (2.31)$$

where $d = d_1 + d_2$.

Usually, $V_R$ is much larger than $V_{\text{diff}}$, so the diffusion potential can be neglected. The reverse voltage can be also expressed as

## 2.1 THE p–n JUNCTION

$$V_R = \frac{\varepsilon_r \varepsilon_0}{2e} E_{max}^2 \left( \frac{1}{N_A} + \frac{1}{N_D} \right) \quad (2.32)$$

It follows from (2.30) that almost all the reverse voltage is dropped across the space charge region in the less heavily doped semiconductor.

The total thickness of the space charge region can be expressed as

$$d = d_1 + d_2 = \left[ \frac{2\varepsilon_r \varepsilon_0}{e} \frac{(N_A + N_D)}{N_A N_D} (V_{\text{diff}} + V_R) \right]^{1/2} \quad (2.33)$$

The reverse-biased p–n junction acts like a capacitance between the n-type and p-type conducting regions. The incremental capacitance of an abrupt plane junction of area $A$ under reverse bias $V_R$ is given by the rate of change with bias voltage of the charge stored in the space charge layer:

$$C_j = \frac{dQ}{dV} = \frac{\varepsilon_r \varepsilon_0 A}{d} = A \left[ \frac{e \varepsilon_r \varepsilon_0}{2} \frac{N_A N_D}{(N_A + N_D)} \right]^{1/2} (V_{\text{diff}} + V_R)^{-1/2} \quad (2.34)$$

This capacitance depends mainly on the concentration profile of the less heavily doped region. It decreases with increasing reverse voltage and increases very steeply in the case of a forward bias, when it adds to the diffusion capacitance.

At low reverse bias voltages (up to a few tens of volts) the linearly graded junction is a better approximation to the doping profile of a junction that is fabricated by the diffusion of donor or acceptor impurities. The analytical treatment may be extended to this and other junction profiles [2.1, 2.2].

### 2.1.3 Electrical Breakdown

In between two successive scattering interactions, the electrons and holes in the space charge region are accelerated in the electric field and their kinetic energy increases. The strength of the field increases with increasing reverse bias voltage. Although the average drift velocity of the carriers reaches the maximum value $v_{\text{dmax}}$ at high field strengths, as shown in Figure 1.12, their velocity distribution develops a high energy tail. If, as a result, the kinetic energy of some of the carriers becomes high enough, an electron–hole pair can be generated in an inelastic collision between an accelerated carrier and a neutral atom. The new free carriers so generated are also accelerated by the electric field and they too generate further carriers. The free carrier concentration thus increases in an avalanche multiplication process. To initiate this effect, the free carriers have to gain an energy of about twice the bandgap energy.

Avalanche multiplication coefficients, $\alpha_e(E)$ and $\alpha_h(E)$, are defined as the number of electron–hole pairs generated per unit length of the carrier trajectory in the direction of the electric field. They are steeply increasing functions of the electric field strength. Room temperature measurements of the multiplication of optically generated carriers at a silicon p–n junction indicate that the multiplication coefficient for electrons is about $10^6 \, \text{m}^{-1}$ ($10^4 \, \text{cm}^{-1}$) at a field strength of $3 \times 10^7 \, \text{V/m}$ ($3 \times 10^5 \, \text{V/cm}$) whereas the coefficient for holes is about an order of magnitude less. Both the electrons and the holes so generated can go on to create more carriers, and this introduces an element of positive feedback into the process. When each carrier creates, on average, one further electron–hole pair in the space charge region, the reverse current increases very steeply with reverse voltage, as illustrated in curve Figure 2.3(e). In theory the current increases without limit when

$$\int_{-d_1}^{d_2} \alpha_e(E) \exp\left[\int_{-d_1}^{x} -(\alpha_e - \alpha_h)\, dx'\right] dx = 1 \tag{2.35}$$

using $x'$ as a dummy coordinate. This condition is known as avalanche breakdown and it occurs when the peak electric field at the junction reaches the critical breakdown field $E_{BR}$, i.e. when the reverse bias voltage reaches $V_{BR}$. Then

$$V_{BR} = \frac{\varepsilon_r \varepsilon_0}{2e} E_{BR}^2 \left(\frac{1}{N_A} + \frac{1}{N_D}\right) \tag{2.36}$$

A useful approximation, originally suggested by Fulop [2.4] for p–n junctions in silicon and widely used for purposes of analysis, is to define a composite multiplication coefficient $\alpha(E)$ for electrons and holes such that

$$\frac{\alpha(E)}{(m^{-1})} = \frac{1.8 \times 10^{-47} E^7}{(V/m)^7} \quad \left(\frac{\alpha(E)}{(cm^{-1})} = \frac{1.8 \times 10^{-35} E^7}{(V/cm)^7}\right) \tag{2.37}$$

Then (2.35) reduces to $\int_{-d_1}^{d_2} \alpha(E)\, dx = 1$, and in the case of a highly asymmetric, abrupt, plane p–n junction in silicon, it is possible to express the breakdown field, the breakdown voltage and the width of the space charge region at breakdown as functions of the effective dopant concentration, $N = |N_D - N_A|$, in the less heavily doped region:

$$\frac{E_{BR}}{V/m} = 7.13 \times 10^4 \left(\frac{N}{m^{-3}}\right)^{1/8} \quad \left(\frac{E_{BR}}{V/cm} = 4010 \left(\frac{N}{cm^{-3}}\right)^{1/8}\right) \tag{2.38}$$

$$\frac{V_{BR}}{V} = 1.68 \times 10^{18} \left(\frac{N}{m^{-3}}\right)^{-3/4} \quad \left(\frac{V_{BR}}{V} = 5.34 \times 10^{13} \left(\frac{N}{cm^{-3}}\right)^{-3/4}\right) \tag{2.39}$$

$$\frac{d_{BR}}{m} = 4.75 \times 10^{13} \left(\frac{N}{m^{-3}}\right)^{-7/8} \quad \left(\frac{d_{BR}}{cm} = 2.67 \times 10^{10} \left(\frac{N}{cm^{-3}}\right)^{-7/8}\right) \tag{2.40}$$

The derivation of these expressions is left as an exercise for the reader. The variation of $V_{BR}$ with $N$ is shown in Figure 2.6 and the variation of $d_{BR}$ with $N$ in Figure 2.7. The maximum extent of the space charge region increases as the impurity concentration in the more lightly doped region decreases, i.e. with increasing breakdown voltage. For the plane, asymmetric, abrupt junction

$$d_{BR} = \sqrt{\frac{2\varepsilon_r \varepsilon_0 V_{BR}}{eN}} = \frac{\varepsilon_r \varepsilon_0 E_{BR}}{eN} = \frac{2V_{BR}}{E_{BR}} \tag{2.41}$$

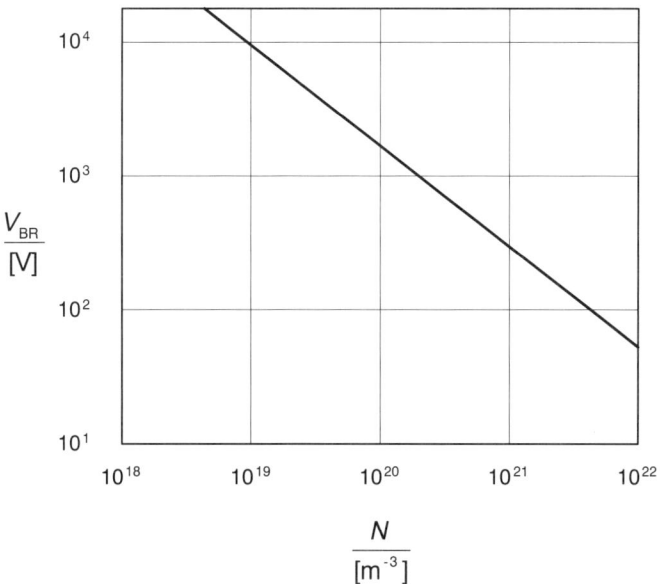

**Figure 2.6** Theoretical dependence of the breakdown voltage of an abrupt p–n junction on the impurity concentration

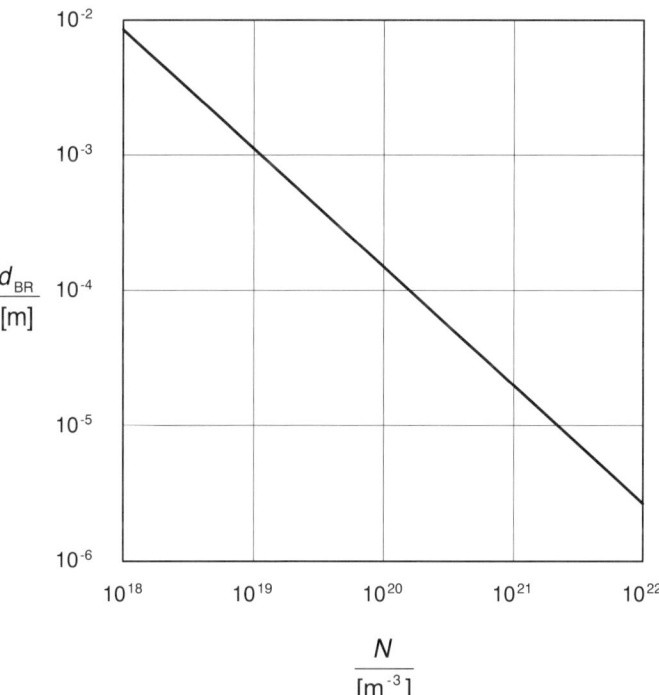

**Figure 2.7** Theoretical dependence of the space charge region thickness at breakdown, $d_{BR}$, on the impurity concentration

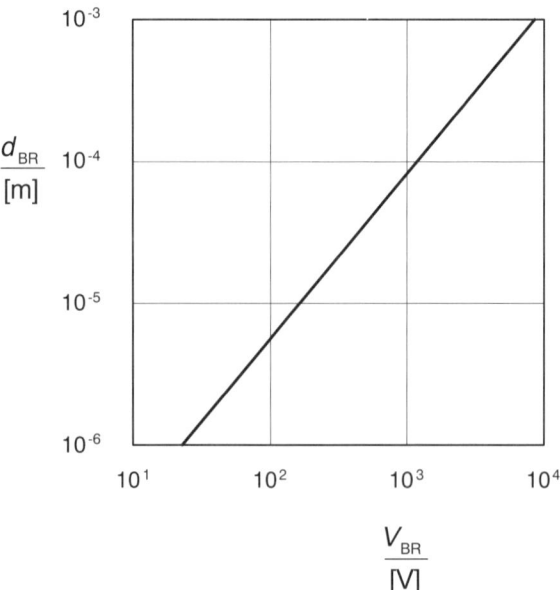

**Figure 2.8** Theoretical dependence of the space charge region thickness, $d_{BR}$, on the breakdown voltage, $V_{BR}$, for a diode with unlimited base widths

Thus for silicon devices

$$\frac{d_{BR}}{m} = 2.6 \times 10^{-8} \left( \frac{V_{BR}}{V} \right)^{7/6} \qquad (2.42)$$

which is plotted in Figure 2.8.

At higher temperatures the average mean free path between scattering collisions decreases and higher values of the electric field are needed if carriers are to reach the energy necessary for the generation of electron–hole pairs. As a result, $\alpha(E)$ is a decreasing function of temperature, and the breakdown voltage increases with temperature. However, the reverse current density $J_R$ increases rapidly with temperature, as given by (2.19), and the combination of these two effects causes the reverse characteristics to vary with temperature as illustrated in Figure 2.9.

As a result of impact ionisation, the reverse current is found to increase at voltages slightly lower than the breakdown voltage. This can be expressed in terms of a multiplication factor $M$, defined by

$$J_R(V_R) = MJ_0 \qquad (2.43)$$

where

$$\frac{1}{M} = 1 - \left( \frac{V_R}{V_{BR}} \right)^{\kappa} \qquad (2.44)$$

with $3 < \kappa < 6$, depending on the material and the technology used to fabricate the junction.

## 2.1 THE p–n JUNCTION

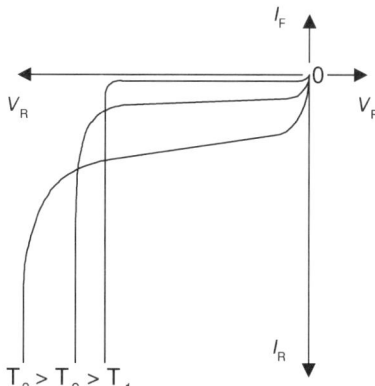

**Figure 2.9** The effect of temperature on the reverse current–voltage characteristics of a p–n junction

The exact solution of the Poisson equation in the case of a plane p–n junction made by diffusion and having the doping concentration profile described by (2.26) predicts a slightly higher breakdown voltage, but the difference is small and the abrupt junction approximation is generally used. The solution becomes much more complicated when the p–n junction is not plane [2.2, Section 2.5]. It is found that the breakdown voltage decreases when the junction is curved, and also when it comes to the surface of a semiconductor structure of finite size. These matters are taken up in later chapters when we discuss details of the construction and fabrication techniques that allow the breakdown voltage of individual types of device to approach that of the infinite plane junction (Sections 3.4 and 5.2).

It has been assumed in this analysis that the impurities in each region are laterally homogeneous. But this may not be the case; suppose there are structural defects in the volume of the semiconductor or on its surface, then the electric field can reach very high values locally and breakdown can occur in very small localised regions at reverse voltages lower than the breakdown voltage of a homogeneous structure. This is known as microplasma breakdown. It causes a change in the reverse current–voltage characteristic and an increase in the diode noise [2.5].

### 2.1.4 Thermal Breakdown and Thermal Runaway

Thermal breakdown results from the temperature rise caused by current flowing through a reverse-biased p–n junction. The power dissipated per unit area is

$$P_J = V_R I_R \tag{2.45}$$

Using (2.19) and (1.6) this becomes

$$P_J = V_R \left[ A_1 T_J^3 \exp\left(\frac{-W_g}{kT_J}\right) + A_2 T_J^{3/2} \exp\left(\frac{-W_g}{2kT_J}\right) \right] \tag{2.46}$$

where $T_J$ is the junction temperature and $A_1$ and $A_2$ are parameters that depend on details of the diode structure such as the junction area, the doping concentrations and the carrier lifetimes. The heat generated at the junction is conducted out of the device at a rate given by

$$P_A = \frac{T_J - T_A}{R_{thJA}} \qquad (2.47)$$

where $T_A$ is the ambient temperature and $R_{thJA}$ is the thermal resistance between the junction and the surrounding ambient into which the heat ultimately flows. Thus, in the steady state, when all the power generated at the junction flows out through the device to the ambient surroundings, $P_A = P_J$, so that

$$V_R \left[ A_1 T_J^3 \exp\left(\frac{-W_g}{kT_J}\right) + A_2 T_J^{3/2} \exp\left(\frac{-W_g}{2kT_J}\right) \right] = \frac{T_J - T_A}{R_{thJA}} \qquad (2.48)$$

The junction temperature can be obtained by solving (2.48), which can be done graphically, as shown in Figure 2.10. The power dissipation increases exponentially with temperature, but the maximum power that can be conducted away increases only linearly. If the two curves intersect in two places, (1) and (2), then point (1), at temperature $T_{op}$, represents a stable solution for thermal balance. A small increase in the junction temperature, for any reason, causes the heat lost by conduction to exceed the additional heat generated. Point (2), at temperature $T_{crit}$, represents an unstable solution, since any increase in the junction temperature causes the power generated to increase (as a result of the increase in the reverse current) by more than the increased heat loss by conduction. This causes thermal runaway, which in turn leads to thermal breakdown.

For thermal stability we require that

$$\frac{\partial P_J}{\partial T_J} < \frac{\partial P_A}{\partial T_A} \qquad (2.49)$$

Simultaneous solution of (2.48) and (2.49) is complicated and solutions exist only when $R_{thJA}$ is less than some critical value, as shown in Figure 2.10.

If, for any reason, the junction temperature locally becomes higher than $T_{crit}$, it starts to increase rapidly in that region. As it rises, the local current density increases and current is drawn into the region where the temperature is highest. Once the temperature at some point reaches the intrinsic temperature $T_i$, defined in Figure 1.3, the local rate of carrier generation can easily rise by several orders of magnitude. The p–n junction is then effectively shunted by a small filament of highly conducting intrinsic semiconductor known as a *mesoplasma*.

The development of a mesoplasma invariably results in an irreversible degradation of the p–n junction. This phenomenon was originally called *second breakdown*, though that term is now normally restricted to bipolar devices where positive feedback causes it to occur at voltages well below the avalanche breakdown voltage. It is better called *thermal runaway* in contrast to ordinary avalanche breakdown in which the temperature remains below $T_{crit}$ and the junction is undamaged. During thermal runaway, a device can be destroyed in several different ways:

## 2.1 THE p–n JUNCTION

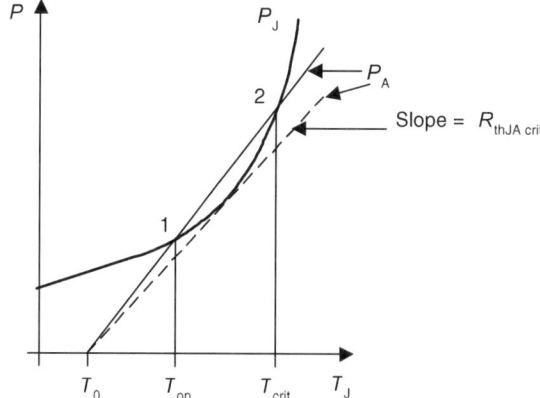

**Figure 2.10** Illustration of the condition for thermal stability of a p–n junction. The dashed line indicates the limiting value of the thermal resistance, above which stable operation is not possible

(a) Thermal shock can damage the crystal lattice and cracks can develop.
(b) The temperature in the mesoplasma region can exceed the melting temperature of the eutectic metal–silicon alloy at the contact.
(c) The temperature in the mesoplasma can exceed the melting temperature of the semiconductor, which for silicon is 1360 °C.

In all cases the formation of a mesoplasma represents a terminal phase that leads to irreversible device failure. A detailed discussion of the development of mesoplasmas and the causes of thermal runaway is given in the literature [2.2].

The onset of thermal runaway sets a limit for the operating conditions of p–n junction devices. For reliability, the condition for thermal stability given by (2.49) must be satisfied. In any application the design of a suitable heat sink is an important consideration and a detailed discussion of the issues involved is given in Section 13.1.

### 2.1.5 The Influence of Illumination on p–n Junction Characteristics

As discussed in Section 1.4, electron–hole pairs can be generated by the absorption of light provided that the photon energy $h\nu$ exceeds the bandgap energy $W_g$. Such excess carrier generation is described by (1.29). It takes place close to the irradiated surface; only radiation whose photon energy is near to or less than the bandgap energy penetrates more deeply into the volume of the semiconductor crystal. From (1.29) it follows that the concentration of the excess carriers generated in this way decreases away from the surface, with the result that a concentration gradient arises and carriers diffuse into the volume of the semiconductor. Those that reach the edge of the space charge region around a p–n junction, together with those generated within the space charge region itself, are subjected to the built-in electric field. Electrons are transferred into the n-type region and holes into the p-type region, where in each case their potential energy is minimum. This is illustrated in Figure

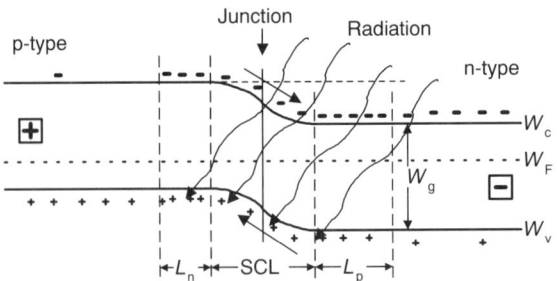

**Figure 2.11** Electron energy level diagram indicating the effect of illumination on carrier generation and flow at a p–n junction

2.11. The current generated by these excess carriers is superimposed on the current of thermally generated carriers. The total current density crossing the p–n junction is thus given by

$$J = J_0 \left[ \exp\left(\frac{eV_{\text{app}}}{kT}\right) - 1 \right] - J_L \qquad (2.50)$$

where $J_0$ is given by (2.17) and $J_L$ is the current density arising from the irradiation of the semiconductor. This depends on the detailed construction of the junction and the intensity and wavelength of the incident radiation [2.1, Section 13.3]. An illustration of the influence of illumination on the current–voltage characteristics of a p–n junction diode at different intensity levels is shown in Figure 2.12.

Under illumination, the current density flowing through a reverse-biased p–n junction increases in direct proportion to the incident optical power. This effect is exploited in devices such as light-triggered thyristors and photodiode detectors. Under open circuit conditions, the n-type region is charged negatively and the p-type region positively. The maximum voltage that can be obtained is limited by the diffusion potential $V_{\text{diff}}$ at the p–n junction. This effect is exploited in solar cells.

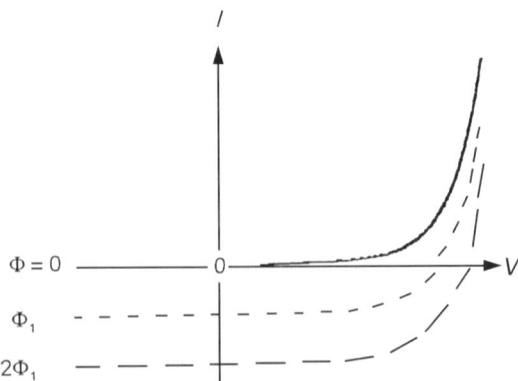

**Figure 2.12** Current–voltage characteristics of a p–n junction, showing the effects of illumination

## 2.1.6 Transient Behaviour of the p–n Junction

All the results derived in the previous sections assume that the p–n junction diode is operating under steady-state conditions, i.e., that all externally applied voltages are constant and a steady-state distribution of carrier concentrations is established. If the applied voltages change, the carrier distributions also change, as a result of the processes of diffusion, drift, generation and recombination that we have discussed. The transient excess carrier distributions can be found by solving the continuity equations (1.73). In the electrically neutral regions adjacent to a p–n junction and in the absence of carrier generation, the evolution of the excess carrier concentration is determined by (1.78). In the one-dimensional case this becomes

$$\frac{\partial \Delta n(x,t)}{\partial t} = -\frac{\Delta n(x,t)}{\tau_n} + \frac{1}{e}\frac{\partial J_n(x,t)}{\partial x} \tag{2.51}$$

The boundary conditions determine the solution of this type of equation. Here it is the constraints imposed by the external circuit on the device containing the p–n junction that set the boundary conditions. In this section we consider some idealised cases in order to establish the basic principles underlying the transient behaviour of p–n junctions. In later chapters these principles are used to obtain an understanding of the waveforms observed when real devices operate in real circuits.

Consider first a forward-biased junction in which the current density carried by the electrons that diffuse into the bulk p-type semiconductor from the edge of the space charge layer is given by

$$J_n(t) = eD_n \frac{\partial \Delta n(0,t)}{\partial x} \tag{2.52}$$

We assume low injection conditions. By analogy, a similar equation can be derived for the holes diffusing into the n-type semiconductor on the opposite side of the depletion layer. One matter that we have to be aware of here is that the boundary we have used to define $x=0$ moves under transient conditions.

In general, the voltage distribution across a device containing a p–n junction can be obtained by solving Poisson's equation in its general form (1.84). However, a simplified, analytical approximation for the voltage across an abrupt, forward-biased p–n junction can be derived from (2.4) as

$$v_J(t) = \frac{kT}{e} \ln\left(\frac{n_p(0,t)}{n_{p0}}\right) \tag{2.53}$$

In (2.53) we are really assuming that the electron concentration throughout the depletion layer remains in thermodynamic equilibrium with the concentration in the n-type region, under both steady-state and transient conditions. Likewise, the concentration of holes in the depletion region may be assumed to remain in thermodynamic equilibrium with the hole concentration in the p-type region. The processes that maintain the equilibrium are assumed to be sufficiently strong and sufficiently rapid to ensure this. Effectively, in the depletion region, the Fermi level for electrons $W_{Fn}$ is separated from the Fermi level for holes $W_{Fp}$ by an amount $eV_J$ corresponding to the voltage across the junction, as shown in Figure 2.1.

If, during a transient process, $n_p(0,t)$ becomes *less* than the equilibrium value $n_{p0}$, additional space charge forms. Then the instantaneous *reverse* voltage that appears across the junction is determined by the instantaneous thickness $d(t)$ of the space charge layer. With an abrupt, asymmetrical junction, where the impurity concentration of the p-type region $(N_A - N_D)$ is either much greater or much less than for the n-type region $(N_D - N_A)$, the reverse voltage can be expressed using (2.31) as

$$v_R(t) = \frac{e|N_D - N_A|}{2\varepsilon_r \varepsilon_0} d^2(t) \qquad (2.54)$$

For illustration, we apply these principles to an idealised, plane $n^+$–p junction that is first subjected to a sudden transition from a steady-state forward bias condition to one of reverse bias and then the converse, from steady-state reverse bias to forward bias. The circuit is shown in Figure 2.13 and the two situations are illustrated in Figures 2.14 and 2.15. In each figure, part (a) shows the current that flows through the diode, part (b) the junction voltage $V_J$, and part (c) the evolution of the minority electron concentration distribution in the p-type region. The curves in (c) hold the key to the observed waveforms.

Consider first the diode being turned off (Figure 2.14). At $t=0$ the situation is representative of steady-state forward conduction. Reversal of the applied voltage causes an almost instantaneous reversal of the current to a value $-(V_2 + v_J)/R_R$, mainly governed by the circuit. In the p-type semiconductor this current is carried by electrons diffusing into the space charge region, and it determines the electron concentration gradient at the boundary of that region. Further into the p-type semiconductor, electrons and holes diffuse in the opposite direction, towards the contact. They recombine either within the bulk material or at the contact. The

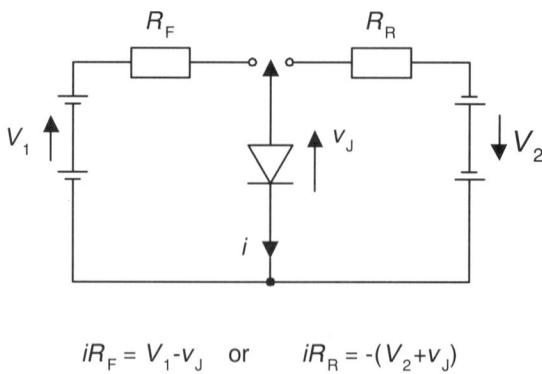

**Figure 2.13** Idealised circuit used to illustrate transient switching effects at a p–n junction

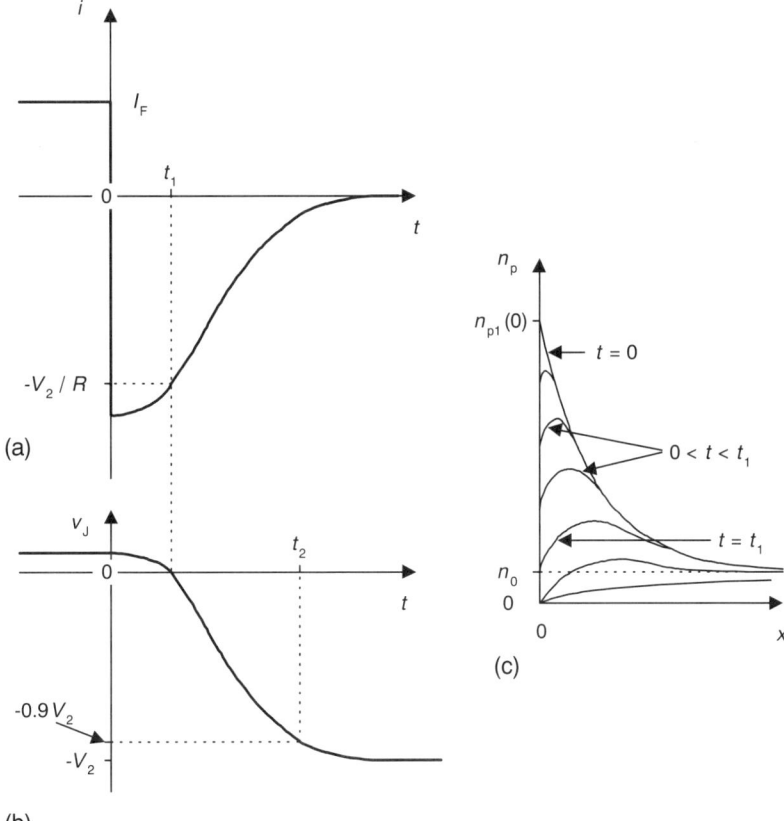

**Figure 2.14** The turn-off process at an abrupt n$^+$–p junction: (a) diode current waveform; (b) junction voltage waveform; (c) evolution of the distribution of excess carriers

electron concentration thus diminishes until, after a time $t=t_1$, $n(0)=n_{p0}$ and $v_J$, which falls slowly during this period, reaches zero and then reverses. A slight expansion of the space charge region accompanies this variation.

A very short time after $t_1$ the electron concentration at the edge of the space charge region becomes almost zero, $n(0)=0$. The continuing outdiffusion of electrons causes the concentration gradient at the edge of the space charge region, and with it the reverse current $(-i)$ to decrease exponentially with time after $t_1$. The reverse voltage across the junction $-v_J = V_2 - R_R(-i)$ thus increases and the expansion of the space charge region is accelerated. This continues until steady-state reverse bias conditions are established. At time $t_2$, shown in Figure 2.14, $-v_J$ has reached $0.9 V_2$. A fuller analysis would have to take account of the effects of circuit inductance and junction capacitance. These are dealt with in the context of practical circuits in later chapters.

Now consider the diode turning on, as shown in Figure 2.15. After the reversal of the applied voltage from $-V_2$ to $+V_1$, the space charge region is discharged and the voltage across the junction $v_J$ rises from $-V_2$ initially to reach zero at time $t_3$. At that point, the electron concentration in the p-type region at the edge of the space charge layer is $n_{p0}$. It continues to rise at a rate limited by the diffusion of the electrons into

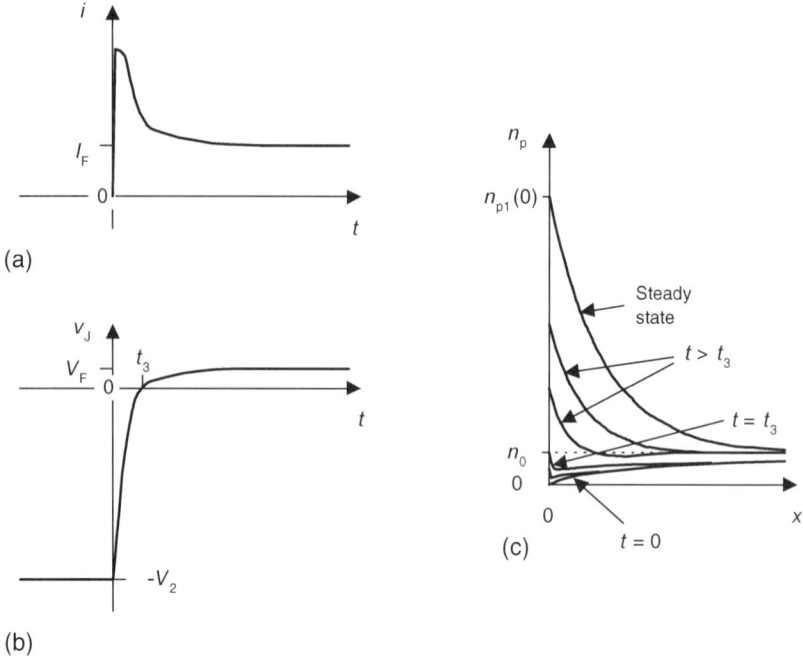

**Figure 2.15** The turn-on process at an abrupt n⁺–p junction: (a) diode current waveform; (b) junction voltage waveform; (c) evolution of the distribution of excess carriers

the bulk material. This is governed by the available current $(V_1 - v_J)/R_F$, which determines the electron concentration gradient at the boundary of the space charge layer and changes very little during this period. Excess carriers continue to build up until the steady-state forward bias condition is established, with a steady current $(V_1 - V_F)/R_F$ flowing.

This analysis is based on solutions of the diffusion equation, which depend on the boundary values and the initial conditions. By specifying the current at $x=0$ as the controlling parameter, we have set the boundary conditions in terms of the gradient of the variable $n(x,t)$ at this boundary. In the less common situation where it is the voltage across the junction that is the controlling parameter, the values of $n(0,t)$, as determined by (2.53), set the boundary conditions. We have ignored the fact that the boundary may move. Exhaustive treatments of the solutions of the diffusion equation are to be found in the literature [2.6, 2.7]. A brief analysis of the transitions described here is given in the appendix. Following a sudden change in the value of $n(0)$, the subsequent variation of $n(x,t)$ over distances short compared to $L_n$ is mainly a function of $x/2\sqrt{D_n t}$. This means that the time required for the concentration at a distance $w$ from the origin to reach a particular value is proportional to $w^2/4D_n$.

For the accurate modelling of semiconductor structures containing p–n junctions, it is necessary to provide a simultaneous solution of (1.69), (1.72) and (1.84). When the electric field changes rapidly, the displacement current density $\varepsilon_r \varepsilon_0 \, \partial E/\partial t$ must be added to the carrier current density (1.71). A general solution is very complicated and the transient behaviour of devices is normally analysed for particular structures under specific initial and boundary conditions, usually by numerical simulation. Equation (1.79), which expresses in integral form the time variation of the total excess charge stored in the neutral semiconductor regions, is the basis of several simplified techniques that are widely used to model the behaviour of junction devices under transient conditions. Further details on numerical modelling methods for semiconductor devices are given in the literature [2.8 to 2.10]. In later chapters the basic principles outlined here are applied to different device structures.

## 2.2 n–n$^+$ and p–p$^+$ Junctions

For reasons discussed in Section 2.4, it is normal practice for the semiconductor regions in the vicinity of a contact to be heavily doped. In order to obtain high breakdown voltages, the active regions of power devices normally include at least one layer of lightly doped material. So it is reasonable to expect there will be one or more junctions between heavily doped and lightly doped material of the same type, and the properties of these junctions can have a significant effect on the detailed device characteristics.

Because of the change in the position of the Fermi energy with the doping concentration, when an n$^+$-region, doped with a donor concentration $N_{D+}$ is in contact with an n-region doped with a donor concentration $N_D$, we should expect a diffusion potential $V_{n^+n}$, to be set up between them; that is,

$$V_{n^+n} = \frac{kT}{e} \ln\left(\frac{N_{D+}}{N_D}\right) \tag{2.55}$$

When the junction is abrupt, as shown in Figure 2.16(a), we might expect a narrow depletion region to be formed on the n$^+$-side and an accumulation layer, also very narrow, to arise on the n-side. A similar diffusion potential, $V_{p^+p}$, occurs at a p$^+$–p junction:

$$V_{p^+p} = \frac{kT}{e} \ln\left(\frac{N_{A+}}{N_A}\right) \tag{2.56}$$

where $N_{A+}$ and $N_A$ are the acceptor concentrations in the two regions.

In practice all such junctions are graded, so the transition takes place over a distance greater than the thickness of the space charge layer that would form at an abrupt junction. The result is that approximate charge neutrality is maintained and a built-in electric field $E$ is set up, as described in Section 1.6. This field exactly balances the tendency of the carriers to diffuse down the concentration gradient. At an n$^+$–n transition,

$$E = \frac{-kT}{eN_D} \frac{dN_D}{dx} \tag{2.57}$$

when $N_D$ varies only in the x-direction. At a similar p$^+$–p transition,

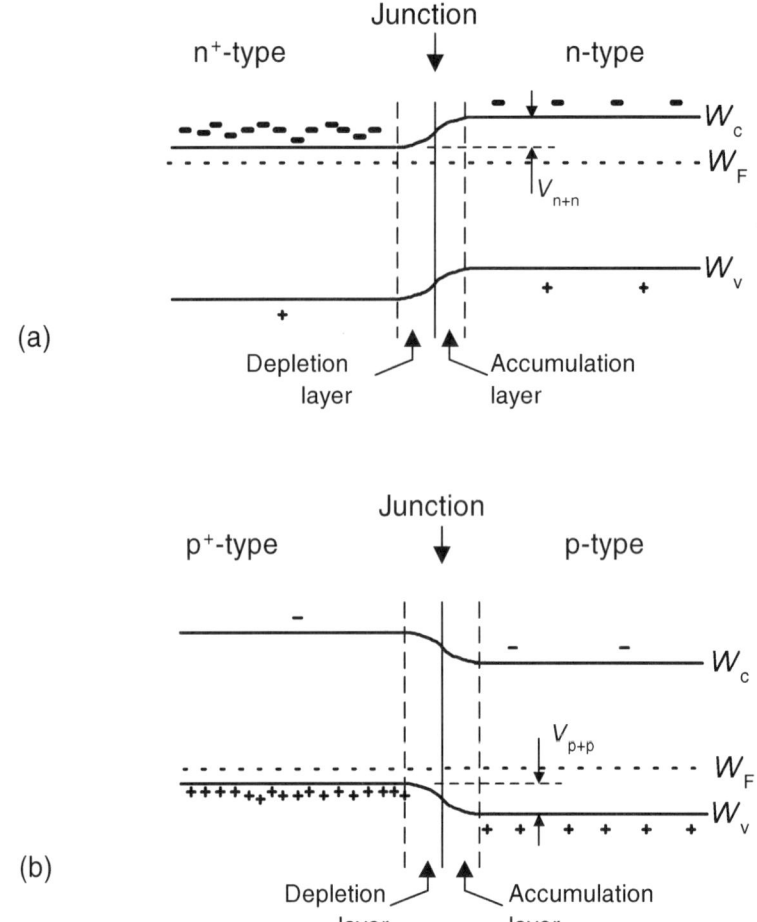

**Figure 2.16** Band diagrams, showing the variation of the electron energy levels across: (a) an abrupt n⁺–n junction; (b) an abrupt p⁺–p junction

$$E = \frac{kT}{eN_A}\frac{dN_A}{dx} \qquad (2.58)$$

Unlike a p–n junction, n⁺–n and p⁺–p transitions do not present a high impedance when they are biased. At an n⁺–n junction, making the n⁺-region positive with respect to the n-region adds to the diffusion potential and helps to accelerate the majority electrons out of the n-region and into the n⁺-region. Any minority holes would be accelerated out of the n⁺-region into the n-region, but as there are likely to be very few, the injection efficiency of this structure for holes may be considered to be zero and for electrons almost one.

Bias of the opposite polarity subtracts from the diffusion potential and reduces the magnitude of the built-in field. There is a net flow of electrons out of the n⁺-region and these diffuse into the n-region. Minority holes are assisted to flow out of the n-region and into the n$^+$-region. When there is an excess of holes flowing by diffusion through the n-region towards the n$^+$–n junction, they are assisted into the n$^+$-region, where they normally recombine. The recombination lifetimes in the heavily doped regions are usually very short, so the diffusion lengths are short compared to the thickness of these layers. In this situation it is often convenient to characterise recombination in the n$^+$-region by a recombination velocity $s_p$, as defined in Section 1.4.3.

Consider the situation shown in Figure 2.17, where a current density $J_p$ of holes leads to an excess hole concentration $\Delta p(0)$ at the boundary of the n$^+$-layer. Within the heavily doped region, the hole diffusion coefficient is $D_p$ and the recombination lifetime is $\tau_p$. Then

$$s_p = \sqrt{\frac{D_p}{\tau_p}} \qquad (2.59)$$

In Section 1.2 we saw that bandgap narrowing takes place at high impurity levels. Usually, the changes in the positions of the conduction and valence bands are not equal, with the result that the built-in fields for the electrons and holes may be different. As with a p–n junction, described in Section 2.1.1, bandgap narrowing also affects the injection efficiency. At an n$^+$–n junction, bandgap narrowing in the n$^+$-region causes the equilibrium minority hole concentration to be increased by the factor $\exp(\Delta W_g/kT)$ and so reduces the injection efficiency for electrons.

## 2.3 Surface Effects and MOS Structures

The boundary surface of a finite crystal represents a considerable disruption to the periodicity of the crystal lattice. The valence electrons of the atoms in the surface layer are unable to bond properly because of the missing layer beyond. The unoccupied positions at the crystal boundary result in additional local electron energy levels within the bandgap. The surface atoms also interact with their

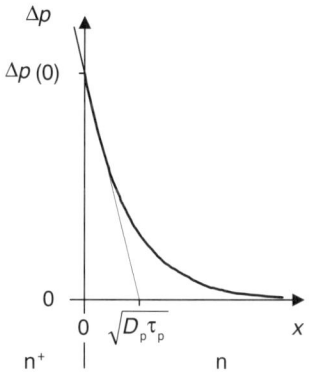

**Figure 2.17** Carrier concentrations at a biased n$^+$–n junction

surroundings, with the result that the surface is usually covered with a thin layer of oxide or other adsorbed atoms or molecules. Such surface layers may be charged, so attracting or repelling the free charge carriers of the bulk semiconductor and altering the band structure at the surface. Consequently, there are variations in the carrier concentrations at the surface. Some of the resulting features are discussed in this section. More detailed theory can be found in the literature [2.11, Ch. 2].

Consider a semiconductor having a surface charge per unit area $Q_{sc}/A$. This attracts carriers of opposite polarity and repels those with the same polarity. A space charge region is induced at the surface by this charge and the potential energy of the free carriers is also changed, causing a variation in the band structure. As before, we set the zero of potential to be that of an electron at the bottom of the conduction band in the bulk semiconductor, in a defect-free region. Then, in thermal equilibrium, the carrier concentrations vary with the local value of the potential $V$ as follows:

$$n = n_0 \exp\left(\frac{eV}{kT}\right) \quad (2.60a)$$

and

$$p = p_0 \exp\left(\frac{-eV}{kT}\right) \quad (2.60b)$$

where $n_0$ and $p_0$ are the equilibrium concentration of free electrons and holes in the volume of the crystal.

The potential distribution $V(x)$ in the surface region is related to the local space charge density $\rho(x)$, given by

$$\rho(x) = e(N_D^+ - N_A^- + p - n) \quad (2.61)$$

where $N_D^+$ and $N_A^-$ are the concentrations of ionised donors and filled acceptors, respectively.

The potential distribution can be found by solving Poisson's equation, which when $\rho(x)$ varies only in one dimension, has the form

$$\frac{d^2 V}{dx^2} = -\frac{dE}{dx} = \frac{-\rho(x)}{\varepsilon_r \varepsilon_0} \quad (2.62)$$

We think of the $x$-coordinate as directed into the bulk semiconductor at right angles to the surface, with its origin set at the surface. Away from the surface region, $\rho = 0$ as required by the condition for electrical neutrality in the bulk semiconductor. Within the surface region, the Poisson equation (2.62) is given by

$$\frac{d^2 V}{dx^2} = \frac{-\rho}{\varepsilon_r \varepsilon_0}\left\{p_0\left[\exp\left(\frac{-eV}{kT}\right) - 1\right] - n_0\left[\exp\left(\frac{eV}{kT}\right) - 1\right]\right\} \quad (2.63)$$

utilising (2.61) and (2.60). Using Gauss's law, the relationship between the charge per unit area in the surface layer, $Q_{sc}/A$, and the surface potential $V_S$ can be expressed as

## 2.3 SURFACE EFFECTS AND MOS STRUCTURES

$$\frac{Q_{sc}}{A} = \varepsilon_r \varepsilon_0 E_S = -\varepsilon_r \varepsilon_0 \left.\frac{dV}{dx}\right|_{V=V_S} \quad (2.64)$$

The potential and carrier concentration at the surface depend on the type and density of the surface charge. The following situations are illustrated in Figure 2.18. For p-type semiconductor:

(a) A negative surface charge, $V_S < 0$. The hole concentration increases at the surface in an *accumulation layer* of thickness $h_a$, as shown in Figure 2.18(a).

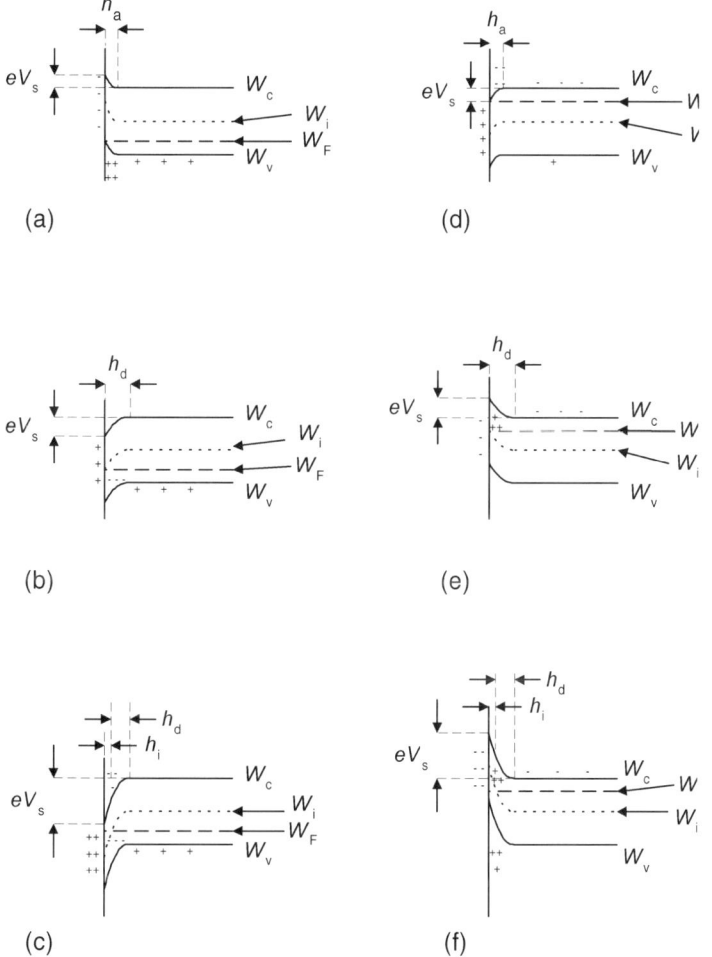

**Figure 2.18** Electron energy level diagrams, showing the effect of surface charge on the band structure: (a) formation of an accumulation layer in p-type material; (b) formation of a depletion layer in p-type material; (c) formation of an inversion layer in p-type material; (d) formation of an accumulation layer in n-type material; (e) formation of a depletion layer in n-type material; (f) formation of an inversion layer in n-type material

(b) A small positive surface charge, so that $(V_F - V_i) > V_S > 0$, where $V_F$ is the Fermi potential. There is a reduced concentration of holes at the surface and a *depletion layer* of thickness $h_d$, forms, as shown in Figure 2.18(b).
(c) A larger positive surface charge, so that $V_S > (V_F - V_i)$. At the surface, the Fermi level $W_F = eV_F$ is closer to the conduction band than the valence band and an *inversion layer*, in which electrons are the majority carriers, is created. This has a thickness $h_i$ and is illustrated in Figure 2.18(c). The inversion layer is separated from the bulk semiconductor by a depletion layer of thickness, $h_d$.

For n-type semiconductor:

(a) A positive surface charge, $V_S > 0$. An accumulation layer with an enhanced concentration of majority electrons is set up at the surface, as shown in Figure 2.18(d).
(b) A small negative surface charge, so that $(V_F - V_i) < V_S < 0$. In this case a depletion layer with reduced electron concentration forms at the surface, as shown in Figure 2.18(e).
(c) A larger negative surface charge, so that $V_S < (V_F - V_i) < 0$. In this case an inversion layer, with holes as the majority carrier, forms at the surface, as shown in Figure 2.18(f).

Accurate computation of the potential and electric field distributions induced by a surface charge is made difficult by the interactive effect of the free carriers, whose concentration varies exponentially with the potential. A number of computer simulation tools for device modelling provide solutions to Poisson's equation under these conditions.

## 2.4 Metal–Semiconductor Contacts

Metal–semiconductor contacts are essential for the connection of semiconductor components to external circuits. They therefore have to be formed at all semiconductor device terminals. The metal–semiconductor contact, as a junction of two different materials, represents another type of inhomogeneity that needs to be studied. Its characteristics are non-linear in general and they vary considerably from those of a fully rectifying contact at one extreme to those of a low resistance ohmic contact at the other.

### 2.4.1 Rectifying (Schottky) Contacts

The energy band diagram of an ideal contact between a metal of work function $\phi_M$ and an n-type semiconductor is shown in Figure 2.19(a). The n-type semiconductor is characterised by its electron affinity $\chi$, which specifies the energy required to raise an electron from the level $W_c$, corresponding to the bottom of the conduction band, to the vacuum level and so release it from the crystal. When a metal and a semiconductor are joined, electrons from the semiconductor cross to the metal until the Fermi level of the whole of the metal–semiconductor system is aligned. In

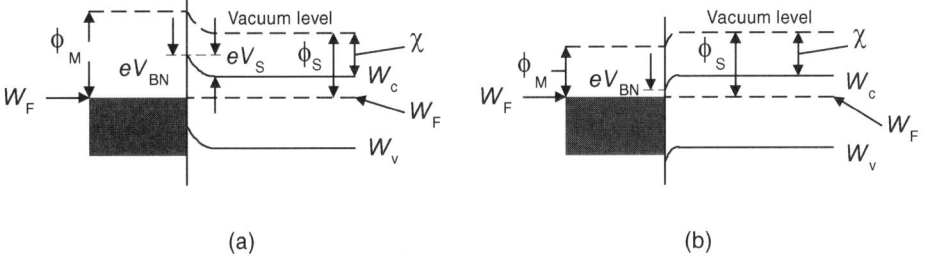

**Figure 2.19** Electron energy level (band) diagrams across different possible combinations of metal–semiconductor contact: (a) a metal and n-type semiconductor, forming a depletion layer at the surface (rectifying); (b) a metal and n-type semiconductor, forming an accumulation layer at the surface (ohmic)

thermal equilibrium the Fermi level has to be uniform everywhere. In the metal it is $\phi_M$ below the vacuum level, whereas at the metal–semiconductor boundary the bottom of the conduction band is $\chi$ below the vacuum level. When $\phi_M$ is large, potential barriers impeding the free flow of electrons are established on both sides of the boundary. On the metal side, the barrier height is given by

$$eV_{BN} = \phi_M - \chi \qquad (2.65)$$

while on the semiconductor side it is

$$eV_S = \phi_M - [\chi + (W_C - W_F)] \qquad (2.66)$$

The barrier in the n-type semiconductor is associated with a space charge region of ionised donors. It inhibits the flow of free electrons into the metal. Applying an external voltage $V_{app}$ to an n-type semiconductor–metal contact, such that the metal is positively biased with respect to the semiconductor (forward bias), reduces the space charge, lowers the barrier $eV_S$ and leads to a forward current that increases exponentially with the applied voltage. With an applied voltage of opposite polarity, the height of the barrier $eV_{BN}$ is unaffected and the only current that can flow is that resulting from the diffusion of holes out of the bulk semiconductor to the interface, where they recombine with electrons from the metal. Such a rectifying contact is known as a Schottky-barrier contact.

The current density $J$ passing through the contact is the sum of the electron and hole current densities flowing from the semiconductor to the metal and can be expressed as

$$J = J_S \left[ \exp\left(\frac{eV_{app}}{kT}\right) - 1 \right] \qquad (2.67)$$

where

$$J_S = A^* T^2 \exp\left(\frac{eV_{BN}}{kT}\right) \qquad (2.68)$$

In (2.68) the parameter $A^* = 4\pi e m_e k^2 / h^3 = 1.2 \times 10^6$ A m$^{-2}$ K$^{-2}$ is called the Richardson constant. Equation (2.67) may be compared with (2.13), which represents the current–voltage characteristic of a p–n junction.

The saturation current density $J_S$ at a metal–semiconductor contact is usually much greater than its equivalent $J_0$ at a forward-biased p–n junction. Another very important difference between the p–n junction and the Schottky-barrier junction is that in the former the forward current is carried by a flow of injected excess minority carriers, whereas in the latter it is carried by majority carriers.

This theory of the energy barriers at ideal metal–semiconductor contacts predicts a dependency on the metal work function, as given by (2.65) and (2.66). The barriers should be very different for different metals and $eV_S$ should become zero when $\phi_M = \chi + (W_C - W_F)$. With still smaller values of $\phi_M$, we might expect the situation shown in Figure 2.19(b) to arise, so that an electron accumulation layer forms at the interface. If $\phi_M$ is small enough, we would expect both forward and reverse bias to cause a free flow of electrons in either direction across the interface. In reality, however, the disruption of the crystal lattice at the semiconductor–metal interface produces a large number of surface (or rather, interface) states. These minimise the effect of metal work function differences on the barrier heights. For example, the work function of platinum (Pt) is nearly 2 eV higher than that of magnesium (Mg), but the barrier height $eV_S$ at a Pt–Si(n) contact is 0.85 eV, whereas at an Mg–Si(n) contact it is 0.40 eV. The barrier heights of most silicon–metal contacts lie between 0.6 eV and 0.8 eV, with the result that most junctions between metals and normally doped semiconductors are rectifying. Metal–semiconductor interface barriers are discussed in more detail in the literature [2.1, Ch. 5].

Although we have discussed Schottky barriers in terms of contacts with n-type material, exactly analogous situations apply to contacts between metals and p-type semiconductors, as shown in Figure 2.20. We would expect a contact with a low work function metal to create a depletion layer in the semiconductor at the interface, as shown in Figure 2.20(a) and hence to be rectifying. The barrier height decreases when the semiconductor is biased negatively with respect to the metal, and holes flow to the interface where they recombine with electrons from the metal. When the p-type semiconductor is positively biased, the height of the barrier increases and the hole current is rapidly cut off. On the other hand, we might expect a junction formed

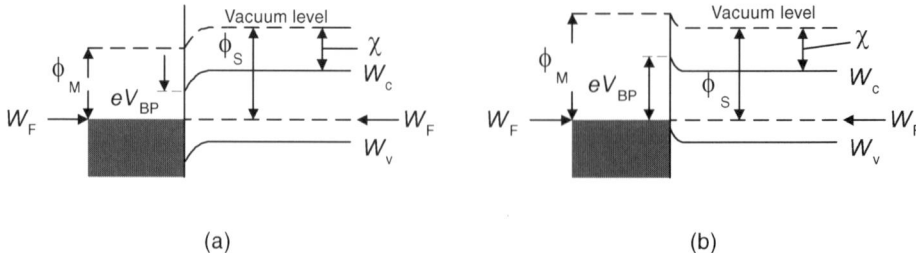

**Figure 2.20** (a) A metal and p-type semiconductor, forming a depletion layer at the surface (rectifying); (b) a metal and p-type semiconductor, forming an accumulation layer at the surface (ohmic)

with a high work function metal to offer little resistance. As with n-type material, the presence of interface states invalidates this, unless the semiconductor is heavily doped.

### 2.4.2 Ohmic Contacts

The rectifying characteristics of the metal–semiconductor contact are associated with the energy barrier at the metal–semiconductor interface and with a space charge region that develops in the semiconductor. When silicide layers are put down onto the silicon surface, the energy barrier is quite low ($<0.2\,\text{eV}$). Then the thermal emission at the contact is so high that the current–voltage characteristic is linear over a very broad range of currents. This is an example of an ohmic metal–semiconductor contact. The specific contact resistance $R_c$ is a parameter that is often used to characterise the contact:

$$R_c = \frac{dV_{app}}{dJ}\bigg|_{J\to 0} \quad (2.69)$$

It has the units $\Omega\,\text{m}^2$ and can vary rapidly with temperature.

The metal–semiconductor ohmic contact can be realised in several other ways. The most important is the contact between a metal and a heavily doped semiconductor. Then the thickness of the space charge region decreases with increasing impurity concentration. For an n-type semiconductor with donor concentration $N_D$, it is given by

$$d = \left[\frac{2\varepsilon_r \varepsilon_0 (V_S - V_{app})}{eN_D}\right]^{1/2} \quad (2.70)$$

where $eV_S$ is the height of the contact potential barrier in thermal equilibrium and $V_{app}$ is the applied voltage. If the electron concentration is higher than $10^{24}\,\text{m}^{-3}$, the barrier region is so narrow that electrons can tunnel through it. Contacts prepared in this way are linear with a low contact resistance and less dependence on temperature. Ohmic contacts can be realised in a similar manner on heavily doped p-type semiconductors.

From this it follows that the rectifying metal–semiconductor contact can be realised with a relatively lightly doped semiconductor where the barrier region is sufficiently wide and the interface barrier is sufficiently high, whereas the ohmic contact is obtained most easily on heavily doped semiconductor. Differences in the energy diagrams in both cases are shown in Figure 2.21.

The heavy doping technique is used in semiconductor device fabrication, where the metal–semiconductor contact at the terminals is usually made to a layer of degenerate semiconductor of the same conductivity type ($n^+n$ or $p^+p$ structures). The structure is thus $Mn^+n$ or $Mp^+p$. A working definition of an 'ohmic' contact is one in which the voltage dropped across the contact is small compared with that dropped elsewhere in the device under all operating conditions. Then it matters not whether the V–I characteristic of the contact itself is linear or non-linear.

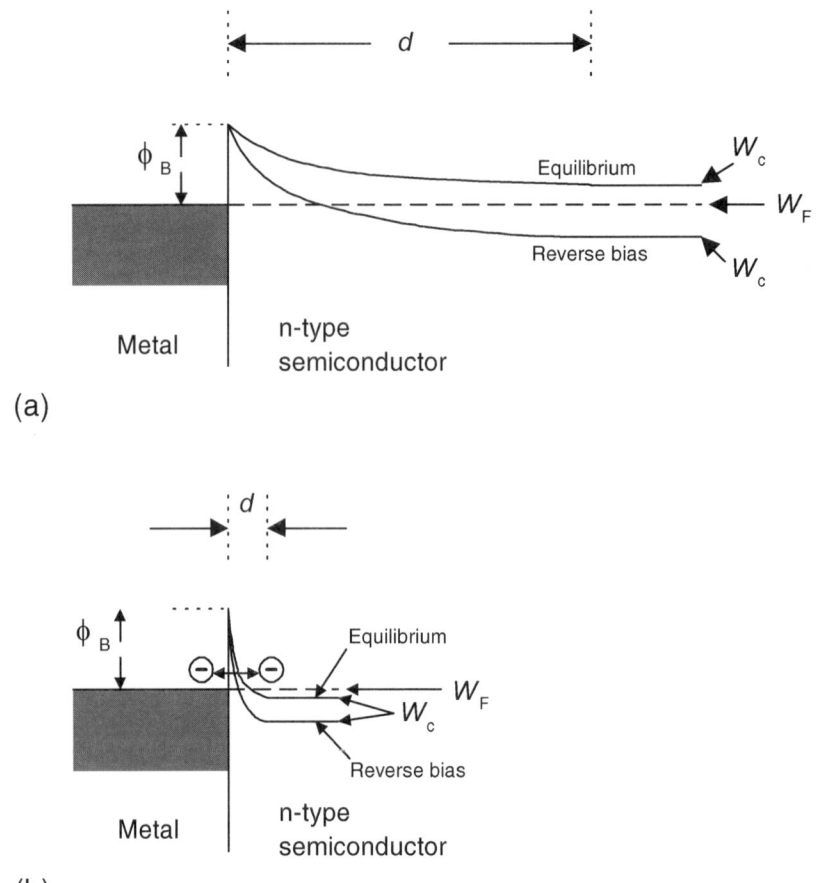

**Figure 2.21** Comparison between the electron energy diagrams across rectifying and ohmic metal–semiconductor junctions: (a) a Schottky contact to lightly doped material; (b) an ohmic contact to heavily doped material

## Summary

Nearly all semiconductor devices depend on the properties of one or more p–n junctions. The current–voltage characteristic of a single p–n junction can be approximated as

$$I = I_S \left[ \exp\left( \frac{eV_{app}}{\alpha kT} \right) - 1 \right]$$

where $\alpha$ normally lies between 1 and 2. Allowance has to be made for the effects of ohmic resistance at high forward current densities and for avalanche breakdown at high reverse voltages. Illumination of the junction with light whose photon energy exceeds the bandgap energy causes the junction current to be reduced by an amount proportional to the optical power.

A space charge layer forms either side of the junction. It extends a distance that varies as the square root of the total junction voltage and inversely as the square root of the doping concentration of the less heavily doped region.

In practice the breakdown voltage varies inversely as the three-quarters power of the doping concentration of the less heavily doped region.

A local increase of the junction temperature above a certain critical temperature leads to thermal runaway and catastrophic failure. In bipolar devices this process can be accelerated by the injection of minority carriers and is known as second breakdown.

Under transient conditions the junction voltage and current have to satisfy the constraints imposed by the external circuit, the space charge layer and the diffusion of minority carriers into the bulk semiconductor. These lead to a set of highly non-linear equations that require numerical solution, except for a few artificially simple cases. The transitions between the conducting and non-conducting states require the establishment or removal of excess carriers in the neutral regions through the processes of diffusion and recombination. In a layer of thickness $w$ these processes take a time that is proportional to $w^2/4D$, where $D$ is the appropriate carrier diffusion coefficient.

The formation of layers of charge on the semiconductor surface can lead to depletion, accumulation or inversion regions under the surface, depending on the polarity and density of the surface charge.

Contacts between a metal and relatively lightly doped semiconductor can be rectifying, forming what is known as a Schottky diode. These have the potential benefit of a low forward conduction voltage and a fast transition to the blocking state. A contact between a metal and heavily doped semiconductor is normally ohmic. The formation of ohmic contacts with a low contact resistance to both n-type and p-type semiconductors is an important feature of semiconductor device technology.

# References

[2.1] Sze, S. M. *Physics of Semiconductor Devices* (John Wiley & Sons, 2nd Edn, 1981).

[2.2] Ghandi, S. K. *Semiconductor Power Devices* (John Wiley & Sons, 1977).

[2.3] Moll, J. L. The Evolution of the Theory of the Current–Voltage Characteristics of p–n Junctions, *Proc. IRE*, **46**, 1076–82 (1958).

[2.4] Fulop, W. Calculation of Avalanche Breakdown of Silicon p–n Junctions, *Solid State Electronics*, **10**, 39–43 (1967).

[2.5] Grove, A. S. *Physics and Technology of Semiconductor Devices* (JohnWiley & Sons, 1967).

[2.6] Crank, J. *The Mathematics of Diffusion* (Oxford University Press, 2nd Edn, 1975).

[2.7] Carslaw, H. S. and Jaeger, J. C. *Conduction of Heat in Solids* (Oxford University Press, 2nd Edn, 1959).

[2.8] Selberherr, S. *Analysis and Simulation of Semiconductor Devices* (Springer-Verlag, 1984).

[2.9] Kurata, M. *Numerical Analysis for Semiconductor Devices* (Lexicon, 1982).

[2.10] Snowden, C. M. *Semiconductor Device Modelling* (Peter Peregrinus, 1988).

[2.11] Grant, D. A. and Gowar, J. *Power MOSFETs: Theory and Applications* (John Wiley & Sons, 1989).

# 3

# DEVICES, FABRICATION AND MODELLING

In Chapters 5 to 11 we describe the main types of power semiconductor device that have been developed. In each case we detail their structure, give examples of their performance characteristics and describe some typical applications. In this chapter we present an introductory overview of the various types of device, give a brief discussion of the techniques used in their manufacture and summarise some of the numerical methods that are available for their analysis and simulation and hence for the prediction and modelling of their behaviour under different circuit conditions. In Chapter 4 we review some of the typical applications in which these devices are commonly used.

## 3.1 The Range of Power Semiconductor Devices

The definition of a *power* device is a matter of judgement. Here we use the term to describe any semiconductor device that is capable of handling currents in excess of 1 A. This still leaves a very wide range of sizes and capabilities, with some devices able to carry currents of more than 10 kA and some able to withstand voltages in excess of 4 kV. The larger power devices are often made using the whole cross-sectional area of a silicon wafer. Smaller devices may be fabricated more conventionally, with tens or hundreds being formed on a single slice of single-crystal silicon. Special techniques for defining the device periphery are required, if high off-state hold-off voltages are to be obtained, as discussed in Section 3.4.

Some of the features of silicon power semiconductor devices are set out in Table 3.1. It can be seen from this table that they can be classified in several distinct ways. First, there are two- and three-terminal devices: diodes and transistors or thyristors. In the three-terminal devices, the characteristics of the conduction path between the two main terminals (the emitter and collector of the bipolar transistors, the source and drain of the field-effect transistors, the anode and cathode of the thyristors) are determined by the control electrode (the base or gate). The control function can be exercised either by the injection of current or through the voltage of the control electrode. In some devices the control electrode makes a direct contact to the

Table 3.1 The range of power semiconductor device structures

| Abbreviated name | Full name | Number of terminals | Number of layers | Number of junctions in current path | Control method | MOS-control electrode | Minority carrier injection |
|---|---|---|---|---|---|---|---|
| p–n diode | p–n junction diode | 2 | 3 | 1 | – | – | yes |
| p–i–n diode | p–i–n diode | 2 | 3/4 | 1 | – | – | yes |
| Schottky | Schottky-barrier diode | 2 | 3 | none | – | – | no |
| JBS | junction barrier Schottky | 2 | 2 | none | – | – | a |
| BJT | bipolar junction transistor | 3 | 4 | 2 | current | no | yes |
| Darlington | bipolar transistor Darlington-pair | 3 | 4 | 2 | current | no | yes |
| SCR | silicon controlled rectifier (thyristor) | 3 | 4 | 2/3 | current | no | double |
| GATT | gate-assisted turn-off thyristor | 3 | 4 | 2/3 | current | no | double |
| GTO | gate turn-off thyristor | 3 | 4 | 2/3 | current | no | double |

| | | | | | | | |
|---|---|---|---|---|---|---|---|
| LTT | light-triggered thyristor | 3 | 4 | 2/3 | light | no | double |
| Triac | triac | 3 | 5 | 2/4 | current | no | double |
| JFET/SIT | power junction field-effect transistor/static induction transistor | 3 | 4 | none | voltage | no | b |
| SITh/FCTh | static induction thyristor/field-controlled thyristor | 3 | 4 | 1/none | voltage | no | yes |
| MOSFET | power MOSFET | 3 | 4 | 2[c] | voltage | yes | no |
| IGBT | insulated-gate bipolar transistor | 3 | 5 | 3[c] | voltage | yes | yes |
| MCT | MOS-controlled thyristor | 3 | 6 | 4 | voltage | yes | double |

[a] Depends on the current level.
[b] Can be operated in bipolar mode by forward biasing the control junction; the device is then called a BSIT.
[c] In these MOS devices the p–n junctions through which much of the on-state current passes are effectively short circuited by an inversion channel.

semiconductor, whereas in others, such as the MOSFET, the IGBT and the MCT, it is separated from the semiconductor by a thin layer of insulator so there is no flow of current under static conditions.

The number of distinct layers of semiconductor material affects the complexity of the fabrication process. In the table, we have counted the heavily doped substrate and contact layers in the total, even when they make junctions with more lightly doped material of the same type ($n^+$–n and $p^+$–p junctions). The number of p–n junctions in the normal main conduction path is also tabulated. Note that in many of the thyristor-type devices, it is usual to include anode or cathode shorts that provide alternative conduction paths avoiding one or more of the junctions. Likewise, in the MOSFET and IGBT, the conduction path formed under the gate oxide when the device is turned on effectively short-circuits the junctions at either end.

The structures of the different device types and, in particular, the number of p–n junctions involved in their operation, have a major influence on their performance characteristics. For example, devices in which conduction is by majority carriers only, such as Schottky diodes and field-effect transistors, are able to switch between the on- and off-states very rapidly but, in high voltage devices, the on-state resistance is likely to be high for a given device area. In the bipolar devices, such as the p–n junction diode, the BJT and the IGBT, the flow of current at some point is carried by the minority carriers. These reduce the forward volt-drop, but they have to diffuse into the active region in order to initiate conduction and be discharged following a period of conduction. They therefore delay the turn-on and turn-off processes and limit the operating frequency of such devices.

In the four-layer devices listed in Table 3.1, double injection of carriers occurs. This is regenerative and causes latching, but it does enable these devices to have a very low on-state voltage, even under conditions of high surge current. However, they normally require special capacitive circuits called *snubbers* that are described in Section 8.3. In silicon controlled rectifiers, snubber circuits protect the device against an excessive rate of rise of anode voltage during switching. In gate turn-off thyristors, the use of a polarised snubber limits the local dissipation that would otherwise occur during turn-off. They are also required to control the rate of rise of the collector voltage of transistors and to keep them within their safe operating area, as described in Section 6.4. Like the excess carrier charge that is stored in the device during the on-state, the charge stored in snubber circuits during the off-state has to be supplied and removed during switching operations and so may limit the maximum frequency of operation.

All these matters are explored in detail in the later chapters for each of the various device types. Their basic characteristics are summarised in the remaining parts of this section.

### 3.1.1 Power Diodes

Figure 3.1 shows the structure of a typical diffused power diode, the standard symbol used to represent it in electrical circuits and a typical current–voltage characteristic. The figure also shows a typical distribution of the active impurities along a section through the active region of the device. The diode comprises a lightly

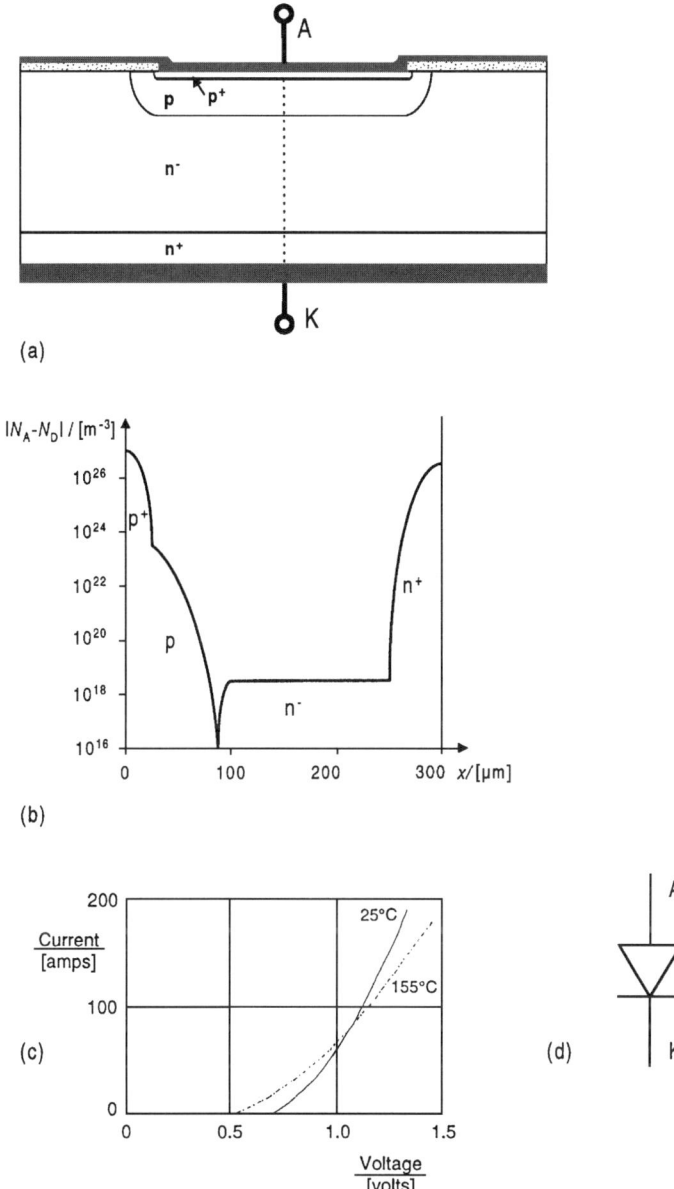

**Figure 3.1** The structure of a power diode: (a) cross-section through the active region; (b) doping profile along the section A–K; (c) current–voltage characteristic; (d) device symbol

doped n–type region (usually the semiconductor material from which fabrication starts) and a more heavily doped p-type region (usually produced by diffusion). These form a p–n junction capable of withstanding the required breakdown voltage $V_{BR}$. Heavily doped p$^+$- and n$^+$-layers are used to ensure a low contact resistance at

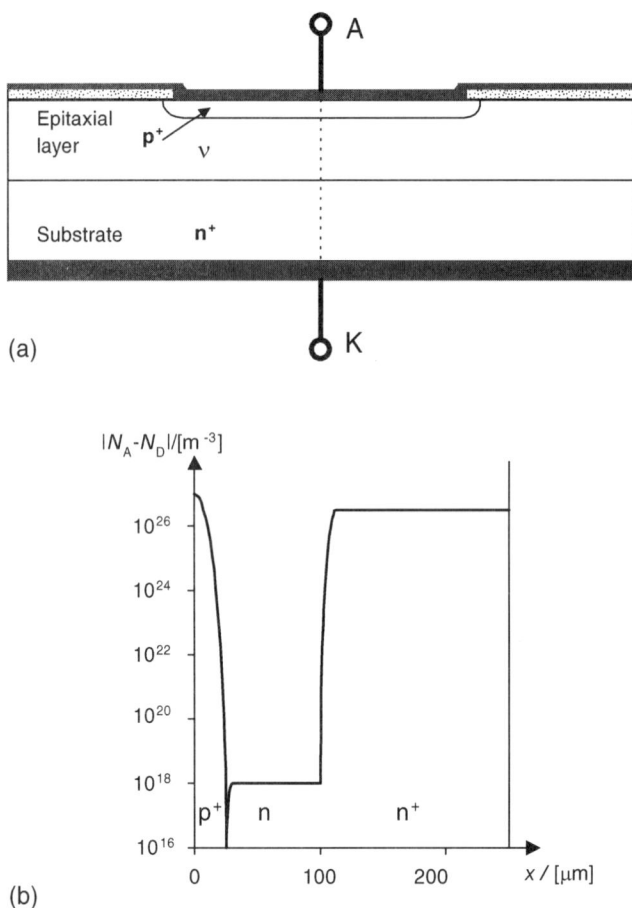

**Figure 3.2** A typical epitaxial power diode designed for fast switching applications: (a) cross-section through the active region; (b) doping profile along the section A–K

the metal–semiconductor interfaces. Thus, a typical power diode has a $p^+pnn^+$ structure.

An alternative design that has advantages when fast switching characteristics are more important than the ability to handle very high powers, is the epitaxial power diode shown in Figure 3.2. Here the starting material is a wafer of heavily doped $n^+$-substrate, onto which a lightly doped $n^-$-layer is deposited epitaxially. A $p^+$-surface layer is diffused into this to form the junction. A moat may be etched around individual devices in order to provide isolation and junction passivation.

Power Schottky diodes are used whenever a low forward volt-drop is needed and a high blocking voltage is not required. The structure is shown in Figure 3.3. The blocking voltage of the Schottky diode itself without suitable field termination is limited to about 60 V, but with p–n junction guard rings, 200–250 V can be obtained. For low forward voltage and high frequency operation, the gallium arsenide

## 3.1 THE RANGE OF POWER SEMICONDUCTOR DEVICES

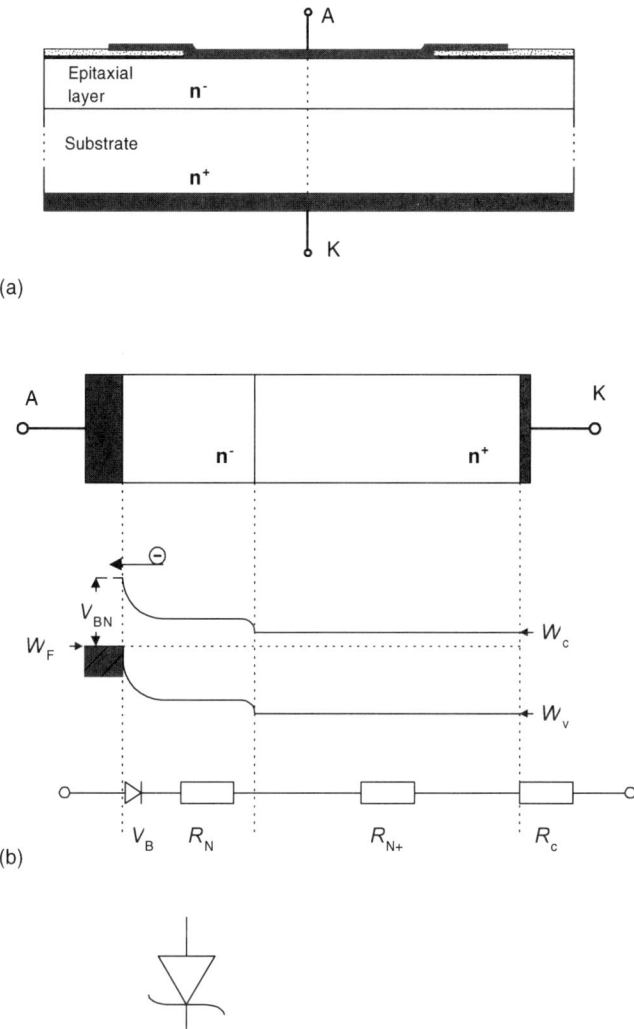

**Figure 3.3** A power Schottky diode: (a) cross-section through the active region; (b) band diagram along the section A–K and an equivalent circuit; (c) device symbol

Schottky diodes described in Chapter 14 benefit from the high electron mobility of that material.

### 3.1.2 Conventional Power Bipolar Transistors

The basic structure of an n–p–n bipolar junction transistor (BJT) is shown schematically in Figure 3.4. A typical doping profile, the normal circuit symbol and a typical set of current–voltage characteristics are also shown. The three layers are

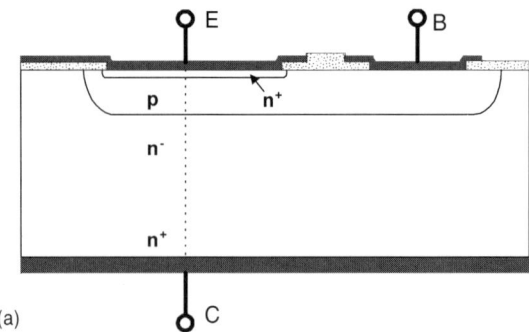
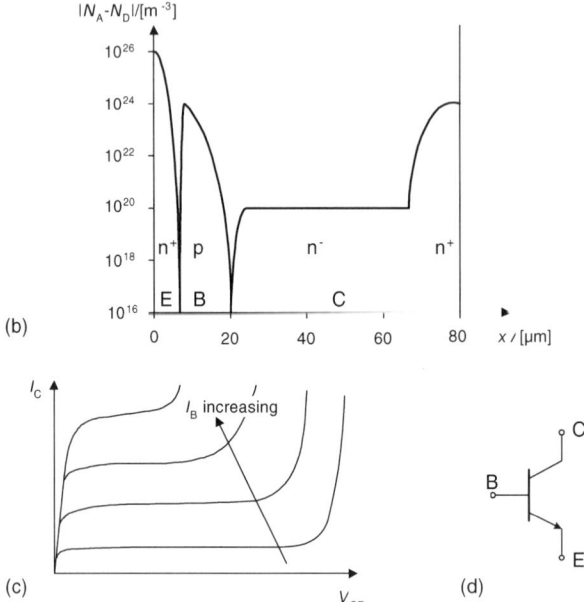

**Figure 3.4** An n–p–n bipolar junction transistor: (a) cross-section through the active region; (b) doping profile along the section E–C; (c) current–voltage characteristics; (d) device symbol

called *emitter*, *base* and *collector*. Either junction may be forward or reverse biased, so there are four possible modes of operation. When both junctions are reverse biased, the transistor is non-conducting and is said to be at *cut-off*. When both are forward biased, the injection of electrons into the base across both junctions causes a very great increase in the base carrier concentrations and renders the transistor highly conducting. It is said to be in *saturation*. By switching between saturation and cut-off, the BJT can be used as an opening switch as well as a closing switch.

With the emitter junction forward biased and the collector junction reverse biased, the transistor is said to be in the normal or *forward active* state. With the polarities reversed, the transistor would be in the *reverse active* state; however, this is not a condition that is ever used in power circuits. In the active states, the current supplied to the base controls the current flowing between collector and emitter.

Although the complementary p–n–p structure has similar properties, for power applications, n–p–n BJTs are always preferred. The higher mobility of electrons gives faster switching and a lower on-state voltage compared with p–n–p transistors of similar dimensions.

Power BJTs fall into three general categories. First are the devices that occupy a complete silicon wafer. They were originally designed for heavy-duty applications such as traction and have largely been displaced by GTOs and IGBT modules. Second are the power BJTs which dominated the market for ac motor-drive inverters and uninterruptable power supplies (UPS) for many years. Modules using BJTs formed on square dies with highly interdigitated base–cathode structures became dominant in inverter and UPS designs. Such devices have the advantage of a 10 μs short circuit capability and a square safe operating area, as described in Section 6.4. However, despite a cost advantage in some instances, during the 1990s they have lost considerable market share to the insulated gate bipolar transistor (IGBT), especially in new applications (Section 3.1.5). In general, IGBTs give a better power gain, are easier to drive and can be switched at higher frequencies. Finally, there are the smaller high voltage devices that can be manufactured very cheaply and find application for switched-mode power supplies and in the line output stage of TV-tube drives. For single-ended switched-mode power supplies working from a 240 V supply and requiring switching frequencies in excess of 75 kHz, small highly interdigitated BJTs continue to hold an advantage.

A disadvantage of the power bipolar junction transistor is its limited current gain. This can be improved, at the cost of a reduction in the maximum operating frequency and an increase in forward voltage drop, by the use of two transistors in the Darlington configuration described in Section 6.5.

There are two main fabrication processes for power BJTs. The first is the triple diffusion process in which the p-base and $n^+$-emitter layers are diffused into the starting material (the $n^-$-collector layer) from one side of the wafer and the $n^+$-collector contact layer is diffused in from the other side. In the second process, the starting material is the $n^+$-collector contact layer. The $n^-$-collector layer is grown epitaxially onto this and the p-base and $n^+$-emitter layers are diffused in from the surface of the grown layer.

### 3.1.3 Thyristor Structures

*Thyristor* ($\theta\upsilon\rho\alpha$ = door) is the common name used for a family of switching devices consisting of four layers of semiconductor of alternating dopant type (pnpn). Many different devices having this basic structure have been described and are in common use. They can be classified in different ways. One way is with respect to the number of electrodes connected to the layers. Thus, there have been diode, triode and tetrode thyristors. Three-terminal thyristor switches are much the most important members of this family.

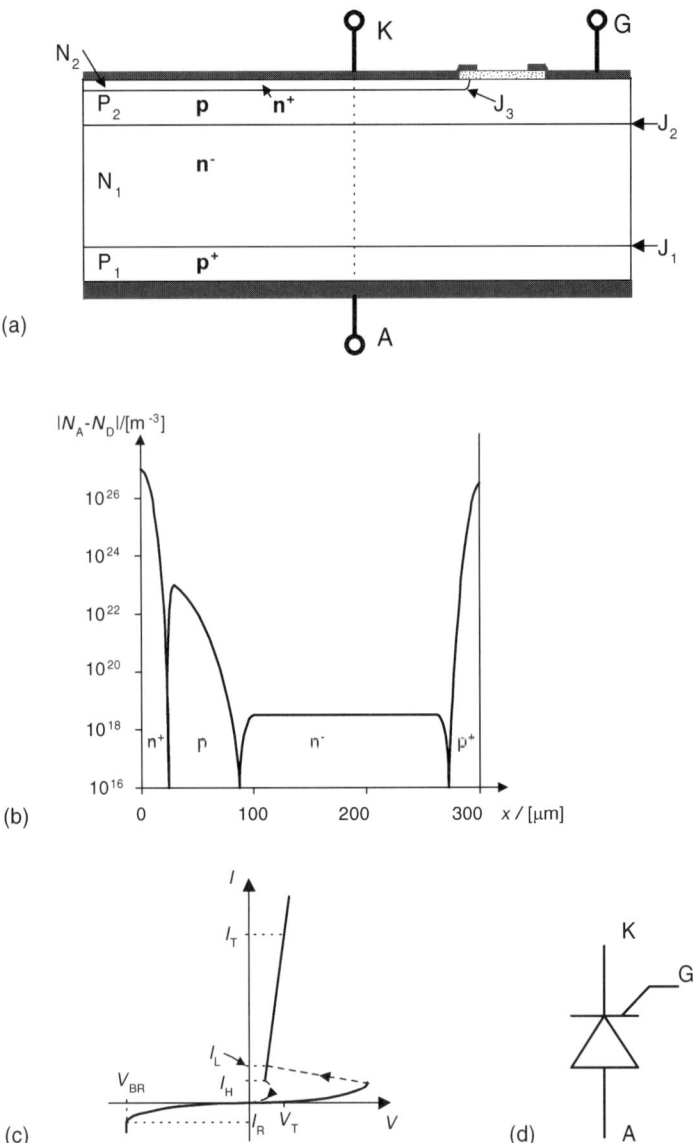

**Figure 3.5** A conventional thyristor: (a) cross-section through the active region; (b) doping profile along the section K–A; (c) current–voltage characteristics; (d) device symbol

The basic triode thyristor structure, its circuit symbol, a typical current–voltage characteristic and a typical doping profile are shown in Figure 3.5. When the anode (A), is negatively biased with respect to the cathode (K), the thyristor is in a high impedance state called the *reverse blocking* mode. With forward bias, i.e. when the anode is made positive with respect to the cathode, the thyristor can be in either a high impedance or a low impedance state. These are called, respectively, the *forward blocking* state and the *forward conducting* state.

## 3.1 THE RANGE OF POWER SEMICONDUCTOR DEVICES

Thyristors differ from transistors in that the gate electrode (G), is not able to control the anode current at all times. It can only bring about the transition between the two forward states. In a conventional thyristor, the gate can only control the turn-on transition from forward blocking to forward conduction. Turn-off depends on commutation; the blocking state can be recovered only after the anode current has fallen below a certain critical value known as the *holding current*. In the important class of thyristors known as gate turn-off thyristors (GTOs), the application of forward bias to the gate initiates turn-on in the usual way but the application of reverse bias, while the thyristor is conducting, can turn it off. This means that the GTO, like the transistor, can act as an opening switch as well as a closing switch.

Ideally the thyristor should have a very high breakdown voltage in the off-states (blocking), a very low impedance in the on-state (forward conducting) and should allow fast turn-on and turn-off transients while switching between the two modes of operation. Although it is not possible to fulfil all of these requirements simultaneously, a study of the device physics indicates how compromises can be made in the technology to give the best combination of device parameters for particular applications. Some variants of the basic thyristor structure have special names:

| | |
|---|---|
| SCR | silicon controlled rectifier |
| ASCR | asymmetric thyristor |
| GATT | gate-assisted turn-off thyristor |
| GTO | gate turn-off thyristor |
| RCT | reverse conducting thyristor |
| RCGTO | reverse conducting gate turn-off thyristor. |
| GCT | gate commutated thyristor |

The SCR was the first semiconductor switch capable of handling high power levels. It requires a period of current reversal to allow it to turn off. Its first application was as a phase control rectifier operating at power frequency, when natural commutation can ensure turn-off. Later, higher frequency inverter applications required forced commutation. Development of the design of the gate–cathode structure led to GATT devices in which turn off is accelerated by withdrawing current from the gate during the commutation period. In the ASCR, the reverse blocking capability is sacrificed so that the forward volt-drop can be minimised. The RCT and RCGTO have integrated antiparallel diodes built into the structure so that they have no reverse blocking capability at all. The detailed designs of these different types of thyristor are discussed in Chapter 8.

A related device, the *triac*, is a five-layer device that can be triggered into both a reverse conduction mode and a forward conduction mode. It is shown in Figure 3.6. It acts as a bidirectional switch and can be considered to be an integration of two antiparallel pnpn thyristors. The special gate layout enables the triac to be turned on in either direction with either a positive or a negative voltage applied to the gate electrode with respect to the main electrode $MT_1$. A detailed discussion is given in Section 8.4.

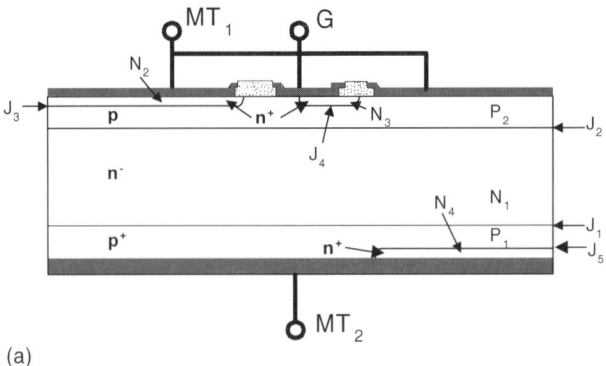

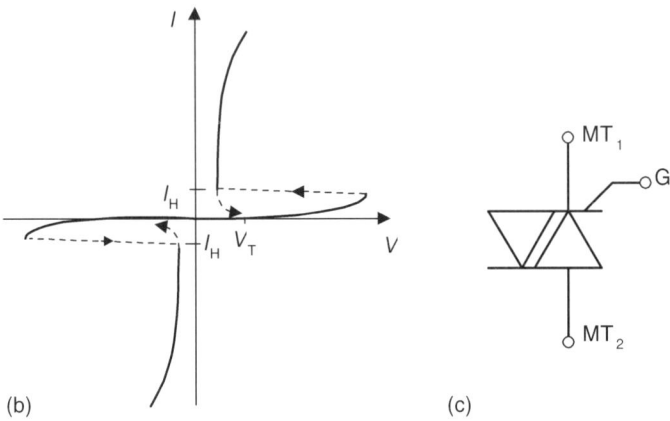

**Figure 3.6** The triac: (a) cross-section through the active region; (b) operating characteristics; (c) device symbol

### 3.1.4 Junction Field-Effect (Static Induction) Devices

A conventional small-signal, junction field-effect transistor (JFET) takes the form shown schematically in Figure 3.7(a). Typical current–voltage characteristics are shown in Figure 3.7(d). The current flows through the channel formed in the n-type region under the $p^+$-gate layer. A negative voltage applied to the gate expands the space charge region at the p–n junction and restricts the area through which conduction can occur, eventually cutting it off altogether. The power levels that can be achieved are limited by this lateral flow, which restricts both the maximum current and the maximum voltage.

To overcome these limitations, designs of JFET have been proposed for power applications, in which the current flows vertically through the device. The cross-section shown in Figure 3.8(a) illustrates one such structure with a buried gate which forms a grid. This is one version of the device known as the static induction

## 3.1 THE RANGE OF POWER SEMICONDUCTOR DEVICES

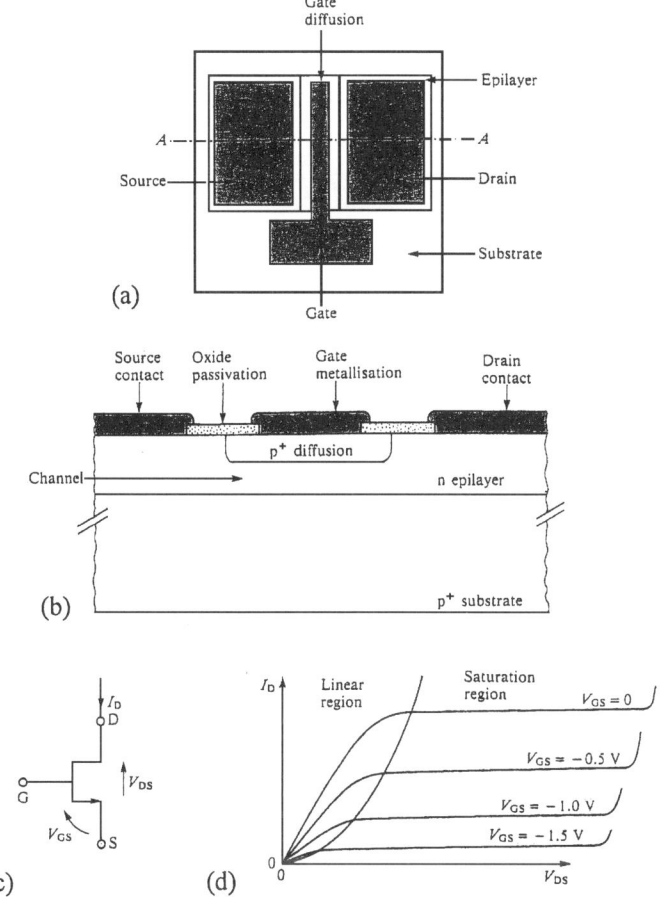

**Figure 3.7** The conventional JFET: (a) a plan view; (b) a schematic cross-section; (c) the device symbol; (d) typical current–voltage characteristics. Reproduced from *Power MOSFETs Theory and Applications* by D. A. Grant and J. Gowar, by permission of John Wiley & Sons, Inc.

transistor (SIT), which has found limited application when very high frequency operation is required. The growth of a high resistivity epitaxial $n^-$-region is halted, the $p^+$-fingers are diffused in, then the epitaxial growth is continued. A final diffusion forms the $n^+$-source layer.

The short length of the $n^-$ conducting channel between the $p^+$ diffusions gives rise to the current–voltage characteristics shown in Figure 3.8(b). The origin of these characteristics and the application of static induction transistors are discussed in Section 9.1.

The static induction thyristor (SITh) is a three-terminal device derived from the SIT by growing the n-type epitaxial layer on a $p^+$-substrate. Although it has thyristor-like characteristics, its structure and mode of operation are quite unlike

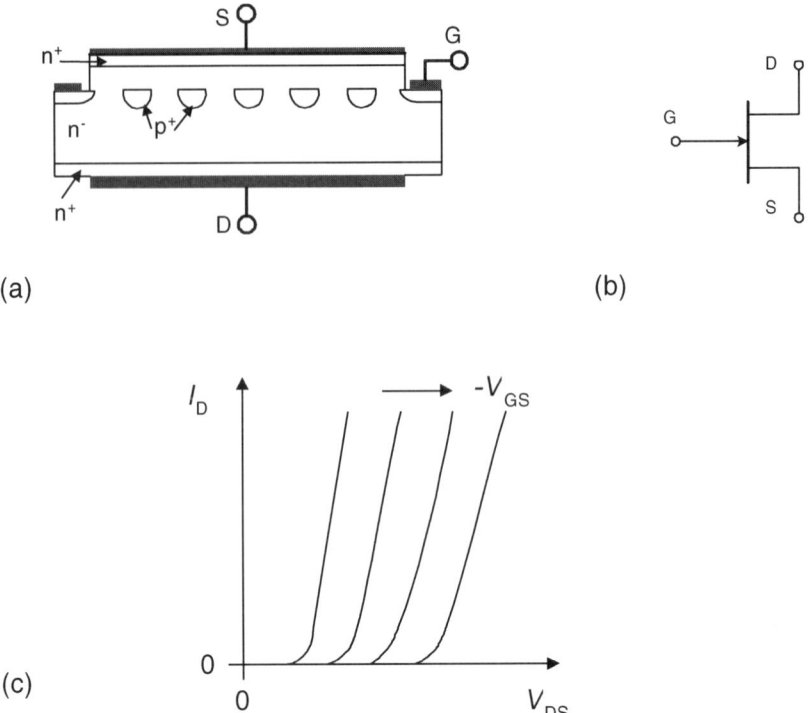

**Figure 3.8** The static induction transistor (SIT): (a) a cross-section through a device with a buried gate; (b) device symbol; (c) typical current–voltage characteristics

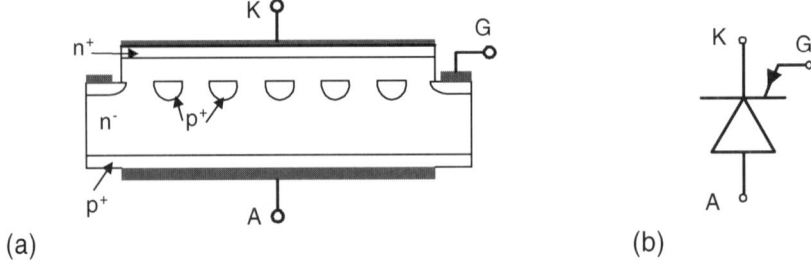

**Figure 3.9** The static induction thyristor, or field-controlled diode, structure: (a) cross-section; (b) device symbol

those of a conventional thyristor and it is perhaps better called a field-controlled diode (FCD) or a field-controlled thyristor (FCT). A device of this type with a buried grid is illustrated in Figure 3.9. Bipolar injection from the $p^+$-layer produces conductivity modulation and greatly reduces the on-state resistance. Because a high gate voltage is needed, the FCD can withstand very high rates of change of anode voltage and is less susceptible than other semiconductor devices to electromagnetic interference. As a majority carrier device, its operating characteristics are less dependent on carrier lifetime.

Both the SIT and the FCD suffer from the serious operational disadvantage that they are normally on and require a negative gate voltage to render them non-conducting. This causes great difficulty in power applications, in ensuring safe switch-on procedures. Nevertheless, they have found limited use and are discussed briefly in Chapter 9.

### 3.1.5 Power MOS Structures 1: The Power MOSFET

In metal–oxide–semiconductor (MOS) field-effect devices, the gate electrode, which controls the flow of current through the main device, is separated from the semiconductor by a thin layer of gate oxide. As a result, it draws no current under static conditions. There are three power devices based on this principle. The power MOSFET is a majority carrier device, like the JFET; the insulated gate bipolar transistor (IGBT), as its name implies, is more analogous to the bipolar junction transistor; and the MOS-controlled thyristor (MCT) behaves very like a conventional thyristor.

The normal lateral, small-signal metal–oxide–semiconductor field-effect transistor (MOSFET) suffers from the same power limitations as the conventional JFET. There are four possible types of MOSFET: n- and p-channel, operating in enhancement and depletion modes (Figure 3.10). The great advantage of the enhancement mode MOSFET over the JFET is that it is normally off. An n-channel device then requires the application of a positive gate voltage to turn it on. Power MOSFETs are usually n-channel devices, because the higher carrier mobility of electrons reduces the on-state losses.

Power versions of these devices have been developed in which, like the SIT, high off-state voltages are supported in the vertical direction. Figure 3.11 shows how this is achieved in the earlier type of V-groove MOSFET and Figure 3.12 illustrates a more modern trench design. For low voltage applications, the trench MOSFET has become pre-eminent because of the very long gate widths that can be achieved in a given chip area. The trenches form a mesh with a cell size of micron dimensions. The vertical, double-diffused MOSFET structure (VDMOS) shown in Figure 3.13 now dominates medium voltage power electronic applications up to about 10 kW. It combines the technology and small feature size of integrated circuit manufacture with a high voltage structure in the vertical direction. The current is controlled by the gate as it flows laterally under the gate oxide, but it then turns vertically and spreads out to take full advantage of the chip area.

All three types of vertical MOSFET are based on successive p-type and n-type diffusions into an n-type epitaxial layer. The ratio of gate width to gate length $w/l$ is a critical parameter in determining the transconductance of these transistors. Like the base width of a bipolar junction transistor, the gate length is precisely controlled by the two diffusion processes to a value in the region of $1–2\,\mu m$. In VDMOS devices the gate width is maximised in a given area of the chip by the use of one of the cellular structures described in Section 10.2.

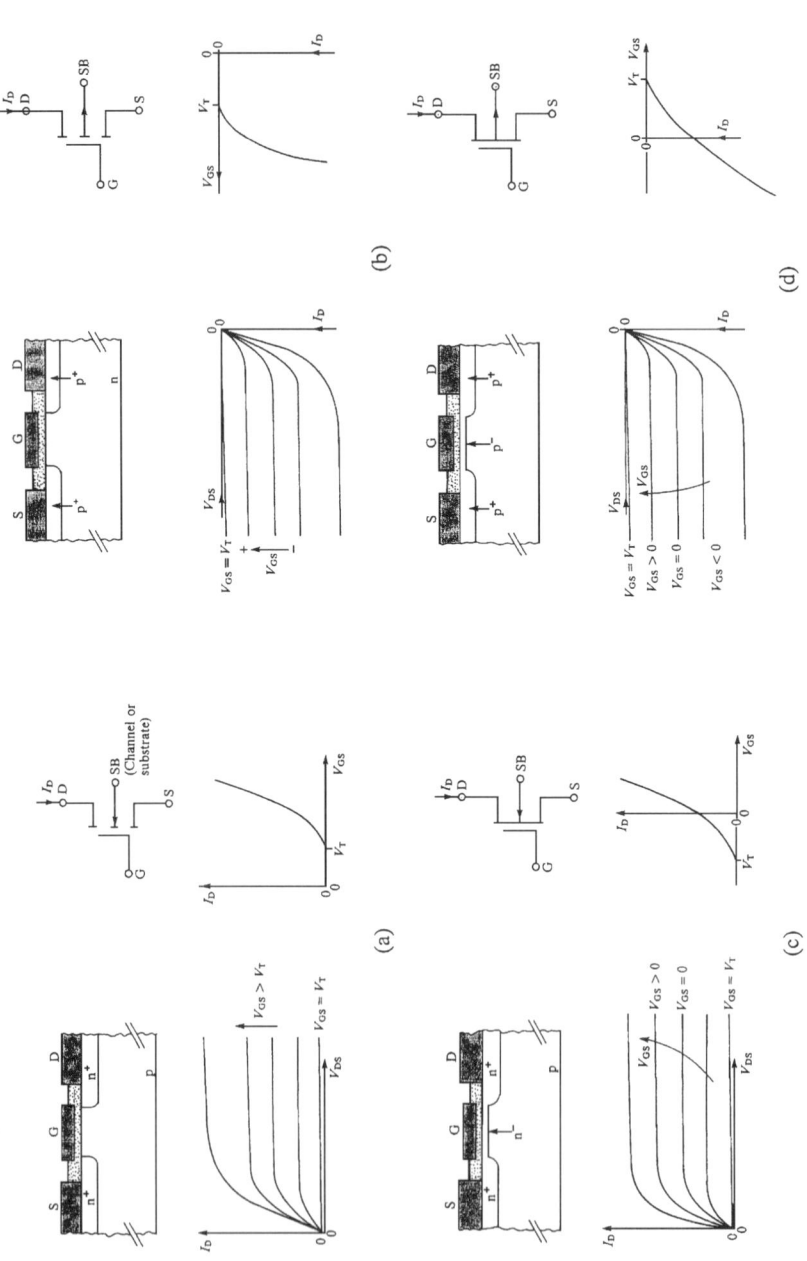

**Figure 3.10** The four types of lateral small-signal MOSFET: (a) n-channel, enhancement mode; (b) p-channel, enhancement mode; (c) n-channel, depletion mode; (d) p-channel, depletion mode. In each case, a schematic cross-section, a set of typical current-voltage characteristics and the device symbol are shown. Reproduced from *Power MOSFETs Theory and Applications* by D. A. Grant and J. Gowar, by permission of John Wiley & Sons, Inc.

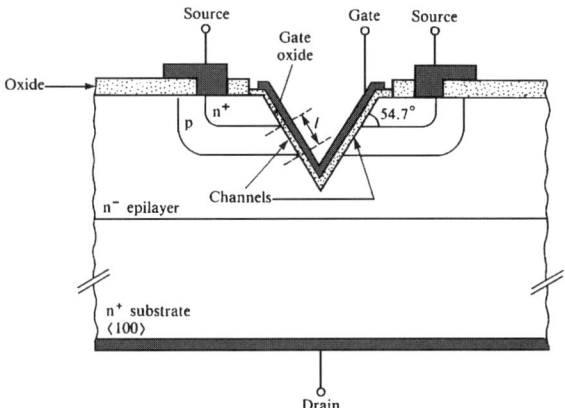

**Figure 3.11** A cross-section through a vertical V-groove MOSFET. Reproduced from *Power MOSFETs Theory and Applications* by D. A. Grant and J. Gowar, by permission of John Wiley & Sons, Inc.

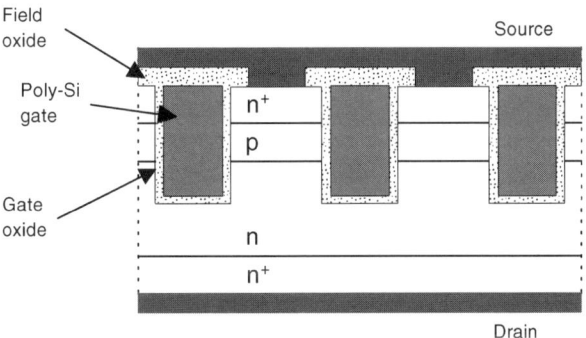

**Figure 3.12** A cross-section through a trench MOSFET

### 3.1.6 Power MOS Structures 2: The Insulated Gate Bipolar Transistor (IGBT)

The insulated gate bipolar transistor (IGBT) has come to dominate medium power applications. Essentially, the n-type substrate of the VDMOS transistor is replaced by p-type material, as shown in Figure 3.14. In this way, bipolar transistor action and conductivity modulation of the on-state resistance are achieved.

The reduced on-state voltage of an IGBT, in comparison with a power MOSFET of similar dimensions, makes it advantageous for applications where a high blocking voltage and a high on-state current are required. The voltage across junction $J_1$ contributes about 0.5 V to the on-state voltage drop, even at relatively low current levels. However, at high current densities, $V_{CE(on)}$ never exceeds about 4 V. Power IGBTs with blocking voltages of 2.5 kV and current-handling capacity greater than 1 kA at the maximum rated temperature are commercially available. Manufacturers

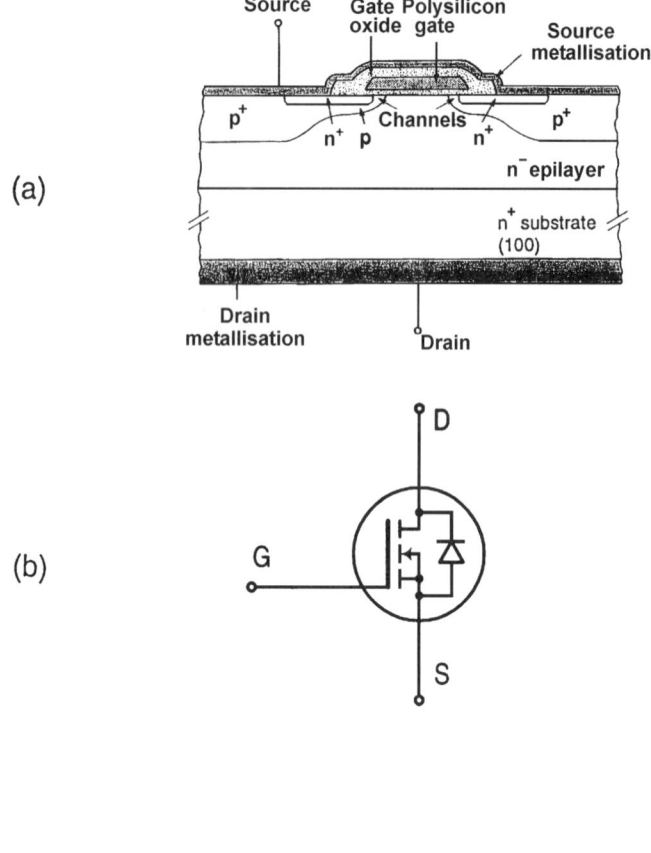

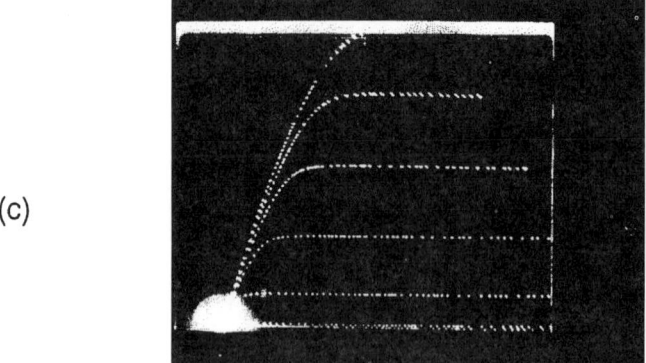

**Figure 3.13** An illustration of a vertical, double-diffused, power MOSFET: (a) cross-section through the active region of the device; (b) device symbol, with the parasitic diode shunting drain and source explicitly included; (c) typical device characteristics. Reproduced from *Power MOSFETs Theory and Applications* by D. A. Grant and J. Gowar, by permission of John Wiley & Sons, Inc.

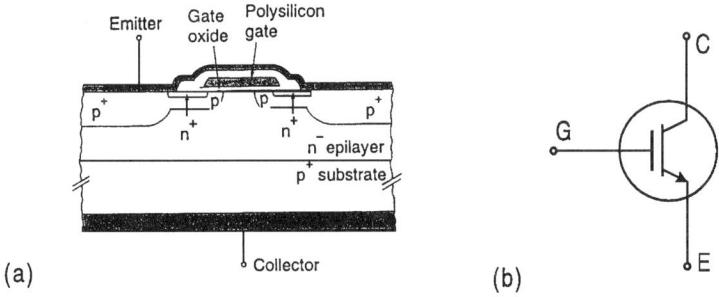

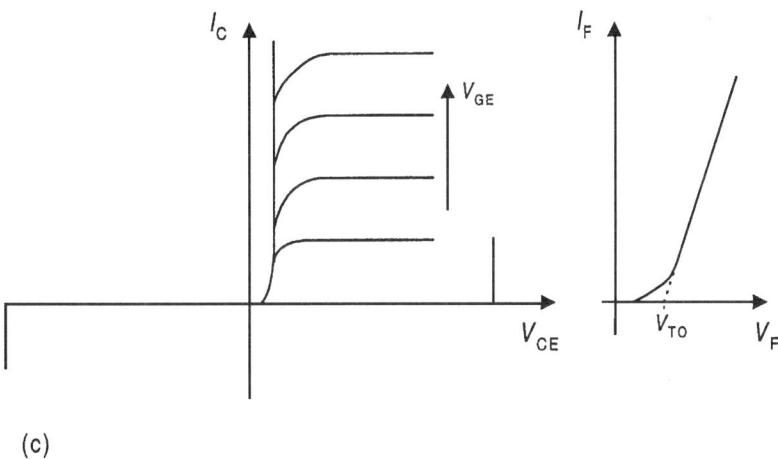

**Figure 3.14** The insulated gate bipolar transistor: (a) cross-section; (b) device symbol; (c) typical device characteristics. Part (a) reproduced from *Power MOSFETs Theory and Applications* by D. A. Grant and J. Gowar, by permission of John Wiley & Sons, Inc.

have development programmes to raise the blocking voltage to 4.5 kV and devices of that rating have been tested [3.1].

### 3.1.7 Power MOS Structures 3: The MOS-Controlled Thyristor (MCT)

An extra layer of opposite doping produces the device structures shown in Figure 3.15. The double injection of carriers lowers the volt-drop in the on-state, but at the expense of a much longer anode current tail during turn-off. These devices demonstrate thyristor-like behaviour but, compared to conventional thyristors, they

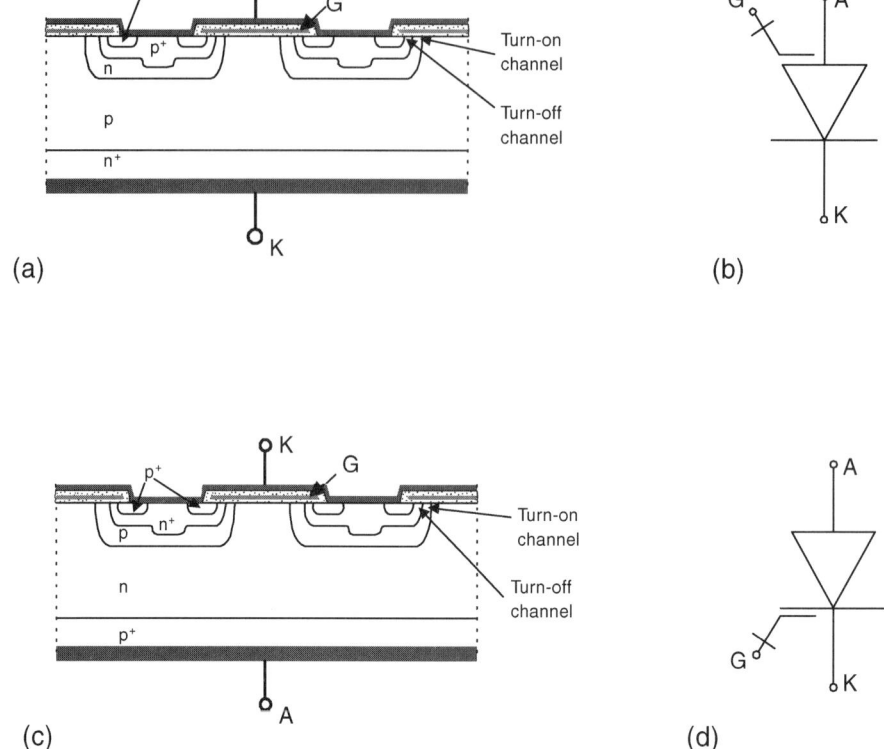

**Figure 3.15** The MOS-controlled thyristor (MCT): (a) an n-channel device; (b) device symbols for an n-channel device; (c) the complementary p-channel device; (d) device symbols for a p-channel MCT

have greatly reduced power requirements in the gate drive circuit. They are known as MOS-controlled thyristors (MCTs). It has proved difficult to design an MCT that can be turned on and off equally well. Such devices remain under development. They are discussed in Section 11.2.

### 3.1.8 A Summary of Device Characteristics

Of the various types of semiconductor switching device discussed in Sections 3.1.2 to 3.1.7, namely the bipolar transistors, the field-effect transistors and the different types of thyristor, the thyristors are able to carry the highest on-state current per unit area of silicon. Bipolar injection of carriers into the inner, high resistivity regions of the structure takes place from both the $n^+$-emitter (cathode) and the $p^+$-emitter (anode), so modulating the conductivity, as described in Section 5.1. The electron–hole plasma extends across all the space between the emitters and a high on-state current density can be obtained, even with a wide $N_1$-layer of high resistivity. Thus, high current, whole wafer devices, up to 150 mm (6 in) in diameter, can be made with

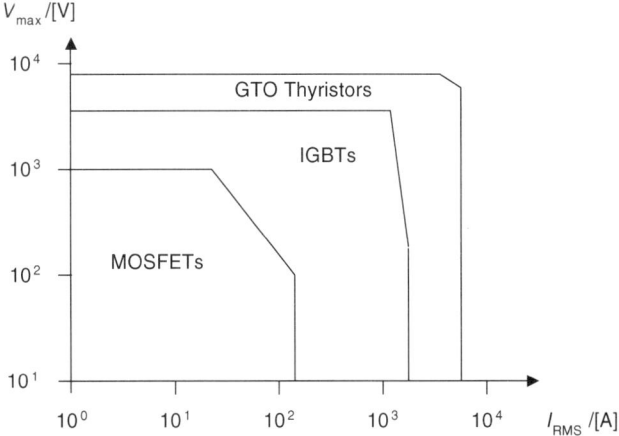

**Figure 3.16** Comparison of typical maximum current and voltage ratings of the basic types of semiconductor device in multi-chip packages: GTO thyristors, IGBTs and power MOSFETs

a high blocking voltage capability. Double injection of carriers occurs in diodes as well as thyristors and is described in detail in Chapters 5 and 7.

Bipolar junction transistors also have a relatively high level of carrier injection, so decreasing the on-state resistance of the device. However, the excess carriers are injected only from the heavily doped emitter layer and the highest injection occurs only at the edge of emitter region. As a result, the current-carrying capacity per unit area of silicon is lower than for thyristors.

MOS field-effect transistors have a narrow conducting channel below the gate oxide layer and an epitaxial layer of relatively high resistivity in series with it. No excess carrier injection modulates the resistivity of the inner layers, and the on-state resistance of devices with high breakdown voltages is relatively high. Consequently, the maximum current density of these devices is considerably lower than that obtainable in bipolar devices having the same blocking voltage.

These basic features determine the working voltage and current ranges of the different types of device, as shown in Figure 3.16. At the higher current levels, for example above about 100 A for IGBTs, devices are supplied in packages containing several chips connected in parallel. For thyristors, the limit shown refers to the maximum average current, rather than the peak current.

The large charge of excess carriers injected into the inner layers of bipolar devices in the on-state requires a relatively long time for it to be removed before the blocking capability can be recovered. In many circuits this charge leads to a significant current flow during the turn-off process, a delay in recovery and the dissipation of power. In consequence, the maximum permissible on-state current density decreases with increased operating frequency. It is important that the residual stored charge is minimised. For different reasons, discussed in detail in Chapter 10, the maximum operating current density of power MOSFETs also decreases at high frequency, but at much higher frequencies than bipolar devices, as shown in Figure 3.17.

MOS transistors are voltage controlled, which enables much simpler gate drive circuits to be used, compared with those needed for the bipolar transistor, which is

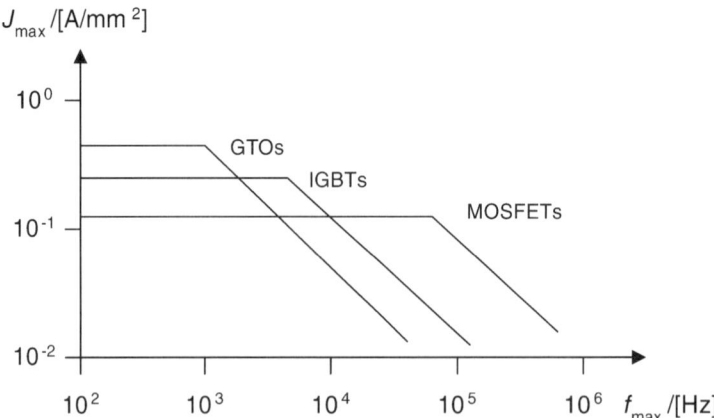

**Figure 3.17** Comparison of the maximum allowed current density as a function of the switching frequency for different types of power semiconductor

current controlled. Figure 3.18 gives a schematic comparison. This simplification impelled the development of bipolar devices with MOS gate structures, such as the IGBT and MCT for use in higher current applications.

A very important characteristic of a device is its safe operating area (SOA). This is a graphical representation of the simultaneous current and voltage values at which it can be used. Time is also involved since a device will often be able to carry a short pulse of current which is larger than its continuous current rating. The power MOSFET probably has the most straightforward SOA and an example is shown in Figure 3.19. Line 1 shows the dc capability of the device. The SOA is the area inside this line and the device may operate continuously with any values of current and voltage which fall within it. Its bounds are set by the voltage rating of the device and by its continuous current rating. In addition there is an overall power dissipation limit and the effect of the device on-resistance limits the amount of current which the device can pass at low values of drain–source voltage.

The power MOSFET has a pulse current rating which is typically four times its continuous current rating. This is represented on the SOA graph by line 2. This shows that, for pulses of up to $10\,\mu s$ duration, the MOSFET can be operated anywhere within a square region bounded by its voltage rating and its pulsed current rating. Because the pulse current rating is so much in excess of the continuous rating, it is usual to use log-log scales for the SOA diagram, as in this example. This means that the power dissipation limit is represented by a straight line. For pulses longer than $10\,\mu s$, the temperature rise in the silicon die limits the safe operating area and further diagonal power limits have to be added.

Figure 3.20 shows typical switching trajectories in the circuit environments of Figure 3.21. With the device used at its maximum voltage and current capabilities, the on-state in any circuit is represented by point A and the off-state by point B. When the device is switched between these points, the voltage and current trace a trajectory on the SOA diagram that depends on the external circuit. Path (a) is the trajectory obtained with the circuit shown in Figure 3.21(a), where the load is purely resistive.

## 3.1 THE RANGE OF POWER SEMICONDUCTOR DEVICES

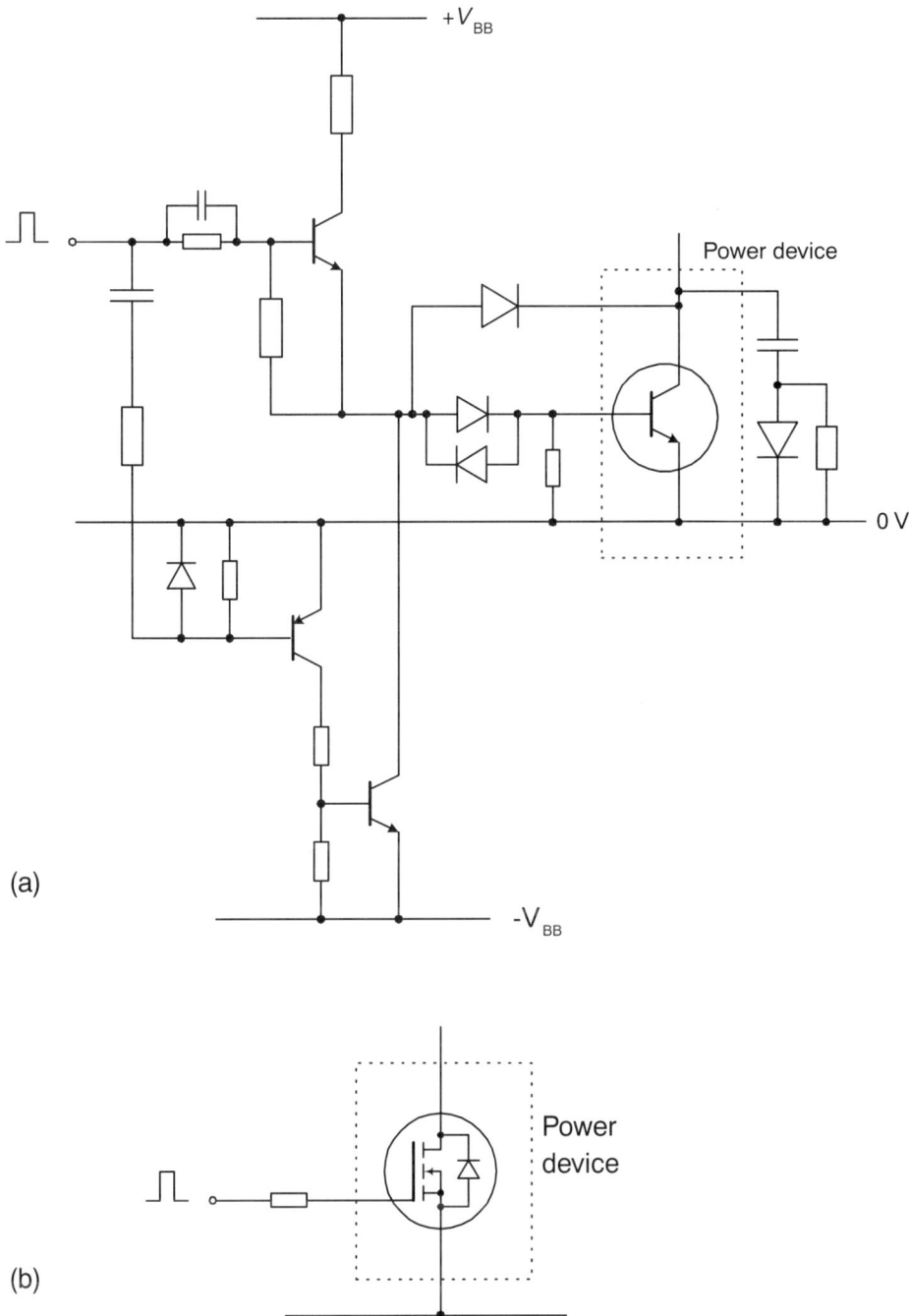

**Figure 3.18** A comparison of the typical drive circuits needed for: (a) a power bipolar junction transistor; (b) a power MOSFET

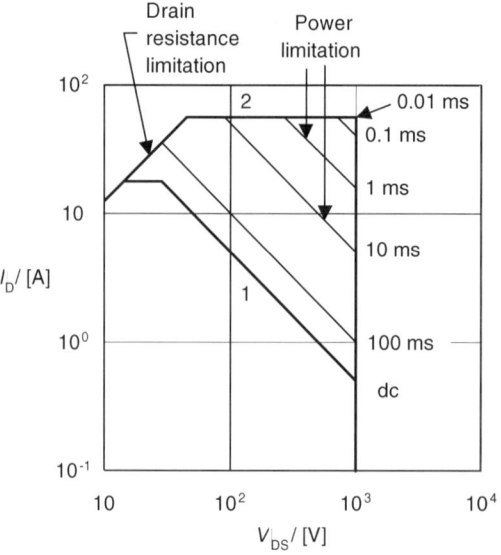

**Figure 3.19** The Safe Operating Area diagram for a power MOSFET

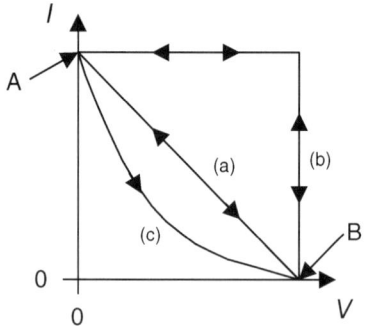

**Figure 3.20** Typical switching trajectories

Loads often take the form of a clamped inductance, as shown in Figure 3.21(b). While the device is switching, the load is predominantly inductive but a clamping diode which provides a circulating path for the current prevents voltage overshoot on turn-off. The turn-off trajectory is then path (b) in Figure 3.20. If current continues to circulate through the freewheeling diode while the device is off, the turn-on trajectory is again given by path (b) traversed in the reverse direction. Both turn-on and turn-off follow a square trajectory, requiring the device to have a square SOA for a time greater than the switching transients. If the switching device cannot tolerate this level of stress, some form of snubber must be added to the circuit to modify the switching trajectories.

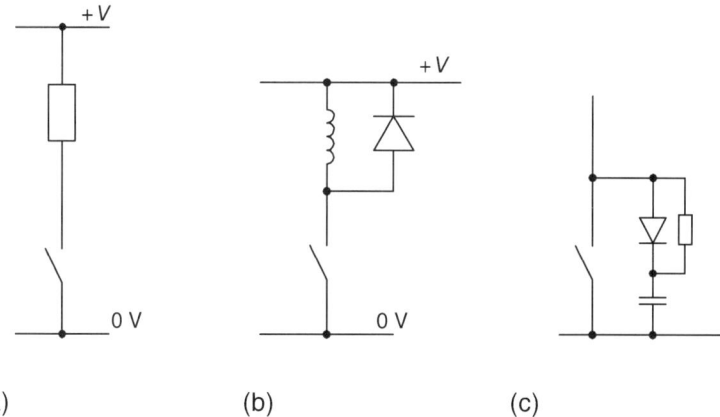

**Figure 3.21** Three switching circuit configurations: (a) a resistive load; (b) a clamped inductive load; (c) an RCD snubber

Path (c) in Figure 3.20 shows the turn-off trajectory produced when an RCD snubber circuit of the kind shown in Figure 3.21(c) is employed. The value of the capacitor is chosen so that the current becomes zero just at the moment the voltage across the device reaches its maximum value. At turn-on the switching device has to take over the load current and also discharge the capacitor C through the resistor R. However, the device may well be more robust in turn-on than turn-off, making this acceptable. Otherwise, a more complex snubber circuit is needed, including series inductance to limit the rise rate of current during turn-on.

Bipolar junction transistors are generally not as robust as power MOSFETs for the reasons discussed in Section 6.4. The SOA characteristics of GTO thyristors and IGBTs are discussed in Chapters 8 and 10, respectively.

## 3.2  Fabrication Processes

From the basic theory presented in Chapters 1 and 2, and from the descriptions of power semiconductor devices given in the last section, it is clear that successful device fabrication depends on the ability to form precisely located p–n junctions in semiconductor material whose composition and properties are carefully controlled. In this section we briefly outline the processes and techniques that enable this to be achieved. Many are identical with those used for the fabrication of integrated circuits, but the larger scale of most power semiconductor devices and the need for them to support high breakdown voltages and high current densities often make for significant differences. These include the use of deep diffusions and the careful control of carrier lifetime; in particular, the use of material with very long carrier lifetime. One of the technical challenges in the fabrication of large-area, high power devices is that of maintaining homogeneity across the structure. In this case homogeneous, high resistivity, single-crystal silicon, with a very low defect density, is the essential starting material. Subsequent processing must ensure that it is not degraded.

### 3.2.1 The Preparation of High Purity Single-Crystal Silicon

Ultrapure, single-crystal silicon, used as starting material in the fabrication of power semiconductor devices, is made in a few specialised production plants around the world. The processing sequence is as follows:

- *Preparation of metallurgical grade silicon*
  In this step the silicon is formed by the reduction of silicon dioxide in a carbon-lined electric arc furnace.

- *Purification by the fractional distillation of trichlorsilane*
  The metallurgical grade silicon is converted to trichlorsilane at 1260 °C in the following reaction:

$$Si + 3HCl \rightarrow SiHCl_3 + H_2$$

  Trichlorsilane is liquid at room temperature and can be purified by fractional distillation to the quality required for semiconductor fabrication. In particular, chlorides of metals such as iron and copper and other impurities have to be eliminated.

- *Preparation of pure polycrystalline silicon by thermal reduction in a hydrogen atmosphere*
  The reaction

$$SiHCl_3 + H_2 \rightarrow Si + 3HCl$$

  takes place in a chamber with electrically heated rods of pure silicon which serve as nucleation surfaces for the deposition of silicon from the decomposition process. Cylinders of polycrystalline silicon can be formed in this way, up to 200 mm in diameter and up to 3 m in length.

- *Growth of a boule of single-crystal silicon*
  Single-crystal silicon is produced either by the Czochralski (CZ) method illustrated in Figure 3.22, or by the float zone (FZ) method shown in Figure 3.23. The Czochralski method is mostly used to prepare the substrate material for the growth of epitaxial layers in which most small-signal devices and integrated circuits are formed. The MOS power devices and some types of power diode are made in this way. However, most high power semiconductor devices are made from very pure, high resistivity starting material prepared by the float zone method. We briefly describe each in turn.

To start the CZ process, polycrystalline silicon is melted in a quartz crucible, together with the required dopants. A small single crystal of silicon is dipped into the melt and slowly rotated and withdrawn. With this starting crystal acting as a seed, layers of silicon are deposited on its facets, maintaining its single-crystal structure as

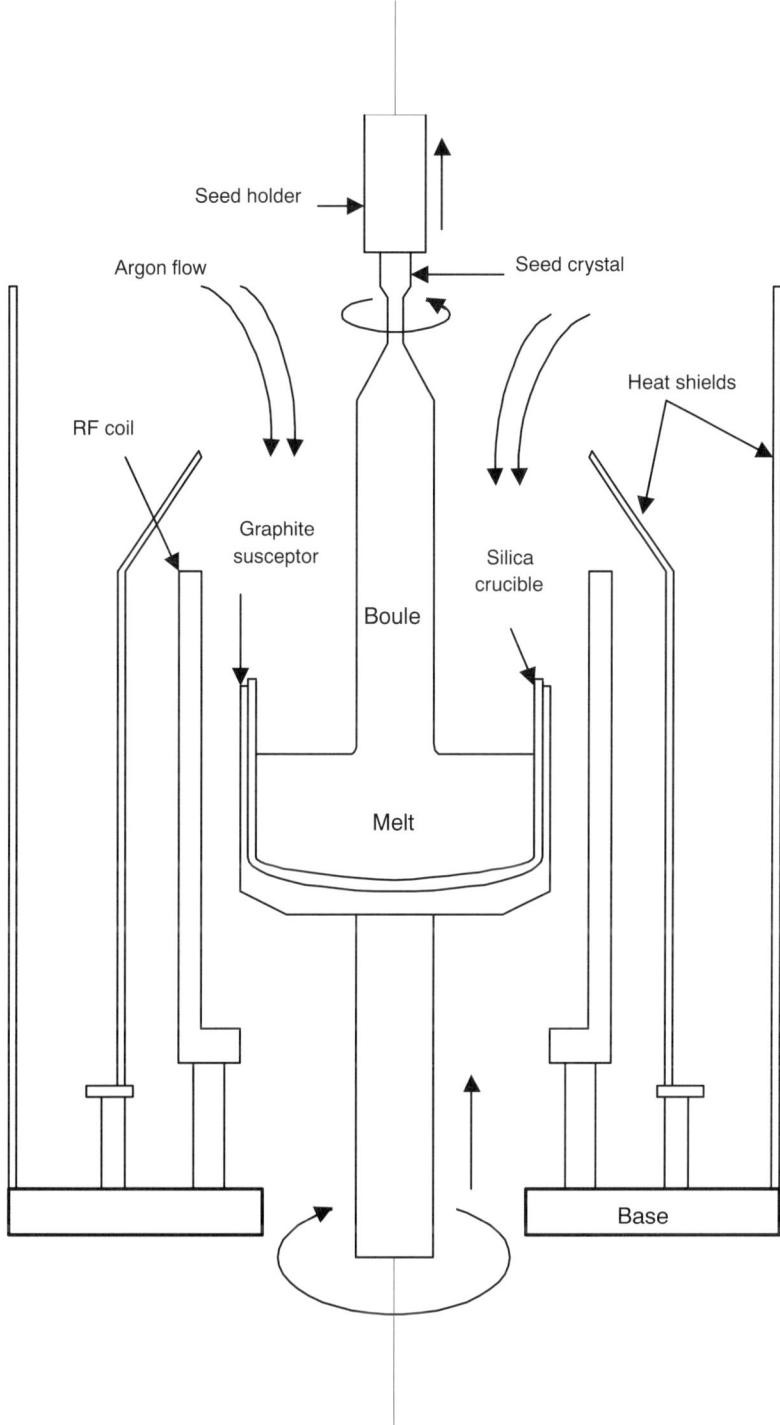

**Figure 3.22** A schematic illustration of the Czochralski (CZ) method of single crystal growth

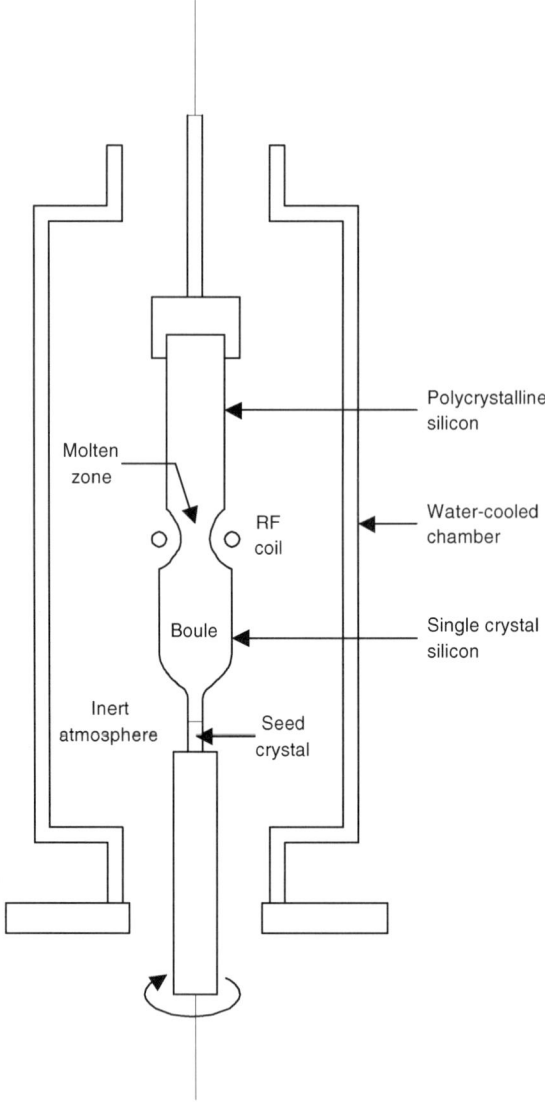

**Figure 3.23** A schematic illustration of the Float Zone (FZ) method of single crystal growth

it is withdrawn. The rates of rotation and withdrawal and the temperature distribution are all carefully controlled.

The CZ method allows cylindrical single-crystal boules of either n-type or p-type silicon to be grown up to diameters of 300 mm and weights of 250 kg. The CZ material contains a relatively high concentration of oxygen and other impurities.

The FZ process shown in Figure 3.23 is carried out in a chamber containing a controlled atmosphere of inert gas. A rod of pure polycrystalline silicon of appropriate diameter is held at the top and rotated. A single-crystal seed of appropriate orientation is clamped in contact with the other end of the rod. An induction heating coil is placed around the rod, melting a small length next to the seed crystal. The molten zone is then slowly moved along the length of the rod by raising the coil from one end to the other. Since the melted region is in contact only with the surrounding inert gas, very few impurities are introduced into the silicon and particularly low levels of carbon and oxygen can be achieved. By making multiple passes, the electrical resistivity can be as high as 10–100 $\Omega$ m (1000–10 000 $\Omega$ cm).

FZ silicon can be made either p-type or n-type. The dopant is added in gaseous form ($PH_3$ for phosphorus or $B_2H_6$ for boron) to the inert gas inside the crystal-growing chamber. A disadvantage of this method of doping is the relatively inhomogeneous incorporation of the dopant into the silicon. This causes striations, especially in the high resistivity material needed to make high voltage devices. These cause periodic variations in the electrical conductivity which often exceed $\pm 10\%$.

Homogeneous, high resistivity n-type silicon can be obtained from high purity FZ silicon using neutron transmutation doping. This technique is based on the nuclear reactions that occur when silicon is subjected to neutron irradiation. These involve the isotope of atomic weight 30, $^{30}Si$, which accounts for approximately 3% of natural silicon:

$$^{30}Si + n \rightarrow {}^{31}Si + \gamma$$

$$^{31}Si \rightarrow {}^{31}P + \beta$$

The beta decay has a half-life of 2.63 hours. Single crystals of silicon, normally with a starting resistivity greater than 10 $\Omega$ m (1000 $\Omega$ cm), are exposed to a flux of thermal neutrons in a nuclear reactor. The neutrons have a penetration range in silicon of about a metre, so the transmutation of isotope $^{30}Si$ into phosphorus occurs homogeneously throughout the boule, which is rotated to prevent any inhomogeneity arising as a result of any non-uniformity in the neutron flux. The doping level is determined by the neutron dose and can be very well controlled. Only n-type silicon can be prepared in this way but this is the preferred starting material for most power devices.

Technological progress has enabled single-crystal boules of silicon to be grown by the CZ method to a diameter that has increased over the years to 300 mm (12 in) as Figure 3.24 shows. Crystals grown by the FZ method can be up to 150 mm (6 in) in diameter.

### 3.2.2 Wafer Preparation

For the manufacture of devices, silicon wafers of the correct thickness and diameter have to be prepared. The first step is the centreless grinding of the grown crystal boule to the correct diameter. When several devices of rectangular shape (chips) are to be fabricated on the wafer, a flat with a predetermined crystallographic orientation, usually the {110} plane, is ground along one edge of the boule. This is important for mechanical handling later in the process.

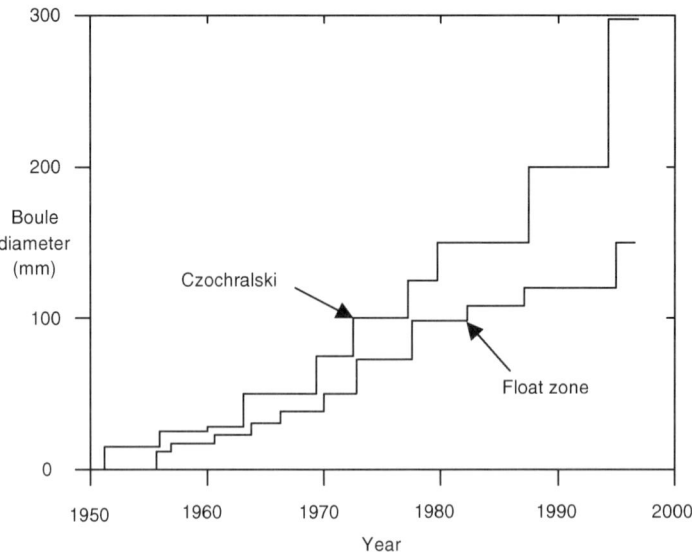

**Figure 3.24** The increase in the maximum available diameter of single-crystal boules of silicon grown by the CZ and FZ methods

Following grinding, the silicon crystal is sliced into wafers of predetermined thickness, usually using an internal circular diamond-edged saw. Cutting damages the wafer surfaces and is usually followed by mechanical lapping to make both sides planar and reduce the wafer thickness to a precise value. The lapping process also causes mechanical damage to the surface layer, which may extend to a depth of 30 $\mu$m. This damage may be fully or partially removed by chemical etching. However, this must not compromise the flatness of the surfaces or the uniformity of the wafer thickness. Wire cutting can also be used. It is slower than using a diamond-edged saw, but causes less surface damage.

Many types of modern power semiconductor device require relatively fine structures (on a micrometre scale) covering a large (centimetre scale) area. For the photolithography used during device fabrication, at least one surface must be damage-free and optically flat. A mirror finish is obtained using a polishing mixture that both chemically etches and mechanically polishes the wafer surface. A dispersion of silica in ammonia, caustic soda and hydrogen peroxide can remove about 1 $\mu$m per minute.

The lapping and polishing processes may introduce contamination into the surface layers of the silicon. Therefore, before any high temperature operations are begun, the surface is cleaned. This may be either a wet or a dry process. When wet cleaning is used, any oxide on the silicon surface is removed using dilute hydrofluoric acid (HF). Heavy metals may be removed by a mixture of hydrochloric acid and hydrogen peroxide. This is followed by rinsing and washing with deionised water, then drying in a flow of dry nitrogen. The dry cleaning process removes the contaminated surface layer by plasma etching, as described in Section 3.2.6.

### 3.2.3 Epitaxial Growth

When a thin layer of high resistivity is required, epitaxial deposition of silicon onto a single-crystal substrate is often used. The deposited layers take up and continue the crystal structure of the substrate. Epitaxial layers are normally used in the fabrication of IGBTs and power MOSFETs. For very high power, high voltage devices, their use is limited because it is difficult to reduce the defect density to the levels that can be achieved in all-diffused devices starting from bulk material.

Two basic processes are used. The first is based on the high temperature decomposition of chlorosilanes in a hydrogen atmosphere, a process similar to that used for the preparation of pure polycrystalline silicon. With silicon tetrachloride,

$$SiCl_4 + 2H_2 \rightarrow Si + 4HCl$$

In order to obtain a good single-crystal layer, deposition temperatures of 1150–1220 °C are normally used. The growth rate is a function of the temperature and the gas composition. Other chlorosilanes can be used, e.g. $SiHCl_3$ at 1100–1175 °C or $SiH_2Cl_2$ at 1025–1100 °C.

Alternatively, deposition can be through the pyrolitic decomposition of silane, at temperatures of 950–1050 °C:

$$SiH_4 \rightarrow Si + 2H_2$$

The deposition rate depends on many parameters such as the mixture of gases, the temperature and the pressure in the reaction chamber; it is normally between 0.1 and 2.0 $\mu$m per minute. In all cases, hydrogen is used as a carrier gas. Dopant gases are used to determine the conductivity type and resistivity of the deposited layer. For n-type layers, phosphine ($PH_3$) or arsine ($AsH_3$) is added to the hydrogen gas flow in a controlled proportion. For p-type layers, diborane ($B_2H_6$) is used.

The silicon wafers are placed on a heated susceptor, usually made from graphite with a thin surface coating of silicon carbide. They are heated in a quartz reaction chamber by means of radio frequency coils or by ultraviolet energy generated from special quartz-envelope lamps. A typical epitaxial deposition cycle consists of the following steps:

- Before the deposition, the substrate is cleaned either by a series of acid washes or by physically scrubbing the wafer surface. This cleaning is of particular importance since any residual particles may give rise to imperfections in the deposited layer.

- At the start of deposition, nitrogen is used to flush unwanted gases from the reaction chamber. When the temperature exceeds 600 °C, the nitrogen is replaced by hydrogen, which is mostly used during the subsequent fabrication steps.

- Before epitaxial deposition, the silicon surface is usually etched with hydrochloric acid in hydrogen at 1180–1240 °C. This helps to ensure a perfect surface and helps to minimise defect concentration; it is particularly important for high voltage, large-area devices, where a very low defect level throughout a large volume of silicon is necessary.

• After the etching, a mixture of hydrogen with a doped silicon source is fed into the reaction chamber. At the end of the deposition sequence, the reaction chamber is cooled and, when its temperature falls below 600 °C, it is flushed with nitrogen until the temperature reaches room temperature, when the wafers can be removed.

### 3.2.4 Thermal Oxidation

Silicon dioxide layers play a vital part in the successful fabrication and operation of silicon devices. They are used as insulating layers to separate and isolate the semiconductor from interconnecting conductors, as masking layers at different stages of fabrication and as a seal at the silicon surface.

Thermal oxidation takes place naturally at the silicon surface when either pure oxygen or a mixture of oxygen with water vapour reacts with silicon. During device fabrication, these reactions are usually carried out in the temperature range 900–1250 °C:

$$Si + O_2 \rightarrow SiO_2$$

and

$$Si + 2H_2O \rightarrow SiO_2 + 2H_2$$

Thermal oxidation is preceded by a cleaning sequence designed to remove all contamination, particularly alkali metals (Na and K) and their compounds, which can give rise to undesirable positive charge in oxide layers. After cleaning, the silicon wafers are loaded in a quartz wafer holder. Oxidation is usually carried out in a quartz tube (sometimes a polycrystalline silicon tube is used), in a furnace with a carefully controlled temperature and gas mixture.

For short oxidation times, while the oxide is very thin, the growth rate is constant. The oxide thickness $d_{ox}$ increases as

$$d_{ox} \approx d_0 + At \tag{3.1}$$

As the thickness increases, the rate of growth is limited by the need for the oxidising gases to diffuse through the oxide that has already formed. The thickness then increases with time as

$$d_{ox} \approx Bt^{1/2} \tag{3.2}$$

In (3.1) and (3.2), both $A$ and $B$ are functions of gas composition and they vary exponentially with temperature.

Oxidation in dry oxygen is a relatively slow process and is used whenever high quality oxide layers of precisely controlled thickness are required. For the gate oxide of MOS structures, where a very low density of oxide defects is essential, chlorine additives (HCl or trichlorethylene) are used. The oxidation rate in steam is much faster and increases with the water vapour pressure, as shown in Figure 3.25, but the quality of the oxide layer is lower. Very often 'wet' oxygen is used as the gas ambient. The oxygen is either bubbled through water (when the percentage of water vapour

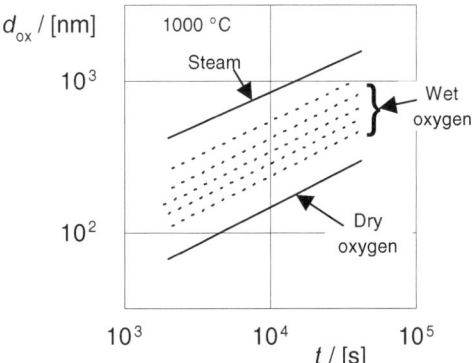

**Figure 3.25** Measured oxide growth rates in oxygen with different partial pressures of water vapour

taken up is determined by the water temperature) or produced by burning a controlled amount of hydrogen in oxygen (pyrogenic oxidation).

Oxide layers may also be formed by chemical vapour deposition, as described in Section 3.2.9.

### 3.2.5 Photolithography

The geometrical patterns of the different layers of a semiconductor device, needed to obtain the desired electrical characteristics, are usually produced by photolithography. This involves the transfer of an image onto the wafer surface by means of a photosensitive polymer material, called photoresist, which should have the following characteristics:

- It should be photosensitive, i.e. it either polymerises or depolymerises when exposed to light of a particular wavelength.
- It should retain a high image resolution so that very fine lines or spaces can be produced.
- It should adhere strongly to Si and $SiO_2$.
- It should have a high chemical resistance to the agents used for etching oxide, silicon and metallic layers.

Photoresists are selected so that they respond to short wavelengths such as the ultraviolet light emitted by mercury arc lamps or excimer lasers, but do not respond to longer wavelengths (yellow and longer). It is then possible to work with them in areas illuminated with yellow light. Two distinct types are used:

1. *Positive photoresist.* Exposure to light increases the solubility of the resist in the developing solution, so that the area illuminated is easily removed from the wafer surface.

2. *Negative photoresist.* The light induces polymerisation that decreases the solubility of the photoresist in the developing solvent and it is the unexposed resist that can easily be removed from the wafer surface in the unilluminated areas.

The photolithographic process consists of a number of steps performed sequentially.

- The photoresist layer is deposited in solution onto the prepared surface to which the process is to be applied. The surface must be clean and dry; an even spread of the resist is usually obtained by dropping a small amount onto the wafer, which is then briefly spun. Other techniques such as dipping or spraying are sometimes used. The layer must be of a suitable thickness and homogeneous over all the area of the wafer.

- After deposition, a 'soft-bake' process, at about 90 °C for about 15 minutes, is applied to remove the remaining solvent.

- Next the photoresist is exposed to ultraviolet light through a mask of the required pattern. In the case of the first masking process, the orientation of the mask with respect to the crystallographic orientation of the wafer is important. If the wafer already has a pattern on its surface from an earlier stage in the processing sequence, the mask must be registered with and aligned to this pattern. Precise optical–mechanical positioning equipment is used for this step. The mask is normally formed by photographic reduction on a glass plate. In processes of increasing sophistication, the mask may be exposed

    (a) in contact with the photoresist on the wafer surface (contact printing)
    (b) in close proximity to, but not in contact with the wafer surface (proximity printing)
    (c) imaged onto the wafer surface, usually with a $\times 10$ reduction (projection printing)
    (d) projected onto the wafer surface in a small area covering just one, or a few, devices at a time and progressively stepped across the whole of the wafer surface (direct step-on wafer, or step-and-repeat)

- After exposure the photoresist is developed by dissolving the regions of high solubility (the exposed regions with positive photoresist, the unexposed regions with negative resist). This may be realised by immersing the wafer in the developer or spraying the developer on the wafer. Then a rinse is usually applied to remove any remaining developer from the surface. The choices of developer and rinsing solution depend on the photoresist being used.

- The developed pattern is examined visually to determine the accuracy of the image transfer process. Wafers that do not pass this inspection are reworked.

- The process of preparing the photoresist mask is completed by hard baking, in which the wafers are maintained at about 120 °C for about 20 minutes to evaporate any solvents left in the resist layer, to increase adhesion and to strengthen the chemical resistance of the photoresist layer. The part of underlying wafer surface that is covered with the photoresist mask is protected, while it remains, against etching and other effects in subsequent processing operations.

- At the end of the sequence of operations connected with photolithography, the photoresist mask has to be removed ('stripped') without damaging the rest of the wafer. Negative photoresist deposited onto an oxide layer can be removed with a sulphuric acid and hydrogen peroxide mixture at about 100 °C. Positive resists are soluble in acetone and similar solvents and can be easily used for preparing metallisation patterns. Both positive and negative resists can also be removed in an oxygen gas discharge, when organic molecules of the photoresist convert to the gases $CO_2$, $H_2O$ and $N_2$, leaving no residue on the wafer surface.

### 3.2.6 Etching Processes

Etching is used to remove layers (oxide, metal, single-crystal or polycrystalline silicon) from the surface of wafers. For certain devices it is also used to cut grooves and trenches into the surface. Etching is used in different stages of device fabrication, very often in combination with photolithography, to prepare a desired pattern of a particular layer of the device. In this case the ideal etchant removes material from regions not protected by the photoresist but does not attack either the photoresist or the underlying layer. In practice it is likely to etch the protected surfaces very slowly – a 10:1 ratio of etch rates is considered the minimum acceptable. The etching process may be either isotropic, where the etch rate is the same in all directions, or anisotropic, where it etches more rapidly in one particular direction.

Two basic etching techniques are used: liquid (wet) etching and dry etching. In liquid etching the wafers are immersed in an etching solution at a chosen temperature for a time determined experimentally. The wet etch is followed immediately by a rinse in deionised water. Examples of etchants commonly used in semiconductor device fabrication are given in Table 3.2. Wet etching processes are normally isotropic, with the result that lateral etching increases the size of the details

Table 3.2  The properties of wet etchants

| Material | Etchant | Temperature (°C) | Etch rate ($\mu$m/min) | Note |
|---|---|---|---|---|
| $SiO_2$ (thermal and CVD) | $NH_4F$–HF (49%) 4:1 to 7:1 | 20–30 | 0.08–0.12 | |
| Silicon | HF-$HNO_3$–$C_2H_5COOH$ | 20–30 | 0.35–0.50 | isotropic |
| | KOH | 20–100 | | anisotropic |
| Aluminium | $H_3PO_4$–$C_2H_5COOH$–$HNO_3$–$H_2O$ (50:10:2:3) | 20–40 | 0.2–0.6 | |

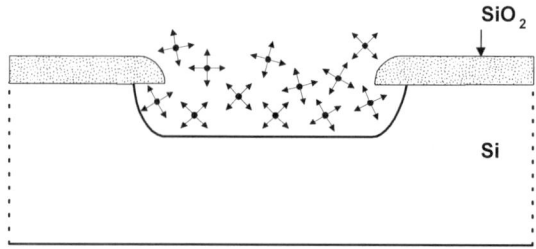

**Figure 3.26** An illustration of the under-cutting caused by an isotropic etch

in the mask pattern. This is illustrated in Figure 3.26 and is an effect that has to be allowed for. Some etchants preferentially attack the less densely packed atomic planes and remove material more slowly from the more closely packed {111} planes. In this way, V-grooves can be formed, as shown in Figure 3.11.

Three dry etching techniques are generally available for use. No liquid is used during the etching process, all etchants are gaseous.

1. *Plasma etching.* Wafers are placed in a chamber which is evacuated and refilled with low pressure gas. A radio frequency electromagnetic field is then applied at a level that ionises the gas. The wafer surface is etched away by the ions. To etch silicon, silicon dioxide or silicon nitride, a gas containing fluorine (usually a type of Freon) is normally used. For metallic layers such as aluminium, a chlorine-containing gas is preferred. When selective etching is required, the surface areas not to be etched can be protected by photoresist.
2. *Sputter etching.* Energetic particles are used to remove material by direct bombardment with inert gas ions such as argon. If part of the surface is covered with a mask, it is protected against etching until the masking layer itself is removed. This process is highly anisotropic because material is only removed along the direction of the ion beam.
3. *Reactive-ion etching.* This technique is a combination of sputtering and chemical etching. It involves the controlled bombardment of the surface to be etched with chemically reactive ions. The etching rate retains a high level of anisotropy and is faster than sputtering. It is used when trenches are required, as in Figure 3.12.

The dry etching methods require a vacuum chamber and the means to ionise the etchant gas. But there are no problems of etchant depletion or the formation of bubbles. The etching products are gaseous, so there is no difficulty in removing the remaining material from the surface at the end of the process.

### 3.2.7 The Introduction and Redistribution of Impurities 1: Diffusion

An essential part of semiconductor device fabrication is the introduction of dopant impurities (donors or acceptors) into specific regions of the wafer in order to create the required device structures. Two techniques are used, diffusion and ion implantation.

Diffusion is the process whereby particles move from regions of higher concentration to regions of lower concentration, as described in Section 1.5 with reference to the electrons and holes in a semiconductor. Here we are concerned with the migration of impurity atoms in single-crystal silicon at high temperature, which enables a controlled amount of chosen impurities to be introduced into selected regions.

The distribution of the impurity atom concentration $N(x,y,z,t)$, evolves according to Fick's laws, so that

$$\frac{\partial N}{\partial t} = \text{div}(D\,\mathbf{grad}\,N) \tag{3.3}$$

where $D$ is the atomic diffusion coefficient, which increases rapidly with temperature. As long as the diffusion coefficient is independent of the impurity concentration, equation (3.3) is linear. Then, with the impurity concentration $N_0$ at the surface of a semi-infinite crystal kept uniform and constant, the solution [3.2] follows the complementary error function:

$$N(x,t) = N_0 \text{erfc}\left(\frac{x}{\sqrt{Dt}}\right) \tag{3.4}$$

where $x$ is the depth below the surface. The normalised concentration profile resulting from a constant-source diffusion is shown in Figure 3.27(a).

In order to maintain a constant surface concentration, sufficient dopant has to be transported onto the surface of the wafer during the diffusion process. In this case, $N_0$ is usually limited by the solubility of the dopant in silicon at the diffusion temperature. This is shown for some important dopants in Figure 3.28.

A different solution of the diffusion equation is obtained when the diffusion source is limited to a fixed total number of atoms $Q$ which diffuse in per unit area from a thin layer at the surface. Then a Gaussian profile is obtained:

$$N(x,t) = \frac{Q}{\sqrt{\pi Dt}} \exp\left(-\frac{x^2}{4Dt}\right) \tag{3.5}$$

This is shown in Figure 3.27(b). A limited-source diffusion is often carried out after an initial very short diffusion under constant source conditions or following the heavy doping of a thin layer by ion implantation. In power devices this process is called *drive-in*.

Solutions of the diffusion equation become more complicated in the case of very thin wafers, or very long diffusion times at high temperatures, when the wafer cannot be treated as semi-infinite. If the diffusion coefficient depends on the dopant concentration, the diffusion equation is non-linear and solution is much more difficult. This occurs with phosphorus in silicon and when the process is controlled by the substitutional–interstitial mechanism (e.g gold, platinum, and some other fast-diffusing metals in silicon).

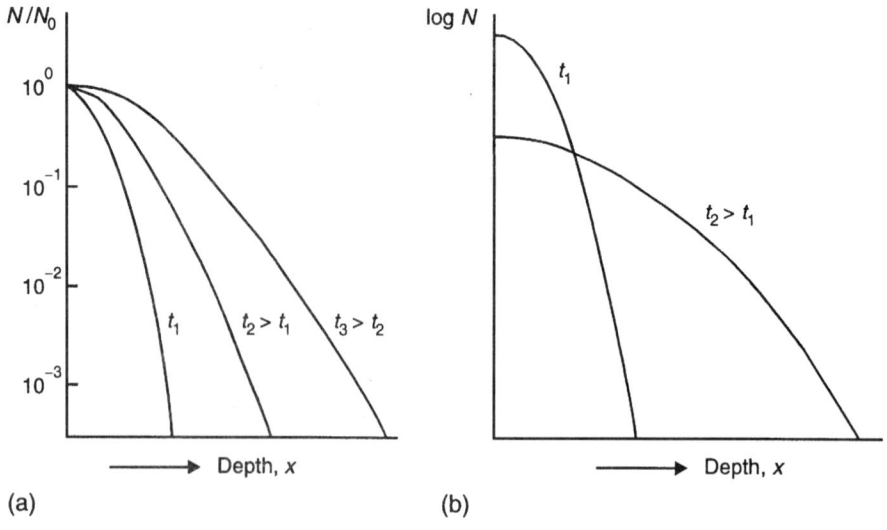

**Figure 3.27** Concentration profiles following constant-source and limited-source (drive-in) diffusions: (a) constant-source diffusion; (b) limited-source diffusion

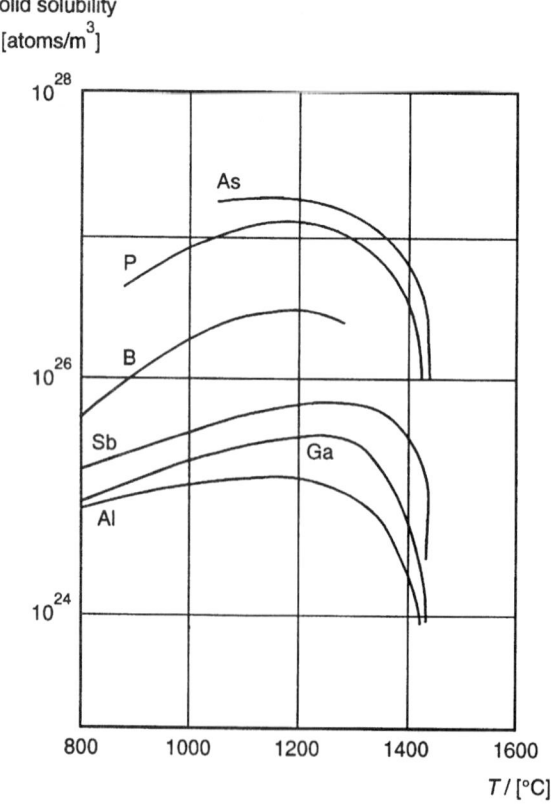

**Figure 3.28** Solid solubility of impurities in silicon

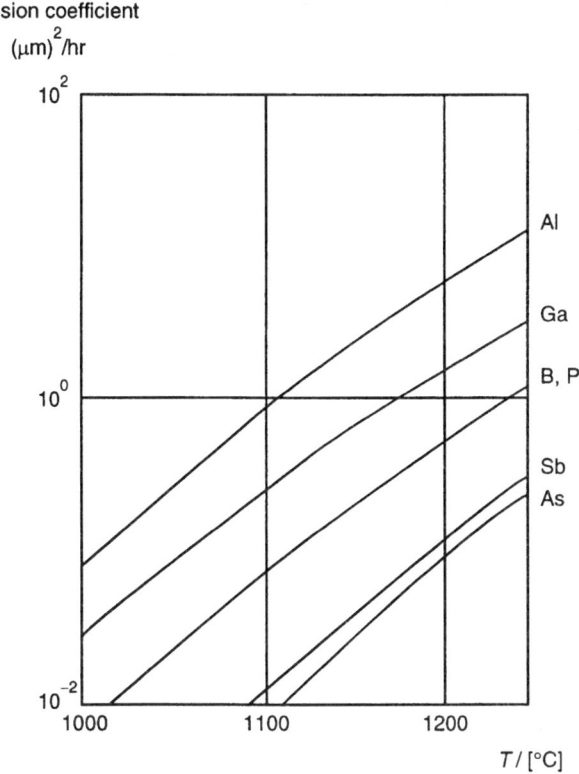

**Figure 3.29** Impurity diffusion coefficients as a function of temperature

Phosphorus, boron, gallium and aluminium are the dopants used most often in the technology of silicon power devices. In Figure 3.29 it can be seen that phosphorus has the highest diffusion coefficient of the donor impurities. It is therefore most commonly used to make n-type material and, because of its high solid solubility in silicon, phosphorus is especially suitable for heavily doped layers. The diffusion coefficients of arsenic (As) and antimony (Sb) are more than an order of magnitude lower, so that long diffusion times and high temperatures are required. They are used to produce n-type layers that are subject to later high temperature processing, because they are less prone to redistribution.

Boron is the most commonly used acceptor impurity in silicon technology. It has the highest solid solubility but a relatively low diffusion coefficient. A disadvantage of boron as dopant is that it has a much smaller atomic radius than silicon; this can produce strain in the crystal lattice.

Both boron and phosphorus diffuse much more rapidly in silicon than in silicon dioxide. This permits the use of layers of $SiO_2$ to mask areas of the silicon wafer surface against the diffusion of these dopants. Successive diffusions through windows, defined by photolithography and etched in oxide layers grown on one surface of the wafer, enable structures of both discrete devices and monolithic

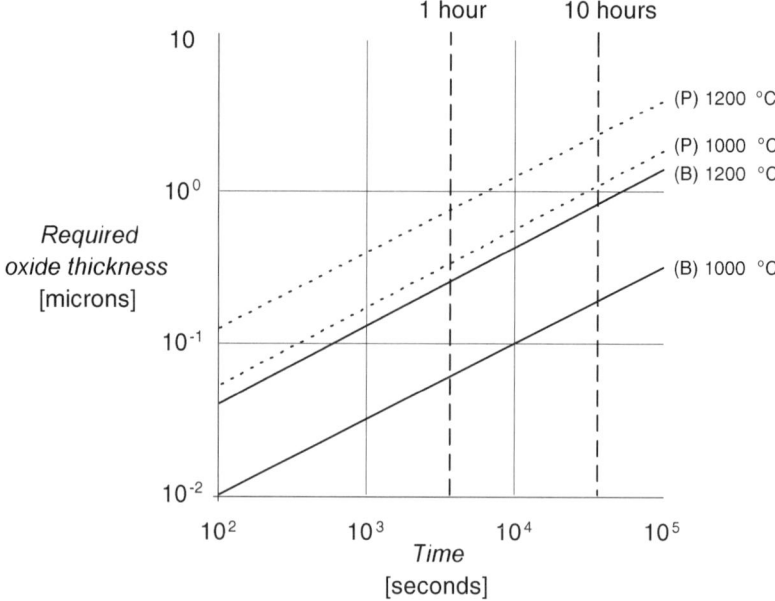

**Figure 3.30** Oxide thickness required for effective masking

integrated circuits to be prepared by what is known as planar technology. The thickness of the oxide layer needed for an effective mask depends on the duration and temperature of the diffusion process, as shown in Figure 3.30.

Gallium and aluminium diffuse faster than boron but have lower solid solubility. Their lattice match to silicon is good. These dopants are often used to form deep p–n junctions with a high breakdown voltage. The diffusion coefficient of gallium in silicon dioxide is relatively high in comparison with those of boron and phosphorus, so an oxide layer is not an effective mask against gallium diffusion. Indeed, gallium is often diffused through an oxide layer in order to protect the silicon wafer surface from the condensation of gallium vapours.

There are several variations of the basic diffusion technique:

1. *Solid to solid.* The diffusant source is a layer of solid material deposited on the surface of the wafers. This layer may be the element to be diffused or one of its compounds. It may be evaporated or sputtered onto the surface, or deposited from solution. For example, a thin layer of doped oxide can be formed by chemical vapour deposition (CVD), whereas a coating of aluminium nitrate ($Al(NO_3)_3$) can be deposited from an ethanol solution. When the compound is soluble in water (e.g. $(NH_4)_2PtCl_6$ as a source of platinum), the hydrophobic nature of the silicon surface complicates the process. Such compounds are put down as a suspension of very fine silicon dioxide, often using a spinner to obtain a homogeneous film over the surface (the so-called spin-on technique), or by means of a spray.

2. *Liquid to solid.* At the diffusion temperature, the diffusant is liquid, perhaps a metal forming a eutectic alloy with silicon or perhaps a borosilicate glass. This method is rarely used.
3. *Gas to solid.* In the most commonly used method, the diffusant or one of its compounds enters the surface in gaseous form, using either a sealed tube process or an open tube process. The sealed tube is usually quartz and the silicon wafers are placed inside along with the dopant, often in its elementary form. The tube is evacuated, filled with a low pressure of inert gas (the pressure at the diffusion temperature should be atmospheric), sealed and placed in a diffusion furnace. After the diffusion is complete, the tube is cut open and the diffused wafers are taken for the next operation. The sealed tube technique is used relatively rarely, e.g. for the diffusion of aluminium and gallium.

The open tube technique is used more frequently. Wafers, held separated in a quartz or silicon boat, are placed in a quartz tube in the diffusion furnace with a flow of a suitable ambient gas atmosphere. This consists of a carrier gas such as nitrogen plus vapours of the required dopant or one of its compounds. The source of the diffusant vapour can be in the solid phase at room temperature (e.g. $B_2O_3$, $Ga_2O_3$, $P_2O_5$), in the liquid phase (e.g. $POCl_3$, $BBr_3$) or a gas (e.g. $PH_3$, $B_2H_6$). The gas flow and the partial pressure of the dopant are carefully controlled as the surface concentration is determined by the gas composition. In the case of phosphorus diffusion, a small amount of oxygen is mixed with the nitrogen carrier gas to produce $P_2O_5$, which reacts with the silicon surface to form a phosphosilicate glass. It is this glass which acts as the source of phosphorus during the diffusion process. Similarly, in the case of boron diffusion, a layer of borosilicate glass is always produced on the surface of silicon and it is this that acts as the source of boron.

### 3.2.8 The Introduction and Redistribution of Impurities 2: Ion Implantation

Ion implantation accelerates ions of the desired dopant into the surface of the wafer, using an electric field. The ions are generated in a low pressure gas discharge. They penetrate into the silicon crystal, losing energy in nuclear and electronic collisions. The depth of penetration increases with the ion energy and so can be controlled by the voltage applied. The accelerated ions are scanned across the wafer to obtain a homogeneous distribution of implanted ions. The number of ions that enter the wafer per unit area (the so-called dose) can be controlled very precisely through the ion beam current. It may vary from $10^{11}$ to $10^{16}$ ions per square centimetre, a range of values that cannot be obtained reliably by other means.

The actual distribution of implanted ions in an amorphous layer approximates to a Gaussian distribution, with a peak at the depth $R_p$ below the surface, as shown in Figure 3.31. This is known as the projection range. It increases approximately linearly with the ion energy and decreases with an increase in the atomic radius of the implanted element. Thus, the projection range for 100 keV boron ions is 0.3 $\mu$m, and for 200 keV ions it is 0.55 $\mu$m. The projection range of 100 keV phosphorus ions is

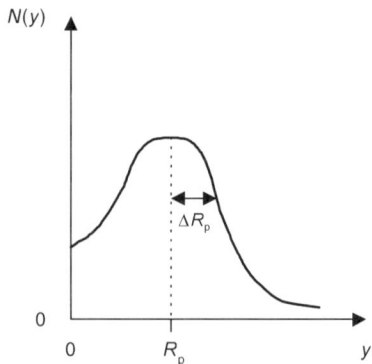

**Figure 3.31** The distribution of impurities with depth, following ion-implantation, at different ion energies

0.12 μm, and for 200 keV ions it is 0.26 μm. With crystalline material, some ions are channelled between the lattice planes and penetrate deeper when that direction coincides with the beam direction. Channelling may be minimised by tilting the silicon surface to give an angle of about 7° between the beam direction and the surface normal. A thin layer of amorphous silicon on the surface also helps to prevent channelling.

The ion implantation can be readily masked from defined areas by a layer covering the surface which prevents the ions from penetrating the silicon. As ion implantation is a low temperature process, it can be masked either by an $SiO_2$ layer (like diffusion) or by a layer of photoresist or a metallic film. The necessary thickness of mask depends on the ions, their energy and the masking material. Usually less than 1 μm is needed to reduce the dose by a factor of $10^6$.

During implantation the dopant ions damage the crystal lattice. Because they are not in substitutional sites, they are not initially electrically active. The ion implantation is therefore followed by a thermal annealing cycle at 600–950 °C to remove the radiation damage and move the dopant atoms into regular substitutional positions.

The ion implantation process gives well-controlled impurity profiles with a high level of uniformity and reproducibility. It can be used to form very thin layers. In MOS devices it is used to make shallow p–n junctions. In other power devices it can be used to provide the material for a well-controlled limited-source diffusion in a subsequent drive-in process. In this high temperature redistribution the dopant profile can be closely controlled.

### 3.2.9 Chemical Vapour Deposition Techniques

Chemical vapour deposition (CVD) is the formation of a layer of a stable compound on a heated substrate by the thermal reaction or decomposition of gaseous compounds. There are many variations of the CVD technique. Epitaxial growth, described in Section 3.2.3, is a specific type of CVD process that enables the crystal

## 3.2 FABRICATION PROCESSES

structure of the substrate to be continued in the layer deposited. In general, a gas mixture flows into the reaction chamber in which the substrate is maintained at a suitable temperature. The gaseous components react with the surface, the desired product of the reaction forms a layer at the surface and the other reaction products have to be exhausted.

The rate of growth of the deposited layer depends on the temperature ($\propto T^{3/2}$), the total pressure ($\propto 1/P_T$) and the partial pressure of the reactant, ($\propto P_n$). The deposition rate also depends on the geometry of the reactor and the velocity of the gas flow. A given deposition rate can be obtained at a lower temperature when the total pressure is lower. For this reason, CVD often takes place at a pressure of about 100 Pa and is known as low pressure CVD (LPCVD). The lowest deposition temperatures (100–400 °C) can be obtained using low pressure, plasma-enhanced CVD.

CVD can be used to deposit many materials. In semiconductor device fabrication, four materials are generally encountered:

- polycrystalline silicon (perhaps deposited on an oxide layer)
- silicon dioxide
- silicon nitride
- phosphosilicate glass (PSG)

Polycrystalline silicon is usually deposited using the reaction

$$SiH_4 \rightarrow Si + 2H_2$$

which occurs at 500–900 °C, using LPCVD, with a carrier gas of hydrogen or nitrogen.

Polycrystalline silicon can be deposited either undoped or doped by including, as required, some phosphine, arsine or diborane in the reaction gas mixture. Undoped polycrystalline silicon is used for silicon-on-insulator (SOI) technology [3.3, 3.4]. The gate electrodes of MOS transistors use heavily doped n-type polysilicon. Semi-insulating polycrystalline silicon (SIPOS), deposited by the LPCVD method using $SiH_4$–$N_2O$–$N_2$ or $SiH_4$–$NH_3$–$N_2$ reaction gas mixtures, can be used for the passivation of high voltage p–n junction terminations.

Silicon dioxide can be obtained by using reactions

$$SiH_4 + 4CO_2 \rightarrow SiO_2 + 2H_2O + 4CO$$

and

$$SiH_4 + 2O_2 \rightarrow SiO_2 + 2H_2O$$

Layers of silicon dioxide can be deposited using LPCVD at temperatures in the range 300–500 °C. This is known as pyrolitic oxide. Low temperature deposition is used when the oxide layer is required to provide dielectric insulation between metallic contacts and interconnections.

Silicon nitride can be obtained using the reaction

$$3SiH_4 + 4NH_3 \rightarrow Si_3N_4 + 12H_2$$

which occurs at 500–900 °C, using LPCVD. Using low pressure, plasma-enhanced deposition, silicon nitride layers can be prepared at temperatures around 300 °C.

### 3.2.10 The Preparation of Contacts

The electrical contacts of power semiconductor devices serve to conduct both electric current to the semiconductor structure and dissipated power away to a heat sink. In the case of integrated structures, metallisation interconnects the individual devices. To serve as an effective contact with silicon, the chosen metal must

- make a contact with high thermal and electrical conductance
- adhere well to the underlying material (silicon, $SiO_2$, etc.)
- be stable and not exhibit electromigration under normal operating conditions
- be compatible with photolithography
- readily bond to external connectors
- be economically competitive

The basic methods used for preparing contacts are

- vacuum deposition
- chemical deposition
- sputtering
- alloying

Aluminium is the most commonly used contact material and is invariably deposited by evaporation in a vacuum. In order to prevent silicon from dissolving in the metal layer, 1–2% silicon is normally added; and to decrease electromigration, 3–5% copper is also often included. Because the contact layers for power devices are relatively thick in comparison with those of integrated circuits, typically about 10 μm, electron beam evaporation, which allows a high deposition rate, is frequently used.

A metal layer can be deposited on the silicon surface by a chemical reaction in a bath containing a suitable compound. In this way, layers of nickel can be put down onto a silicon surface by the catalytic reduction of ions from a solution of $NiSO_4$ and $NaH_2PO_2$. This has frequently been used in the fabrication of small to medium area diodes and thyristors where the connections are made by soft soldering. This technique allows Ni, Au and Cu layers, several tens of microns thick, to be deposited. After deposition, adhesion to the silicon is improved by heating to about 600–700 °C to form silicides at the interface. A further chemical deposition of Ni provides a metal contact that can be used to make a soldered connection.

Alloyed contacts are widely used to join electrodes to the device casing, either directly or via discs of molybdenum or tungsten. The technology requires a metal–

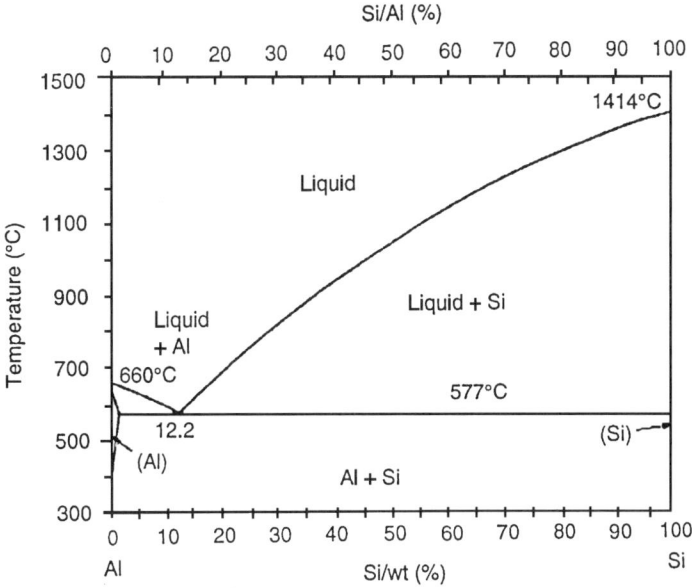

**Figure 3.32** The aluminium–silicon phase diagram. Reproduced from 'Silicon/Heat-sink Assemblies for High Power Device Applications: Present Technology and Developments' by D. E. Crees, G. Humpston, D. M. Jacobson and D. Newcombe, in the *GEC Journal of Research*, **6**, 71–9 (1988), by permission of The General Electric Company, plc

silicon eutectic composition with a relatively low melting temperature. A layer, or foil, of the metal or alloy in contact with the silicon is heated in vacuum or in a reduced atmosphere to a temperature higher than the melting point of the metal. The melt wets the silicon surface and dissolves some silicon in the melt. The temperature is slowly decreased and silicon regrows from the melt. At the freezing point, a layer of composition close to that of the eutectic is plated on the silicon surface.

The most commonly used composition is Al–Si whose eutectic (11.3% Si) has a melting point of 577 °C. In this case the regrown silicon is doped with aluminium, creating a $p^+$-layer in contact with the eutectic. The Al–Si phase diagram is shown in Figure 3.32. This technique is frequently used for joining the anode contacts of thyristors and diodes to molybdenum or tungsten backing discs, as described in Chapter 13.

Another possibility is to use the gold–silicon composition with a eutectic melting point of 370 °C. In this case, adding 1% of antimony results in the regrowth of $n^+$-silicon layers, whereas 1% of gallium provides $p^+$-layers, in each case covered with the gold–silicon eutectic (31% Si). The cost of gold contacts is very high. It is interesting to note that alloying was the first technique used to make p–n junctions. Sputtering is rarely used.

## 3.3 The Control of Carrier Lifetime

Carrier lifetime is a parameter that influences all the important parameters of bipolar power devices, as discussed in detail in subsequent chapters. For high voltage devices, such as diodes, thyristors and IGBTs, a long carrier lifetime is desirable, in order to keep on-state losses acceptably low. On the other hand, devices required to operate at high frequencies need the carrier lifetime to be short, in order to obtain fast turn-off and to minimise the reverse recovery charge.

As discussed in detail in Section 1.4, the carrier lifetime in silicon is controlled by the concentration of recombination centres. These are levels, deep in the band gap, caused by lattice defects and certain types of impurity. Carrier lifetime control in bipolar power semiconductor devices has two basic requirements. The first is to avoid the uncontrolled creation of recombination centres during high temperature processing operations such as diffusion, thermal oxidation and epitaxial growth. The second is the controlled reduction of the carrier lifetime in selected regions of the device by the introduction of suitable recombination centres. We deal first with the techniques used to ensure a long carrier lifetime when this is needed and then with the techniques used to reduce the carrier lifetime. It should be noted that at very high carrier concentrations the lifetime is influenced by Auger recombination, which is unaffected by processing technology.

### 3.3.1 Techniques for Obtaining Long Carrier Lifetimes

In the as-grown single-crystal silicon, the carrier lifetime can be as long as 1 ms. However, during high temperature processing, lattice defects are generated in the volume of the silicon wafers and unwanted impurities, such as iron, copper and gold, can enter the silicon from the surface and contaminate the interior. In both cases, recombination centres are created and the carrier lifetime is shortened. Possible sources of unwanted contamination include

- residual surface impurities from the lapping and polishing operations
- contamination adsorbed from chemicals used in the etching and cleaning processes
- impurities from quartz tubes and boats

In order to prevent this contamination, very pure (electronic grade) chemicals are used for cleaning the silicon; the diffusion sources, gases, diffusion tubes and boats are all of the highest purity; good environmental control is maintained.

The concentration of lattice defects can be minimised by cooling the wafers slowly (at a rate not exceeding about 1 °C per minute) after the final high temperature process. This is particularly effective after a phosphorus diffusion, when the surface is covered with a layer of phosphosilicate glass (PSG). The types of metallic impurity which create deep trapping levels normally diffuse rapidly in silicon by the interstitial–substitutional mechanism. That is, they migrate through interstitial positions and become substitutional either by filling a lattice vacancy or by replacing a silicon atom (the 'kick-off' mechanism). They diffuse to the surface covered with the PSG layer, where they form compounds either with the glass or in the n$^+$-layer under the glass. A similar effect occurs after a boron diffusion when the p$^+$-surfaces are covered with borosilicate glass. It is known as *gettering*. It is a powerful technique for reducing the concentration of defects and impurities in critical regions. Another method of gettering is to create mechanical damage on one side of the wafer surface, perhaps by grinding. The trapping impurities quickly diffuse to occupy the vacancies created underneath the damaged layer.

Using these techniques individually or in combination, it is possible to fabricate devices with carrier lifetimes exceeding 100 $\mu$s, even after high temperature processing operations. This is especially important in the fabrication of high voltage thyristors and diodes.

### 3.3.2 Techniques for Reducing Carrier Lifetimes

In many other circumstances, especially for devices operating at higher frequencies, it is important to reduce the carrier lifetime in a controlled and predetermined manner and so control critical parameters such as the recovered charge and the turn-off time.

In power device fabrication technology, the carrier lifetime is reduced by introducing efficient recombination centres either by the diffusion of metallic impurities such as gold and platinum immediately before metallisation, or by high energy irradiation, which is usually the last step in the processing sequence. The high and non-linear diffusion coefficients of these metals give rise to concentration profiles very different from those of impurities like B, P, Al, As or Ga, as described in Section 3.2.7. Very often a U-profile is formed, with a high impurity concentration at both surfaces of the wafer and a flat plateau in the middle, as shown in Figure 3.33.

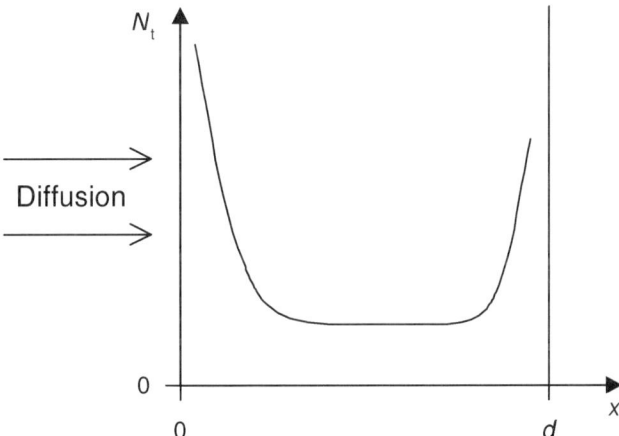

**Figure 3.33** A typical profile of the concentration of trapping levels across a wafer

Copper has the highest diffusion coefficient, but tends to form clusters in areas of high dislocation density. These degrade the breakdown characteristics of devices. Gold is most frequently used as an effective recombination centre. It is diffused into silicon at temperatures in the range 800–1000 °C until the silicon is saturated. Platinum is an alternative. Because the platinum energy levels lie further from the middle of the bandgap, as shown in Figure 1.2, the leakage current is reduced. However, the lifetime under low injection conditions is longer than under high injection conditions. For thyristors, the trade-off between the turn-off time and the on-state voltage drop is poorer.

Although the diffusion of lifetime-reducing impurities is straightforward and does not require precise control, it may occur preferentially in regions of higher defect density and this may result both in non-uniform carrier lifetime distribution and hot spot formation. As it has to be performed before metallisation, the device characteristics cannot be checked before and after doping, as they can if irradiation is used.

In irradiation methods, high energy particles penetrate the crystal lattice, losing their energy in interactions with the lattice atoms, which become displaced from their normal positions and create defects. These include vacancies, divacancies, impurity–vacancy pairs, interstitials and impurity–vacancy–interstitial complexes, all of which introduce recombination centres into the silicon. Electron, proton, alpha and gamma radiation can be used in this way to influence the carrier lifetime. For a given energy, the penetration depth is approximately in inverse proportion to particle mass. There is most radiation damage at the end of the particle range, as illustrated in Figure 3.34 [3.10].

Gamma radiation is able to penetrate very deeply, allowing the carrier lifetime in metallised and encapsulated devices to be modified. However, it can damage the p–n junction surface termination and degrade its reverse blocking characteristics.

Electron irradiation is commonly used. After metallisation but before encapsulation, devices are bombarded with electrons of several MeV energy. These penetrate the silicon to a depth of several millimetres (e.g. > 6 mm for 3 MeV electrons) and create defects quite homogeneously through the device structure. The concentration of deep levels is proportional to the radiation dose applied, so the carrier lifetime depends on the radiation dose as

$$\tau = \frac{\tau_0}{1 + K\Phi\tau_0} \tag{3.6}$$

where $\tau_0$ is the carrier lifetime before irradiation, $K$ is a constant that varies slightly with the electron energy and $\Phi$ is the radiation dose, normally measured in coulombs per m$^2$. The reproducibility of the process is excellent because the dosage can be monitored accurately.

Protons have a much shorter penetration depth than electrons. It varies, as a function of the particle energy, from a few microns to several hundred microns. This is shown in Figure 3.35. An increase in radiation damage at the end of the range allows a well-defined thin layer of reduced lifetime to be created at a depth below the surface that is controlled by the bombardment energy. In this way, the trade-off between the static and dynamic parameters of bipolar power devices can be

3.3 THE CONTROL OF CARRIER LIFETIME 117

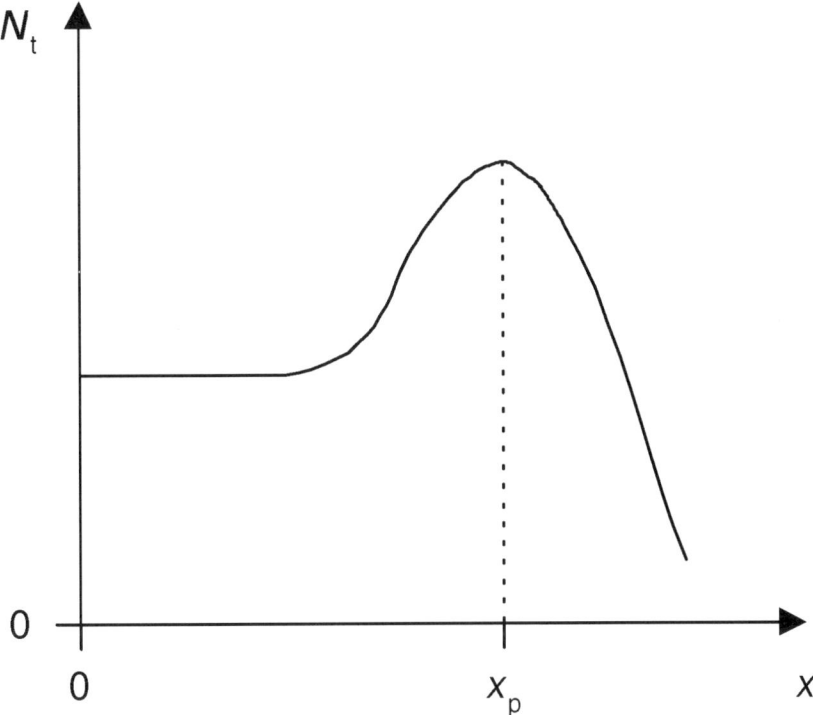

**Figure 3.34** A schematic illustration of the distribution of crystal lattice damage with depth following high energy particle irradiation. The figure shows a typical variation with depth below the surface of the defect concentration. The depth of the peak value is a function of the particle mass and energy

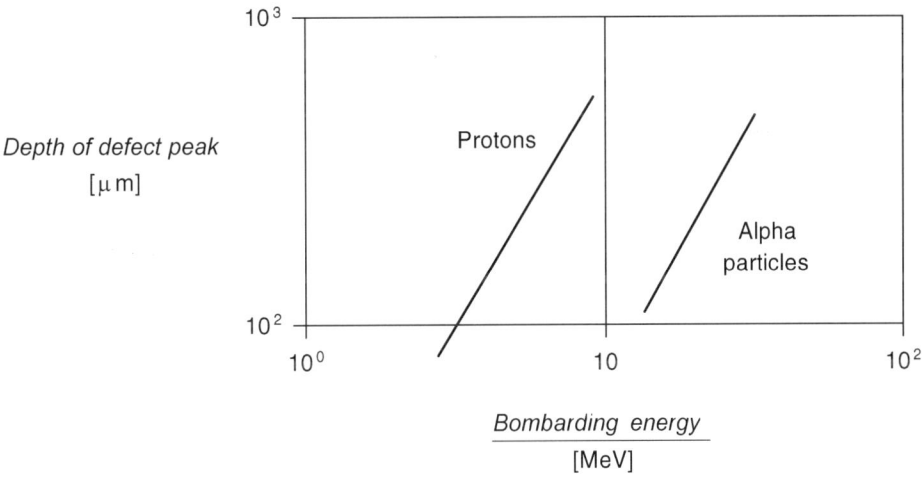

**Figure 3.35** Proton penetration depth in single crystal silicon, as a function of energy

optimised. The trapping levels mainly arise from vacancy complexes but also from implanted hydrogen.

Irradiation with alpha particles, $He^{2+}$ ions, causes similar effects to proton bombardment. Because alpha particles have greater mass, higher particle energies (typically $\sim 10$ MeV) are needed to give the same penetration depth. Both proton and alpha irradiation have to be performed under vacuum with the result that equipment costs are high.

Some of the defects created by irradiation are unstable and can be annealed out at temperatures close to the working temperatures of devices. As this can lead to undesirable changes in parameters, devices are annealed at 250–300 °C for up to several hours after irradiation, to ensure long-term operational stability.

## 3.4 High Voltage Structures

A p–n junction designed to have a high reverse breakdown voltage is usually fabricated using diffusion technology. The junction is formed at a depth $x_j$ that is often more than 50 µm below the surface of the silicon wafer. This region should be free from defects resulting from any mechanical working of the initial single-crystal silicon. As shown in Section 2.1.3, the breakdown voltage of a planar junction in bulk material is determined by the doping concentration on the more lightly doped side of the junction.

However, for real semiconductor structures of limited area, the maximum available breakdown voltage $V_{R(BR)}$ is determined not only by the impurity concentrations, but also by the junction shape and by the detailed structure of the region where the junction comes to the semiconductor surface. In real structures it may also be necessary to consider the effects of finite boundaries such as the edge of the wafer or the diffused area. At the periphery of the plane junction, the depletion layer becomes curved and consequently the electric field in the space charge region is altered. It is usually increased locally, with the result that the junction breakdown voltage and its repetitive reverse voltage capability are decreased.

Junction curvature is important and has to be taken into account when analysing nominally planar junctions. In the case of a p–n junction formed by diffusion into an area delineated by a photolithographically prepared oxide mask, the junction is plane except at the edge of the oxide window. There, as a result of lateral diffusion, the junction is cylindrical in shape and, at the corners, it becomes spherical or ellipsoidal, as shown schematically in Figure 3.36. The radius of curvature $r_j$ in these regions is approximately equal to the junction depth $x_j$. The breakdown voltage decreases with $r_j$ as shown in Figure 3.37.

To increase the breakdown voltage requires both an increase in the radius of curvature of the p–n junction and a widening of the space charge region at the semiconductor surface. One way of achieving this is to diffuse an auxiliary floating region (a diffused guard ring) around the circumference of the junction, as shown in Figures 3.38 and 3.39. An example of the improvement in $V_{R(BR)}$ that can be obtained is shown by the dashed curve in Figure 3.37. The gap between the ring and the main diffusion has to be optimised. For high voltage devices, up to six concentric guard rings may used.

## 3.4 HIGH VOLTAGE STRUCTURES

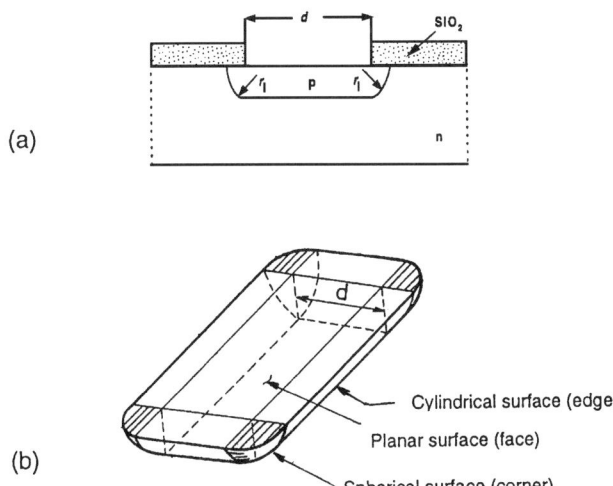

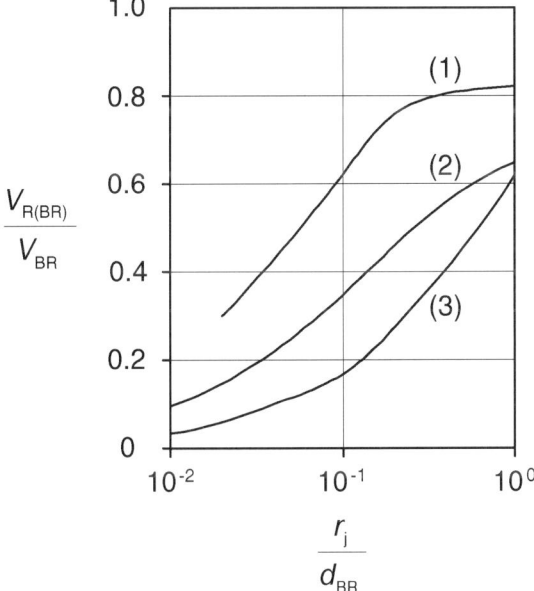

**Figure 3.36** Planar, spherical and cylindrical regions of a diffused p–n junction: (a) cross-section; (b) perspective view. Reproduced from *Physics of Semiconductor Devices* by S. M. Sze, by permission of John Wiley & Sons, Inc.

**Figure 3.37** The influence of the radius of curvature of a spherical and cylindrical p–n junction on the breakdown voltage, compared to a junction with a single guard ring: (1) single guard ring; (2) cylindrical junction; (3) spherical junction. The radius of curvature is normalised against the width of the space-charge layer at breakdown

**120**   3 DEVICES, FABRICATION AND MODELLING

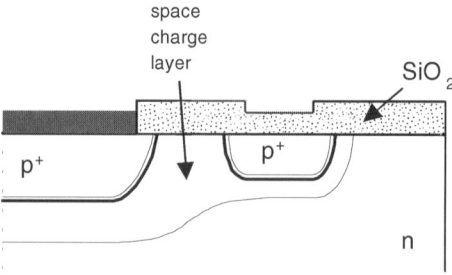

**Figure 3.38**  A planar p–n junction with a diffused field-limiting guard ring

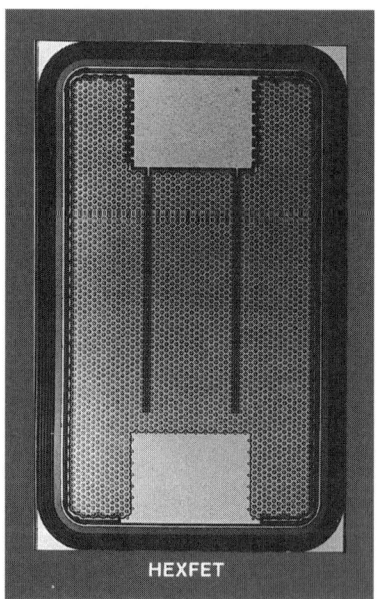

**Figure 3.39**  Photograph showing the guard-ring structure around the periphery of a high voltage power device. Reproduced from *Power MOSFETs Theory and Applications* by D. A. Grant and J. Gowar, by permission of John Wiley & Sons, Inc.

Another way to increase the width of the space charge region at the junction periphery is to implant a thin, very lightly doped layer around the perimeter of the diffused region, as shown in Figure 3.40.

A third means of increasing $V_{R(BR)}$ is to use a field plate, as shown in Figure 3.41(a). This involves extending the contact metallisation over the surrounding oxide

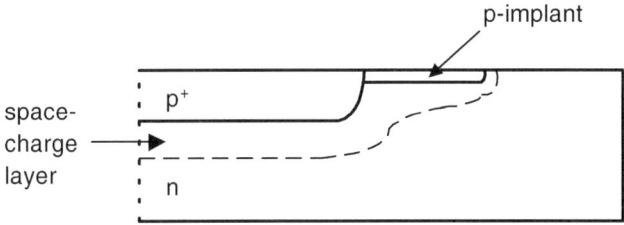

**Figure 3.40** A planar junction with an implanted high-resistivity peripheral layer

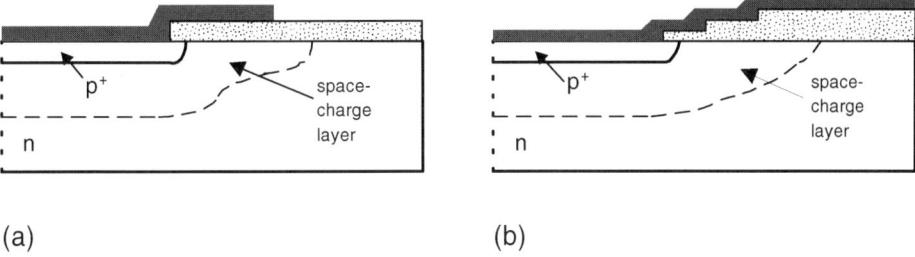

(a)                                    (b)

**Figure 3.41** The use of field plates: (a) with a homogeneous oxide layer; (b) with an oxide layer of increasing thickness

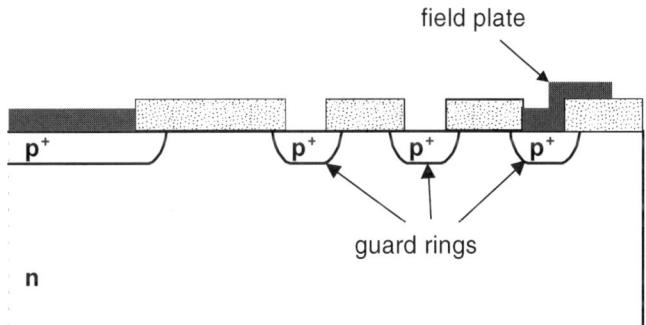

**Figure 3.42** A combination of a field plate and guard rings

layer. Under reverse bias, the depletion of the n-type region beneath the oxide is extended laterally. This technique is especially effective in structures with a thin oxide layer. However, the breakdown field in the oxide layer itself can then become a critical factor. Using a simple field plate, it is possible for the breakdown voltage to reach 60% of the theoretical value for a plane junction $V_{BR}$. Putting a field plate onto a stepped oxide layer, as shown in Figure 3.41(b), allows $V_{R(BR)} = 0.9 V_{BR}$ to be achieved.

It is possible to combine the use of guard rings and field plates, as shown in Figure 3.42. With a sufficient number of each, the breakdown voltage can be made arbitrarily close to that of the ideal plane junction, but it takes up silicon that cannot be used for conduction. One of the aims of the device designer is to obtain a

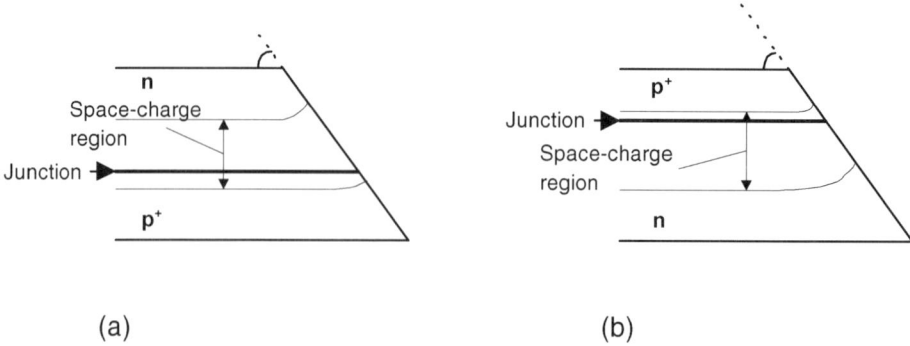

**Figure 3.43** Bevelling of the p–n junction surface: (a) positive bevel; (b) negative bevel

breakdown voltage as close as possible to $V_{BR}$ using the minimum additional silicon area.

The electric field in the region where the p–n junction comes to the surface can be decreased by coating a thin layer of surface oxide with one or more high resistivity conducting (i.e. semi-insulating) layers. The semi-insulating layer is usually a deposited layer of polycrystalline silicon, heavily doped with oxygen. It is known as SIPOS. It has the effect of making the electric field distribution at the surface more uniform. The SIPOS technique can be combined with a field plate termination and junction breakdown voltages close to $V_{BR}$ can be obtained.

The surface electric field is likely to reach peak values at the edges of the contacts and the diffusions, i.e. where the p–n junctions come to the surface. The distribution of the field is sensitive to the depth of the plane junction areas below the surface and, in lateral power devices and integrated circuits, its peak values can be minimised by controlling this depth. This technique is called RESURF (reduced surface field) [3.5].

In many power semiconductor devices the p–n junction comes to the surface, not on the top surface, but at the edge of the dice or wafer. Then the electric field at the surface can be controlled by contouring the device edge, often by a suitable angle of bevelling. Some bevelled junctions are illustrated in Figure 3.43. An analysis of the effects of surface contouring on the electric field of a reverse-biased p–n junction requires the solution of Poisson's equation in two dimensions [3.6]. The thickness of the space charge layer on either side of the junction changes until a charge balance is reached in the surface region. When the bevelling is such as to reduce the device area on the less heavily doped side of the junction, it is said to be positive, as shown in Figure 3.43(a). In this case the depletion layer at the surface expands, the peak electric field at the surface is reduced and a breakdown voltage close to that of a planar junction can be achieved.

With a negative bevel, where the device area of the more heavily doped side is reduced, as shown in Figure 3.43(b), the depletion layer thickness at the surface is initially reduced, so the surface field is increased. However, at very acute bevelling angles (less than 10°), the depletion layer extends sufficiently far along the surface to reduce the surface field. The peak field below the surface is always higher than in the

## 3.4 HIGH VOLTAGE STRUCTURES

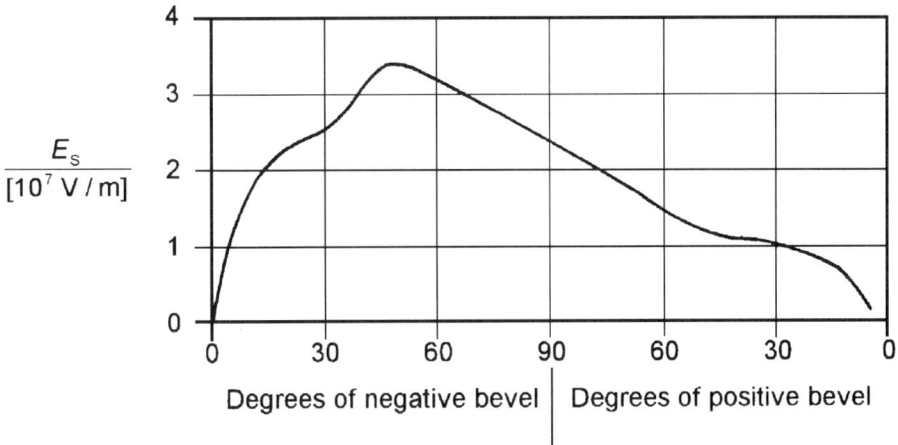

**Figure 3.44** The dependence of the surface electric field on the bevel angle

middle of the device so, at best, a breakdown voltage of about $0.9V_{BR}$ can be obtained.

The variation of the surface electric field with the bevel angle is shown in Figure 3.44. It is necessary to reduce the electric field at the surface to about half that in the bulk material, if an equivalent breakdown voltage is to be obtained. This is because of surface defects which increase the probability of carrier ionisation. Thus, the optimum positive bevel angle normally lies in the range 30–60°.

A positive bevel is usually produced by sandblasting the finished wafer and removing the surface damage by chemical etch. Negative bevelling may be achieved mechanically, e.g. by grinding. In both cases the surface layer damage has to be etched away. After etching and cleaning, the surface is covered with a layer of passivating dielectric, whose function is to minimise the surface defect density and to protect the surface against contamination. The passivating layer is usually made from very pure silicone rubber or from polyimide. No type of relaxation dielectric polarisation is admissible.

The disadvantages of bevelling are the reduction of the effective device area and the increase in the generation–recombination component of the reverse current that occurs, as a result of the increase in the thickness of the space charge layer. It is only

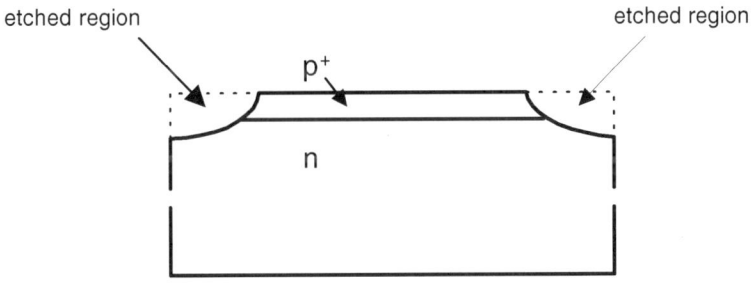

**Figure 3.45** A mesa, or 'moat-etched' structure

practical for whole-wafer devices that are handled individually in the later stages of fabrication.

Surface contouring may also be realised by deep local etching, producing a mesa, or moat-etched profile, as illustrated in Figure 3.45. By this method it is possible to reach $V_{R(BR)} = 0.8 V_{BR}$.

For the passivation of small-area devices, surface chemical contouring is often combined with the use of a surface layer of glass (e.g. Pb–Al–B glass) which is more adhesive and stable than silicone rubber. Instead of the glass, a passivating layer of polycrystalline silicon which is heavily doped with oxygen (SIPOS) may be used. In the case of devices fabricated by the planar technology, a SIPOS layer or a layer of thermally grown oxide may be used for surface passivation. When there are inhomogeneities in the space charge layer, either at the surface or in the bulk, the electric field may locally exceed the critical field $E_{BR}$, so a local breakdown occurs. Even an inhomogeneity in the passivating dielectric layer, perhaps an air bubble in the silicone rubber, may be enough to cause a local surface breakdown.

The design of high voltage structures that enable the breakdown voltage of real devices to approach that of a plane junction in bulk material remains an active area for research. Many novel structures have been suggested, some of which develop or combine the techniques described in this section. Among them, we may mention a non-floating spiral guard-ring structure [3.7] and a combination of the bevel and planar techniques for symmetrically blocking devices with two high voltage junctions [3.8]. Surface passivation is also under investigation [3.9] with materials such as diamond-like carbon, aluminium nitride and boron nitride, which have good breakdown strength and high relative permittivity and which bond strongly to silicon.

## 3.5 Computer Modelling and Simulation Techniques

During the past twenty years or so, a great range of software has been developed for the analysis of semiconductor devices and the prediction of their behaviour in particular circuit configurations. Mostly, this has been directed towards the modelling of integrated circuits. But often it can be adapted and occasionally directly applied to power devices. Some of the software packages have been developed in research groups and are freely available. Many have been developed and are marketed commercially. They are often known as technology computer aided design (TCAD) tools.

There are several levels at which modelling is appropriate. At what may be called the lowest level comes the modelling of the fabrication processes and the device structures that they yield. Thus, modelling the diffusion of impurities through a surface mask into bulk silicon, with the temperature and other conditions specified, tells us the three-dimensional doping profile that is to be expected. Processes such as ion implantation and oxide growth, deposition and wet or dry etching can be modelled in a similar way, giving a predicted device geometry for a particular sequence of fabrication processes.

The next level of numerical modelling aims at the solution of the semiconductor equations so that the terminal characteristics of the device can be predicted. These come in varying levels of sophistication, one-dimensional, two-dimensional and three-dimensional models, either limited to steady-state conditions or able to deal with transients and with temperature variations in time and space.

For the results to be trusted and used to influence device design, it is essential that both the models and the physical parameters on which the results depend should be reliable in all the circumstances modelled. The difficulty in validating the parameters is well illustrated by the discussion on carrier mobilities (a critical parameter) in Section 1.3.2.

The next level of modelling aims to represent the terminal characteristics in terms of an equivalent circuit model that can be inserted into one of the standard circuit simulators such as SPICE. Finally, in switching circuits, comes the development of a functional representation in which the important considerations are the switching conditions and the delay times.

## Summary

The main families of power semiconductor devices are diodes, bipolar junction transistors, power MOSFETs, insulated gate bipolar transistors, silicon controlled rectifiers and gate turn-off thyristors. There are many variants and special devices that satisfy niche applications. Diodes are important components in virtually every power electronic circuit. Of the controlled devices, power MOSFETs dominate the lower voltage, lower power and higher frequency applications, SCRs and GTO thyristors the high voltage and high power applications, while BJTs and IGBTs occupy the expanding middle ground, with IBGTs becoming increasingly dominant.

Most of the processing techniques used for the fabrication of integrated circuits are also used in making power semiconductor devices. However, their large size and the need to support high breakdown voltages and high current densities means that the starting material must be homogeneous, high resistivity, single-crystal silicon with a very low defect density. Deep junctions require long diffusion processes. Carrier lifetime must not be degraded, but controlled reduction in certain regions may be needed. Devices requiring fine structures are normally fabricated in epitaxially grown material.

The working voltages of power devices are determined by the control of the peak field in the device. This usually occurs at a corner of a diffused or implanted region and most often at the surface in the region of a p–n junction. Many techniques are used to minimise the peak electric field, so that it approaches that of a plane p–n junction in bulk material. They are normally used in combination for high voltage devices.

Many device modelling software packages have been developed or adapted for application to power devices. These include packages to model processing and predict device structure, packages to model carrier behaviour and predict terminal characteristics, and packages to match the terminal characteristics to circuit constraints. The underlying data needed to give validity to these models is under constant review and significant advances have been made recently.

# References

[3.1] Bauer, F., Dettmer, H., Fichtner, W., Lendenmann, H., Stockmeier, T. and Thiemann, U. Design Considerations and Characteristics of Rugged Punchthrough (PT) IGBTs with 4.5 kV Blocking Capability, *Proc. ISPSD'96*, 327–30 (1996).

[3.2] Carslaw, H. S. and Jaeger, J. C. *Conduction of Heat in Solids* (Oxford University Press, 2nd Edn, 1959).

[3.3] Gise, P. and Blanchard, R. *Modern Semiconductor Fabrication Technology* (Prentice Hall, 1986).

[3.4] Middleman, S. and Hochberg, A. K. *Process Engineering Analysis in Semiconductor Device Fabrication* (McGraw-Hill, 1993).

[3.5] Appels, J. A., Collet, M. G., Hart, P. A. H., Vales, H. M. J. and Verhoeven, J. F. C. M. High Voltage Thin Layer Devices (RESURF Devices), *Philips Journal of Research*, **35**, 1–13 (1980).

[3.6] Pathak, V. K. and Gowar, J. Numerical Solutions for Surface Electric Field Distributions in Avalanching p–i–n Power Diodes, *Proc. IEE*, **I-130**, 17–23 (1983).

[3.7] Krizaj, D., Amon, S., Mingues C. and Charitat, G. Spiral Junction Termination, *IEEE Trans. on Electron Devices*, **ED-44**, 2002–10 (1997).

[3.8] Mitlehner, H. and Schulze, H.-J. Current Developments in High-Power Thyristor, *EPE Journal*, **4**, 36–42 (1994).

[3.9] Lisik, Z., Mitura S. and Szmidt, J. Application of Diamond-like Layers as Passivation and Isolation Layers in Power Semiconductor Devices, *Proc. 6th European Conference on Power Electronics and Applications* (Seville, Spain, Sept. 1995).

[3.10] Hazdra, P. and Vobecky, J. Application of High Energy Ion Beams for Local Lifetime Control in Silicon, *Material Science Forum*, **248/9**, 225–8 (1997).

# 4

# POWER SEMICONDUCTOR DEVICE APPLICATIONS

Power electronics, at its most general, is concerned with the conversion of ac power taken from the mains to a dc output or an ac output. The dc may be of variable voltage or it may be required to supply a controlled level of power to a varying load. In general, the ac output will require the voltage and frequency to be controlled. In other situations it may be a dc supply that is converted to provide the dc or ac output. In this chapter we give an overview of the basic circuits used most commonly for these purposes, so that the essential features of the power semiconductor devices required for their successful operation can be identified. The ways in which the individual types of device can be designed to embody the required characteristics are discussed in Chapters 5 to 11.

## 4.1 Uncontrolled Rectification

Figure 4.1 shows a basic single-phase rectifier employing a single diode D1 as the rectifier. When the load is inductive, a second diode D2 may be required to provide a freewheeling path for the load current. Without the second diode, current continues to flow in D1 after the supply voltage has reversed and the voltage across the load may then become negative during a part of each cycle. This makes the average output voltage dependent on the load. If the supply is of mains frequency, the diodes do not need to have a fast recovery time or a particularly low stored charge. If, on the other hand, the load contains parallel capacitance, a surge of current passes through D1 when the supply is connected. In this situation, rectifier diodes require a good surge current capability.

Figure 4.2 shows a single-phase, full-wave, bridge rectifier circuit. A freewheeling path for the load current is automatically provided by this circuit. A rectifier bridge of this kind may be operated from the low voltage winding of a transformer. It is then protected to a certain extent from surges on the mains supply by the impedance of the transformer. Increasingly, however, diode bridges are used to rectify the untransformed mains for subsequent processing in a switched-mode power supply (SMPS) or a power factor correction (PFC) circuit. In these cases the diode bridge

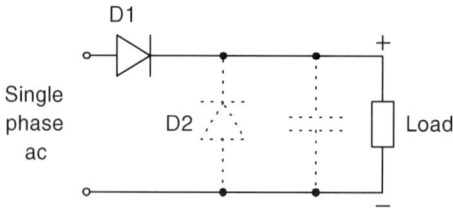

**Figure 4.1** The single-phase, half-wave rectifier circuit

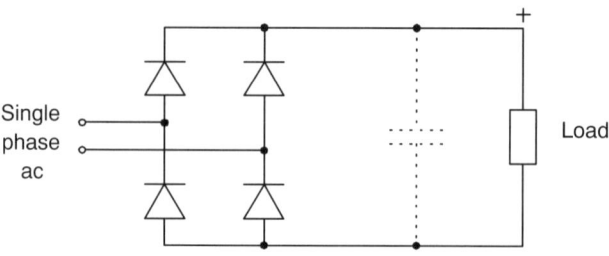

**Figure 4.2** The single-phase, full-wave bridge rectifier circuit

has to be designed to withstand the peak of the mains voltage plus any transients which may appear on the mains. These may occur as a result of lightning strikes somewhere on the power system or as surges caused by the opening and closing of contactors. Filtering may be placed between the bridge and the supply in order to limit the effects of interference conducted along the power supply leads, especially when the bridge is used in conjunction with a high frequency power conditioning circuit such as an SMPS or PFC circuit.

If the bridge is followed by a reservoir capacitor to smooth its dc output, there is a surge of current when it is switched onto the mains. A resistor, especially a temperature-dependent resistor, may be connected in series with the ac supply to limit the magnitude of this surge. Nevertheless, the diode bridge requires a high surge current rating in this situation. If the bridge is followed by a reservoir capacitor, the current drawn from the ac supply consists of peaks of current that flow whenever the instantaneous voltage of the ac supply becomes greater than that on the reservoir capacitor. The amplitudes of such peaks are in part related to the impedance of the ac supply, so it can be difficult to predict the exact shape of the ac input current waveform and hence the losses in the rectifier diodes.

Diode bridges are available as modules. However, single diodes can often be obtained at such a low price that four single diodes are used for relatively low power equipment being produced in high volume at low cost.

The power that can normally be supplied by the single-phase mains is limited to about 2.5 kW. For equipment requiring a dc supply of greater power, a three-phase supply and a three-phase rectifier circuit, like the one shown in Figure 4.3, are required. The dc output is smoother than the output from a single-phase, full-wave rectifier and it is more easily filtered. If a reservoir capacitor alone is used to smooth the output, the diode current waveform again consists of pulses of current which

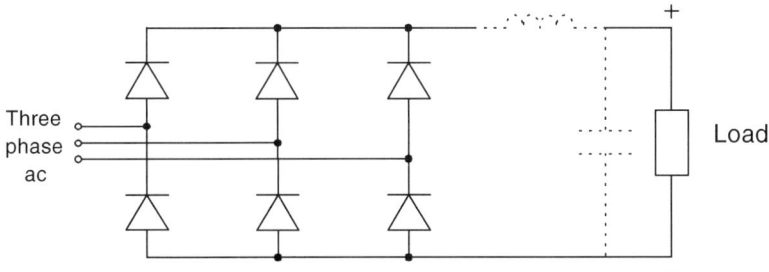

**Figure 4.3** The three-phase, full-wave bridge rectifier circuit

occur whenever the instantaneous value of the ac input voltage exceeds the dc voltage on the capacitor. However, in high power, three-phase rectifiers a choke is often added in the dc link, before the reservoir capacitor, to aid in the filtering of the dc supply. The current drawn from the bridge is then much more constant and the current carried by each of the diodes has a rectangular waveform. Each diode conducts for one-third of each cycle of the supply and the losses are more easily predicted. The inductance also limits the surge current experienced at switch-on.

Each diode experiences the line-to-line voltage of the supply and so requires a higher voltage rating than a diode used in a single-phase bridge which sees only the line-to-neutral voltage. Furthermore, the diode voltage rating has to take account of possible surges on the supply due to lightning strikes and switching phenomena, and these considerations tend to dominate the device selection. With a low frequency supply, the diode recovery losses are negligible, so the diodes can be optimised for low forward voltage drop to minimise dissipation during conduction.

## 4.2 Controlled Rectification

In the three-phase bridge rectifier circuit shown in Figure 4.4, the diodes in the upper row have been replaced by silicon controlled rectifiers (SCRs). This permits the average output voltage of the bridge to be varied from zero to full voltage by varying the point in each cycle at which the SCRs are turned on. The bridge can supply only positive current and positive voltage to the load and it therefore operates in one of the four possible quadrants of operation (positive or negative voltage with positive or negative current). Because of the large instantaneous difference between the line voltage and the dc output voltage when the SCRs are turned on, an inductance is normally interposed between the bridge and the smoothed dc supply. If the load itself is inductive, as in the case of a dc motor, an additional choke may not be required.

The SCRs can be of the phase-control type, in which the on-state voltage drop is minimised and less emphasis is placed on reducing the time required for turn-off. Snubber circuits are needed to limit the rise rate of the anode voltage experienced by the SCRs. Simple RC snubbers are usually sufficient in this application and, since the frequency of switching is low, the additional power loss is not large. Commutation of the output current from one SCR to another usually takes place

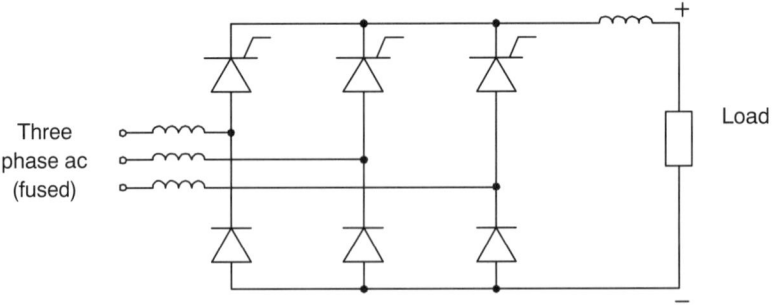

**Figure 4.4** The three-phase, half-controlled bridge rectifier circuit, offering single-quadrant operation only

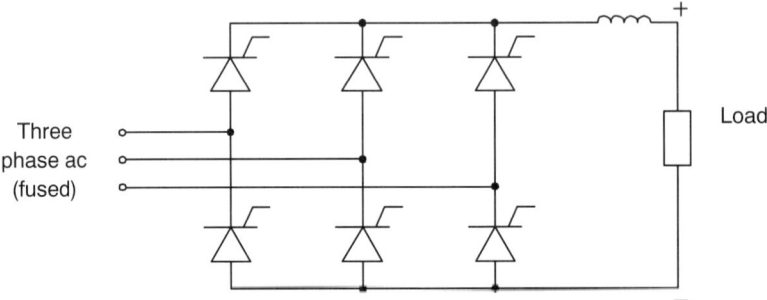

**Figure 4.5** The three-phase, fully-controlled bridge rectifier circuit, offering two-quadrant operation

at a relatively slow rate because of the inductance in each supply phase. The SCRs therefore do not require a particularly high rating for the rise rate of the current. They are usually protected by fast-acting fuses that can cut off the power supply in about 10 ms.

Figure 4.5 shows a three-phase bridge rectifier circuit in which all the rectifying devices are SCRs. Although the bridge can supply only positive current, this allows the dc output voltage to be negative as well as positive, so the bridge is capable of two-quadrant operation.

## 4.3 Conversion of ac to ac

Figure 4.6 shows a circuit which, operated as a cycloconverter, is often used to provide very large, controlled levels of ac power, perhaps for mining machinery or large rotary kilns used in cement works and other plants. A pair of two-quadrant bridges is connected together in antiparallel. The left-hand bridge provides current to the load in one direction and the right-hand bridge provides it in the other direction. Both bridges can supply current to a load carrying either a positive or a negative voltage.

## 4.3 CONVERSION OF AC TO AC

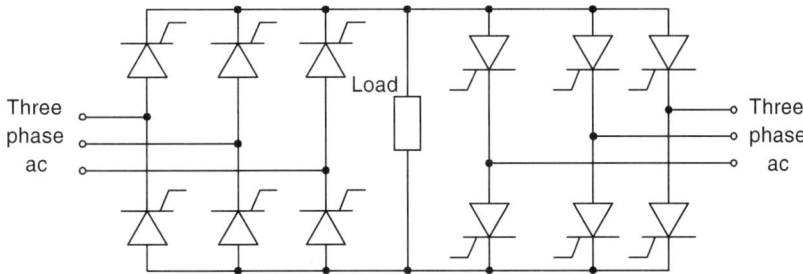

**Figure 4.6** A cycloconverter offering four-quadrant operation with inhibited firing

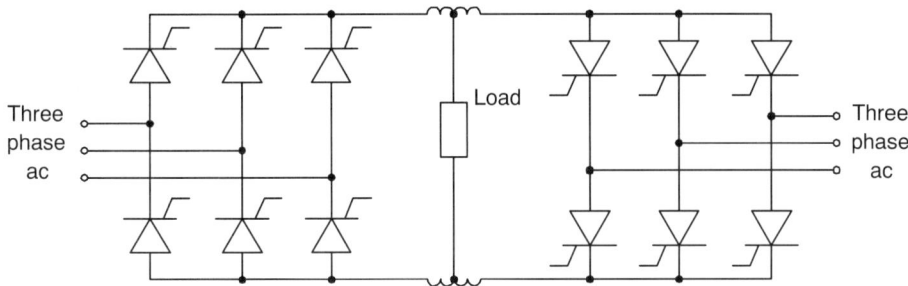

**Figure 4.7** A circulating-current cycloconverter

In Figure 4.6 the bridges can operate only one at a time, and the SCRs in the bridge that is not operating must be inhibited from firing. Both bridges can be operated simultaneously if centre-tapped inductors are included at the connections of the two bridges and the load, as shown in Figure 4.7. The load current then flows naturally in either the positive or the negative direction, according to the load conditions.

In either case the dc output of each converter can be varied cyclically to produce an ac output. The circuit then becomes a cycloconverter, capable of providing a controlled, variable-frequency, ac supply. In order to produce a three-phase variable-frequency supply, three such cycloconverters are required. The maximum output frequency is typically limited to one-third of the input frequency. In practice the use of cycloconverters is now limited to low frequency applications of very high power. Phase-control thyristors (SCRs) with the lowest possible on-state volt-drop are required.

The circuits shown in Figures 4.6 and 4.7 employ natural commutation of the SCRs and this severely limits the output frequency. Forced commutation may be used to improve the quality of the output waveform [4.1]. The ultimate force-commutated cycloconverter design is shown in Figure 4.8. This circuit is now usually known as the *matrix converter* [4.2]. It creates a three-phase set of output voltage waveforms by connecting each output phase to one of the three input phases in a rapidly changing manner. Thus, each output waveform is made up of short sections of the input waveforms.

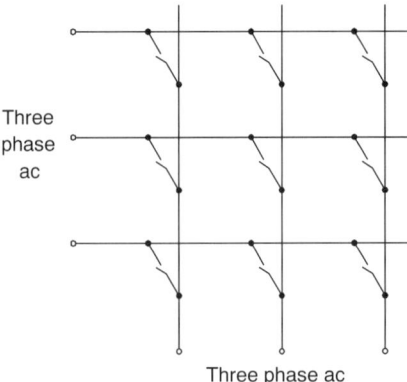

**Figure 4.8** The matrix converter

The power semiconductor switches used at each of the nodes must be able to switch at high frequency, block voltage of either polarity and conduct current in both directions. It is said that the matrix converter will become a practical proposition when a power semiconductor switch which combines all these requirements has been developed. However, recent research has shown that individual control of conduction in each direction is desirable and this points to the use of two separate switches operated in antiparallel. The device best suited for this application is the IGBT, which normally does not have a reverse blocking capability, for reasons discussed in Chapter 11. This means that diodes have to be added in series with each IGBT to prevent exposure to reverse voltage. However, non-punch-through IGBTs could be made with symmetrical blocking capabilities which would enable each switching node in the matrix to be made up of a pair of these devices operating in antiparallel.

The matrix converter is attractive in that it offers the possibility of an 'all-silicon' ac to ac conversion process. It is particularly advantageous when a constant frequency supply has to be obtained from a variable-speed generator, as in many aircraft systems. In that situation it would replace a mechanical system and is the subject of intensive research. In practice, power quality and electromagnetic compatibility (EMC) considerations demand substantial filtering of the ac input so that large passive components are still needed. To some extent the input filters protect the switches from voltage spikes on the input lines but they must still withstand voltage surges on the supply and must therefore be rated accordingly.

## 4.4 Inverters

More conventionally the conversion of a fixed-frequency ac supply to a variable-frequency, variable-voltage ac supply is carried out in two stages, as shown in Figure 4.9. The ac mains is first rectified to create a dc link, as described in Section 4.2. This dc supply is then converted back to ac using one of several possible types of inverter circuit. In its simplest mode of operation, each leg of the inverter circuit creates a

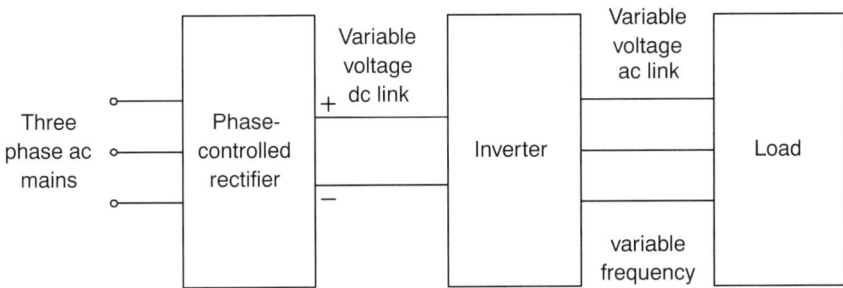

**Figure 4.9** A schematic illustration of the normal approach to ac to ac conversion

rectangular waveform of the required frequency. Voltage control is achieved by use of a controlled rectifier circuit to create the dc link.

The main application for inverters is in variable-speed drives based on induction motors. The three-phase induction motor is simple, reliable and inexpensive but its normal operating speed is a function of the frequency of the supply. Speed control thus requires a variable-frequency ac supply. At the same time, the supply voltage has to be varied, approximately in proportion to the frequency. Inverters in which the power semiconductor devices are switched to provide pulse width modulated (PWM) waveforms are able to fulfil both of these requirements. Inverters are also used in an uninterruptible power supply (UPS). A dc source of backup power, such as a bank of lead–acid batteries, is converted to fixed-frequency ac using an inverter. In the event of a mains failure, the UPS can be switched into operation and take over the supply within a mains cycle.

Figure 4.10 shows a standard auxiliary-commutated three-phase thyristor inverter circuit. The main switching thyristors are SCR1, SCR2, SCR3 and SCR4. The auxiliary thyristors are SCR5, SCR6, SCR7 and SCR8. In order to understand the operation of this circuit, suppose that, in the left-hand-side of the bridge, SCR1 is conducting and SCR2 is off. In order to turn off SCR1, SCR5 is triggered on. From the previous commutation, the commutation capacitor $C_R$ will have been left with a charge on it such that the end connected to the inductance $L_R$ is negative. A current builds up in $L_R$, which eventually exceeds the load current flowing out of SCR1 and the current through SCR1 tries to reverse, turning the device off. The excess current then flows through the diode D1. The current in $L_R$ and $C_R$ reaches a peak then begins to decline. Before this current falls to zero, it builds up a charge on the capacitor of opposite polarity to that present at the beginning of the commutation process. While D1 is conducting, SCR1 is able to recover its blocking capability. When the commutation cycle is complete, SCR2 is fired. A small pulse of current in the resonant circuit tops up the charge on the capacitor $C_R$ to the full value and reverse biases SCR5, which turns off. The circuit is now ready to repeat the commutation cycle, this time with SCR7 being triggered to turn off SCR2.

All circuits which force-commutate SCRs have one objective in common: to reverse the current flow through the SCR and reverse bias it for long enough to allow

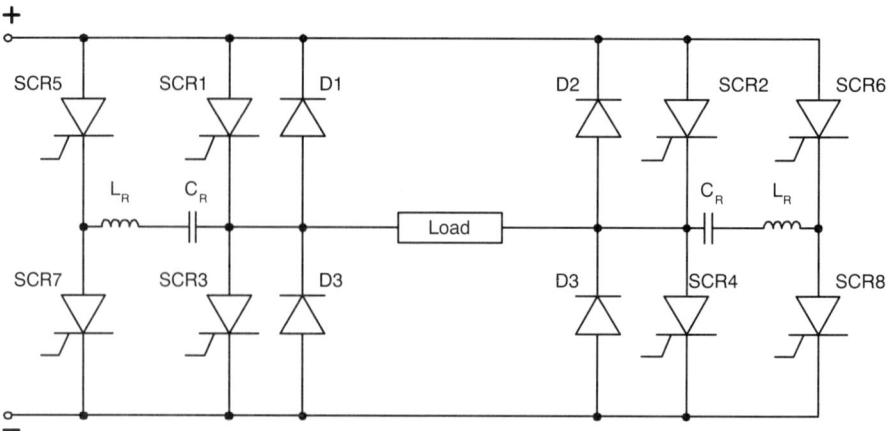

**Figure 4.10** An auxiliary commutated inverter circuit using thyristors as switches

it to recover. During this time the commutation circuit must carry the load current. This usually involves a capacitor supplying the current. For example, in the auxiliary commutated circuit, the current in the resonant circuit must be greater than the load current for long enough to enable the SCR to turn off. The size and cost of the commutation components is therefore very dependent on the recovery time or turn-off time of the SCR. The SCRs used in such applications are fast-recovery devices known as *inverter grade* thyristors. Minority carrier lifetime killing and heavy cathode shorting help to ensure the rapid recovery of the device, as discussed in Section 8.2.

The SCR data sheet specifies the length of time for which the device must be reverse biased in order to ensure turn-off and also the critical rate of rise of anode voltage which the device can support when forward voltage is reapplied at the end of the recovery period. The thyristors used in this type of circuit also have to be able to take over the load current very quickly when they are turned on and thus need to cope with a relatively high rise rate for the current at turn-on. They are usually made with an intricate interdigitated gate structure, which provides the long gate periphery needed to turn on the device rapidly. Many devices designed for this purpose use the amplifying gate structure (Figure 7.17) to aid fast turn-on.

When IGBTs or gate turn-off thyristors (GTOs) are used as the switches in an inverter circuit, the auxiliary commutation arrangements just described are unnecessary. However, GTOs do require a large polarised snubber circuit in order to limit the rise rate of the anode voltage at turn-off. A typical circuit is shown in Figure 4.11. The losses associated with the snubber and the losses caused by the tail of anode current which flows during turn-off both restrict the frequency at which GTO inverters can operate. They are more suited to high power applications, where a high device voltage rating is required. Wherever possible, IGBTs tend to be used in preference to GTOs because they are easier to control and have more moderate snubber requirements.

Voltage control in inverter circuits may be achieved using pulse width modulation (PWM) waveform generation techniques. The use of this technique is limited in

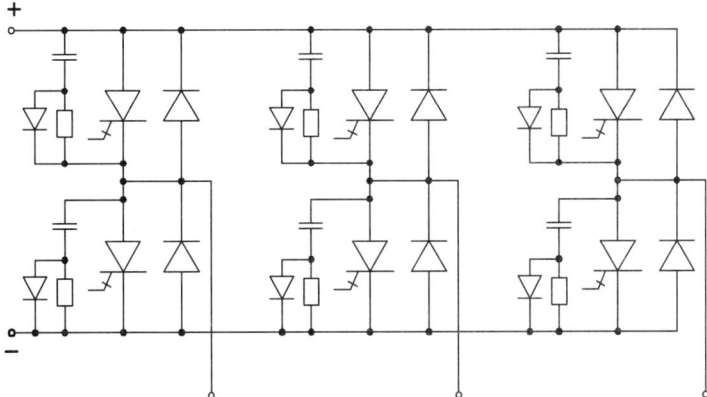

**Figure 4.11** A three-phase inverter circuit using GTO thyristors as switches

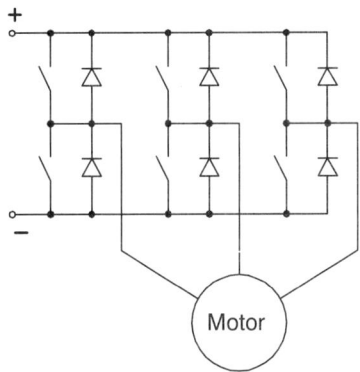

**Figure 4.12** A generalised three-phase inverter circuit

SCR-switched inverter circuits because of the relatively long turn-off time of SCRs. It is more suitable for inverters using MOSFET, BJT or IGBT switching devices, which are capable of faster turn-off.

Figure 4.12 shows a three-phase inverter circuit in which the power semiconductor switching device is represented simply as a switch. There is competition between several different devices to fill this position. The inverter can be operated in two principal ways. If the output of the inverter is a rectangular wave at the desired frequency, the switches are operated at this relatively low frequency and, in the presence of a partially inductive load, the load current commutates naturally from the freewheeling diode to the power switch after the switch is turned on, so there are no diode recovery losses. If the output of the bridge is a PWM waveform, with switching at a frequency much higher than the desired output frequency, the load current is forcibly commutated out of the freewheeling diode whenever a switch turns on. The stored charge in the diode gives rise to losses in the switching device, so the choices of switch and diode are important in determining the overall operating efficiency of the inverter.

At lower power levels, say up to 1 kW, the power MOSFET may be an economic proposition, particularly if the dc supply is generated from a single-phase rectifier. Since the on-state resistance of a MOSFET rises rapidly with the voltage rating, ($R_{DS(on)} \propto V^{2.5}$), MOSFETs are not normally used for inverters supplied from the three-phase mains. The use of MOSFETs permits a high switching frequency, certainly greater than 20 kHz, and thus outside the audible range. No snubbers are required and the switching losses are small. Conduction losses can be made as small as desired by use of large-area devices. Devices may be connected in parallel to lower the overall on-state resistance of the switch. When operating from low voltage supplies, MOSFETs become increasingly attractive because of the lower $R_{ds(on)}$. One serious problem with MOSFETs in this application is associated with the antiparallel body-drain diode, inherent in the device structure. This is a relatively slow device having a large stored charge compared with that of a discrete fast diode of comparable rating. Consequently, arrangements may have to be made to isolate the body-drain diode (Section 10.4) if switching losses are to be kept to an acceptable level.

The device which has come to dominate mains-powered inverter applications is the IGBT, since it tends to have the best mix of characteristics. It is voltage controlled and therefore easy to drive. It is conductivity modulated so that high voltage devices have a relatively low on-state voltage drop. However, the switching speed is fast enough and the switching losses low enough for the inverter to be operated at frequencies high enough to permit the generation of good quality waveforms using PWM techniques. A 100 Hz waveform requires a switching frequency of at least 3 kHz, and this may be the source of a serious level of acoustic noise. High power drives using IGBTs optimised for low forward voltage drop may operate at this level. Smaller drives using IGBTs optimised for fast switching can be operated at 20 kHz.

The IGBT has the advantage over the MOSFET that its structure does not include an integrated body-drain diode, so the circuit designer is free to chose an antiparallel diode in the way that best suits the application. For minimum switching losses and low electromagnetic interference, a diode with a small reverse recovery charge and a soft recovery characteristic (Section 5.3.2) is optimum. As might be expected, because they have the most desirable characteristics, they tend to command the highest price.

Inverter circuits need to withstand their output terminals being short circuited. The usual means is to incorporate load current sensing in the bridge with a command fed back to the IGBTs to turn off when a fault is detected. Discriminating a fault condition from normal operation takes time, and during this time the IGBT has to support the full dc link voltage. IGBTs are available which can sustain this condition for 10 $\mu$s, a time that is usually considered adequate to detect the fault and turn the devices off.

When an inverter bridge is operated in the PWM mode, the commutation of the load current from the diode to the switching device and vice versa imposes a switching trajectory on the device, and if snubbers are not to be employed, a square safe operating area (SOA) is required. The safe operating area is thus of prime importance. Power MOSFETs have a square safe operating area at all voltages and currents up to the surge current rating. IGBTs usually have this property for the rated current and voltage, but may need a snubber circuit if the device is to be switched at the maximum allowable current. The snubber may also reduce the switching losses in the IGBT and the electromagnetic interference generated by rapid switching and the fast rate of rise of the output voltage when the freewheeling diode snaps off.

Before the advent of IGBTs, the power bipolar junction transistor (BJT) had become the device of choice in PWM motor drive applications. In the 1980s BJTs, with square safe operating areas, became available at an economic price, particularly in module form. The short-circuit capability of these devices can be as long as 10 $\mu$s. Somewhat surprisingly, they are capable of parallel connection, at least of individual dice within the module, notwithstanding the tendency of the forward voltage drop of a BJT to fall with increasing temperature. This can be aided by built-in resistance in the emitter area, as described in Section 6.4. The relatively slow switching of a BJT, compared with that of an IGBT, means that the freewheeling diodes in the inverter do not need to be quite so fast. The main problem with BJTs is the complexity of the base drive circuits required for optimum performance, as illustrated in Figure 3.17. The gain of power BJTs in the saturated condition is quite low and therefore the base driver must be capable of providing a significant level of current all the time that the transistor is switched on. During turn-on a higher than usual level of base drive current may be provided to accelerate the process. Similarly at turn-off, in order to achieve fast switching, the base drive is required to remove charge from the base rapidly, implying a large negative pulse current capability. To reduce the level of continuous base current needed to keep the device turned on, the main switching transistor may be operated with a second BJT in the Darlington configuration (Section 6.5). This has the disadvantage of increasing the overall forward drop of the combination, and steering diodes in the base connections may also be required to permit the extraction of charge from the main transistor at turn-off. These are often incorporated within the module.

Bipolar transistors are usually operated in saturation to keep the forward drop, and thus the conduction losses, as low as possible. However, this does mean a significant delay at turn-off and, perhaps more important, large turn-off losses. Turn-off losses can be reduced, at the expense of increased forward drop, by preventing the transistor from entering the saturation region by the use of an antisaturation diode (Section 6.3.2). The BJT structure does not inherently include an antiparallel diode, but just like the IGBT it does not usually have a high reverse voltage capability and care must be taken to avoid spikes of reverse voltage. One possible source of such negative voltage spikes is the rapid change of current which occurs in the freewheeling diode path during commutation. It is therefore good practice to locate the freewheeling diode physically close to the IGBT or BJT. For this reason, manufacturers commonly offer modules which include the freewheeling diode together with the IGBT or BJT.

A factor which has a significant influence on the cost of a power semiconductor switch is its voltage rating. Transients caused by parasitic inductance in the circuit, or by external phenomena such as lightning strikes and circuit-breaker operation, have to be considered when deciding how large a margin to allow between the voltage to which the device is exposed during normal operation and the voltage rating of the device used. Some power MOSFETs and IGBTs have the ability to withstand avalanche breakdown in the forward direction, provided that the energy dissipated in the device is limited. The device may be given two ratings. One is a repetitive avalanche energy capability and the other a larger single-shot avalanche capability. The presence of these characteristics in a device may give the equipment designer the confidence to work with a lower safety margin in choosing the voltage rating.

Figure 4.12 shows a three-phase inverter in which the output is a three-phase set of voltage waveforms. However, there are many applications in which a single-phase ac supply or a variable-voltage dc supply with four-quadrant capability is required. These requirements can be met with the H-bridge circuit shown in Figure 4.13. When producing an ac output, the bridge operates in the same manner as the three-phase inverter described above. When operating as a dc power source with PWM control, a zero voltage condition can be obtained by turning on only one of the switches, with the current freewheeling through the switch and a diode.

## 4.5 Non-Isolated dc to dc Converters

Power supplies form a major application area for power semiconductor devices. Power supply circuits are basically dc to dc converters, although the input dc in a mains-powered supply is commonly derived by rectifying the ac supply. Elimination of ripple and the regulation of the average output voltage, which was previously performed by a dissipative voltage regulator, is now commonly achieved by the use of a switching converter, or switched-mode power supply (SMPS). Pulse width modulation of a power semiconductor switch is employed to keep the output voltage constant at the desired value. Switched-mode converter circuits can be of the non-isolated type or they can include a transformer to provide galvanic isolation between the input and output sides of the supply. There are three basic switched-mode power supply circuits, each imposing different requirements on the semiconductor devices.

Figure 4.14 shows a non-isolated buck regulator circuit which steps down the voltage. The power semiconductor switch S1 is turned on and off at a fixed frequency to produce a waveform across D1 whose average value approximates the desired output voltage. The output voltage is approximately given by the product of the input voltage and the S1 duty cycle. This type of regulator is usually operated in

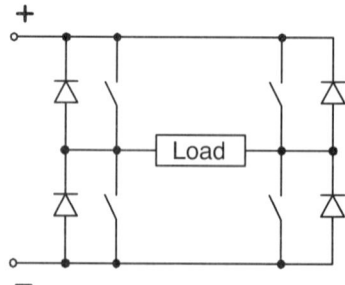

**Figure 4.13** The H-bridge circuit

## 4.5 NON-ISOLATED DC TO DC CONVERTERS

the continuous mode, with the current in the inductor L1 never falling to zero under normal load conditions. When S1 is off, the inductor current circulates through the freewheeling diode D1. When S1 switches on, the diode is forcibly recovered from the conducting state. It must therefore be a fast diode with a low reverse recovery charge. S1 is likely to be a power MOSFET since the required value of the inductor is approximately inversely proportional to the switching frequency. Such circuits are commonly operated well above 100 kHz. The body-drain diode in the MOSFET is always reverse biased and does not conduct.

Figure 4.15 shows a buck-boost circuit. This is not a commonly used circuit although its transformer-isolated equivalent, the flyback converter, is widely used. Figure 4.16 shows a boost converter which is used to step up a dc supply. In this circuit, when S1 closes, current builds up in the inductor. When S1 opens, the current in L1 continues to flow but its magnitude declines because the output voltage is greater than the input voltage. The circuit may be operated in the continuous or discontinuous mode. In the continuous mode, diode D1 is forcibly recovered when S1 turns on and it must therefore have good recovery properties. If the current in the

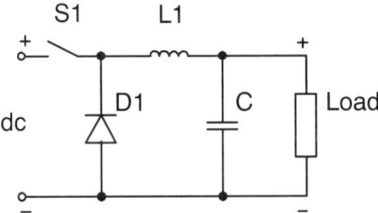

**Figure 4.14** The buck regulator circuit, a non-isolated converter

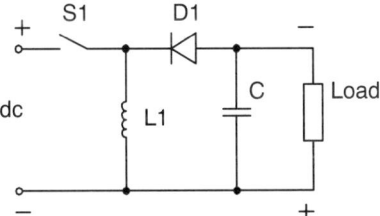

**Figure 4.15** The non-isolated buck-boost converter circuit

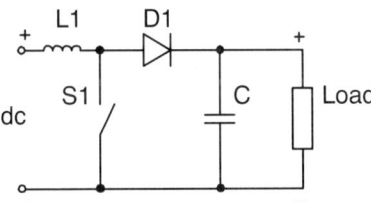

**Figure 4.16** The non-isolated boost converter circuit

# 4  POWER SEMICONDUCTOR DEVICE APPLICATIONS

inductance is allowed to fall back to zero each cycle and the circuit operates in the discontinuous mode, there is not the same need for diode D1 to have a particularly low reverse recovery charge or a soft recovery characteristic. When this circuit is supplied from a low voltage supply, i.e. well below rectified mains voltage, the device of choice for S1 is likely to be a power MOSFET.

## 4.6  Transformer-Isolated dc to dc Converters

Figure 4.17 shows a transformer-isolated version of the buck regulator. The use of a transformer permits a wide variation between input and output voltage. The input voltage is commonly derived from the rectified and smoothed ac mains. S1 is operated at a constant frequency with its duty cycle varied in order to keep the output voltage constant at the desired value.

When S1 opens, energy stored in the magnetic field of the core of the transformer is recovered into the reservoir capacitor C2 through the clamp winding and diode D1. On the forward stroke when S1 is closed, power is delivered to the secondary side of the circuit. Diode D2 rectifies the output voltage while D3 provides a freewheeling path for the inductor current, which is most likely to be continuous. If D2 and D3 are p–n junction diodes, they will need to be of the fast recovery type with low reverse recovery charge, since current is force-commutated between them. When the output voltage of the supply is low, say 5 V or less, the forward voltage drop in D2 and D3 can be a significant proportion of the output voltage and Schottky diodes are frequently used. They have the advantage of zero reverse recovery charge, but there can be significant capacitance associated with the Schottky junction and the depletion region. This can give rise to losses in the same way as the reverse recovery charge at a p–n junction. At turn-off the Schottky capacitance and parasitic inductance may cause oscillations, requiring snubbers for suppression.

The choice of device for S1 is not obvious. The size of the magnetic components in the circuit is minimised when as high a switching frequency as possible is used. This suggests the use of a MOSFET. However, on turn-off the clamp circuit does not operate until D1 is forward biased. This means that the voltage across S1 rises to twice the dc input voltage by transformer action. The peak input voltage is

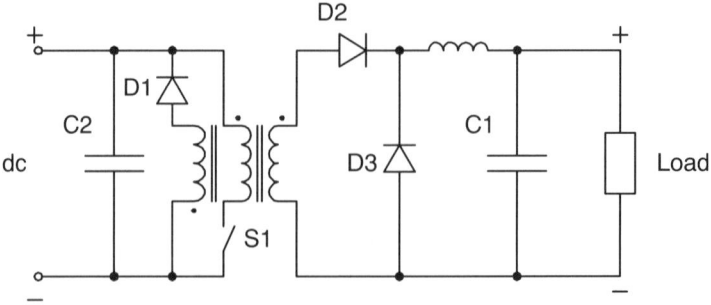

**Figure 4.17**  A single-ended, transformer-isolated forward converter

## 4.6 TRANSFORMER-ISOLATED DC TO DC CONVERTERS

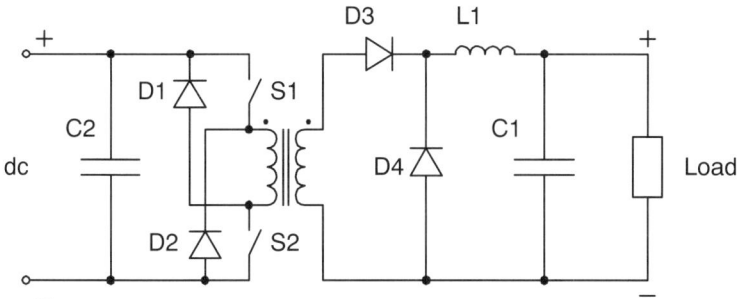

**Figure 4.18** A double-ended, transformer-isolated forward converter

approximately 325 V if it is derived by the rectification of a 230 V single-phase ac supply and the voltage across S1 can rise to 650 V. If the coupling between the clamp winding and the main transformer winding is less than perfect, this voltage can rise briefly to higher levels, even when a snubber or clamping device is used across S1.

With a margin for safety, this suggests that a MOSFET with a voltage rating approaching 1000 V is required. These MOSFETs are relatively expensive because they have a high on-state resistance. A high voltage bipolar transistor is therefore often used for S1 in this circuit. Although its turn-off losses are greater than those of a MOSFET, it is conductivity modulated in the on-state and a low forward voltage drop is easily achieved. The switching frequency of the bipolar transistor is likely to be limited to less than 100 kHz by switching losses but its cost is relatively low.

The advent of fast-switching IGBTs with reduced turn-off losses means that the IGBT has become a contender for this application. It combines the benefit of voltage control with conductivity modulation and a relatively low forward voltage drop. However, its maximum practical operating frequency is lower than for the MOSFET and it is likely to be more expensive than the bipolar transistor. Hence the choice of device for S1 in this application is not an easy one.

Figure 4.18 shows a double-ended transformer-isolated converter. Both transistors are turned on simultaneously on the power stroke and both are turned off during the recovery part of the cycle. No clamp winding is required. When S1 and S2 are turned off together, the flux in the transformer core starts to collapse and this induces a negative voltage in the primary winding. The magnitude of this induced voltage is clamped to the dc rail voltage by D1 and D2. The voltage across S1 and S2 during this period is equal to the dc supply voltage. Hence there is no voltage-doubling effect as in the single-ended converter. Furthermore, because there is no clamp winding and therefore no voltage overshoot due to leakage flux, the voltage seen by each transistor in a well laid out circuit is little more than the dc supply voltage and MOSFETs are more acceptable in this circuit. It is true that two transistors are required instead of one, but the cost of each MOSFET is likely to be significantly lower than the cost of the high voltage MOSFET needed for the single-ended converter.

Figure 4.19 shows a flyback converter that is the transformer-isolated version of the buck-boost converter. Switch S1 is turned on, current builds up in the

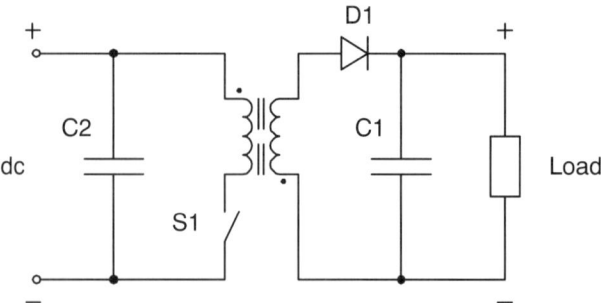

**Figure 4.19** A flyback converter, a transformer-isolated buck-boost converter

transformer primary and energy is stored in the transformer core. When S1 turns off, the collapsing flux in the transformer core induces a voltage in the secondary and D1 conducts. The energy stored in the transformer core is transferred to the reservoir capacitor C1 and thus to the load. The transformer turns-ratio must be chosen to cause S1 to operate with a duty cycle somewhere in the region of 50%. Then the voltage across the primary winding when the energy in the transformer core is being recovered is approximately equal to the dc supply voltage, and the voltage across S1 is twice the supply voltage.

Again, leakage flux can increase the value of voltage impressed on S1 when it turns off. Therefore a bipolar transistor might be the preferred device for S1 when the input dc supply is derived from a rectified 230 V ac mains supply. If ease of control is sought, and the switching frequency is not too high, an IGBT might be used, This type of circuit is generally used for supplies of about 150 W or less, because the more complicated single-ended or double-ended forward converters generally prove more economic at higher power levels.

At lower power levels, the relatively high price of high voltage MOSFETs might be offset by the cost of the bipolar transistor gate drive circuit, by its snubber requirements, and by the increased cost of the transformer that results from having to operate at a lower switching frequency. The flyback circuit may be operated in the continuous mode or the discontinuous mode. In the discontinuous mode the current in the secondary winding of the transformer falls to zero each cycle. Thus diode D1 turns off naturally each cycle. However, the form factor of the current waveform in S1 is poor, leading to high losses in this device. In the continuous mode the current in the secondary of the transformer never falls to zero. This implies a better form factor of the current in S1 but it also means that D1 is forcibly recovered, so this diode must have a suitably low reverse recovery charge.

## 4.7 Power Factor Correction

A high voltage version of the boost converter is shown in Figure 4.20. This circuit is commonly used for power factor correction circuits used in front of switched-mode power supplies to ensure that an almost sinusoidal waveform of current at almost

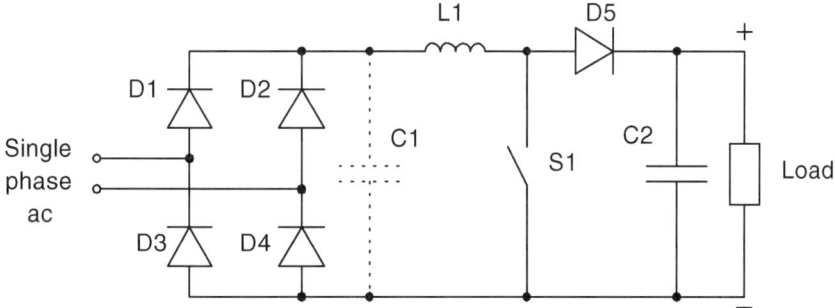

**Figure 4.20** A high voltage, non-isolated boost converter for power factor correction

unity power factor is drawn from the ac mains supply. Diodes D1, D2, D3 and D4 form a full-wave rectifier. These can be conventional low-forward-drop rectifier diodes. Capacitor C1 is not an electrolytic reservoir capacitor but a small foil capacitor which circulates high frequency switching currents within the converter and which ensures that the instantaneous voltage of the rectified supply does not change significantly when pulses of current are drawn by the boost circuit.

S1 opens and closes in a manner which ensures that the average current drawn by the inductor L1 follows the same waveform as the rectified ac mains. After filtering by C1 this current is steered by the rectifier bridge so that it appears as a sinusoid of current at the ac input terminals in phase with the voltage waveform. A control circuit senses the average voltage at the output terminals of the boost converter and adjusts the duty cycle of S1 to increase or decrease the magnitude of the current waveform in L1, hence the average current delivered to C2 and the output, thereby maintaining a balance between the charge entering and leaving C2 for all load conditions.

The circuit may be operated in discontinuous mode or continuous mode. In the discontinuous mode, the current in L1 falls to zero each cycle, enabling D5 to recover naturally, but the form factor of the current in S1 is poor. In the continuous mode, the current in L1 has much less ripple component. D5 is forcibly recovered, necessitating the use of a fast recovery diode, but the form factor of the current in S1 is relatively good. The voltage seen by S1 is clamped by D5 and the output reservoir capacitor C2 so that, if the output voltage is not much above the peak value of the rectified ac mains input, the use of a power MOSFET for S1 is practical. Again, if a high switching frequency is not desired, an IGBT might be a suitable choice for S1.

## 4.8 Resonant Circuits

No survey of the environments in which power semiconductors are used would be complete without mentioning soft-switching and resonant or quasi-resonant circuits. Most of the circuits dealt with so far are hard-switching circuits in that when they are operated in the continuous current mode, the switch has to switch current on and off while the full supply voltage is present across the switch. This requires the switch to have a square safe operating area unless a snubber is used. In a fully resonant circuit, such as those shown in Figures 4.21 and 4.22, an inductor and capacitor dominate the load characteristics and the converter switches at about their resonance frequency. In Figure 4.21 they are in parallel, leading to parallel resonance, whereas in Figure 4.22 they are in series and series resonance is the mode of operation. Resonant loads have the advantage that either the current or the voltage can be arranged to fall to zero during turn-on or turn-off.

Resonant circuits such as those shown in Figures 4.21 and 4.22 offer the possibility of operation in several ways, depending on whether they are switched at above or below resonance. But the objective is always to reduce the switching losses in the semiconductor devices by ensuring that either the current through the device or the voltage across the device is zero at the moment of switching. The disadvantage is that the peak value of the current drawn by the resonant load is considerably greater than the average value and therefore greater than the current which would be drawn by a hard-switching converter of similar specification. Thus the current rating of the power semiconductor switches has to be increased.

Quasi-resonant circuits employ partial resonance. In zero voltage switching and zero current switching quasi-resonant circuits, the circuit capacitance and inductance is such that a half-cycle of resonance, at a frequency well above the switching frequency, occurs when the semiconductor switch is either turned off or turned on.

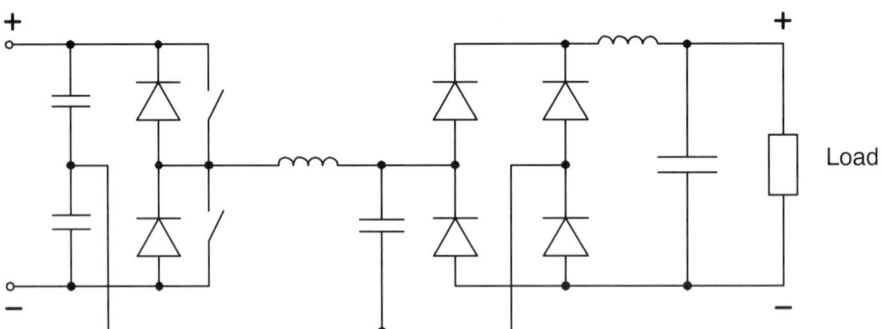

**Figure 4.21**  A converter employing parallel resonance

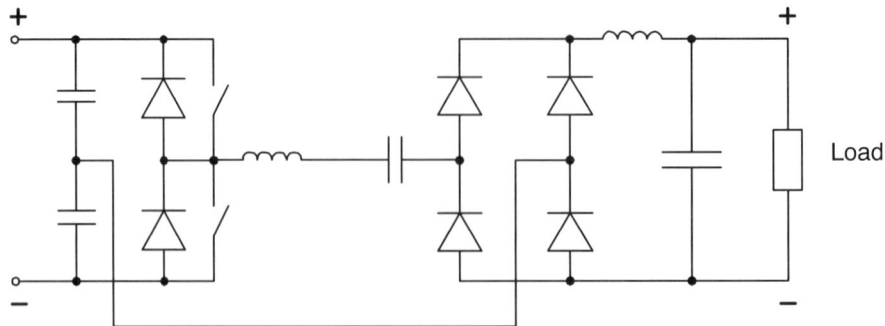

**Figure 4.22**  A converter employing series resonance

By this means the switching losses at the most critical part of each cycle of operation are avoided. There are many possible circuit configurations which can be employed to make use of this principle. The use of quasi-resonance normally imposes some restriction on the way in which the circuit may be operated. For example, either a fixed on-time or a fixed off-time of the switching device may be demanded. In resonant and quasi-resonant circuits, losses in components other than the switching device may be reduced. Diode recovery losses may be avoided by ensuring that the diode is in a resonant circuit in which resonance reduces the current in the diode to zero naturally during each cycle.

## Summary

Power electronics involves converting ac mains power to supply a controlled level of power to a load as either ac or dc. Diodes and diode bridge circuits can provide uncontrolled rectification of the mains. Thyristor (SCR) and mixed thyristor/diode bridges provide a controlled dc output. Cycloconverters can provide very high levels of low frequency ac power, using phase-control thyristors.

In inverters, a variable-voltage dc supply, usually obtained by rectifying the mains, is converted to a variable-voltage, variable-frequency ac supply, often for motor control. Such circuits require fast-switching devices. MOSFETs are used for low power levels, IGBTs at medium power levels and GTO thyristors for the highest power levels.

Other applications include transformer-isolated and non-isolated dc to dc converters and power factor correction. In some cases, high frequency resonant circuits may be preferred to those involving hard switching.

## References

[4.1] Gjugyi, L. and Pelly, B. *Static Frequency Changers* (John Wiley & Sons, 1976).
[4.2] Wheeler, P. and Grant, D. A. Optimised Input Filter Design and Low-Loss Switching Techniques for a Practical Matrix Converter, *IEE Proc. Electrical Power Applications*, **144**, 53–60 (1997).

# 5
# POWER DIODES

Power diodes are used in applications which require current to flow in one direction only. The main application is rectification, in which alternating current is converted to direct current. However, it is clear from Chapter 4 that diodes play an important role in almost all power electronic circuits. The current ratings of individual devices range from less than one ampere to several thousand amperes and their blocking voltages from tens of volts to several thousand volts. As well as a rating for continuous current, devices also need to be able to withstand surges of current that, under fault conditions, may last for a few tenths of a second.

Most often power diodes make use of the rectifying action of a p–n junction, although a rectifying Schottky contact can also be used. As well as being capable of blocking the required reverse voltage in the off-state, power diodes require a low forward voltage drop in the on-state. In many applications a rapid transition from the forward conducting to the reverse blocking state is required. The rate at which the device changes in the other direction, from blocking to conducting, may also be important.

As described in Section 3.1.1, the internal structure of a diffused power diode typically takes the form $p^+pnn^+$, whereas an epitaxial device is better described as having a $p\nu n$ structure. Our first aim is to establish the relationships between the electrical parameters of the diode and the details of its construction. With this information, the device design can be optimised to meet the requirements of particular circuit applications.

## 5.1 The Forward-Biased Diode

When a p–n junction is forward biased (with the p-type region made positive with respect to the n-type region), holes are injected into the n-type region and electrons into the p-type region, as described in Section 2.1. The fact that these regions are bounded by $n^+$- and $p^+$-layers, in which the carrier concentrations are several orders of magnitude higher, means that the concentrations of excess carriers so injected can be many times greater than the equilibrium majority carrier concentrations. The excess carriers modify the conductivity of the n-type and p-type regions, with the result that it is only at low current densities that a $p^+pnn^+$ power diode behaves like a simple p–n junction. At higher levels of current density, theoretical analysis is made much simpler by the use of the p–i–n diode approximation, illustrated in Figure 5.1. When the impurity concentrations of the p- and n-regions are much less than the carrier concentration, both may be considered to be very lightly doped, so they may be treated as equivalent to intrinsic material. Hence the use of the letter 'i' in 'p–i–n' diode. The epitaxial p–ν–n-diode already approximates to this structure.

Typical distributions of the free carrier concentrations across a forward-biased power diode are shown in Figure 5.1. At the $p^+$–i junction, we should expect that the current is carried almost entirely by holes injected from the $p^+$-layer into the 'i'-region. Only a small current of minority electrons flows to the $p^+$-contact. At the $n^+$–i junction, the converse is true: the current is predominantly carried by the electrons injected from the $n^+$-layer into the 'i'-region, with only a small current of holes flowing to the $n^+$-contact. The difference between the hole current injected from the $p^+$-contact and that continuing into the $n^+$-layer is accounted for by recombination within the 'i'-region. The same applies to the difference between the electron current entering the 'i'-region from the $n^+$–i junction and that leaving at the $p^+$–i junction. The carrier concentrations in the 'i'-region rise, until the rate of recombination balances the difference between the electron and hole currents flowing in and those leaving. At the same time, a minute imbalance in the carrier concentrations is sufficient to provide the variation of the electric field that is needed to control the distributions of the electron and hole currents in the 'i'-region.

The potential variation across the forward-biased diode is shown schematically in Figure 5.1. The forward voltage drop $V_F$ can be expressed as the sum of three terms:

$$V_F = V_P + V_N + V_I \qquad (5.1)$$

Here $V_P$ is the voltage dropped across the $p^+$–i junction, $V_N$ is that dropped across the $n^+$–i junction and $V_I$ is that across the 'i'-region. To a first approximation, $V_P$ and $V_N$ vary logarithmically with the current density $J$, as would be expected for the case of a p–n junction. Following (2.22), we put

$$V_P + V_N = K_0 + \frac{\alpha k T}{e} \ln J \qquad (5.2)$$

where the constant $K_0$ depends on the temperature and the doping profiles, and the parameter $\alpha$ may depend on the current density.

The other component of $V_F$, the voltage $V_I$ across the middle region of the diode, varies in quite a complicated way with both the current density and the width $w$ of this region. What matters is the ratio of $w$ to the *ambipolar diffusion length*, $L_a = \sqrt{D_a \tau_H}$, where $D_a$ is the bipolar diffusion coefficient defined in (1.74) and $\tau_H$ is the trap-dominated carrier lifetime described by (1.62), under conditions of high injection. The ambipolar diffusion length is analogous to the electron and hole diffusion lengths introduced in (2.8) and (2.9), which apply under conditions of low injection. With high injection the electrical conductivities $\sigma_n$ and $\sigma_p$, are dominated by the injected carrier concentrations $\Delta n$ and $\Delta p$, which we assume to be equal. Equation (1.74) then becomes

## 5.1 THE FORWARD-BIASED DIODE

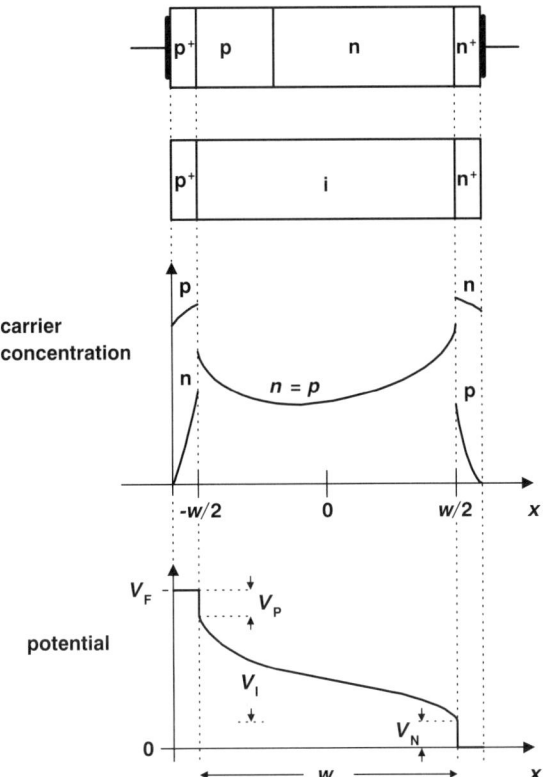

**Figure 5.1** Carrier and potential distributions in a forward-biased power diode, demonstrating the p–i–n diode approximation. The figure shows the basic structure, the pin diode approximation, the free carrier concentrations under normal, high injection, forward bias conditions and the variation of potential across the forward-biased diode. Note that the carrier concentrations greatly exceed the dopant concentrations in the n- and p-regions

$$D_a = \frac{2D_n}{1 + \mu_n/\mu_p} = \frac{2D_p}{1 + \mu_p/\mu_n} \quad (5.3)$$

And $\mu_{\text{diff}}$, as defined in equation (1.75), is zero.

Using the values of carrier mobility appropriate to lightly doped silicon at room temperature, $\mu_n = 0.142 \, \text{m}^2/\text{V s}$ and $\mu_p = 0.047 \, \text{m}^2/\text{V s}$, we obtain $\mu_n = 3.0 \mu_p$ and

$$D_a = 0.50 D_n = 1.5 D_p = 1.8 \times 10^{-3} \, \text{m}^2/\text{s}$$

When either the doping levels or the carrier concentrations are very high, say in the region of $10^{26} \, \text{m}^{-3}$ ($10^{20} \, \text{cm}^{-3}$), equation (1.21) indicates that the carrier mobilities fall to values of $\mu_n = 0.007 \, \text{m}^2/\text{V s}$ and $\mu_p = 0.0045 \, \text{m}^2/\text{V s}$. Then $\mu_n = 1.5 \mu_p$ and at room temperature,

$$D_a = 0.8 D_n = 1.2 D_p = 1.4 \times 10^{-4} \, \text{m}^2/\text{s}$$

These two sets of asymptotic values are used in the analysis that follows.

In order to analyse the voltage distribution across the middle part of the diode, we define the x-coordinate to be at right angles to the planes of the junctions, with its origin midway between them. The p$^+$–i junction is then at $x=-w/2$ and the n$^+$–i junction at $x=w/2$. The first step is to solve the continuity equation (1.77) in the steady state and under high injection conditions. With $\partial \Delta n/\partial t=0$, putting $\tau=\tau_H$ and neglecting any variation of $D_a$ with carrier concentration, equation (1.77) becomes

$$\frac{d^2 \Delta n}{dx^2} = \frac{\Delta n}{L_a^2} \tag{5.4}$$

The boundary conditions are determined by the current density $J$, which in turn is fixed by the external circuit. If we assume for the moment that the injection efficiency for electrons at the n$^+$–i junction is 100%, we are asserting that the hole current at that point is zero. However, there is an equal concentration of both types of carrier, hence an equal concentration gradient, as shown in Figure 5.1. The electric field strength needed to reduce the flux of holes to zero at $x=w/2$ is obtained by applying (1.83) to our one-dimensional situation:

$$J_p(w/2) = pe\mu_p E(w/2) - eD_p \frac{dp}{dx}\bigg|_{x=w/2} = 0$$

Thus

$$E(w/2) = \frac{D_p}{\mu_p}\frac{1}{p}\frac{dp}{dx}\bigg|_{x=w/2} = \frac{kT}{e}\frac{1}{p}\frac{dp}{dx}\bigg|_{x=w/2} = \frac{kT}{e}\frac{1}{n}\frac{dn}{dx}\bigg|_{x=w/2} \tag{5.5}$$

using the Einstein relation (1.70a). If we substitute for $E(w/2)$, using (5.5), and again apply the Einstein relation, the current density, which is due to the electrons, is given by

$$J = J_e = ne\mu_n E(w/2) + eD_n \frac{dn}{dx}\bigg|_{w/2} = 2eD_n \frac{dn}{dx}\bigg|_{w/2} \tag{5.6}$$

Hence

$$\frac{dn}{dx}\bigg|_{w/2} = \frac{d\Delta n}{dx}\bigg|_{w/2} = \frac{J}{2eD_n} \tag{5.7}$$

Applying a similar argument at $x=-w/2$, where the current density is assumed to be carried entirely by the holes, we obtain for the boundary condition at that point:

$$\frac{dp}{dx}\bigg|_{-w/2} = \frac{d\Delta n}{dx}\bigg|_{-w/2} = \frac{J}{2eD_p} \tag{5.8}$$

The solution of (5.4) that satisfies the boundary conditions imposed by (5.7) and (5.8) can be expressed as

$$\Delta n(x) = \frac{J\tau_H}{2eL_a}\left[\frac{\cosh(x/L_a)}{\sinh(w/2L_a)} - \delta\frac{\sinh(x/L_a)}{\cosh(w/2L_a)}\right] \tag{5.9}$$

where $\delta=(\mu_n-\mu_p)/(\mu_n+\mu_p)$. The second term in the square brackets on the right-hand side of (5.9) gives rise to the asymmetry in the carrier concentration distribution, revealed in Figure 5.1, as a result of the difference in the electron and hole mobilities.

At any point between $-w/2$ and $+w/2$, the current density $J$ is given by the sum of the drift and diffusion currents of each type of carrier:

$$J = J_n + J_p = e(n\mu_n + p\mu_p)E(x) + e\left(D_n\frac{dn}{dx} - D_p\frac{dp}{dx}\right)$$
$$= e\Delta n(\mu_n + \mu_p)E(x) + kT(\mu_n - \mu_p)\frac{d\Delta n}{dx} \qquad (5.10)$$

So the electric field strength can be expressed as

$$E(x) = \frac{J}{e(\mu_n + \mu_p)\Delta n(x)} - \frac{kT}{e\Delta n(x)}\delta\frac{d\Delta n}{dx} \qquad (5.11)$$

We have again assumed that the carrier concentrations $n$ and $p$ can be represented by the injected excess carrier concentration $\Delta n$.

The voltage $V_I$ dropped across the middle region of the diode, can be obtained by substituting the carrier concentration profile given by (5.9) into (5.11) and integrating the electric field $E(x)$ from $x=-w/2$ to $x=+w/2$. The solution [5.1] is complicated and of limited use because of the assumptions made; it is

$$\frac{eV_I}{kT} = \frac{8b}{(b+1)^2}\frac{\sinh W}{[1-\delta^2\tanh^2 W]^{1/2}}\tan^{-1}\{[1-\delta^2\tanh^2 W]^{1/2}\sinh W\}$$
$$+\delta\ln\left\{\frac{1+\delta\tanh^2 W}{1-\delta\tanh^2 W}\right\} \qquad (5.12)$$

where $W = \dfrac{w}{2L_a} = \dfrac{w}{2\sqrt{D_a\tau_H}} \qquad b = \dfrac{\mu_n}{\mu_p} \qquad \delta = \dfrac{\mu_n-\mu_p}{\mu_n+\mu_p}$

Equation (5.12) is plotted in Figure 5.2 using $b=3.0$ and $\delta=0.50$ to represent low injection levels, and $b=1.5$ and $\delta=0.22$ to represent high injection levels. Note that $V_I$ is not a function of the current density. This is because the conductivity of the middle region is proportional to the injected carrier concentration, which itself is proportional to the current density. Such *conductivity modulation* is characteristic of bipolar devices. It enables them to be made with the wide, lightly doped layers required for high blocking voltages and still to have a low on-state voltage.

When $w/L_a \gg 1$ the diode is described as a 'long-base' diode. Then $\tanh W \to 1$. Also, the $\tan^{-1}$ term in (5.12) tends to $\pi/2$ and the second term becomes negligible. As a result, equation (5.12) reduces to

$$V_I = \frac{2\pi b}{(b+1)^2}\frac{kT}{e}\exp\left(\frac{w}{2L_a}\right) = B\frac{kT}{e}\exp\left(\frac{w}{2L_a}\right) \qquad (5.13)$$

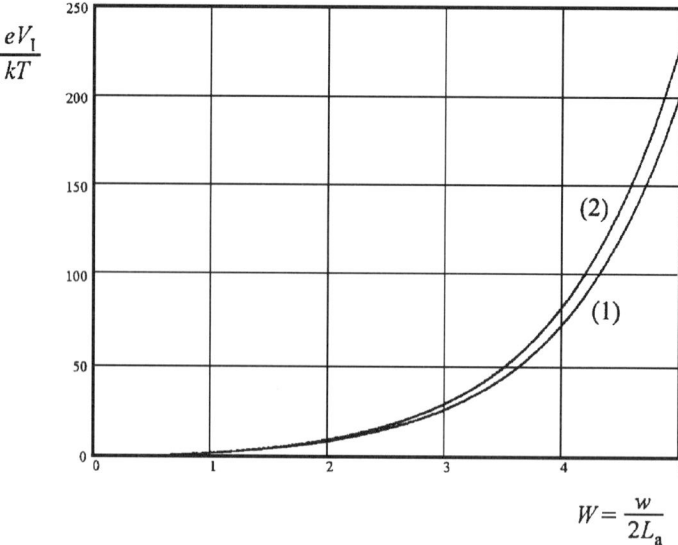

**Figure 5.2** Theoretical calculation of the forward voltage drop across the intrinsic region of a $p^+in^+$ diode as a function of the ratio of its thickness to the carrier diffusion length. Curve (1) has been calculated using carrier mobilities appropriate to low injection conditions, Curve (2) using high injection values. Note that the diffusion length is proportional to the square root of the carrier lifetime, which is reduced at high carrier concentration by Auger recombination. Thus, $W$ is inversely proportional to the square root of the lifetime and increases at high current density

The voltage drop increases exponentially with the width of the middle region. With $b = 3.0$ the constant is $B = 1.17$. With $b = 2$, the constant is $B = 1.5$. Note that the carrier concentration falls by the factor $\cosh W$ between the injecting junctions and the centre. For $W = 3$, this is a factor of 10 and $V_I$ is rather less than 1 V. At $W = 10$ these values are increased a thousandfold, setting a severe constraint on the values of $W$ that are acceptable in practice.

When $w/L_a \ll 1$ the diode is described as having a 'short' base. However, this implies there is negligible recombination in the 'i'-region of the diode, so that all the injected electrons pass right through from the $n^+$-layer into the $p^+$-layer and all the injected holes pass through from the $p^+$-layer into the $n^+$-layer. Our assumption of 100% injection efficiencies at the $n^+$–i and $p^+$–i junctions is now clearly invalid and the current density should be considered as having three components, $J_p$, $J_n$ and $J_r$: $J_p$ is the current density of holes entering the $n^+$-region across the $n^+$–i junction; $J_n$ is the the current density of electrons entering the $p^+$-region across the $p^+$–i junction; and $J_r$ is the current density of carriers recombining in the base region. Thus,

$$J = J_p + J_n + J_r \tag{5.14}$$

The recombination current density can be written as

$$J_r = \frac{ew\langle \Delta n \rangle}{\tau_H} \tag{5.15}$$

### 5.1 THE FORWARD-BIASED DIODE

where $\langle \Delta n \rangle$ signifies the average excess carrier concentration in the base.

Under these conditions, $V_I$ can be obtained by substituting $(\mu_n + \mu_p) = \mu_a/K$ into (5.11) and integrating, so that

$$V_I = -\int_{-w/2}^{w/2} E \, dx = -K\frac{J}{e}\int_{-w/2}^{w/2} \frac{dx}{\mu_a \Delta n} - \delta\frac{kT}{e}\int_{-w/2}^{w/2} \frac{1}{\Delta n}\frac{d\Delta n}{dx}dx \quad (5.16)$$

Substituting $eD_a/kT$ for $\mu_a$, $\langle \Delta n \rangle = J_r \tau_H/ew$ for $\Delta n$, on the understanding that the variation of $\Delta n$ is small in this case, and putting $L_a^2 = D_a \tau_H$, we obtain

$$V_I = K\frac{kT}{e}\frac{J}{J_r}\left(\frac{w}{L_a}\right)^2 + \delta\frac{kT}{e}\ln\left(\frac{\Delta n(-w/2)}{\Delta n(w/2)}\right) \quad (5.17)$$

Equation (5.17) shows how a reduced injection efficiency at the two junctions causes an increase in $V_I$ through the increase in the factor $J/J_r$. The factor $K$ has the value 0.37 when $b=3.0$ and rises to 0.48 when $b=1.5$.

At first sight, a low voltage diode, with a relatively narrow base region, might be expected to behave as a short-base diode. This may be true at low values of the forward current density, but two effects that arise at high current densities cause even low voltage devices with thin base regions to take on the characteristics of a long-base device. These are carrier–carrier scattering, as discussed in Section 1.3, and Auger recombination, as discussed in Section 1.4. They become particularly important under surge current conditions, when current densities may exceed 1 A/mm² (100 A/cm²), where they limit the beneficial effects of conductivity modulation.

As indicated by (1.21), carrier–carrier scattering becomes significant at carrier concentrations in the region of $10^{23}$ m$^{-3}$ ($10^{17}$ cm$^{-3}$), and leads to a reduction in the carrier mobilities. We have seen that their ratio is believed to fall from a room temperature value of $b=3.0$ for low carrier concentrations to $b=1.5$ at very high injection levels. More important, the ambipolar diffusion coefficient and hence $L_a$ decrease as the carrier concentrations rise. In material where the carrier lifetime $\tau_H$ is 10 μs, this reduced value gives $L_a = 39$ μm. With $J=1$ A/mm², and putting $w=2L_a$, the average carrier concentration at the injecting boundaries is given by (5.9) as

$$\frac{J\tau_H}{2eL_a}\coth 1 = 10^{24} \text{ m}^{-3}$$

The resultant increase in $(w/L_a)$ causes $V_I$ to increase with increasing current density. However, in this situation the continuity equation is no longer linear and rigorous analytical solution becomes very difficult. Several attempts at treatments using different simplifying assumptions have been published [5.1 to 5.3] and these have suggested sublinear power-law relationships such as $V_I \propto J^{0.5}$ and $V_I \propto J^{0.75}$.

At current densities greater than about 2 A/mm² (200 A/cm²), Auger recombination starts to dominate over the other recombination processes, so the carrier lifetime decreases as the square of the carrier concentration. This has several important effects. In the p$^+$- and n$^+$- regions, the high concentrations of majority carriers increase the probability of Auger recombination, so that it first becomes significant in these regions. The resulting reduction in the carrier lifetime reduces the injection efficiencies at the n$^+$–i and p$^+$–i junctions because a higher proportion of minority

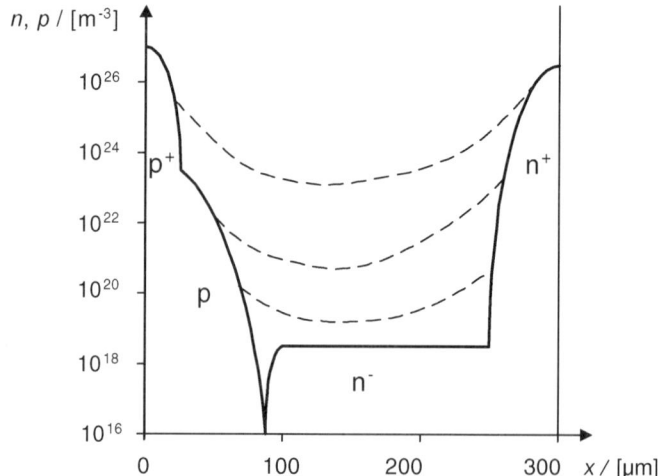

**Figure 5.3** An illustration of the increase at high current densities in the effective base width of a diode with diffused junctions

carriers is drawn across the junctions to supply the necessary recombination currents. This increases the ratio $J/J_r$, and hence $V_I$, as indicated in (5.17). In the 'i'-region, the reduction of the carrier lifetime reduces $L_a$ and causes $(w/L_a)$, hence $V_I$, to increase.

In the discussion so far, we have assumed the $n^+$–i and $p^+$–i junctions to be abrupt. With diffused-junction devices, the impurity concentration gradients cause the effective spacing $w$ between the junctions to expand as the concentration of injected carriers rises. This is illustrated in Figure 5.3. Once again, the effective increase of $(w/L_a)$ causes $V_I$ to increase with the current density.

Note that under conditions of high injection the theory we have presented here is not sensitive to the doping profile in the central regions of the device, or even to the presence of additional p–n junctions there. The theory therefore applies equally well to thyristors and other multijunction devices in which the double-injection of carriers occurs. In all cases the forward current–voltage characteristic does depend on the impurity concentration levels and profiles in the heavily doped $p^+$ and $n^+$ contact layers. For example, an increase in the acceptor concentration in the $p^+$-region of a diode results in an increase in the diffusion potential at the $p^+$–i junction and hence in $V_P$ at low current density. However, it also leads to the injection of a higher concentration of carriers into the 'i'-region and thus to lower values of $V_I$ at high current density. These effects, illustrated in Figure 5.4, are partially compensated by the bandgap narrowing that also occurs, as described in Sections 1.2 and 2.2.

In general, the forward current–voltage characteristic of a power diode can be expressed as

$$V_F = K_0 + K_1 \ln J + K_2 J^m \tag{5.18}$$

## 5.1 THE FORWARD-BIASED DIODE

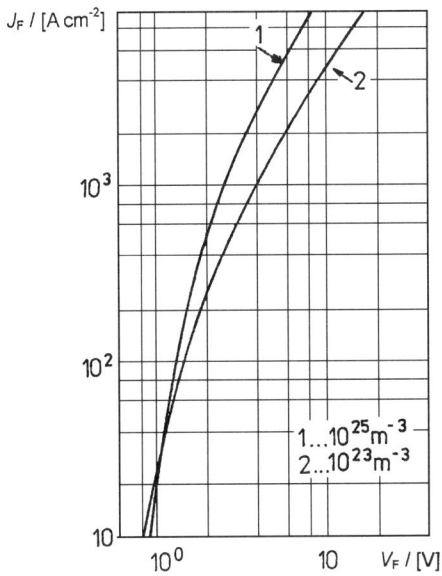

**Figure 5.4** An example of the effect of the dopant concentrations in the contact regions on the forward current–voltage characteristics of a power diode. Adapted from Reference [5.4]

where the index $m$ typically lies between 0.6 and 0.8. The parameters $K_0$, $K_1$ and $K_2$ depend on temperature and on features of the diode structure such as the carrier lifetime, the doping concentrations and the thickness of the various layers. The analysis presented here shows that $V_F$ is strongly dependent on the excess carrier lifetime and on the distance between the n$^+$- and p$^+$- layers. It may also increase significantly at very high current densities, with serious consequences for the surge current rating.

For the purposes of circuit analysis, the current–voltage characteristic is often represented by a threshold voltage $V_{Th}$ and a differential resistance $r_d$, as shown in Figure 5.5. This approximation facilitates the calculation of parameters such as the power dissipation and it enables simple comparisons between different types of diode to be made.

The effect of a rise in temperature on the forward voltage can be quite complex. Because $n_i$ is a rapidly rising function of temperature, $V_P$ and $V_N$ decrease with increasing temperature. Carrier mobilities also decrease, so we should expect $V_I$ to increase. However, carrier lifetime normally increases and this may offset some of the effect of the reduced mobility. In some diodes doped with iridium, it has been found that the lifetime can increase so rapidly that the diffusion length increases and the forward voltage at a given current density decreases with increasing temperature over the range $-30$ to $+150\,^\circ\text{C}$. With gold- and platinum-doped devices, the lifetime may reach a maximum at a certain temperature, and the net result is normally that at low current densities $V_F$ decreases with increasing temperature whereas at high current densities it increases. In the approximation of Figure 5.5 this may be

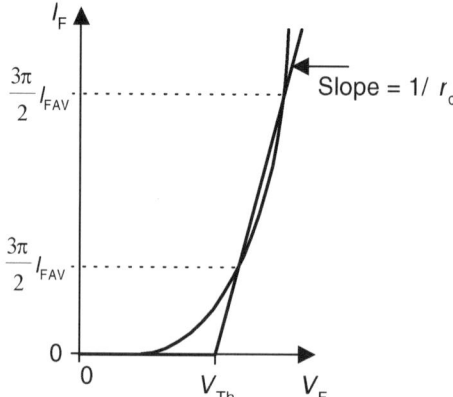

**Figure 5.5** Definition of the threshold voltage $V_{Th}$, and the differential resistance, $r_d$, of a power diode

represented by a decrease in the threshold voltage $V_{Th}$ as the temperature rises, and an increase in the differential resistance $r_d$. The effect is shown in Figure 5.6.

Although the results of the theoretical analysis presented here apply to an idealised structure, they illustrate very clearly some of the problems facing the designer of a power diode. In Section 5.2 it is shown that a high reverse breakdown voltage requires a wide 'i'-region. This causes $V_F$ to be large when the forward current is high. Likewise, in Section 5.3 it is shown that a short carrier lifetime assists the reverse recovery process but it too leads to large values of $V_F$ at high forward currents.

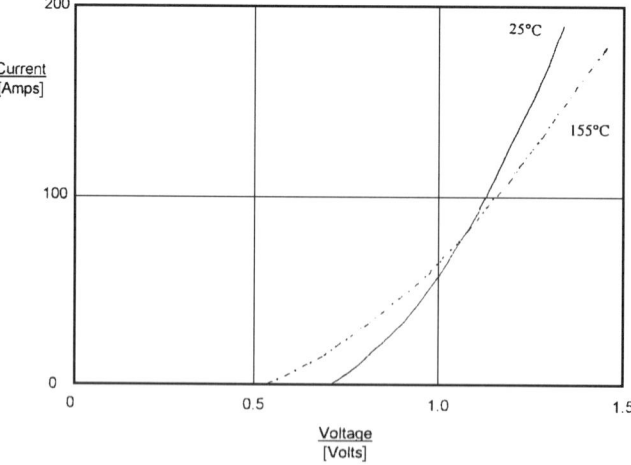

**Figure 5.6** An example of the effect of temperature on the forward characteristics of a power diode. These measurements were made on a 2.5 kV double-diffused diode with a wafer diameter of 16 mm and thickness 250 μm. The maximum current represents a current density of about 1 A/mm²

## 5.2 Reverse Characteristics of Power Diodes

The p–n junction of a high voltage power diode is usually formed 50 μm or more below the wafer surface using diffusion technology. As is shown in Section 2.1.3, in dealing with reverse voltages greater than a few tens of volts, it is usually possible to approximate the diffused junction by assuming an abrupt transition from the p-region to the n-region. The breakdown voltage $V_{BR}$ is determined mainly by the doping concentration on the more lightly doped side of the junction. In a diode made by diffusing acceptors into n-type material, this is the n-region. If its thickness $w_N$ is greater than the thickness $d$ of the space charge region, equation (2.36) for the reverse breakdown voltage of a plane p$^+$–n junction can be simplified to

$$V_{R(BR)} = \frac{\varepsilon_r \varepsilon_0 E_{BR}^2}{2eN_D} \tag{5.19}$$

and (2.38) to (2.42) apply, with $N=N_D$. This structure is illustrated in Figure 5.7 for a typical diode. It can be seen that the reverse breakdown voltage $V_{R(BR)}$ can also be expressed as

$$V_{R(BR)} = E_{BR}d - \frac{eN_D}{2\varepsilon_r\varepsilon_0}d^2 < E_{BR}w_N - \frac{eN_D}{2\varepsilon_r\varepsilon_0}w_N^2 \tag{5.20}$$

In both equations we have neglected the small voltage drop in the p-region.

It follows from (5.19) that to obtain a high breakdown voltage $V_{R(BR)}$ in a p$^+$–n diode, it is necessary to start with silicon having a low donor concentration and hence a high resistivity. A thick layer of such material is needed to enable a wide space charge region to develop when the p–n junction is reverse biased. A low forward voltage drop $V_F$ is then possible only if a long carrier lifetime is preserved during the fabrication of the device. This requires careful control over the quality of the chemicals used and the wafers must be cooled slowly after all high temperature operations.

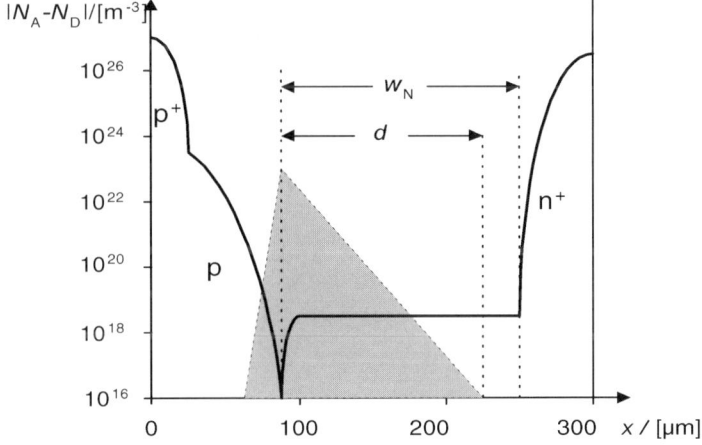

**Figure 5.7** A conventional high-voltage diode structure showing the doping profile and the distribution of the electric field under full reverse bias

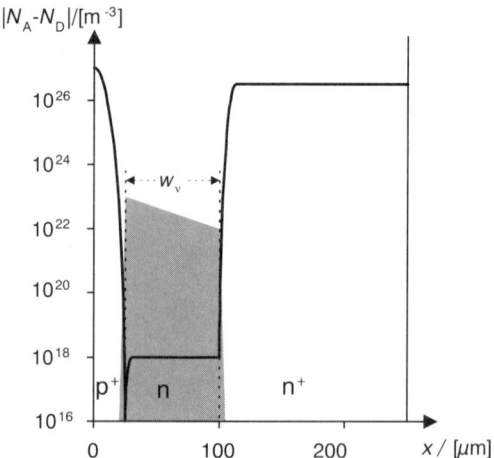

**Figure 5.8** A p⁺νn⁺ (compressed field) structure showing the doping profile and the distribution of the electric field under full reverse bias

It can be seen that there is a conflict between the requirements to achieve both a high reverse breakdown voltage and a low forward voltage drop. A certain compromise can be achieved using the type of construction shown in Figure 5.8. In that case the diode is realised as a p⁺νn⁺ structure, where the space charge region reaches through to the n⁺-region under reverse bias conditions. This is sometimes called a compressed-field structure.

If the donor concentration in the lightly doped region is reduced from $N_D$ to $N_{D\nu}$, so that the space charge spreads through the whole width $w_\nu$ and into the n⁺-region, the reverse breakdown voltage may be expressed as

$$V_{R(BR)} = E_{BR} w_\nu - \frac{eN_{D\nu}}{2\varepsilon_r \varepsilon_0} w_\nu^2 \qquad (5.21)$$

The voltages dropped in the p⁺- and n⁺-regions are neglected as small.

With the same thickness of lightly doped region, $w_N$ or $w_\nu$, and the same carrier lifetime, similar forward current–voltage characteristics would result. But a comparison of (5.20) and (5.21) shows that the compressed-field structure would give a higher reverse breakdown voltage than the conventional structure. Alternatively, similar reverse and forward characteristics can be achieved with material of the compressed-field structure having a shorter carrier lifetime.

Diodes with a compressed-field structure are commonly known as power p–i–n diodes. The dependence of the maximum reverse breakdown voltage $V_{R(BR)}$ on the impurity concentration in the lightly doped region is illustrated in Figure 5.9, with the thickness of that region as a parameter.

## 5.2 REVERSE CHARACTERISTICS OF POWER DIODES

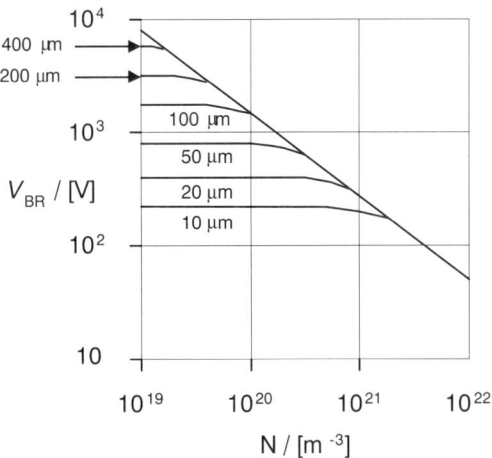

**Figure 5.9** Dependence of the maximum breakdown voltage on the parameters of the $p^+vn^+$ structure

Although p–i–n diodes bring advantages, there are accompanying disadvantages. In particular, there is a problem with surface contouring. In the case of positive-bevelled diodes, the surface electric field increases strongly at the n–n$^+$ junction and breakdown can occur at a relatively low reverse voltage. An improvement of the breakdown voltage can be achieved by a number of means. These include an overhang of the cathode contact [5.5] (Figure 5.10); a change of the concentration profile in the surface region [5.6, 5.7] (Figure 5.11); an auxiliary layer of the same

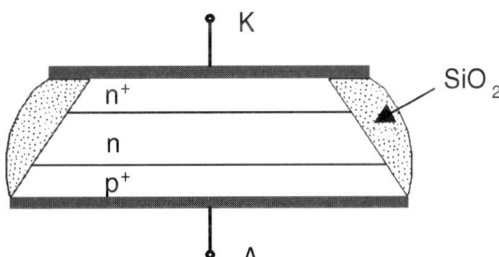

**Figure 5.10** A p–i–n diode with an overhang of the cathode contact

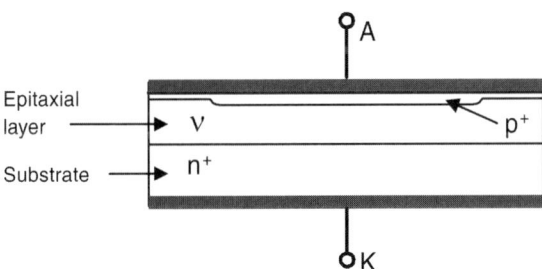

**Figure 5.11** A p–i–n diode with an increased base thickness in the surface region

type of conductivity with a relatively low concentration gradient between the lightly doped layer and the n$^+$-emitter [5.8]. However, the use of negative bevelling, or a planar junction with diffused guard rings or with a field plate, as discussed in Section 3.2.4, can be simpler solutions.

## 5.3 Transient Processes in Power Diodes

When the rate of change of current and voltage in a power diode are relatively slow, their instantaneous values are well described by the steady-state current–voltage characteristic. However, as is shown in Section 2.1.6, when either the voltage or the current changes abruptly, the carrier distributions in the device may be very different from those prevailing under steady-state conditions. Important dynamic processes during the transitions between the on-state and the off-state lead to appreciable differences between the steady-state characteristics and the dynamic characteristics. A theoretical solution of the equations describing the transient processes requires the distribution of excess carriers at any given time in each layer of the semiconductor structure to be determined. The continuity equation must be solved with the correct initial conditions and the proper boundary conditions, using the methods described in Section 2.1.6 and the appendix.

### 5.3.1 The Transition from Reverse Bias to Forward Bias

The most basic method of analysing transient phenomena in diodes is to assume the diode is subject to an abrupt change of current or voltage. In power semiconductor device applications, changes of current are generally the more important. Following equation (5.1), the forward voltage drop of a power diode under time-varying conditions may be expressed as

$$v_F(t) = v_P(t) + v_N(t) + v_I(t)$$
$$= v_P(t) + v_N(t) + v_{Ohm}(t) + v_{Mob}(t) \tag{5.22}$$

where $v_P$ and $v_N$ are the voltages across the p$^+$–i and n$^+$–i junctions, and $v_I$ is that dropped across the middle part of the diode. This last component is further split into the two voltages shown on the right-hand side of (5.17). The first of these terms, $v_{Ohm}$, is representative of the instantaneous carrier distribution, while the second, $v_{Mob}$, arises out of the difference in the electron and hole mobilities when there is a carrier concentration gradient.

If the forward current increases from zero, with a rate of rise of $di_F/dt$, excess carriers are initially injected into the regions closest to the p$^+$–i and n$^+$–i junctions. From there they diffuse into the centre of the middle region, as illustrated in Figure 5.12. Initially, the increasing current causes an increasing voltage to be dropped across the resistance of the lightly doped region. As the number of excess carriers in the middle region of the diode increases, its resistivity diminishes, hence its resistance diminishes. Thus, the forward voltage dropped across the diode normally rises to a maximum, after which it decreases with time to a value that corresponds to the

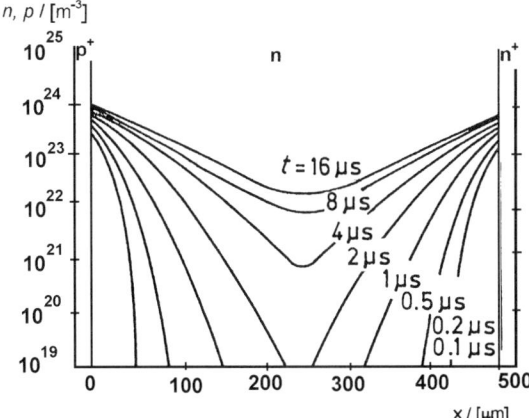

**Figure 5.12** Excess carrier concentration profiles during the turn-on process in a p–i–n diode

steady-state value of $I_{FM}$. Typical current and voltage waveforms are shown in Figure 5.13. The maximum voltage $V_{FM}$ increases with increasing $di_F/dt$. Its value depends on how high the current has risen before the conductivity modulation of the middle region is fully effective. This takes a time of the order of $w^2/8D_a$, which is typically between 0.1 and 10 μs, for values of $w$ between 50 and 500 μm. When the current rise time is less than $w^2/8D_a$, the ratio of $V_{FM}$ to $V_F$ increases. However, attempts to measure $V_{FM}$ accurately are complicated by very small amounts of

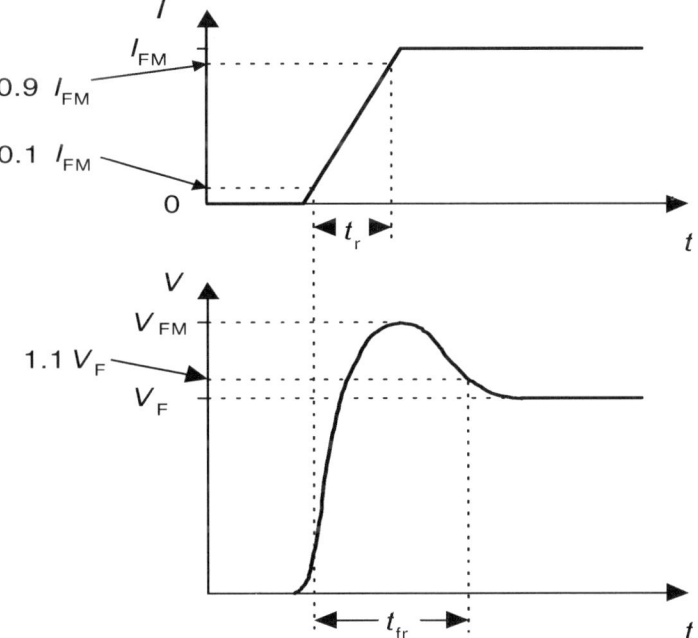

**Figure 5.13** Variation of the forward current and the forward voltage drop during the turn-on transient

packaging inductance, with the result that it is often thought to be greater than it really is.

For practical purposes, the speed of transition from the reverse state to the forward state is normally measured by the parameter $t_{fr}$, as defined in Figure 5.13. It is the interval from the time when the current reaches 10% of its final value to the moment the forward voltage decreases to a level 10% above its steady-state value. When the rate of rise of current is not constant, its rise time $t_r$ may be specified as the time taken for the forward current to increase from 10% to 90% of its final value.

Transient phenomena may result in a significant power dissipation, especially in high voltage diodes operating at high frequencies. The overshoot of forward voltage that occurs at turn-on when the rate of rise of the forward current is high is particularly undesirable in a freewheeling diode or one used in a snubber circuit (Section 5.5).

### 5.3.2 The Transition from Forward Bias to Reverse Bias

In many power electronic circuits, a diode is called upon to switch quickly from the forward conducting state to the blocking state. Usually a semiconductor switch is turned on, effectively reconfiguring the circuit and causing the diode to change state. At the beginning of the turn-off process, the middle region of the diode between the $n^+$- and $p^+$-emitters is assumed to be flooded with excess carriers. The exact carrier distribution is a function of the forward current density, as described in Section 5.1. Initially, the high excess carrier concentration in the vicinity of the p–n junction prevents the establishment of a space charge region, so the voltage across the diode remains small and, in the first part of the transition process, the current falls to zero and reverses at a rate determined by the external circuit.

Only when sufficient charge has recombined, or has diffused out as reverse current, do the carrier concentrations at the p–n junction reach the levels of thermodynamic equilibrium, permitting the formation of a space charge region. At this point, the reverse current starts to fall in a way that is controlled by the diffusion and recombination processes in the base region. Now it is the voltage across the diode, rather than the current, that is determined by the external circuit. In a circuit in which there is significant inductance in the diode leads, the change in $di/dt$ at the peak of the reverse current can give rise to a large spike of reverse voltage.

The hard-switching converters shown in Figures 4.14 to 4.16 are examples of power electronic circuits that cause such a change of state in the diodes. The accompanying transients produce losses that have a significant effect on the maximum operating frequency. They can be emulated in the circuit shown in Figure 5.14(a), which may be used as a diode test circuit. It operates in the following manner. With S1 in the on-state, current is built up in the inductance $L$. When the desired value is reached, S1 is turned off. The flux in the inductance attempts to collapse, causing the voltage at diode terminal K to fall rapidly and the diode to become forward biased. The inductor current is allowed to circulate through the diode for long enough to establish steady-state conditions, at which point S1 is turned on again. This action attempts to reverse bias the diode but, as it is flooded

## 5.3 TRANSIENT PROCESSES IN POWER DIODES

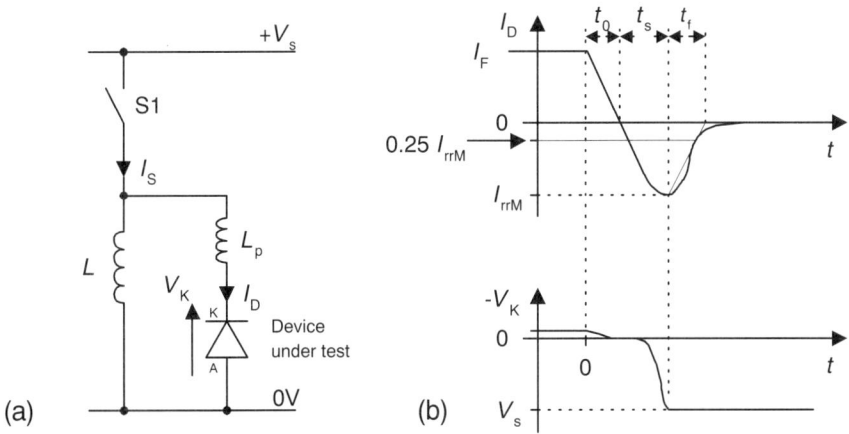

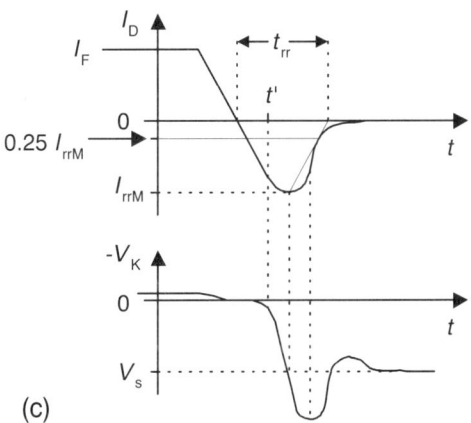

**Figure 5.14** Reverse recovery waveforms: (a) circuit used to emulate diode switching; (b) with the fall of the diode current controlled by the external switch; (c) with the rate of change of current controlled by the diode circuit inductance

with carriers, a short circuit is created between the supply rails. The resulting short circuit current is limited in one of two ways.

In the first, the turn-on of S1 is controlled so that the current rises gradually, initially taking over the inductor current and then drawing reverse current out of the diode. As a space charge layer is established in the diode, the reverse voltage settles at the supply voltage with no significant overshoot, as shown in Figure 5.14(b). The reverse diode current adds to the total current carried by S1 and causes a transient peak.

Alternatively, if S1 is turned on abruptly, the rise of the short circuit current is limited by inductance. In some circuits this may be introduced deliberately, but more commonly it is parasitic. In Figure 5.14(a) it is represented as a lumped inductance $L_p$ in the cathode lead of the diode. With S1 turned on and the diode still conducting in the reverse direction, $L_p$ supports the supply voltage $V_s$. The initial rate of fall of the forward diode current and the later rate of rise of the reverse current is $(-di/dt) = (V_s/L_p)$, as shown in Figure 5.14(c).

Once a space charge layer forms, the reverse current is determined by the rate at which excess carriers continue to diffuse out from the base region of the diode. The reverse voltage $V_R$ starts to rise and the space charge region expands. While the junction voltage is rising, the expansion of the space charge layer adds to the carrier diffusion current and a capacitive component adds further to the total current. When the reverse current reaches its maximum value $I_{rrM}$, the reverse voltage is $V_R = V_s$. The reverse current then falls to zero in a manner that reflects the distribution of any remaining carriers. The further change in $(di/dt)$ induces an overshoot of the reverse voltage as the energy stored in $L_p$ is transferred into the diode.

Practical circuits usually operate somewhere between these two extremes. There is always some parasitic inductance and therefore some reverse voltage overshoot. This is undesirable, so the inductance is minimised by good circuit design. However, it may still be necessary to reduce the rate of rise of current in S1 and the equal rate of fall of current in the diode. In practice the circuit layout, the recovery characteristics of the diode and the turn-on characteristics of the switch have to be harmonised to optimise performance.

The evolution of the excess carrier concentration distribution during the reverse recovery period is shown in Figure 5.15. The process can be analysed using the methods described in the appendix. However, the initial condition is given by (5.9) and the boundary conditions require the carrier concentration gradients at $x=0$ and $x=w$ to satisfy (5.23). As a result, analytical solutions become quite complicated [5.3]. An easier approach is offered by the stored charge approximation described in Section 1.5.

For purposes of analysis, we simplify the diode current waveform by assuming that $i(t)$ decreases linearly from $I_F$ to some negative value:

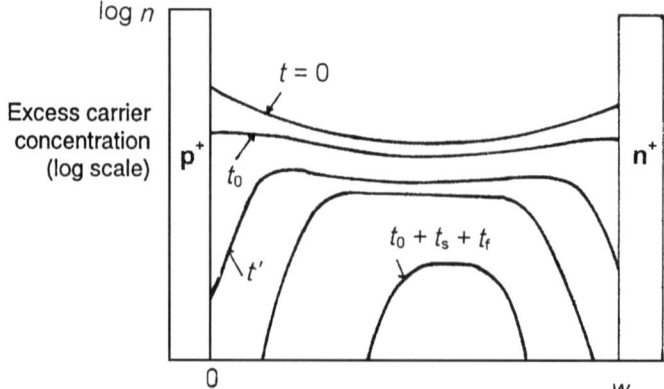

**Figure 5.15** The decay of the excess carriers during the reverse recovery period

## 5.3 TRANSIENT PROCESSES IN POWER DIODES

$$i(t) = I_F \left(1 - \frac{t}{t_0}\right) \tag{5.23}$$

where $t_0$ is the time taken for the current to fall to zero.

The subsequent reverse recovery time $t_{rr}$ is made up of two periods. The first period, $t_s$, is the interval from the moment the current crosses zero to the time that the peak reverse current $I_{rrM}$ is reached. The second period, $t_f$, characterises the fall of the reverse current. It is defined by drawing a straight line from the peak reverse current through the point when $i(t) = 0.25 I_{rrM}$ to intersect the time axis. When the reverse current decreases exponentially with a time constant $\tau_{eff}$, then $t_f = 1.85 \tau_{eff}$. Another definition that has been used is the time taken for the reverse current to fall to 10% of its peak value. For an exponential decay this definition gives $t_f = 2.3 \tau_{eff}$. However, the exact shape of this waveform does depend on the detailed physical construction of the diode; it shows quite wide variations among the different types of diode, and sometimes with the circuit voltage relative to the device rating.

The rate of change of the excess mobile carrier charge in the volume of the diode structure can be written as

$$\frac{dQ}{dt} = i(t) - \frac{Q}{\tau_H^*} \tag{5.24}$$

The time constant $\tau_H^*$ is the effective time constant of (1.79) under high injection conditions. At the beginning of the recovery process, with a constant forward current $I_F$ flowing through the diode, $dQ/dt = 0$ and the excess carrier charge that will have accumulated during the forward conduction period to support $I_F$ is therefore

$$Q(0) = I_F \tau_H^* \tag{5.25}$$

It is necessary to solve (5.24) with $i(t)$ given by (5.23) and with the initial condition given by (5.25). As long as $\tau_H^*$ remains constant, the solution is

$$Q(t) = I_F \tau_H^* \left\{ \frac{\tau_H^*}{t_0} \left[ 1 - \exp\left(-\frac{t}{\tau_H^*}\right) \right] + 1 - \frac{t}{t_0} \right\} \tag{5.26}$$

As the current increases in the reverse direction, the carriers diffuse out of the central region into the $n^+$- and $p^+$-layers. This continues until $t = t'$, as shown in Figure 5.15. This is the time needed for the excess carrier concentrations at the p–n junction to decrease to levels comparable to those of thermodynamic equilibrium, at which point a space charge region starts to form. Up to then the voltage across the diode remains small and the current is determined by the parameters of the external circuit.

Provided that a significant concentration of mobile carriers remains elsewhere in the diode base region, the reverse current comes gradually to a maximum then decays away as the residual carriers recombine and diffuse out. The reverse voltage across the diode passes through its steady-state value $V_s$ as the current reaches its maximum value $I_{rrM}$. It continues rising to a peak whose magnitude $V_{rrM}$ is governed by the rate of fall of the reverse diode current and hence by the rate at which the remaining carriers are swept out from the base region.

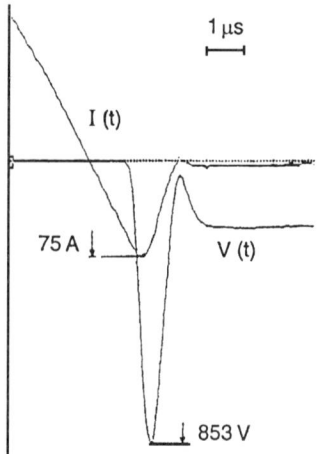

**Figure 5.16** Reverse recovery waveforms of a hard-recovery power diode operating in a rectifying circuit with series inductance

The magnitude of the overvoltage can be expressed as

$$V_{rrM} = V_R \left(1 + \frac{(-di_F/dt)}{di_R/dt}\right) \approx V_R \left(1 + \frac{t_s}{t_f}\right) \qquad (5.27)$$

In the p–ν–n diode structure there is a tendency for the carrier concentration at the p–ν junction to remain higher than elsewhere during the recovery period. Then, once the space charge layer starts to form, it expands rapidly across the ν-layer, sweeping out any residual carriers. The reverse current then falls very abruptly, causing a large spike in the reverse voltage, as shown in Figure 5.16.

A diode in which the rate of decrease of the reverse current is very high, i.e. $t_f \to 0$, is said to 'snap off' or to exhibit hard recovery. The resulting high overvoltage can cause detrimental oscillations in the circuit and give rise to electromagnetic interference, unless an external snubber capacitance is provided. On the other hand, a more gradual decrease of the reverse recovery current (soft recovery) is obtained when the base region is relatively wide and there is a high concentration of excess carriers still present at the time $t'$. The *quality* of the diode recovery is commonly described in terms of the *softness factor* $t_f/t_s$. (In data sheets the symbols $t_b$ and $t_a$ are sometimes used for $t_f$ and $t_s$, respectively.)

This reverse recovery phase of diode operation has many important implications for the circuit designer. The reverse recovery waveforms and, in particular, the free charge that has to be removed before the diode can regain its high impedance state are each subject to many parameters. These include the previous forward current and its rate of fall, the circuit conditions (especially the series inductance) and the carrier lifetime profile within the diode base region.

Further insight can be gained by pursuing the charge-control analysis into the recovery period. Let us assume that the diode forward current falls linearly to zero, reverses, and then the reverse current rises linearly to its peak value $I_{rrM}$ at time $t_s$.

## 5.3 TRANSIENT PROCESSES IN POWER DIODES

From this point, the reverse current can be considered, as a first approximation, to decay exponentially with a time constant $\tau_{\text{eff}}$. That is, for $t > t_0 + t_s$,

$$i(t) = -I_{\text{rrM}} \exp\left(\frac{t_0 + t_s - t}{\tau_{\text{eff}}}\right) \quad (5.28)$$

Substituting (5.28) into (5.24) indicates that the stored charge $Q$ also decays exponentially and

$$Q(t) = Q_1 \exp\left(\frac{t_0 + t_s - t}{\tau_{\text{eff}}}\right) \quad (5.29)$$

where $Q_1$ is the excess carrier charge remaining in the middle part of the diode at the time $t = t_0 + t_s$. From (5.26) this is

$$Q_1 = Q(t_0 + t_s) = I_F \tau_H^* \left\{ \frac{\tau_H^*}{t_0}\left[1 - \exp\left(-\frac{t_0 + t_s}{\tau_H^*}\right)\right] - \frac{t_s}{t_0}\right\} \quad (5.30)$$

It can be seen that $\tau_{\text{eff}}$ is an effective carrier lifetime that now includes the effects of carrier recombination and the loss of carriers by diffusion to the boundaries. From (5.24),

$$\frac{1}{\tau_{\text{eff}}} = \frac{I_{\text{rrM}}}{Q_1} + \frac{1}{\tau_H^*} \quad (5.31)$$

A more detailed analysis [5.9], based on the continuity equation (1.77), shows that

$$\frac{I_{\text{rrM}}}{Q_1} = C \frac{\pi^2 D_a}{w_R^2} \quad (5.32)$$

where $w_R$ is the thickness of the region still flooded with carriers; the constant $C$ has a value rather greater than unity and varies slowly with the base thickness and the applied voltage. Figure 5.17 shows the experimentally determined dependence of $\tau_{\text{eff}}$ on $w_R$ and $\tau$. Note that in practice $w_R$ varies with time during the later stages of turn-off, so $\tau_{\text{eff}}$ can no longer be considered constant:

$$\tau_{\text{eff}} = \frac{w_R^2 \tau_H^*}{\tau_H^* C \pi^2 D_a + w_R^2} \quad (5.33)$$

Thus, the exponential decay of current and charge that we have assumed no longer applies.

The reverse recovery charge $Q_{\text{rr}}$ is defined as the time integral of the reverse current during the reverse recovery period; that is,

$$Q_{\text{rr}} = \int_0^{t_{\text{rr}}} i(t)\, dt \quad (5.34)$$

Approximating the waveform by straight line sections, the reverse recovery charge can be expressed as

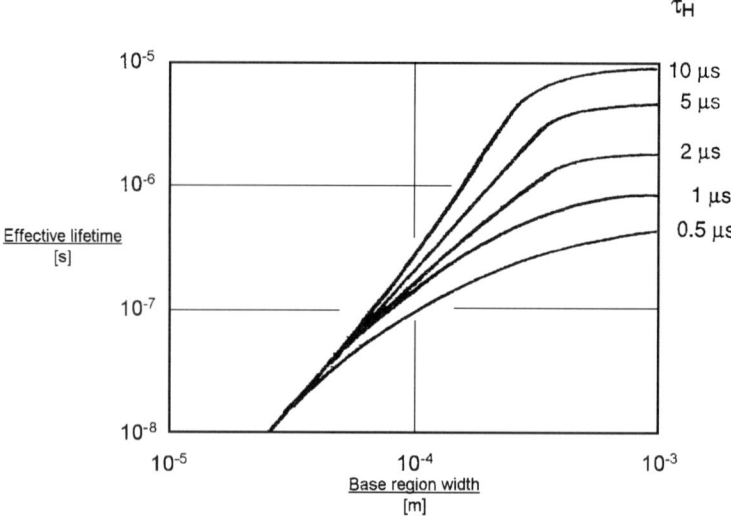

**Figure 5.17** The dependence of the effective carrier lifetime on the width of the base region and the recombination time constant of a typical diode

$$Q_{rr} = \frac{1}{2} I_{rrM} t_{rr} = \frac{1}{2} t_s \left|\frac{di}{dt}\right| t_{rr} \tag{5.35}$$

Since $t_s$ and $t_f$ are each proportional to the carrier lifetime, the reverse recovery charge varies approximately with the square of the carrier lifetime ($Q_{rr} \propto \tau_H^{*2}$). The reverse recovery charge is found to increase with the rate of decrease of forward current $di_F/dt$, as shown in Figure 5.18 for a typical power diode.

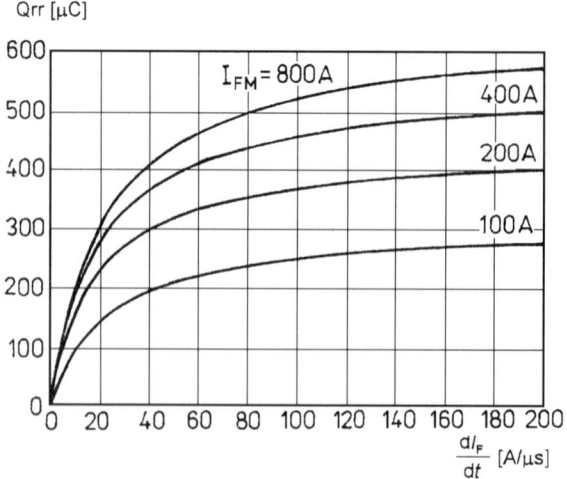

**Figure 5.18** An example of the dependence of the reverse recovery charge, $Q_{rr}$, on the forward current and its rate of decay. Reproduced by permission of Polovodiče, a. s., Prague

From the relationships incorporated in (5.30), (5.33) and (5.35), it follows that the reverse recovery time is long and the reverse recovery charge is large when the excess carrier lifetime under high injection conditions is long. A short reverse recovery time is only obtained by having a short carrier lifetime. This can be achieved by the techniques discussed in Section 3.3. Reducing the lifetime of the carriers can result in a deterioration of the forward current–voltage characteristic and, in the case of high voltage diodes, it is necessary to reduce the distance between the $n^+$- and $p^+$-emitters, in order to obtain reasonably low values for the on-state voltage drop $V_F$. For very fast diodes it is common to use p–i–n structures such as those described in Section 5.2, but with very abrupt junctions.

It is evident from (5.33) that the effective carrier lifetime $\tau_{\text{eff}}$ becomes very small when $w_R \to 0$, and the reverse current then decreases very rapidly. This is most likely to occur when the space charge region occupies the whole of the middle region between the $p^+$- and $n^+$-emitters, as is the case with the compressed-field structure of p–i–n diodes.

Fast diodes with soft recovery characteristics would be desirable for many applications. Although it is difficult to produce a short reverse recovery time simultaneously with a low forward voltage drop and a long effective carrier lifetime, the construction and technology of the diode may be modified in several possible ways that lead to an increase in the softness factor $t_f/t_s$.

Simulations of the reverse recovery process have shown [5.10, 5.11] that the turn-off characteristics can be influenced by the use of a graduated concentration of recombination centres. Figure 5.19 shows a comparison of the reverse recovery current waveforms of a range of diodes which have the same general structure and the same average concentration of recombination centres but which differ in their distribution. By shortening the carrier lifetime in the region of the $p^+$–i junction, the excess carrier concentration is reduced there. This shortens the storage time $t_s$ because less time is needed for the electron concentration to fall below its equilibrium level, so a space charge layer can start to form. However, the techniques that have to

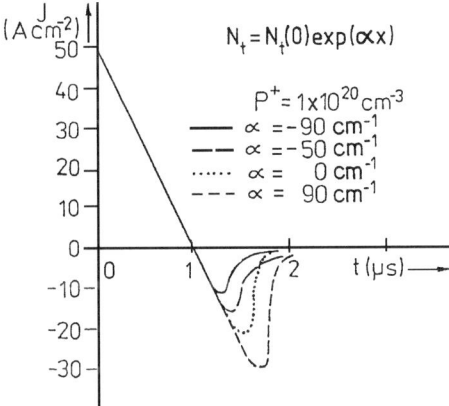

**Figure 5.19** Simulation results showing the influence of a non-uniform distribution of recombination centres on the reverse recovery waveform of a diode. The concentration of trapping centres is assumed to decrease or increase exponentially with distance away from the $p^+$ layer surface, with the characteristic distances shown

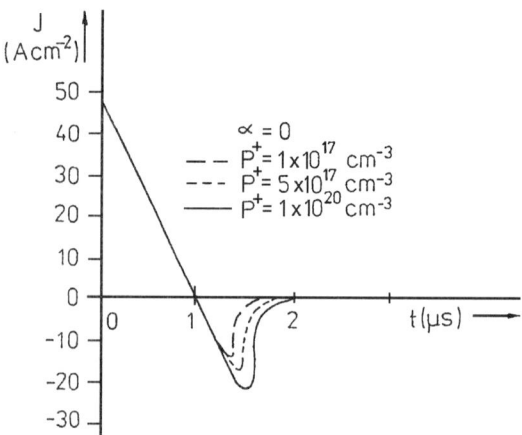

**Figure 5.20** Simulation results showing the influence of the p$^+$-emitter concentration on the reverse recovery waveform of a power diode. The concentration of trapping levels is uniform

be employed to create a carrier lifetime gradient are complicated. In particular, the required gradient cannot be achieved easily with the gold or platinum doping normally used to reduce carrier lifetime because these metals diffuse so rapidly through the silicon that they are normally uniformly distributed across the base region. Likewise, electron bombardment also creates recombination centres uniformly throughout the device structure.

Iridium has a much lower diffusion coefficient than Au or Pt and is also an impurity that generates recombination centres. It may therefore be suitable for setting up a concentration gradient of trapping levels, although its effectiveness for lifetime control is limited to temperatures below 120 °C. Alternatively, irradiation with ionised hydrogen or helium atoms (protons or alpha particles) of suitable energy (about 20 MeV) enables a narrow layer of enhanced recombination to be created in the region of the p$^+$–i junction, as described in Section 3.3.2. Although the equipment costs are high, this technique offers the greatest flexibility in controlling the carrier lifetime profile in power devices.

Another technique that has been considered is to reduce the acceptor concentration in the p$^+$-emitter to something less than $10^{23}$ m$^{-3}$ ($10^{17}$ cm$^{-3}$). The simulation results shown in Figure 5.20 illustrate how this can improve the reverse recovery process. It reduces the excess carrier concentration at the p$^+$–n junction, but it also reduces the injection efficiency, thereby causing an increase in the forward voltage drop at higher current densities.

A structure that attempts to reconcile these conflicting requirements is the self-adjusting p$^+$-emitter efficiency diode (SPEED) [5.12] shown in Figure 5.21. Essentially, this is an integrated structure consisting of p–n–n$^+$ and p$^+$–n–n$^+$ diodes connected in parallel. A network of p$^+$-emitters of diameter 50–80 μm occupy about one-third of the active area of the device. As shown in Figure 5.4, at low current densities the forward voltage drop in the p–n–n$^+$ part is lower than in the p$^+$–n–n$^+$ part. This reduces the storage time $t_s$ by about 40% with a consequent 'softening' of the reverse recovery process. At high current densities the p$^+$–n–n$^+$

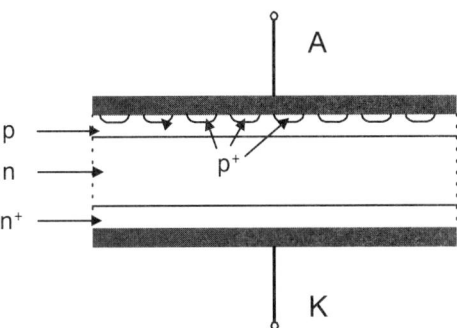

**Figure 5.21** A combined p–n–n⁺ and p⁺–n–n⁺ (SPEED) structure

part of the structure takes over the conduction of current by virtue of its lower voltage drop. By this means it is possible for the diode to have good static properties and good dynamic properties. A further improvement in the dynamic characteristics of the SPEED structure can be obtained by creating a concentration gradient of recombination centres in the base regions of the diode so that the carrier lifetime at the p–n junction is shorter than the lifetime at the n⁺–n junction [5.13].

Fast, high voltage power diodes are very important in modern power electronic applications in which high frequency operation is often advantageous. There is therefore a continual need to develop new devices and to improve the technology of power diodes. This is reflected in the continuing research and development effort that is being applied to these problems.

## 5.4 Power Schottky Diodes

A metal–semiconductor contact with a Schottky barrier, as described in Section 2.3.1, can be used to rectify relatively high currents. Schottky contacts are usually fabricated by evaporating a suitable metal onto the surface of an n–n⁺ epitaxial structure, as shown in Figure 3.3.

As discussed in Section 2.3.1, charge transport in the Schottky diode is by majority carriers. Phenomena connected with minority carrier injection, and the extraction and recombination of excess carriers, are therefore not present during the turn-on and turn-off processes. Because the relaxation time of the majority carriers is less than $10^{-13}$ s, the time constants associated with the turn-on and turn-off processes are limited only by the time required to charge and discharge the capacitance of the space charge region at the metal–semiconductor contact. As a result, Schottky diodes can be used advantageously at high frequencies.

The Schottky diode can be modelled as an ideal rectifying metal–semiconductor contact in series with a resistance. All the major parameters of the diode are functions of the resistivity of the n-type layer $\rho_N$. These are the maximum breakdown voltage ($V_{BR} \propto \rho_N^{3/4}$), the series resistance ($R_S \propto \rho_N$) and the barrier capacitance ($C \propto \rho_N^{-1/2}$). It follows that, with increasing breakdown voltage, the forward characteristics deteriorate and the time constant $CR_S$ gets longer. The

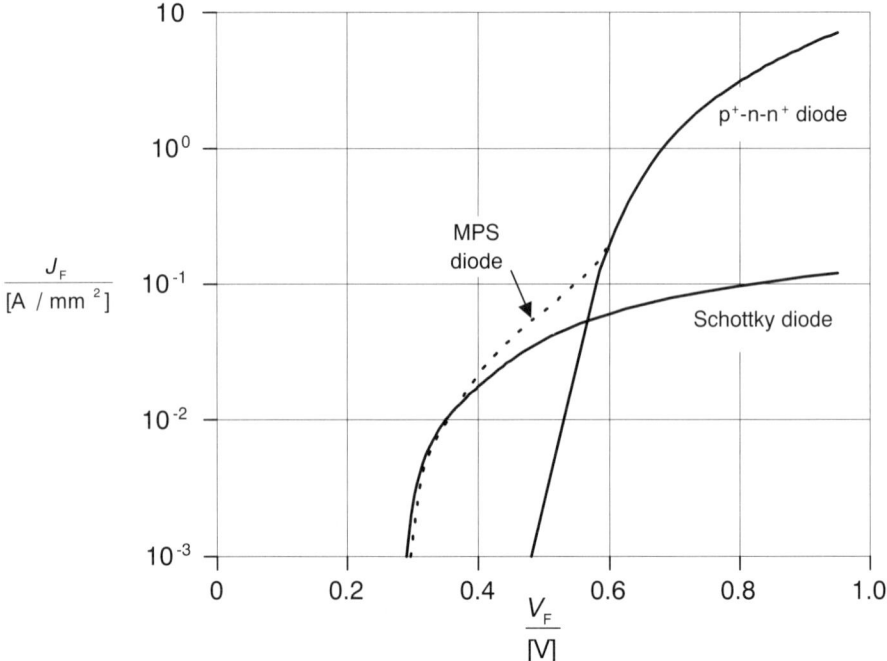

**Figure 5.22** A comparison of the forward characteristics of a Schottky diode, a $p^+$–n–$n^+$ diode and the MPS-diode

threshold voltage depends on the metal used to form the rectifying contact, but because of the relatively high saturation current density $J_s$, it is always lower than for a p–n junction. Because the n-base resistivity is not modulated by excess carrier injection, as it is for the p–n junction diode, the forward voltage drop of a Schottky diode is lower than for a p–n junction diode at low current densities and higher at high current densities. This is illustrated in Figure 5.22. It is possible to raise the height of the Schottky barrier with a thin implant at the semiconductor surface. Although this may increase the forward voltage drop, it also reduces the reverse saturation current.

The reverse characteristics of Schottky diodes are determined by the spreading of the space charge region into n-type semiconductor in the same way as in an abrupt p–n junction. The reverse current density $J_s$ is given by (2.68). The breakdown voltage of a plane Schottky barrier is described by (2.39). At a Schottky contact of finite area the breakdown voltage $V_{R(BR)}$ is limited to less than 100 V by the field concentration caused by the curvature of the space charge region at the edge of the contact area. This can be improved by using one of the edge termination techniques described in Section 3.4. The preferred method is the diffused $p^+$ guard ring shown in Figure 5.23. Although this forms a p–n junction diode in parallel with the Schottky diode, the lower forward voltage drop of the Schottky junction ensures that the p–n junction is only forward biased sufficiently to inject carriers into the active region under surge conditions.

## 5.4 POWER SCHOTTKY DIODES

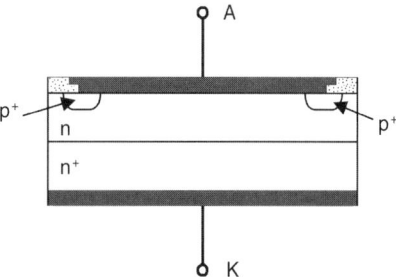

**Figure 5.23** A Schottky diode with a diffused guard-ring

When the reverse breakdown voltage of the guard-ring diode is made low, it can protect the Schottky diode against reverse voltage breakdown. The junction barrier Schottky (JBS) structure [5.14], shown in Figure 5.24, is an advanced device that develops this principle. A fast power diode is formed by the parallel integration of an M–n–n$^+$ Schottky diode and a p$^+$–n–n$^+$ junction diode. The geometric layout is

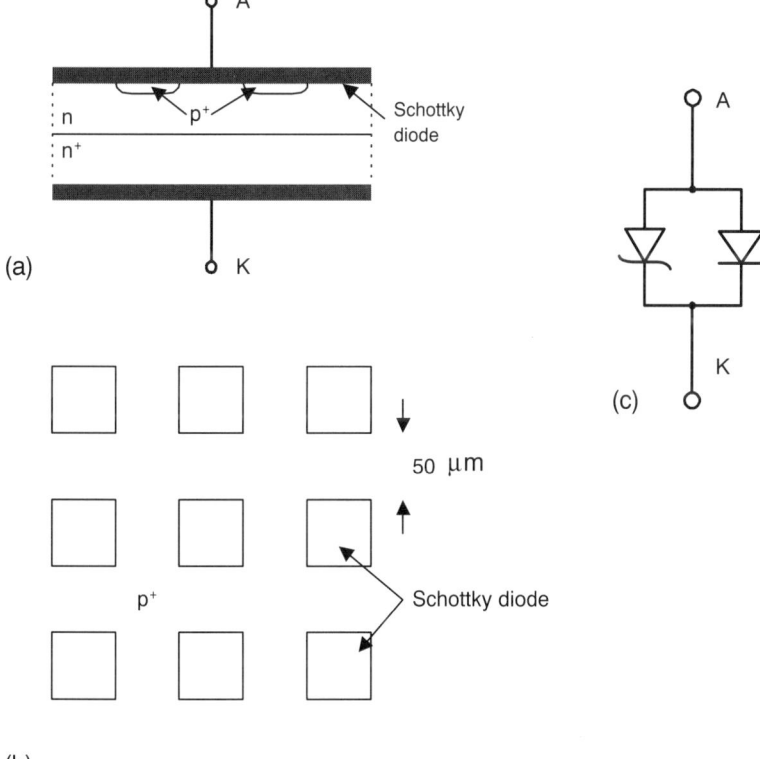

**Figure 5.24** The integration of a Schottky diode and p$^+$–n–n$^+$ diode (the JBS-diode structure): (a) cross-section through the active region; (b) plan view of structure; (c) equivalent circuit

such that the $p^+$–n–$n^+$ diode takes the form of a grid. Under reverse bias, the space charge region expands to shield the Schottky junction from the applied voltage by JFET action. Reverse breakdown voltages greater than 200 V can be achieved.

When the $p^+$–n junction is designed to allow the diode to be forward biased under surge conditions, the device is known either as a Schottky p–i–n (SPIN) diode, or as a merged power Schottky (MPS) diode [5.15]. The current–voltage characteristic is shown by the dashed line in Figure 5.22.

## 5.5 Silicon Power Diode Applications

Although the power diode is the simplest of the range of semiconductor devices, it is the most widely used. Virtually every power electronics circuit requires one or more diodes if it is to function at all, and the performance of the diodes is often critical to its success. Improvements to the other active devices, especially in their speed of switching, put increased demands on the functioning of the associated diodes. The previous sections of this chapter have shown that the detailed theory of the operation of a power diode is exceedingly complex and that many parameters have to be considered in seeking an optimum design for any particular application.

The discussion of diode applications may be separated into those where the rapid transition between on- and off-states is important and those associated with low frequency, usually power frequency (50 or 60 Hz), circuits.

### 5.5.1 Power Frequency Applications

In most power frequency applications the choice of diode is dominated by the need to minimise the forward on-state voltage drop. Carrier lifetime is maximised by the use of diffused or sometimes alloyed devices. Defects are kept to a minimum by the careful choice of starting material and by avoiding rapid temperature changes during processing. As with all components, cost and reliability are important. This is particularly true for rectifier diodes, where many are needed in the 6- or 12-phase circuits used to reduce ripple and improve harmonic distortion. Some examples of typical bridge rectifier circuits are shown in Figures 4.1 to 4.3. The load may include smoothing components.

During the start-up of a smoothed rectifier circuit, the inrush current can be limited by the insertion of a temperature-dependent resistance or a contactor/resistance arrangement.

Applications requiring very high d.c. power, and hence the largest devices that can be produced, include chemical processing (including smelting and electrolytic deposition), battery charging, welding and motor drives. An application at much lower power levels that is still very demanding is the automotive alternator. This may require the diode to operate at frequencies up to 1 kHz in a high temperature environment. Very high reliability and very low cost are essential.

The packaging of power frequency diodes is important because they have to withstand power cycling, with its attendant thermal and mechanical stresses, and have a good surge current capability. Particular attention must be paid to the

### 5.5.2 Fast-Switching Applications

A low on-state forward voltage drop is also important in high frequency switching circuits, but a compromise has to be reached between this and the need to minimise the reverse recovery time and the recovered stored charge. A soft recovery characteristic is also important, in order to minimise electromagnetic compatibility problems, and all these features should be maintained over the full working range of the device specification. Examples of such application include switched-mode power supplies, uninterruptible power supplies and a.c. motor drives.

## Summary

A power diode normally has a $p^+p\nu n$ structure. Under forward bias, the injection of carriers from the heavily doped contact regions modulates the conductivity of the inner layers. This enables wide, high resistivity layers to be used to obtain high blocking voltages without incurring large forward voltage drops. However, the on-state voltage does increase rapidly if the width of the inner layers becomes large compared to the carrier ambipolar diffusion length. Surge currents are limited because this parameter reduces at high current densities as a result of carrier–carrier scattering and Auger recombination. The forward current–voltage characteristic can be expressed as

$$V_F = K_0 + K_1 \ln J + K_2 J^m \tag{5.18}$$

The effect of temperature is shown in Figure 5.6. During rapid turn-on the forward voltage normally rises to a peak value before returning to the steady-state value. During turn-off the current normally reverses and, after reaching a maximum, falls to zero at a rate governed by the distribution of residual carriers in the diode. When the fall of the reverse current is very rapid, the recovery is said to be hard and the diode is said to snap off. Together with the parasitic inductance in the diode circuit, this can cause large reverse voltage spikes across the diode and it can also cause circuit oscillations. Various design techniques have been proposed to achieve the goal of a device with a fast, soft turn-off characteristic.

Power Schottky-barrier diodes offer a lower on-state volt-drop than p–n junction diodes at lower current density. The reverse blocking capability is limited to about 60 V, although mixed Schottky/p–n junction designs can raise this value.

Power diodes are used in virtually every power electronics circuit and are often critical to performance. Many parameters have to be considered in seeking an optimum design for any particular application.

# References

[5.1] Howard, N. R. and Johnson, G. W. $p^+$–i–$n^+$ Silicon Diodes at High Forward Current Densities, *Solid State Electronics*, **8**, 275–84 (1965).

[5.2] Otsuka, M. The Forward Characteristics of Thyristors, *Proc. IEEE*, **55**, 1400–8 (1967).

[5.3] Benda, H. and Spenke, E. Reverse Recovery Processes in Silicon Power Rectifiers, *Proc. IEEE*, **55**, 1331–54 (1967).

[5.4] Homola, J. Thyristor on-state (Propustny stav tyristoru) PhD Thesis, Charles University in Prague (1979).

[5.5] Pathak, V. K. and Gowar, J. Numerical Solutions for Surface Electric Field Distributions in Avalanching p–i–n Power Diodes, *Proc. IEE*, **I-130**, 17–23 (1983).

[5.6] Ghatol, A. A. and Sundarsingh, V. P. Breakdown in Concentration Profile Diodes, *International Journal of Electronics*, **55**, 639–46 (1983).

[5.7] Benda, V., Zamastil, J. and Chytil, M. Czech Patent AO 244077 (1985).

[5.8] Homola, J., Benda, V. and Pína, B. Czech Patent AO 149842 (1972).

[5.9] Benda, V., Kang, C. H. and Klabacka, E. A Note on Charge Analysis Approach of Power Diode Reverse Recovery, *Proc. 2nd International Seminar on Power Semiconductors, ISPS'94*, 53–60 (Prague, Aug./Sept. 1994).

[5.10] Assalit, H. B., Ericson, H. B. and Wu, S. J. High Power Controlled Soft Recovery Diode Design and Application, *IEEE-IAS Conference Record*, 1056–61 (1979).

[5.11] Benda, V. Design Considerations for Fast Soft Reverse Recovery Diodes, *Proc. EPE'93 Conference*, 288–92 (Brighton, 1993).

[5.12] Schlangenotto, H., Serafin, J., Sawitzki, F. and Maeder, H. Improved Recovery of Fast Power Diodes with Self-Adjusting p-Emitter Efficiency, *IEEE Electron Device Letters*, **EDL-10**, 322–4 (1989).

[5.13] Benda, V., Dvorak, M. and Brabec, L. A Note on the Problem of Fast Soft Reverse Recovery Diodes, *Proc. 1st International Seminar on Power Semiconductors, ISPS'92*, 18–24 (Prague, Sept. 1992).

[5.14] Baliga, B. J. Analysis of Junction Barrier Controlled Schottky Rectifier Characteristics, *Solid State Electronics*, **28**, 1089–93 (1985).

[5.15] Hower, P. L. and Weaber, L. E. The SPIN Rectifier, a New Fast-Recovery Diode, *IEEE PESC'88 Record*, 709–17 (1988).

# 6

# BIPOLAR JUNCTION POWER TRANSISTORS

The invention of the bipolar junction transistor [6.1] and the development of the technology necessary for its mass production initiated a revolution in electronics. The transistor has become a key element in this revolution and the bipolar junction transistor structure forms the basis of several families of monolithic integrated circuits. The power bipolar junction transistor (BJT) has also been important as a switching device at frequencies greater than 1 kHz. Although it has been replaced by the insulated gate bipolar transistor in many high power applications, it retains market share in some special areas.

## 6.1 Basic Characteristics of the Bipolar Junction Transistor Structure

The basic structure of the bipolar junction transistor is shown in Figure 3.4. Its operation can be understood most easily in what is known as the *common-base* configuration. This is shown in Figure 6.1, together with a schematic illustration of the electron energy level distribution within the transistor structure in the normal or forward-active state. The emitter–base junction is forward biased, whereas the base–collector junction is reverse biased, thus setting up a thick space charge layer with a strong built-in electric field. At the forward-biased junction, electrons are injected into the p-type base region, where they diffuse towards the reverse-biased junction. Some recombine in the p-type region; the rest diffuse to the edge of the space charge region, where they are swept into the collector under the influence of the electric field.

The relationship between the emitter current $I_E$, the collector current $I_C$, and the parameters of the p-base can be found by solving the continuity equation (1.73). In a one-dimensional approximation under steady-state, low injection conditions, the electron concentration $n_p$ in the neutral regions of the base is given by

$$D_n \frac{d^2 n_p}{dx^2} - \frac{n_p - n_{p0}}{\tau_n} = 0 \qquad (6.1)$$

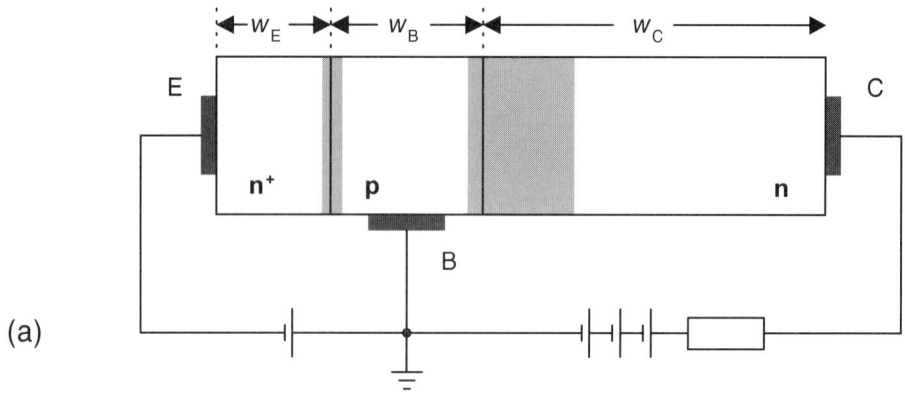

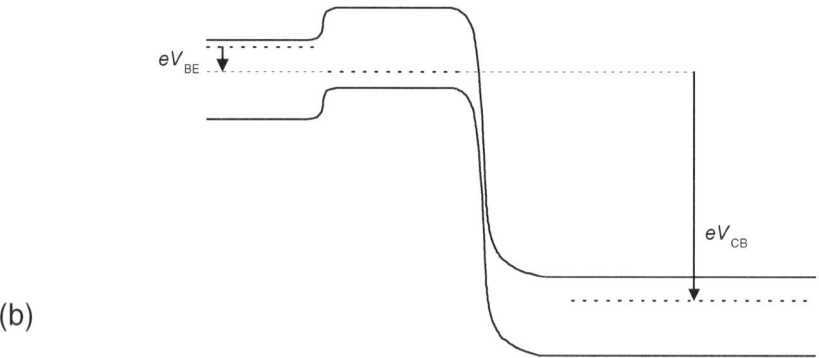

**Figure 6.1** The n–p–n bipolar junction transistor structure in the common-base configuration: (a) a schematic description; (b) energy band diagram in the forward active mode of operation

where $D_n$ is the electron diffusion coefficient and $\tau_n$ is the lifetime of the excess electrons.

We take the origin of the x-coordinate to be at the edge of the space charge region in the base, so the boundary conditions with the base–emitter junction forward biased and with the base–collector junction short circuited ($V_{BC}=0$) are

$$n_p(0) = n_{p0}\exp\left(\frac{eV_{BE}}{kT}\right) \qquad (6.2a)$$

$$n_p(w) = n_{p0} \qquad (6.2b)$$

where $w$ is the thickness of the neutral semiconductor of the base.

The solution of (6.1) then takes the form

## 6.1 BIPOLAR JUNCTION TRANSISTOR STRUCTURE

$$n_p(x) - n_{p0} = \{n_p(0) - n_{p0}\}\frac{\sinh\{(w-x)/L_n\}}{\sinh\{w/L_n\}} \tag{6.3}$$

where $L_n = \sqrt{D_n \tau_n}$ is the electron diffusion length. This distribution of the excess electrons across the base region of the transistor is shown in Figure 6.2.

Using (2.10) and (6.3), the electron current density at the emitter junction can be expressed as

$$J_n(0) = eD_n \left(\frac{dn_p}{dx}\right)_{x=0} = \frac{eD_n}{L_n}\{n_p(0) - n_{p0}\}\frac{\cosh(w/L_n)}{\sinh(w/L_n)} \tag{6.4}$$

and, on the other side of the base, at the edge of the space charge region of the collector–base junction, as

$$J_n(w) = eD_n \left(\frac{dn_p}{dx}\right)_{x=w} = \frac{eD_n}{L_n}\{n_p(0) - n_{p0}\}\frac{1}{\sinh(w/L_n)} \tag{6.5}$$

If the electron injection efficiency at the emitter is $\tilde{\gamma}_n$, as defined in (2.23a), the total emitter current density is given by $J_E = J_n(0)/\tilde{\gamma}_n$. The collector current density is made up entirely of the flow of electrons, $J_C = J_n(w)$. So it is possible to define a common-base current 'gain' as

$$\alpha_F = h_{21B} = \frac{J_C}{J_E} = \frac{\tilde{\gamma}_n J_n(w)}{J_n(0)} = \frac{\tilde{\gamma}_n}{\cosh(w/L_n)} \tag{6.6}$$

An exactly analogous expression can be deduced for a pnp transistor

$$h_{21B} = \frac{\tilde{\gamma}_p}{\cosh(w/L_p)} \tag{6.7}$$

where $\tilde{\gamma}_p$ is the hole injection efficiency across the base–emitter junction and $L_p$ is the hole diffusion length in the base. Where we wish to generalise the discussion to cover both types of transistor, the subscripts n and p are omitted.

It is evident that $h_{21B}$ is highly dependent on the ratio $w/L$, as illustrated in Figure 6.3. If we include the additional leakage current, $I_{CBO}$ that flows across the collector

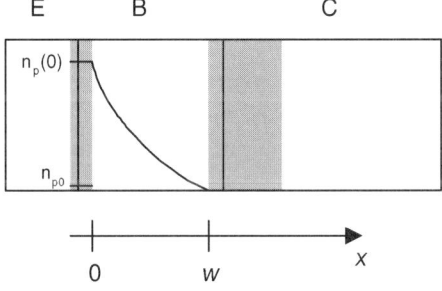

**Figure 6.2** The distribution of excess carriers across the transistor base. The shaded regions are the space-charge layers at the collector and emitter junctions

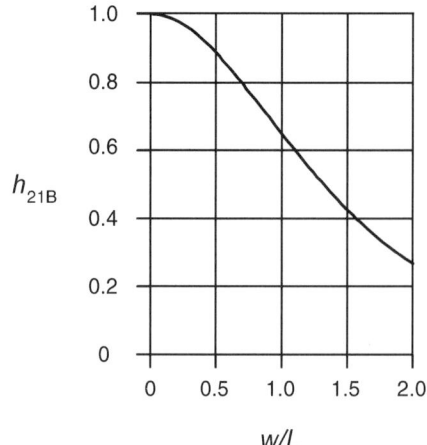

**Figure 6.3** The dependence of the current gain coefficient $h_{21\mathrm{B}}$ on the ratio of the base width to the diffusion length

junction when it is reverse biased rather than short circuited, the total collector current becomes

$$I_\mathrm{C} = h_{21\mathrm{B}}I_\mathrm{E} + I_\mathrm{CBO} \tag{6.8}$$

Thus, $h_{21\mathrm{B}}$ is really the incremental ratio of the collector and emitter currents $\partial I_\mathrm{C}/\partial I_\mathrm{E}$. Equation (6.6) indicates how a transistor should be designed if the value of $h_{21\mathrm{B}}$ is to be as high as possible: it is necessary to maximise the emitter injection efficiency and to minimise the value of $\cosh(w/L_\mathrm{n})$. To obtain a high injection efficiency, the majority carrier concentration in the emitter should be at least two orders of magnitude greater than that in the base. To minimise the value of $\cosh(w/L_\mathrm{n})$, the ratio $w/L_\mathrm{n}$ should be as small as possible. This means that either the base region has to be very thin or the excess carrier lifetime very long.

It is also important that the space charge region at the collector junction should not spread far into the base region as the collector voltage increases. This avoids the current gain becoming voltage-dependent and minimises the risk of *punch-through*, which is discussed further in Section 7.1. For power applications the breakdown voltage at the collector junction has to be high. In order to satisfy these two requirements, the concentration of active impurities in the collector region must be low, whereas their concentration in the base should be considerably higher.

As discussed in Section 1.3.2, the electron mobility in silicon (and hence the electron diffusion coefficient) can be as much as a factor of three greater than the hole mobility. Thus, for material with the same carrier lifetime, $L_\mathrm{n} = \sqrt{3}L_\mathrm{p}$. It follows from this that npn transistors have a higher current gain than pnp transistors with the same base width and carrier lifetime.

In a well-designed transistor, the emitter efficiency is almost 100% and $w$ can be approximated by the distance between the two junctions $w_\mathrm{B}$, which can be assumed to be much less than the minority carrier diffusion length $L$ in the base. Then the current gain coefficient can be expressed as

## 6.1 BIPOLAR JUNCTION TRANSISTOR STRUCTURE

$$h_{21B} = 1 - \frac{w_B^2}{2L^2} \tag{6.9}$$

The ratio $(w_B^2/2L^2)$ represents the fraction of the excess carriers that recombine in the base region. The base current is

$$I_B = I_E - I_C = I_E(1 - h_{21B}) - I_{CBO} = I_E \frac{w_B^2}{2L^2} - I_{CBO} \tag{6.10}$$

and may be considered to be the current of carriers injected from the emitter that recombine during their transport through the base region (less any leakage current across the collector–base junction).

So far we have considered the impurity concentration in the base region to be uniform. When the transistor is fabricated using diffusion technology, which is normal for power transistors, the base impurity concentration decreases from the emitter junction to the collector junction. Associated with this concentration gradient is a built-in electric field, as described in (1.82) and (1.83), that is beneficial to transistor action. In one dimension and for the base region of an npn transistor, equation (1.83) becomes

$$E = -\frac{kT}{e} \frac{1}{\{N_A(x) - N_D\}} \frac{dN_A(x)}{dx} \tag{6.11}$$

where $N_D$ is assumed to be uniform. The built-in field assists the diffusion of minority electrons from emitter to collector (the downstream direction) and inhibits it in the reverse (upstream) direction. This has the effect of increasing the downstream diffusion length to an effective value which can be expressed as

$$L_{eff} = \frac{2L_n^2}{\sqrt{4L_n^2 + (\mu_n E \tau)^2} - \mu_n E \tau} \tag{6.12}$$

where $L_n = \sqrt{D_n \tau}$ is the electron diffusion length and $\tau$ is the recombination lifetime in the base. The internal field $E$ in a diffused-base transistor can reach $10^5$ V/m, so that $L_{eff}$ can be up to five times longer than $L_n$, and $h_{21B}$ is increased according to (6.9).

A bipolar junction transistor can be operated in one of three circuit configurations, as shown in Figure 6.4, depending on which connection is common to both input and output circuits. The common-base configuration that we have used to describe the basic transistor function is not normally suitable for power electronics applications, neither is the common-collector configuration. The common-emitter configuration is generally used since it provides good current and power amplification. The common-emitter current gain coefficient $h_{21E}$ is defined as

$$h_{21E} = \frac{\partial I_C}{\partial I_B} = \frac{h_{21B}}{1 - h_{21B}} \tag{6.13}$$

It follows from (6.13) that any modification of the design or technology of a transistor that improves (increases) the common-base current gain $h_{21B}$ results in a high common-emitter current gain $h_{21E}$.

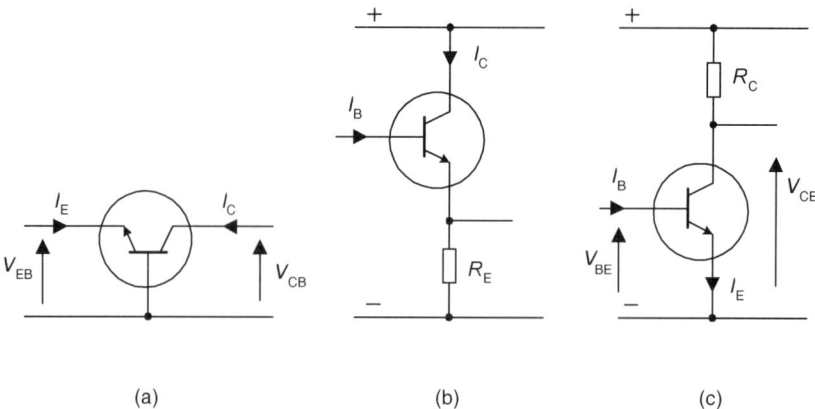

(a)             (b)             (c)

**Figure 6.4** The three possible circuit configurations in which an n–p–n transistor may be operated: (a) common base; (b) common collector; (c) common emitter

Both the gain coefficients $h_{21B}$ and $h_{21E}$ are functions of the emitter current density. As described in Section 2.1.1, at very low current densities the current crossing a p–n junction is mostly generation–recombination current and the injection efficiency is low. At higher current densities the fraction of the current that consists of injected minority carriers increases to a level very close to unity and the current gain coefficients are given by (6.7) and (6.13). At still higher current densities the concentration of carriers injected into the base affects the concentration of majorities and causes the emitter efficiency to decrease. This is discussed in Section 6.2, together with another effect that also reduces the current gain coefficients at high current densities, the base widening effect. An example of the variation of $h_{21E}$ with collector current in a typical power transistor is shown in Figure 6.5.

The current gain coefficients also increase slightly as $V_{CE}$ increases, as a result of the expansion of the collector space charge region into the base. This is known as the Early effect. It shortens the effective base width and so increases the current density

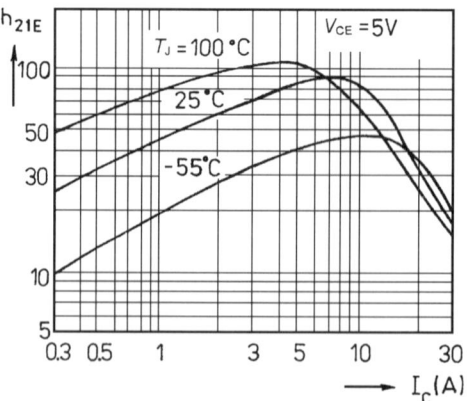

**Figure 6.5** An example of the variation of the current gain parameter, $h_{21E}$, with the collector current $I_C$. Data taken from *Development of High Power Transistor for Power Use* by M. Otsuka (Toshiba, 1975)

of electrons diffusing across the base. The total free-carrier charge in the base is reduced. The rate of recombination of carriers, and hence the base current for any given value of $V_{BE}$, are reduced in proportion.

Increased temperature has two opposing effects on the current gain coefficients. At low current densities the reduced bandgap gives an increase in the emitter injection efficiency. However, the carrier mobility is expected to increase, hence the diffusion coefficients and the diffusion lengths are also expected to increase. A reduction in temperature from 300 K to 77 K was found [6.2] to decrease the current gain by a factor greater than 10, to increase the on-state voltage drop and to reduce the storage time by a factor of 10 and the fall time by a factor of 6. The collector–emitter breakdown voltage increased slightly and emitter crowding was much more severe.

## 6.2 Basic Characteristics of Power Transistors

### 6.2.1 High Voltage Considerations

In the previous section we discussed the current gain coefficients of the transistor structure. Another important parameter, which limits the application of transistors in power electronic circuits, is the maximum available collector–emitter voltage. In order to analyse this, we have to consider the avalanche multiplication factor $M$ for the carriers crossing the collector–base junction, defined in (2.43) and (2.44). The breakdown voltage $V_{BR}$ is that of the isolated collector–base junction (i.e. with the emitter left open circuit), which we now write as $V_{(BR)CBO}$. The current crossing the collector–base junction is multiplied by $M$, so (6.8) becomes

$$I_C = M(h_{21B} I_E + I_{CBO}) \tag{6.14}$$

In the common-emitter configuration, which we now assume, with the base connection left open circuit ($I_B = 0$), the collector emitter current is given by

$$I_E = I_C = I_{CEO} = \frac{M I_{CBO}}{1 - h_{21B} M} \tag{6.15}$$

The collector–emitter breakdown voltage is defined as the voltage at which $I_{CEO}$ increases without limit as far as the device is concerned. It is written as $V_{(BR)CEO}$ and occurs when $h_{21B} M = 1$. Substituting this voltage into (2.44), for $M$, we obtain

$$M = \frac{1}{h_{21B}} = \left[ 1 - \left( \frac{V_{(BR)CEO}}{V_{(BR)CBO}} \right)^\kappa \right]^{-1} \tag{6.16}$$

Hence

$$V_{(BR)CEO} = V_{(BR)CBO} (1 - h_{21B})^{1/\kappa} \tag{6.17}$$

For an npn transistor, $\kappa = 5$ would be a typical value. Note that $V_{(BR)CEO}$ is always lower than $V_{(BR)CBO}$; the difference increases as $h_{21B}$ and $h_{21E}$ increase. If the base region is lightly doped or very thin, $V_{(BR)CEO}$ can also be lowered as a result of punch-through: the depletion layer extends through the base to the emitter, effectively shorting out the transistor. This effect is discussed in Section 7.1 as it affects thyristor structures. It does not normally occur in power transistors.

The influence of the current gain on the maximum value of $V_{CEO}$ is quite important. For example, with $h_{21B}=0.9$ (so that $h_{21E}=9$) and for $\kappa=5$, $V_{(BR)CEO}=0.63 V_{(BR)CBO}$, whereas with $h_{21B}=0.99$ ($h_{21E}=99$), $V_{(BR)CEO}=0.40 V_{(BR)CBO}$.

At low current densities, a high proportion of the emitter current is made up of generation–recombination current (majority carriers). In consequence, as we have seen, the injection efficiency and in turn $h_{21B}$ are low. When the base is left open circuit, $I_B=0$, the emitter current consists only of the reverse current crossing the collector–base junction and the breakdown voltage $V_{(BR)CEO}$ is determined by (6.16) with a relatively low value of $h_{21B}$. However, the increasing current density that flows as a result of carrier multiplication when the breakdown voltage is approached has the effect of increasing the emitter injection efficiency and hence increasing $h_{21B}$. This leads to an increase in $I_{CEO}$, according to (6.15), and consequently a region of negative differential resistance appears on the I–V characteristics. This is illustrated in Figure 6.6.

In the common-emitter configuration, connecting a resistance between the emitter and the base has the effect of reducing the emitter junction injection efficiency and so increasing the breakdown voltage to a value designated $V_{(BR)CBO}$. Short-circuiting the emitter and base or reverse biasing the base–emitter junction causes further rises in the breakdown voltage to the values $V_{(BR)CES}$ and $V_{(BR)CEX}$, shown in Figure 6.6. An example of the effect on the transistor breakdown voltage $V_{(BR)CER}$ of a resistance connected between emitter and base is shown in Figure 6.7.

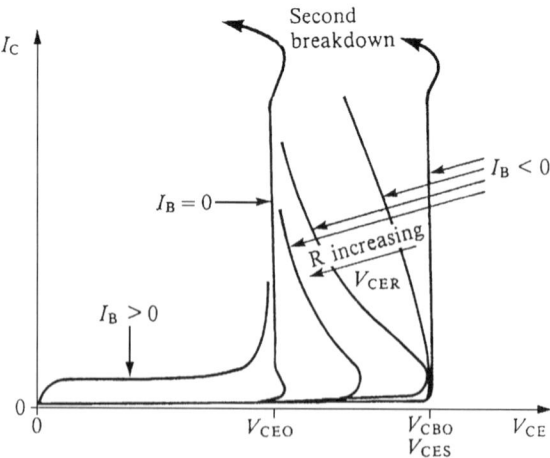

**Figure 6.6** Breakdown characteristics in the common-emitter configuration. The curves show the effect of forward and reverse bias of the base/emitter junction. Reproduced from *Power MOSFETs Theory and Applications* by D. A. Grant & J. Gowar, by permission of John Wiley & Sons, Inc.

## 6.2 BASIC CHARACTERISTICS OF POWER TRANSISTORS

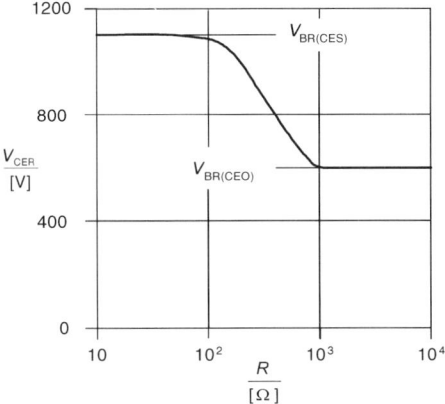

**Figure 6.7** An example of the variation of the breakdown voltage, $V_{CER}$, with the resistance connecting the base and emitter of a 100 A transistor. Data taken from *Development of High Power Transistor for Power Use* by M. Otsuka (Toshiba, 1975)

### 6.2.2 Transistor Operating Regions

The transistor output characteristics $I_C(V_{CE})$ for different values of $I_B$ in the common-emitter configuration are very important and are illustrated schematically in Figure 6.8. We have divided the characteristics into four regions. Region I is the region of *saturation*, in which both junctions are forward biased and a high concentration of carriers makes the transistor highly conducting. In region II the transistor is said to be *cut-off*: both junctions are reverse biased and only a very small leakage current flows. Region III is called the *active* region, in which the base–emitter junction is forward biased and the collector–base junction is reverse biased. Region IV is the *breakdown* region, where a small increment of collector voltage results in a large increase in collector current. This is not a region suitable for operating power transistors.

Region I represents the on-state and region II the off-state for a transistor operating as a switch in a circuit like Figure 6.4(c). The load line corresponding to

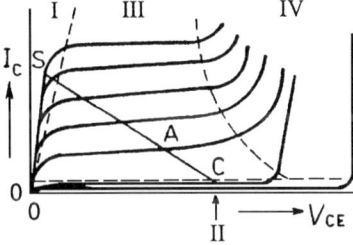

**Figure 6.8** The output characteristics, $I_C$ versus $V_{CE}$, of a transistor in the common-emitter configuration: (I) saturation region; (II) cut-off region; (III) active region; (IV) breakdown region. The points, S, A and C, are operating points on a load line and represent the saturation, active and cut-off regions of device operation, respectively

the load resistor $R_L$ is shown on Figure 6.8. At $I_B=0$, or with the base–emitter junction reverse biased, the collector current is small, the voltage dropped across $R_L$ is also small and practically all the supply voltage appears across the collector–base junction. Increasing $I_B$ causes the collector current to increase and with it the voltage dropped across $R_L$. This results in a corresponding decrease in $V_{CE}$ to a point such as A in Figure 6.8, in the active region. However, if the base current is high enough, point S is reached in region I, there is a low forward voltage drop across the transistor, $V_{CE} < V_{BE}$, and practically all the supply voltage appears across the load resistance.

In the active region, region III, and therefore in the absence of carrier multiplication,

$$I_C = h_{21E} I_B + I_{CEO} \tag{6.18}$$

where $h_{21E}$ is given by (6.13).

Region III is the normal operating region for linear amplifiers. Provided the collector voltage is high enough, the space charge region at the collector junction is thick enough and its electric field is strong enough to sweep out all the excess carriers that diffuse across the base. When the collector voltage decreases so that both the electric field and the thickness of the space charge region decrease, it is not possible for all the excess carriers to be swept out across the collector junction and the transistor becomes saturated.

In region III the collector current is only slightly dependent on the collector voltage. This results from the expansion of the space charge layer at the collector junction into the base region, as discussed at the end of the last section. It is straightforward to show that the $I_C$ ($V_{CE}$) characteristics increase with $V_{CE}$, with a slope that is approximately proportional to $I_C$.

### 6.2.3 High Current Considerations: Collector Conductivity Modulation and Base Widening Effects

The set of characteristics shown in Figure 6.8 is typical of a low voltage transistor, one with $V_{(BR)CEO}$ less than about 150 V, where the thickness and resistivity of the collector region have only a small effect on the on-state voltage drop.

If a high working voltage is to be achieved, the donor concentration in the collector region adjacent to the junction must be low enough and the lightly doped region has to be wide enough for the desired high collector breakdown voltage to be obtained. A typical concentration profile is shown in Figure 3.2. Notice that the base–collector region has a $p\nu n^+$ or compressed-field structure similar to that of the diode discussed in Section 5.2. A field-reducing surface termination is also required.

As indicated in Figure 3.2, the donor concentration in the $\nu$-region of the collector may be less than $10^{20}$ m$^{-3}$, so its resistivity may approach 1 Ω m. Its thickness may range from 20 to 100 μm. Thus, its specific resistance would lead to a significant voltage drop at high current densities, were it not for the fact that under these conditions the excess carriers modulate the collector conductivity. Typical output characteristics can be seen in Figure 6.9. They differ from those of Figure 6.8 in that

## 6.2 BASIC CHARACTERISTICS OF POWER TRANSISTORS

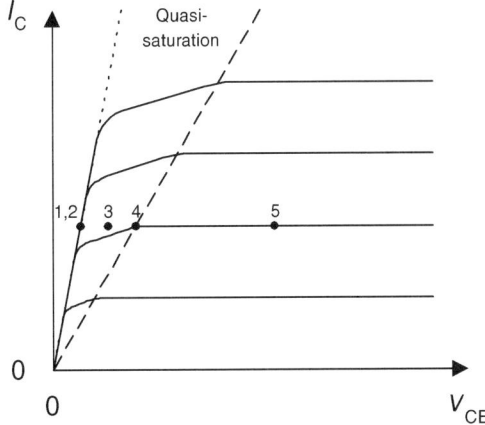

**Figure 6.9** Output characteristics of a power transistor showing a region of quasi-saturation

a region of quasi-saturation appears between the two dashed lines. This can be explained from a consideration of the carrier concentration distributions.

We start by considering a point such as (1) on the characteristics of Figure 6.9, where a very large base current maintains a free carrier concentration that exceeds the doping concentration throughout the base and lightly doped collector regions. The free electron and hole concentrations are everywhere approximately equal to one another, differing only by the local net dopant concentration and the tiny

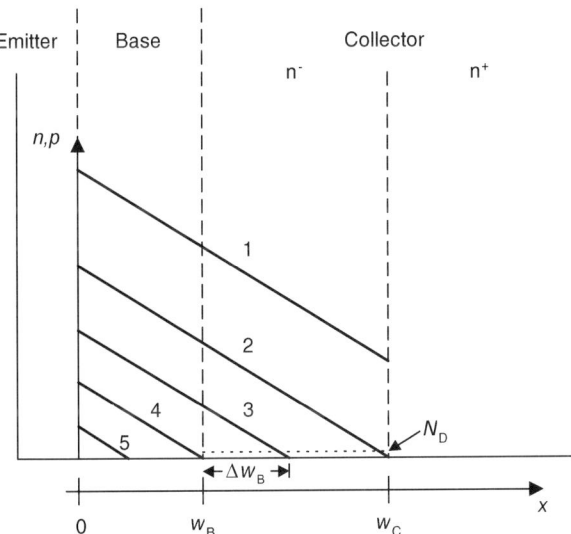

**Figure 6.10** The variation of the excess carrier distribution in a power transistor in the different regions of operation. Each curve refers to the corresponding point shown in Figure 6.9. The constant slope of the curves indicates that the collector current remains constant, while the base current and the collector–emitter voltage are allowed to vary

imbalance required to provide the local electric field. They are represented schematically by curve 1 in Figure 6.10.

The collector–emitter voltage is made up of the ohmic losses at the contacts and in the heavily doped emitter and collector regions, and a small potential difference that develops across the base and the lightly doped collector layer. This last component arises as follows. Apart from the small current of holes needed to provide for the recombination that occurs in the collector region, the current crossing from emitter to collector can be considered to be carried entirely by electrons. Any tendency of the holes to diffuse into the collector is balanced by an internal electric field, so for the purpose of analysis, we may set the hole current density $J_{pC}$ equal to zero:

$$J_{pC} = pe\mu_p E - eD_p \frac{dp}{dx} = 0 \tag{6.19}$$

Thus

$$E = \frac{D_p}{\mu_p} \frac{1}{p} \frac{dp}{dx} = \frac{kT}{e} \frac{1}{p} \frac{dp}{dx} \tag{6.20}$$

using the Einstein relation (1.70a).

Under these conditions of high injection, we can assume that $dp/dx = dn/dx$ and that, with the hole current negligible, the total current density $J_C$ crossing the base–collector junction is that of the electrons, flowing by field-assisted diffusion:

$$J = -J_{nC} = ne\mu_n E + eD_n \frac{dn}{dx} = eD_n \left(2 + \frac{N(x)}{p}\right) \frac{dp}{dx} \tag{6.21}$$

We have again used the Einstein relation and put $N(x) = n - p$ for the net doping concentration: $N(x)$ is the net donor concentration in the collector region $(x > w_B)$ and $-N(x)$ is the net acceptor concentration in the base region $(x < w_B)$. In the lightly doped region of the collector we assume that $N(x) = N_D = $ constant.

With $J$ constant, the excess carrier concentrations decrease approximately linearly with distance:

$$p(x) = p(0) - \frac{J}{2eD_n} x + \frac{1}{2} \int_0^x \frac{N(x)}{p} \frac{dp}{dx} \approx p(0) - \frac{J}{2eD_n} x \tag{6.22}$$

As long as the net hole current is zero, so that (6.20) remains valid throughout the base and the lightly doped region of the collector, the voltage dropped across those regions is

$$V_C = -\int_0^{w_C} E\,dx = -\frac{kT}{e} \int_{p(0)}^{p(w_C)} \frac{dp}{p} = \frac{kT}{e} \ln\left\{\frac{p(0)}{p(w_C)}\right\} \tag{6.23}$$

It is normally no more than a few tenths of a volt.

For purposes of illustration, we follow the transition into the active region by assuming that the base current $I_B$ is steadily reduced, while we use the external circuit to keep $I_C$ constant, allowing $V_{CE}$ to increase, as required. Curve 2 in Figure 6.10 brings us to the point where $p(w_C)$ is comparable to $N_D$ and the region of modulated conductivity just extends across the lightly doped collector layer. In Figure 6.9, point (2) is indistinguishable from point (1).

Further reduction of $I_B$ takes us into the quasi-saturation region (3) of the two figures, in which only part of the high resistivity collector region is conductivity-modulated and the voltage dropped across the unmodulated part causes a significant increase in $V_{CE}$. The voltage across the conductivity-modulated part is obtained by substituting $N_D$ for $p(w_C)$ in (6.24); the voltage across the remainder is obtained from Ohm's law, provided it applies. The overall collector–emitter voltage then comprises three terms which can be approximated as

$$V_{CE} = K\ln\left(\frac{J_E}{J_0}\right) + \frac{kT}{e}\ln\left\{\frac{n(0)}{N_D}\right\} + J_c\frac{(w_v - \Delta w_B)}{e\mu_n N_D} \tag{6.24}$$

The first term on the right-hand-side is the voltage across the emitter–base p–n junction with $K$ a constant, $J_E$ the emitter current density and $J_0$ the saturation current density under reverse bias, as given in (2.13). The second term is the voltage across the conductivity-modulated part of the collector region and the third term is the ohmic voltage dropped across the remainder of the lightly doped collector region, with $w_v = w_C - w_B$ and $\Delta w_B$ defined in Figure 6.10. Note that the collector current density $J_C$ is less than $J_E$ as a result of the base current supplied, and note that any voltage dropped across the n$^+$-collector and n$^+$-emitter layers is neglected in this approximation.

A reduction in the base current to the point where the carrier concentrations at the collector–base junction fall to the equilibrium levels of the majority carriers allows a space charge region to form there, and the transistor enters the active mode of operation. This happens at point (4) in Figure 6.9 and is represented by curve 4 in Figure 6.10. Neglecting the small offset voltage caused by the first two terms in (6.24), the slope of the dashed line in Figure 6.9, marking the transition from the quasi-saturation to the active state, represents the conductance of the lightly doped collector layer:

$$\frac{dJ_c}{dV_{CE}} = \frac{e\mu_n N_D}{w_v} \tag{6.25}$$

The transistor is taken into the normal forward active operating state (5) in Figures 6.9 and 6.10 by allowing $V_{CE}$ to rise. The additional voltage is dropped across the space charge layer that is now able to develop at the base–collector junction and which soon extends right through the lightly doped collector layer to $w_C$. The collector current is now carried in the usual way: by diffusion through the base, by drift across the base–collector space charge region and by the ohmic conduction of majority carriers in the emitter and collector layers.

Collector conductivity modulation occurs when the carrier concentration in the collector exceeds that of the donor impurities. This requires a large number of free carriers to be stored in the base and collector regions. Because they have to be supplied and removed during the turn-on and turn-off processes, switching times and transient dissipation are increased. During switching, the transistor has to pass through the quasi-saturation state and this, too, increases the transition time and the dissipation.

The need for a wide, lightly doped collector layer, in order to support high voltages, gives rise to another effect in power transistors, known as the Kirk effect. This occurs at high current densities in the active mode of operation. It leads to a rearrangement in the distribution of the electric field across the device with the result that the effective base width increases and the gain and frequency response of the transistor decrease.

In the active state, the electric field in almost all of the space charge layer is so high that the electrons travel through at their saturation drift velocity $v_{sat}$, approximately $10^5$ m/s. If the collector current density is $J_C$, the free electron concentration is

$$n(x) = \frac{J_C}{ev_{sat}} = \text{constant} \tag{6.26}$$

If the value of $J_C$ is increased through increasing the base current, $n(x)$ starts to have a significant effect on the space charge density $\rho$ in the lightly doped collector region, even though its effects elsewhere remain negligible. For $x > w_B$,

$$\rho = e\{N_D - n(x)\} \tag{6.27}$$

thus
$$\frac{dE}{dx} = \frac{\rho}{\varepsilon_r \varepsilon_0} = \frac{e}{\varepsilon_r \varepsilon_0}\left(N_D - \frac{J_C}{ev_{sat}}\right) \tag{6.28}$$

and
$$E(x) = E(w_B) + \frac{ex}{\varepsilon_r \varepsilon_0}\left(N_D - \frac{J_C}{ev_{sat}}\right) \tag{6.29}$$

Note that $E(w_B)$ is negative and that the effect of the free carriers is to decrease the rate at which the electric field reduces in magnitude from $w_B$ to $w_C$.

We assume that the collector–emitter voltage is large enough for the space charge layer to extend all the way across the lightly doped collector layer, i.e. from $x = w_B$ to $x = w_C$. Then

$$-V_{CE} \approx -V_{CB} = \int_{w_B}^{w_C} E \, dx = E(w_B)w_v + \frac{ew_v^2}{2\varepsilon_r \varepsilon_0}\left(N_D - \frac{J_C}{ev_{sat}}\right) \tag{6.30}$$

where $w_v = w_C - w_B$.

If the collector current density is increased further through a further increase in $I_B$, when it reaches the value

$$J_C = N_D ev_{sat} \tag{6.31}$$

the electric field remains constant across the lightly doped collector layer and

$$E = E(w_B) \approx -\frac{V_{CB}}{w_v} \tag{6.32}$$

With $N_D = 10^{20}$ m$^{-3}$ and $v_{sat} = 10^5$ m/s, this situation would arise at a current density of $1.6 \times 10^6$ A/m$^2$ (1.6 A/mm$^2$). Further increase in $J_C$ causes a further reduction in $E(w_v)$, for a given value of $V_{CE}$, and the magnitude of the electric field now *increases*

## 6.2 BASIC CHARACTERISTICS OF POWER TRANSISTORS

away from the junction and into the collector, as shown in Figure 6.11. The peak value of the field now occurs at the n–n$^+$ junction.

When the collector current density reaches the value

$$J = ev_{\text{sat}}\left(N_D + \frac{2\varepsilon_r\varepsilon_0 V_{CB}}{ew_v^2}\right) \tag{6.33}$$

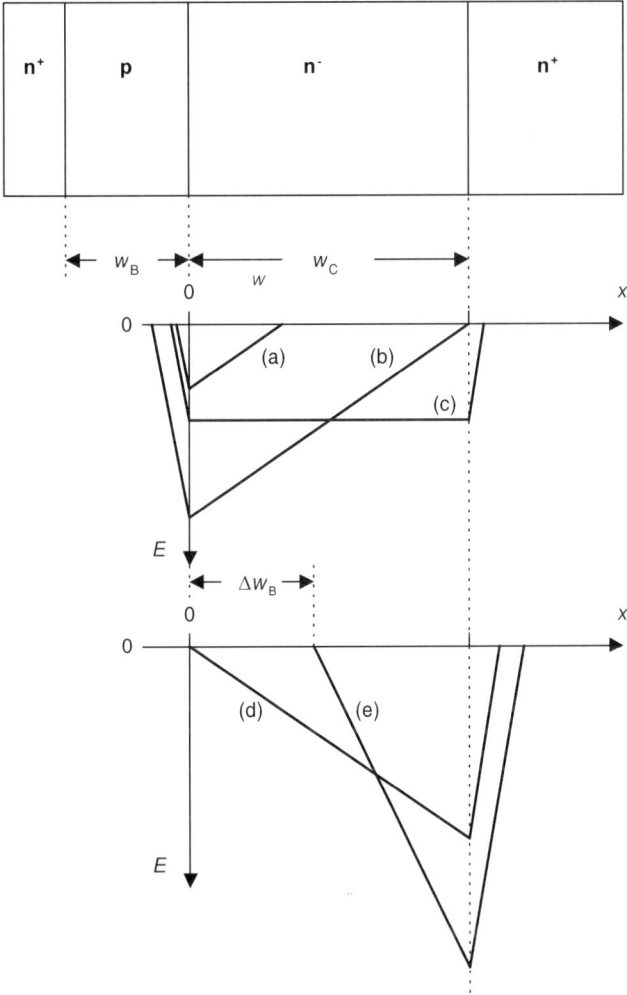

**Figure 6.11** The electric field distribution in a power transistor as the collector current density is increased, keeping $V_{CE}$ constant: (a) in the normal active mode at moderate voltage and current density; (b) when $V_{CE}$ is increased so that the space-charge layer reaches $w_B$; (c) when the current density reaches the value given by Eq. (6.31); (d) when the current density is given by Eq. (6.33); (e) at very high current densities

the electric field at the base–collector junction is $E(w_B)=0$. There is now no space charge region on the base side of the collector junction and the effective base width is the full width $w_B$.

At still higher levels of $J_C$ the region of neutral semiconductor extends a distance $\Delta w_B$ beyond the collector–base junction. Because the collector current has to be carried by electrons diffusing through this region, it has the effect of increasing the effective base thickness and the total amount of stored mobile charge. In consequence, the current gain and the high frequency performance of the transistor are degraded.

Because the base–collector junction is no longer reverse biased, it can be argued that we have again entered a quasi-saturation region of the characteristics. But now the conduction in the high resistivity collector layer is no longer ohmic and the current density is so high that it can only be carried by a concentration of free electrons greater than $N_d$, travelling at their saturation drift velocity. The normal forward active mode of operation is not possible at current densities exceeding the value given by (6.33). However, there is still transistor action, in that there is current gain, with the base current able to modulate the collector current. This is also true for the quasi-saturation region that occurs at lower current densities as a result of conductivity modulation. Indeed, both regions might be better named 'quasi-active' modes rather than quasi-saturation modes.

Using a one-dimensional model [6.3], the current gain coefficient is found to be inversely proportional to the collector current, so a gain parameter $G$ may be defined as

$$G = h_{21E} I_C = \frac{4e^2 n_i^2 D_n^2}{J_0} \frac{A_E}{w_C^2} \tag{6.34}$$

when the transistor is fully saturated and where $A_E$ is the active area of the emitter.

Note that the maximum working voltage $V_{CEW} \leqslant V_{CEO}$ and that for a typical construction

$$V_{CEO} = K_1 w_C \tag{6.35}$$

where $K_1$ is approximately 10 V/μm. Consequently,

$$G \propto \frac{1}{V_{CEO}^2} \tag{6.36}$$

This is one of the basic limitations of power transistor performance.

### 6.2.4 Other High Current Density and Temperature Effects

When the carrier concentrations in the base are comparable with the acceptor concentration at the emitter junction, the emitter injection efficiency is reduced. This lowers the current gain coefficients, and the gain and high frequency performance of the transistor are degraded. It normally occurs at current densities lower than those needed to produce the base widening effect discussed in the previous section. The

## 6.2 BASIC CHARACTERISTICS OF POWER TRANSISTORS

gain coefficients are further degraded at these high current densities when additional recombination through Auger processes becomes significant.

Up to now transistor behaviour has been treated using a one-dimensional model which implies that the emitter and collector currents can be increased indefinitely simply by increasing the emitter and collector areas. In real structures the base current $I_B$ has to flow laterally under the emitter to reach the base contact, as illustrated in Figure 6.12. Because the base layer has a finite resistivity $\rho_B$, this lateral current gives rise to a lateral potential gradient and hence to a non-uniform emitter current density. Two-dimensional modelling of this effect [6.4] indicates that the common-emitter current gain $h_{21E}$ varies inversely as the square of the collector current when this is high.

If we regard the high conductivity emitter as a region of uniform potential, the potential gradient across the base results in a variation of the base–emitter voltage $V_{BE}$ that is a function of the distance $y$ from the edge of the emitter–base junction:

$$V_{BE}(y) = V_{BE}(0) - \int_0^y \rho_B J_B \, dy \qquad (6.37)$$

where $J_B$ is the current density flowing laterally through the base region. The emitter current density increases exponentially with increasing base–emitter voltage. We therefore expect it to decrease approximately exponentially with $y$, giving rise to what is known as current crowding. This is illustrated schematically in Figure 6.12. It follows that it is the part of the emitter adjacent to the base contact that is most effective and that, to increase the current capability of the transistor, it is necessary to extend this part of the emitter periphery. An interdigitated layout, like the one shown in Figure 6.13, achieves this.

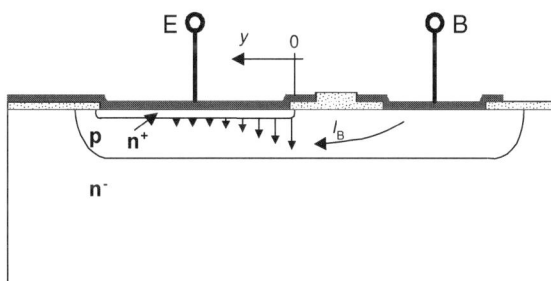

**Figure 6.12** The distribution of the emitter current underneath the emitter contact

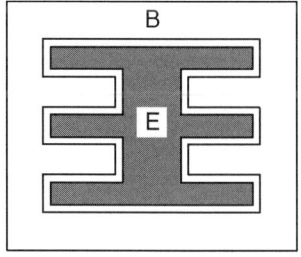

**Figure 6.13** An inter-digitated base-emitter layout

The transistor output characteristics are functions of temperature, a feature that is very important in electronic circuit design. For a given value of $V_{BE}$, the emitter current $I_E$ increases with temperature. This results in an increase of $h_{21E}$. The resulting increase in collector current is offset by the increased resistance of the high resistivity collector region.

The temperature dependence of $h_{21E}$ can give rise to an important potential failure mechanism in transistors, called *second breakdown*. This is a local thermal breakdown that leads to the formation of a mesoplasma, as described in Section 2.1.4. When there is a significant non-uniformity in the emitter current density distribution, which can arise as a result of a local inhomogeneity in the base, or through emitter crowding, any local increase of temperature where the current density is higher can induce a further increase in both current density and temperature. This is illustrated in Figure 6.14. If the local temperature comes to exceed a critical value, this positive feedback leads to the formation of a mesoplasma and to device failure through thermal breakdown. The likelihood of second breakdown increases with current density and is one of the effects that limit the maximum permissible collector current $I_{CM}$. Particular care to avoid second breakdown has to be taken in pulse conditioning circuits, as discussed in Section 6.4.

The temperature dependence of the current gain coefficients also complicates the connection of transistors in parallel. A higher current flowing through one of several transistors connected in parallel results in higher dissipation and an increase in temperature. This induces a further increase in $h_{21E}$ and hence in $I_C$. In order to prevent all the current from being concentrated into the transistor with the highest current gain (which could result in second breakdown), some negative feedback is needed. This can be obtained in one of the two ways illustrated in Figure 6.15. Either a small resistance can be put in series with the emitter of each transistor or a rather

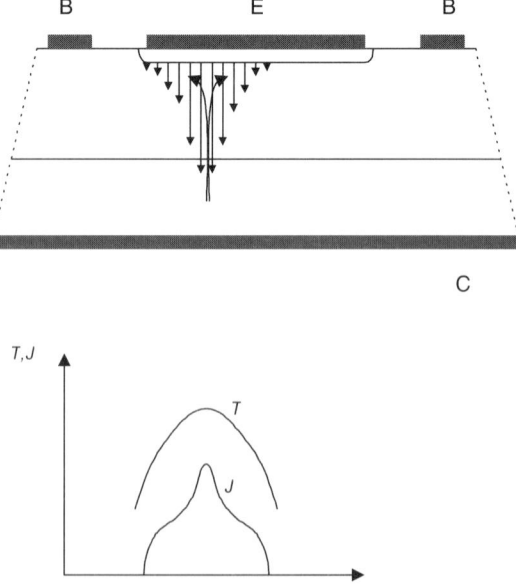

**Figure 6.14** An illustration of local thermal breakdown in a bipolar junction transistor

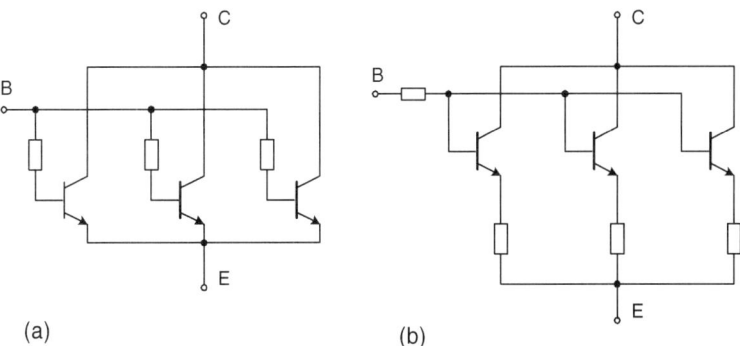

**Figure 6.15** Current balancing in parallel-connected transistors using series resistance: (a) in the base lead; (b) in the emitter lead

higher resistance in series with the base. The latter arrangement helps to maintain the base current constant, but it complicates the drive circuit. If the transistors are not fully saturated, the internal resistance of the collector can sometimes provide the necessary feedback, provided the temperature does not rise sufficiently to cause the collector region to become 'intrinsic'.

## 6.3 The Dynamic Behaviour of Power Transistors

The most important applications of transistors in power electronics are as switches in power converters. In these applications the dynamic processes (turn-on and turn-off) are of great importance. They are characterised by the switching times: the turn-on time $t_{on}$ and the turn-off time $t_{off}$. They also give rise to additional power losses and so affect the safe operating area for transistors in different types of circuit.

### 6.3.1 Turn-on

We consider first the very important turn-on process for a transistor in the common-emitter configuration. This is the transition from the off-state ($V_{CE}$ high, $I_C$ low) to the on-state ($V_{CE}$ low, $I_C$ high). Typical base and collector current waveforms are shown in Figure 6.16.

Initially, $I_B=0$ and there is no excess carrier concentration in the base region. At $t=0$, the base drive circuit causes $V_{BE}$ to increase and supplies a current pulse that rises rapidly to a value that is high enough to keep the transistor in saturation (or quasi-saturation) under steady-state conditions. The excess electron concentration at the base edge of the base–emitter depletion layer increases exponentially with $V_{BE}$, according to (2.5), with $V_{BE}=V_J$. Excess electrons are injected from the emitter into the base and diffuse towards the base–collector junction, while the excess holes needed to maintain space charge neutrality are supplied simultaneously by $I_B$. The time-varying emitter and collector currents, $i_E(t)$ and $i_C(t)$, can be obtained using an argument similar to the one that leads to (6.4) and (6.5) in Section 6.1:

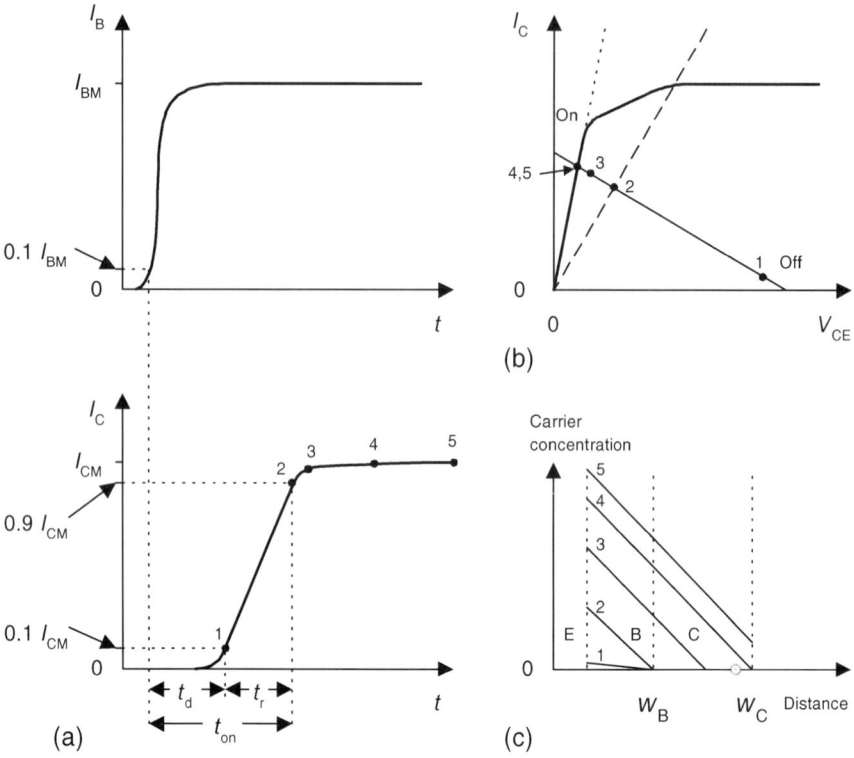

**Figure 6.16** The transistor turn-on transition: (a) current waveforms for the base and collector; (b) corresponding points on the collector circuit load line; (c) the development of the distribution of excess carriers during the turn-on process

$$i_E(t) = \frac{AJ_n(0,t)}{\tilde{\gamma}_n} = \frac{eAD_n}{\tilde{\gamma}_n}\frac{dn_p}{dx}(0,t) \qquad (6.38)$$

$$i_C = AJ_n(w_B,t) = eAD_n\frac{dn_p}{dx}(w_B,t) \qquad (6.39)$$

$$i_E = i_B + i_C \qquad (6.40)$$

The evolution of the excess carrier distribution in the base is illustrated in Figure 6.16(c). The collector current starts to rise only when the excess carriers reach the reverse-biased base–collector junction, i.e. after the delay time $t_d$, defined in Figure 6.16(a). The collector current rise time $t_r$ is normally defined from the 10% and 90% points on the collector current waveform, as shown in the figure. The sum of $t_d$ and $t_r$ is the total turn-on time $t_{on}$. Any exact solution of the continuity equation during the turn-on period is complicated but the stored-charge approach gives a good insight into the processes. The transport of carriers through the base can be characterised by the transit time $\tau_T$, defined from the effective carrier velocity:

## 6.3 THE DYNAMIC BEHAVIOUR OF POWER TRANSISTORS

$$v(x) = \frac{J}{en(x)} \qquad (6.41)$$

Then

$$\tau_T = \int_0^{w_B} \frac{dx}{v(x)} = \int_0^{w_B} \frac{en(x)}{J} dx \qquad (6.42)$$

To a first approximation

$$n(x) = n(0)\left(1 - \frac{x}{w_B}\right) \qquad (6.43)$$

So that

$$\tau_T = \frac{en(0)w_B}{2J} \qquad (6.44)$$

The current density is given by (2.52) and in our approximation

$$\frac{dn}{dx} = -\frac{n(0)}{w_B} \qquad (6.45)$$

and so

$$\tau_T = \frac{w_B^2}{2D_{eff}} \qquad (6.46)$$

where $D_{eff}$ is an effective diffusion coefficient which incorporates the effect of any built-in electrical field arising from a doping concentration gradient across the base. The transit time $\tau_T$ is approximately the time taken by the diffusion wavefront to propagate across the base, starting at time $t=0$.

Following the initial turn-on phase, the excess carrier concentration in the base region increases, as illustrated in Figure 6.16(c). The rate of change of the excess carrier charge in the base region, as expressed in (1.79), is now given by

$$\frac{dQ_B}{dt} = i_B(t) - \frac{Q_B}{\tau_{eff}} \qquad (6.47)$$

When the base current can be considered to rise instantaneously to a steady value $I_{B0}$ at $t=0$, the solution of (6.47) is

$$Q_B(t) = I_{B0}\tau_{eff}\left\{1 - \exp\left(\frac{-t}{\tau_{eff}}\right)\right\} \qquad (6.48)$$

During the accumulation of the excess carrier charge $Q_B$ in the base region, the collector current rises and it continues to increase until it reaches the value $I_{C0}$. Putting $I_C(t) = Q_B(t)/\tau_T$, the turn-on time can be obtained from

$$I_{C0} = \frac{I_{B0}\tau_{eff}}{\tau_T}\left\{1 - \exp\left(\frac{-t_{on}}{\tau_{eff}}\right)\right\} \qquad (6.49)$$

Equation (6.49) is only an approximation because of the gradual expansion during turn-on of the effective base region, as described in Section 6.2.3. Together with the

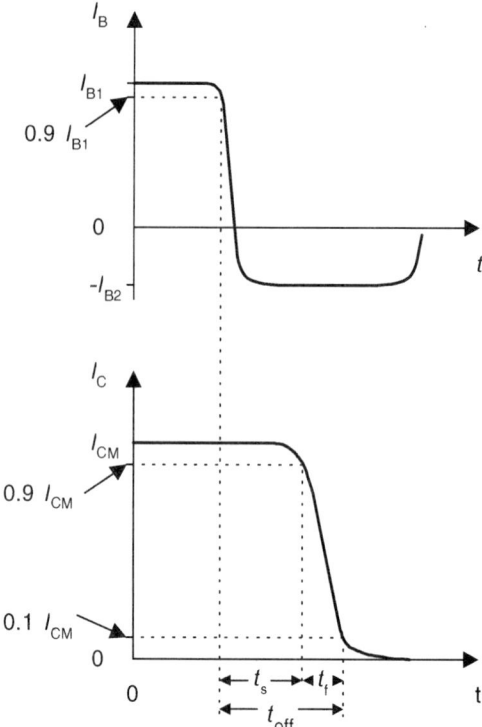

**Figure 6.17** Base and collector current waveforms during the transistor turn-off transition

reduction of the diffusion coefficient under high injection conditions, this prolongs the transit time compared with (6.46). The transistor gradually changes its state from quasi-saturation to full saturation, as indicated by points (3), (4) and (5) in Figure 6.16(b), with a considerable increase in the stored charge.

### 6.3.2 Turn-off

Also very important is the transient behaviour of power transistors during the turn-off process. This is the dynamic transition from the state of saturation or quasi-saturation, with a very low resistance, to the off-state with a very high resistance.

During the turn-off process, the excess carriers stored in the base and the collector regions during the on-state have to be removed in order that the blocking function of the collector–base p–n junction can be recovered. In the on-state, a base current $I_B$ is required to maintain the injection of excess carriers. Interrupting the base current cuts off this injection, but the base region and the high resistivity part of the collector are still flooded with excess carriers. Over a period of time, some of these carriers recombine and others are swept out by the reverse-biased p–n junctions into the collector and emitter regions. After a time interval $t_s$, called the storage time, the excess carrier concentration decreases to such a low level that a space charge region can be rebuilt at the collector junction. As this occurs, there is an increase in the loss rate of excess carriers and the collector current decreases rapidly. Its fall time $t_f$ is defined as the time taken for the current to decrease from 90% to 10% of its maximum value. The overall turn-off time is defined as $t_{off} = t_s + t_f$.

Typical current and voltage waveforms during turn-off are shown in Figure 6.17. We assume that the polarity of the base supply is reversed at time $t=0$, causing an abrupt change of the base current from $+I_{B1}$ to $-I_{B2}$. The excess carrier charge in the base and the high resistivity region of the collector can be shown [6.5] to decrease as

$$Q(t) = \tau_{\text{eff}}(I_{B1} + I_{B2})\exp\left(\frac{-t}{\tau_{\text{eff}}}\right) - \tau_{\text{eff}}\left(I_{B2} + \frac{I_C}{h_{21E}}\right) \tag{6.50}$$

where $\tau_{\text{eff}}$ is the sum of the high injection carrier lifetime $\tau_n^*$ and the carrier transit time through the region flooded with excess carriers (the base plus the conductivity-modulated region of the collector):

$$\tau_{\text{eff}} = \frac{w_C^2}{4D_n} + \tau_n^* \tag{6.51}$$

It is reasonable to set $Q(t_s)=0$. Then it follows from (6.50), that

$$t_s = \tau_{\text{eff}} \ln\left(\frac{I_{B1} + I_{B2}}{I_{B2} + I_C/h_{21E}}\right) \tag{6.52}$$

Equation (6.52) indicates how the storage time is influenced by the reverse base current $I_{B2}$, as is the fall time. While the storage time depends on the carrier transit time through the base and the conductivity-modulated collector regions, the fall time depends only on the transit time through the normal base width. Examples for a typical transistor are shown in Figure 6.18.

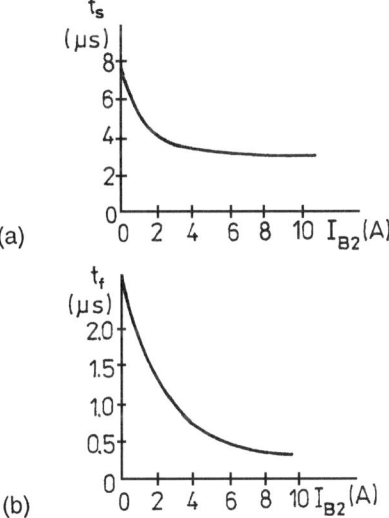

**Figure 6.18** The influence of the reverse base current, $I_{B2}$, on: (a) the storage time, $t_s$; (b) the fall time, $t_f$.

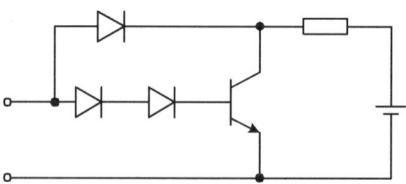

**Figure 6.19** An example of a circuit designed to prevent a transistor from going into saturation (the Baker clamp)

The effective carrier lifetime is the sum of the recombination lifetime and the transit time through the widened base. When a short turn-off time is required, the carrier lifetime can be shortened either by the diffusion of recombination centres or by irradiation with high energy electrons. However, a shortened carrier lifetime results in a lower current gain of the transistor. Another possibility is to operate with the transistor only in quasi-saturation in the on-state, so that no conductivity modulation or base widening occurs. An example of a circuit designed to inhibit saturation is shown in Figure 6.19. A disadvantage of this approach is that there is a higher on-state voltage and consequently higher dissipation, limiting the current at which the transistor can be operated.

In the early part of the turn-off process, as the transistor moves from saturation to quasi-saturation while the collector current remains largely unchanged, there is a slight rise in $V_{CE}$. The main rise in $V_{CE}$ occurs later with the formation of the space charge layer at the collector–base junction. The collector voltage and current waveforms following the storage time are functions of the type of load, as described in Section 3.1.8.

Plotting $I_C(t)$ and $V_{CE}(t)$ on the transistor output characteristics provides an *operating point trajectory* on the $I_C$–$V_{CE}$ plane during turn-off. Trajectories for inductive, resistive and capacitive loads are shown in Figure 6.20. The highest instantaneous power dissipation is generated with an inductive load as the collector current starts to decrease immediately after the collector voltage reaches its maximum value. The lowest power dissipation occurs with a capacitive load.

The load character strongly influences the magnitudes of both the instantaneous and average power dissipation $P_{off(AV)}$ during turn-off:

$$P_{off(AV)} = f \int_{t_1}^{t_n} V_{CE} I_C \, dt \qquad (6.53)$$

where $f$ is the switching frequency and $t_1$ and $t_n$ are defined in Figures 6.20. The power dissipated during the turn-off process can be a major component of the total power dissipation at higher frequencies and hence a limiting factor determining either the maximum collector current or the maximum working frequency.

An important figure of merit is the maximum frequency at which the transistor can be used. In the linear regime this is normally characterised by the frequency $f_T$ at which the current gain $h_{21E}$ decreases to unity. This depends on the carrier transit time through the base. In power transistors which suffer the base widening effect, it can be approximated as

## 6.3 THE DYNAMIC BEHAVIOUR OF POWER TRANSISTORS

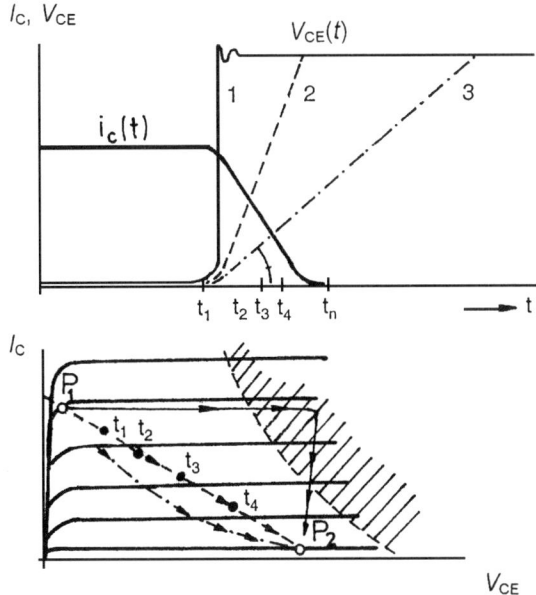

**Figure 6.20** The trajectory of the operating point during turn-off for a transistor supplying the three different types of load shown in Figure 3.21: (1) clamped inductive load; (2) resistive load; (3) capacitive load.

$$\frac{1}{f_T} = 2\pi \frac{(w_B + \Delta w_B)^2}{4D_n} \tag{6.54}$$

For a power transistor, $f_T$ is typically in the range 10–100 kHz.

In switching operation, the maximum working frequency is limited by the turn-on and turn-off times and by the power dissipation. Thus

$$f_{max} \leqslant \frac{1}{\pi(t_{on} + t_{off})} \tag{6.55}$$

and the maximum switching frequency $f_{max}$ is much lower than $f_T$.

In practice the maximum working frequency of a power transistor is limited by dissipation. During the turn-on and turn-off processes, the dissipation rises to instantaneous values that are several orders higher than in the on-state. Let the average values of the power dissipation during turn-on and turn-off be $\overline{P}_{on}$ and $\overline{P}_{off}$ respectively; let the average on-state dissipation be $\overline{P}$. If the transistor is in the on-state for a period $\Delta t$ during each cycle and there is negligible dissipation in the off-state, the average power dissipation $P_{AV}$ at a switching frequency $f$ is

$$P_{AV} = (\overline{P}_{on}t_{on} + \overline{P}_{off}t_{off} + \overline{P}\Delta t)f \tag{6.56}$$

Increasing the switching frequency does not affect the turn-on and turn-off times but $\Delta t$ decreases. Because $\overline{P}_{on}$ and $\overline{P}_{off}$ are both much greater than $\overline{P}$, the average power dissipation increases with frequency and, for given current and voltage waveforms, the maximum working frequency is governed by the need to satisfy the conditions

## 6.4 Safe Operating Area

In Section 6.2 we discussed the possible occurrence of second breakdown during the operation of a transistor. It is marked by an abrupt decrease in $V_{CE}$ and the loss of control of the collector current by the base. Device destruction follows. The origin of second breakdown involves two main phenomena.

First, as discussed in Section 2.1.4, the rating of semiconductor devices is limited by the stability criterion (2.49) and consequently by the maximum allowable power dissipation. One of the limits is the maximum operating temperature $T_{jmax}$ of the p–n junctions. If this is exceeded, the stability of device parameters cannot be guaranteed and thermal breakdown may follow. Local thermal instability associated with the formation of a mesoplasma can be caused either by a high emitter current, as described in Section 6.2, or by excessive dissipation.

Another phenomenon that may initiate second breakdown is avalanche injection. This occurs when a relatively high collector voltage and current density are present simultaneously. If, in any part of the transistor structure, especially in the high resistivity part of the collector region, the electric field exceeds $10^7$ V/m, electron–hole pairs are generated by avalanche ionisation. The holes so generated are accelerated towards the emitter and act as an additional base current. They therefore lead to increased electron injection from the emitter and an increased current density in the collector region. This results in a local increase in the electric field. And this positive feedback causes a local abrupt increase of current density, resulting in thermal breakdown and device destruction.

Conditions for avalanche injection can occur during turn-off, especially in a circuit with an inductive load. During the on-state, while the transistor is saturated, the base region expands into the collector, as described in Section 6.2. Following the reversal of the base current, carriers recombine and are swept out of the expanded base region, which contracts laterally towards the centre of the emitter area, as shown in Figure 6.21. The current density there increases and the space charge of the carriers causes the redistribution of the electric field shown in Figure 6.11(e). The field at the

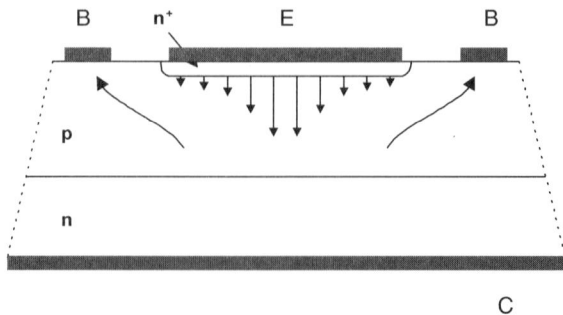

**Figure 6.21** The concentration of current into the centre of the emitter region during transistor turn-off

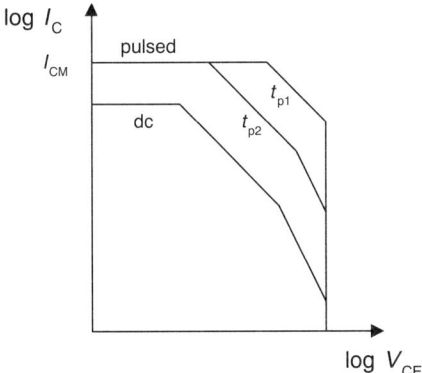

**Figure 6.22** A schematic illustration of a bipolar junction transistor safe operating area during turn-on (FBSOA). Curves are given for continuous and pulsed operation. In the latter case, $t_{p2} > t_{p1}$

n–n$^+$ junction can become high enough to initiate avalanche injection with resulting thermal runaway and second breakdown.

The locus of the $I_C$–$V_{CE}$ characteristic that marks the boundary between stable (safe) and unstable (dangerous) operation is a most important parameter for power transistor applications. Any simultaneous combination of collector current and voltage that lies within this locus, and for which there is no fear of second breakdown, is called the *safe operating area* (SOA). Its shape is determined by the operating conditions. Typical SOA diagrams are shown in Figure 6.22 for turn-on and the on-state (forward-biased base, FBSOA) and in Figure 6.23 for the turn-off phase (reverse-biased base, RBSOA). Logarithmic scales are normally used for these diagrams.

The FBSOA diagram shown in Figure 6.22 illustrates how, at low $V_{CE}$, the maximum allowed collector-current is determined by the upper limit of current-carrying capacity. In switching circuits this can depend on the length of the current pulse because second breakdown is linked to the energy dissipated. With higher collector voltages the FBSOA is determined by the power dissipated. As $V_{CE}$ approaches $V_{CEsus}$, it is determined by the onset of avalanche breakdown.

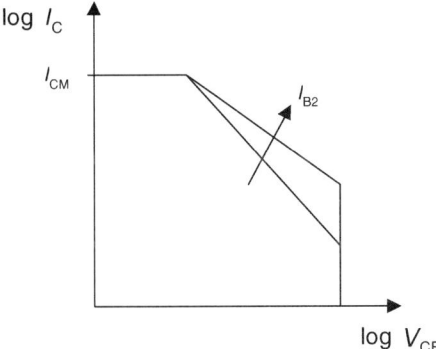

**Figure 6.23** A schematic illustration of a bipolar junction transistor safe operating area during turn-off (RBSOA). The effect of increasing the reverse base current is shown

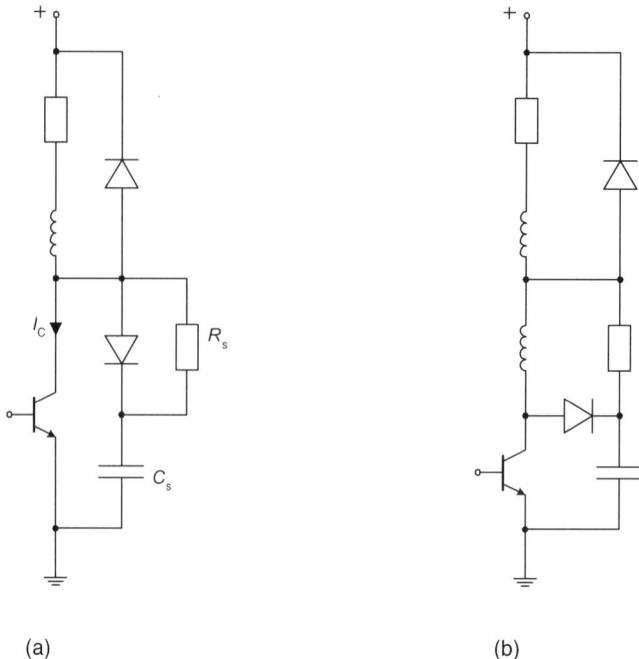

**Figure 6.24** Examples of subber circuits used with power transistors: (a) ideal; (b) including unclamped circuit inductance

The RBSOA, shown schematically in Figure 6.23, sets the limit to all possible trajectories of the $I_C$–$V_{CE}$ working point during turn-off and must not be crossed under any conditions. Its shape depends in some degree on the magnitude of the current $I_{B2}$ drawn out of the base.

Device manufacturers indicate safe operating areas in catalogued data. The circuit designer must ensure that the collector current and collector voltage never lie outside either SOA. Figure 6.20 illustrates how this can occur during turn-off with an inductive load. In this case, decreasing the rise rate of the collector voltage during turn-off by means of a snubber circuit offers some protection. Examples of snubber circuits comprising a capacitor, a diode and a resistor (RCD snubbers) are shown in Figure 6.24. To be effective, the parasitic inductance of the diode and capacitor must be kept as low as possible, typically less than 1 $\mu$H.

The circuits shown in Figure 6.24 also incorporate an inductance in series with the transistor to protect it against too high a rise rate of the current during turn-on. Decreasing the rise rate of the collector current during turn-on and decreasing the rise rate of the collector voltage during turn-off both reduce the transient power dissipation and help to keep the device within the SOA. However, there is dissipation in the snubber components and the reliability of the transistor can become dependent on the reliability of these passive components. It is important to design the snubber so as to minimise the overall dissipation of the transistor plus the snubber. This is discussed further in Section 13.3.1.

Snubber circuits increase the total power dissipation and the total equipment cost, and they can occupy quite a lot of space. It is therefore desirable to maximise the

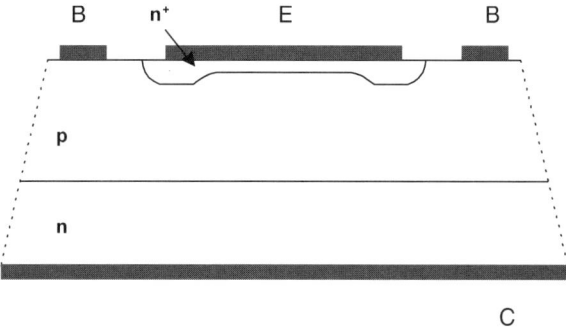

**Figure 6.25** Enlarging the RBSOA by decreasing the injection efficiency in the middle of the emitter strip

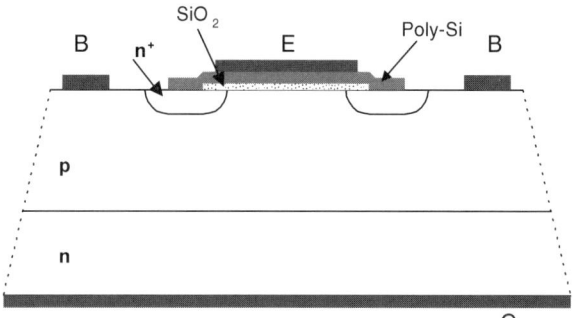

**Figure 6.26** The ring-emitter transistor (RET) structure

device SOA by using new design principles and technology. Reducing the width of the emitter stripe decreases the probability of second breakdown during turn-off. Another possibility is to reduce $h_{21E}$ in the centre of the emitter by reducing the depth of the base–emitter junction there, as illustrated in Figure 6.25. The ring emitter transistor (RET) achieves this by the parallel connection of many elementary ring emitter cells in parallel, as shown in Figure 6.26. The centre of each cell is isolated by oxide so that no electron injection occurs in the region. The small series resistance introduced by the polycrystalline silicon emitter connection helps to offset any variation in the characteristics of the individual parallel-connected cells. Another variation is the use of a perforated emitter structure.

As well as optimising the emitter region, it is possible to improve the dynamic characteristics of transistors by means of a two-step (or multistep) collector region using epitaxial technology [6.6]. Figure 6.27 shows the enlargement of the RBSOA obtained with a RET device. Several manufacturers offer transistors with an optimised RBSOA using this type of technology.

## 6.5 The Power Darlington Configuration

As described in Section 6.2.3, the current gain $h_{21E}$ of a power transistor in the turned-on state (in saturation) is relatively low, typically less than 10. The need in power applications is to be able to switch high power levels rapidly with a simple, low power control signal. A successful way to reduce the base drive power

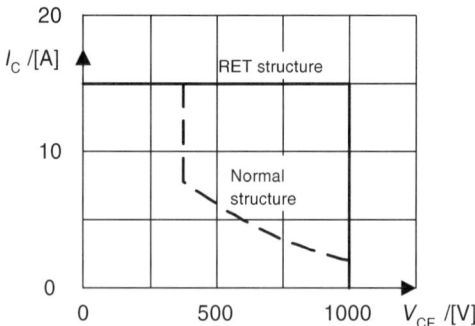

**Figure 6.27** A comparison of the RBSOA of an RET with that of a conventional transistor with an interdigitated emitter structure

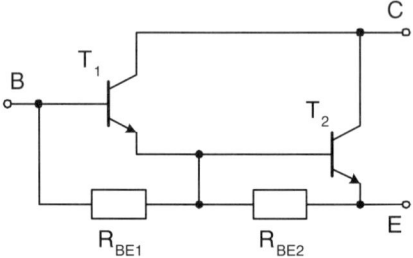

**Figure 6.28** Two bipolar junction transistors in the Darlington configuration

requirements of bipolar transistors is to use the Darlington configuration of two or more cascaded transistors, shown in Figure 6.28. The emitter current of transistor T1 provides the base current for transistor T2 and the overall gain is much higher. With two transistors, having individual current gains of $h_{21E1}$ and $h_{21E2}$, the overall gain is

$$h_{21E(D)} = h_{21E1} h_{21E2} \qquad (6.57)$$

The Darlington connection of an auxiliary transistor and a main power transistor can be realised as a monolithic integrated structure, as shown schematically in Figure 6.29. An integrated antiparallel diode may also be included in the structure to protect the emitter junction against reverse breakdown. The optimum ratio of the areas of the auxiliary and main transistor is found [6.7] to be 1:3 and the resulting gain is

$$h_{21E(D)} = \frac{G_1 G_2^2}{I_C^3} \qquad (6.58)$$

where $G_1$ and $G_2$ are the gain factors of the individual transistors. With two power transistors, this arrangement is usually called a *power Darlington*.

The gain improvements are not obtained without penalties. Integrated Darlington configurations have a higher voltage drop in saturation, $V_{CEsat}$, than a single transistor and therefore a higher power dissipation. The current gain coefficient $h_{21E(D)}$ decreases rapidly at high collector current, as shown by (6.58) and illustrated in Figure 6.30. Without special modifications the decay of the excess carrier

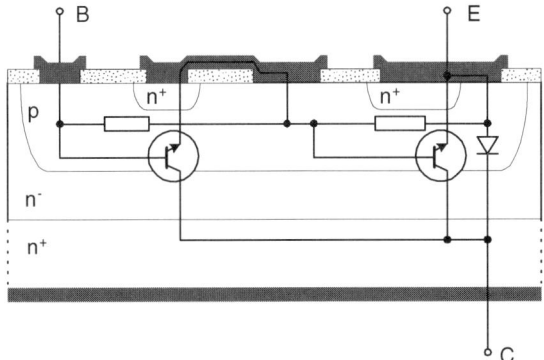

**Figure 6.29** An integrated Darlington structure with an antiparallel diode

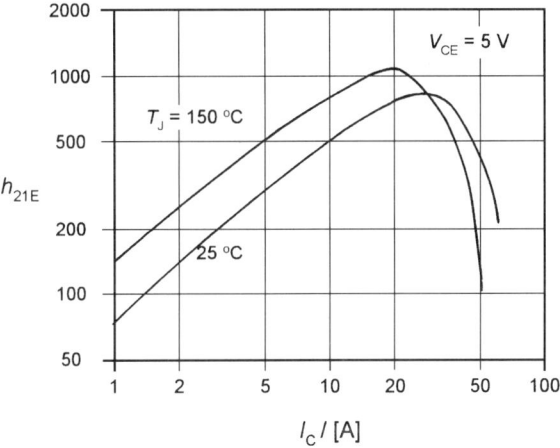

**Figure 6.30** An example of the variation of the current gain coefficient of a Darlington-coupled power transistor module with the collector current. Data taken from *Development of High Power Transistor for Power Use* by M. Otsuka (Toshiba, 1975

concentration in the base of the main transistor during turn-off is much slower than occurs in a single transistor. This results in longer switching times; in particular, the storage time increases and therefore the turn-off time increases too. In comparison with a conventional transistor structure of similar construction, the turn-off time for a power Darlington is typically three times longer. In spite of these penalties, many power transistors with a maximum current rating greater than 50 A are packaged as a module containing a hybrid-integrated double or triple Darlington. Input diodes may be included to assist turn-off by permitting charge extraction from the bases of the second or third transistor. A triple Darlington module of this kind has values of $h_{FE} = 100$, $V_{CE(Sat)} = 3.5$ V and $\tau_s = 12\mu s$. It can operate at 3 kHz.

## 6.6 Power Transistor Applications

Initially, the power bipolar junction transistor was the only viable semiconductor on/off switch for high powers. One of the first applications that still continues was for

TV line output drives. Later uses were for switched-mode power supplies, motor control, uninterruptible power supplies and automotive applications such as electronic ignition. Power BJTs still find application in off-line, single-ended SMPS, although they are now challenged by power MOSFETs and IGBTs. In the 1980s, as the SOA of power bipolar transistors improved, the development of high current (often Darlington-coupled) modules extended the range of ac drives. However, in this application, IGBTs have largely taken over. Operated as linear devices, power BJTs continue to find application as series regulators in power supplies and for audio and ultrasonic amplifiers.

## Summary

Although the bipolar junction transistor has an important history as a power semiconductor device, being the first semiconductor on/off switch, it is now largely superseded by the IGBT at higher power levels and the power MOSFET at lower power levels. Because of the higher electron mobility, power BJTs are invariably npn devices. They are often used in the form of Darlington-configured modules which give higher current gain but reduced frequency response.

Operation in the on-state at high current densities can lead to a redistribution of the electric field as a result of the space charge of the free electrons. Inhomogeneity may cause current crowding and lead to the formation of a hot spot and thermal runaway. This is referred to as second breakdown and is particularly likely to occur during turn-off. It is avoided by ensuring that at no time the locus of the voltage and current simultaneously present lies outside the *safe operating area* defined by the manufacturer. The use of external resistance/diode/capacitance snubber circuits allows the safe operating area to be better exploited but increases dissipation and reduces the operating frequency.

During turn-on it is normal for the forward volt-drop to rise above the steady-state value. During turn-off the current reverses, rises to a maximum in the reverse direction then falls to zero. The changes in $di/dt$ can give rise to large voltage spikes and circuit oscillations as a result of parasitic inductance in the circuit.

## References

[6.1] Shockley, W. The Theory of p–n Junctions in Semiconductors and p–n Junction Transistors, *Bell Systems Technical Journal*, **28**, 435–89 (1949).

[6.2] Singh, R. and Baliga, B. J. Cryogenic Operation of Power Bipolar Transistors *Proc. IEEE 6th International Symposium on Power Semiconductor Devices and ICs, ISPSD '94*, 243–8 (1994).

[6.3] Hower, P. L. Application of a Charge-Control Model to High-Voltage Power Transistors, *IEEE Trans. on Electron Devices*, **ED-23**, 863–70 (1976).

[6.4] Sheng, W. W. The Effect of Auger Recombination on the Emitter Injection Efficiency of Bipolar Transistors, *IEEE Trans. on Electron Devices*, **ED-22**, 25–7 (1975).

[6.5] Baker, A. N. Charge Analysis of Transistor Operation, *Proc. IRE*, **48**, 949–50 (1960).

[6.6] Aloisi, P. A. The Innovation in the Field of High Voltage Bipolar Transistor, *EPE Journal*, **1**, 47–54 (1991).

[6.7] Leturcq, P. Power bipolar devices, *Microelectronics and Reliability*, **24**, 313–37 (1984).

# 7

# THYRISTORS: BASIC OPERATING PRINCIPLES

A schematic illustration of the structure and operating characteristics of a typical three-electrode thyristor is shown in Figure 3.4. Thyristors have very wide application as high power switches and, in consequence, they are made with a wide range of voltage and current ratings. They normally operate in one of three states: reverse blocking, forward blocking or forward conducting. In this chapter we describe each of these conditions and the transitions between them. In Chapter 8 we examine in more detail the design and operation of some of the variants on the basic phase-control thyristor or, to use its original name, the silicon controlled rectifier (SCR). Of these other types of thyristor, the triac, which can block and conduct in both reverse and forward directions and the gate turn-off (GTO) thyristor are particularly important.

The transition from the blocking to the conducting state is usually initiated by the application of a current pulse to the gate. In a conventional thyristor the reverse transition from the conducting state to one of the blocking states is normally achieved by natural commutation. However, the gate turn-off thyristor enables a forward current to be switched off by the application of a negative gate voltage. The principles of gate turn-off are considered in Section 7.4.3; GTO design details are discussed in Section 8.3.

## 7.1 Steady-State Operation

### 7.1.1 The Reverse Blocking State

In the reverse blocking mode illustrated in Figure 7.1(a), the anode (the p-type outer layer, $P_1$) is at a negative potential with respect to the cathode; the p–n junctions $J_1$ and $J_3$ are reverse biased; and the thyristor is in a high impedance state. Junction $J_2$ is forward biased. The donor concentration in the n-base layer $N_1$ is normally many times lower than the acceptor concentration in the gate layer (the p-base layer, $P_2$). As a result, the breakdown voltage of junction $J_3$ is much lower than that of junction $J_1$. Furthermore, except in GTOs, junction $J_3$ is usually shorted by a local connection

# 7 THYRISTORS: BASIC OPERATING PRINCIPLES

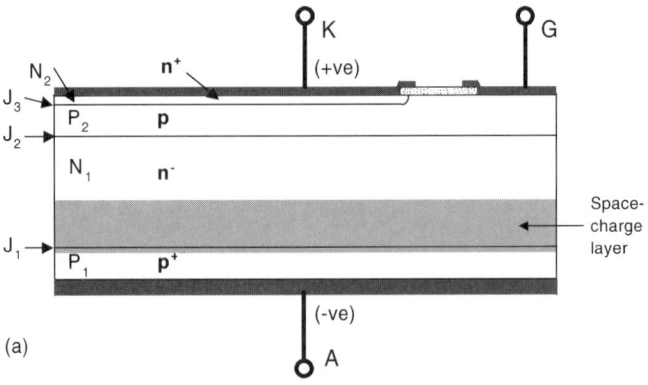

(a)

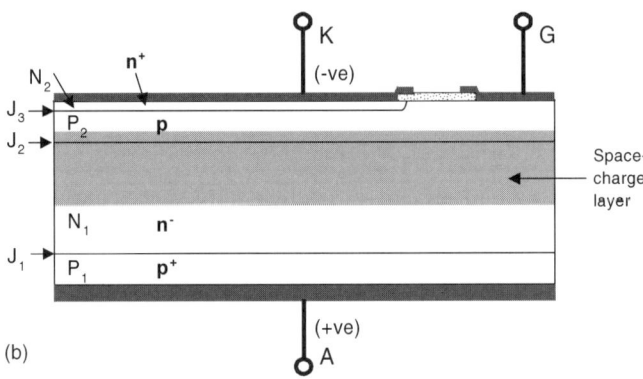

(b)

**Figure 7.1** A schematic illustration of the triode thyristor in its blocking states: (a) reverse blocking; (b) forward blocking

from the $P_2$-layer to the cathode contact, as described in detail in Section 7.2. We can therefore assume that all the reverse voltage is dropped across $J_1$.

At first sight we might expect a thyristor in the reverse blocking mode to break down in one of two ways. The first is through avalanche breakdown at junction $J_1$. The second is as a result of *punch-through*. This occurs when the $J_1$ space charge region extends right through the n-base layer to junction $J_2$. Then, with $J_2$ forward biased, holes are drawn out of region $P_2$ and across region $N_1$ into the anode layer $P_1$ under the action of the electric field, and the anode current can increase without limit.

Assuming uniform doping and an abrupt junction, the punch-through voltage $V_{PT}$ is given by

$$V_{PT} = \frac{eN_{D1}w_N^2}{2\varepsilon_r\varepsilon_0} \tag{7.1}$$

where $N_{D1}$ is the donor concentration in the lightly doped n-base region (the $N_1$-layer) and $w_N$ is its thickness. If $J_1$ were an isolated, abrupt, asymmetric junction, according to the analysis of Section 2.1.3, its avalanche breakdown voltage $V_{BR}$ would be given by

$$V_{BR} = \frac{\varepsilon_r \varepsilon_0 E_{BR}^2}{2 e N_{D1}} \quad (7.2)$$

where $E_{BR}$ is the critical electric field at which carrier generation becomes significant.

Equations (7.1) and (7.2) imply that, for a fixed value of $w_N$, the breakdown voltage is maximum when

$$N_D = \frac{\varepsilon_r \varepsilon_0 E_{BR}}{e w_N} \quad (7.3)$$

However, this analysis ignores the fact that junction $J_1$ is not isolated. Rather, the $P_2 N_1 P_1$ structure can be regarded as that of a transistor connected in common-emitter configuration, with the base left open circuit. From the discussion in Section 6.2.1 it can be seen that the breakdown voltage is influenced by the injection efficiency $\tilde{\gamma}_p$ for holes across junction $J_1$, and the diffusion length $L_p$ of the holes in the n-base. The effective base thickness of this parasitic transistor is the thickness $w_R$ of the part of the $N_1$-region that is not occupied with space charge. Therefore, using (6.16) and (6.7), the breakdown voltage is given by

$$V_{R(BR)} = V_{BR} \left[ 1 - \frac{\tilde{\gamma}_p}{\cosh(w_R/L_p)} \right]^{1/\kappa} \quad (7.4)$$

Figure 7.2 shows the variation of $V_{BR}$ and $V_{PT}$ with the thickness and the donor impurity concentration of the $N_1$-base; it illustrates the influence of the diffusion length $L_p$ on $V_{R(BR)}$ when $w_N = 200\,\mu\text{m}$ and $\tilde{\gamma}_p = 0.5$. A value of 5 has been assumed for $\kappa$.

### 7.1.2 The Forward Blocking State

In the forward blocking mode, when the anode is positive with respective to the cathode, junctions $J_1$ and $J_3$ are forward biased, junction $J_2$ is reverse biased and the thyristor again has a high impedance. This condition is shown in Figure 7.1(b). If the applied forward voltage exceeds a value known as the breakover voltage $V_{D(BO)}$, the thyristor switches into its low impedance state, as discussed in Section 7.2. The reverse breakdown voltages of $J_1$ and $J_2$ are both determined primarily by the doping concentration in the $N_1$-base region. Thus, to a first approximation, we expect that $V_{D(BO)} \approx V_{R(BR)}$, although the influence of electron injection over $J_3$ may cause $V_{D(BO)}$ to be lower.

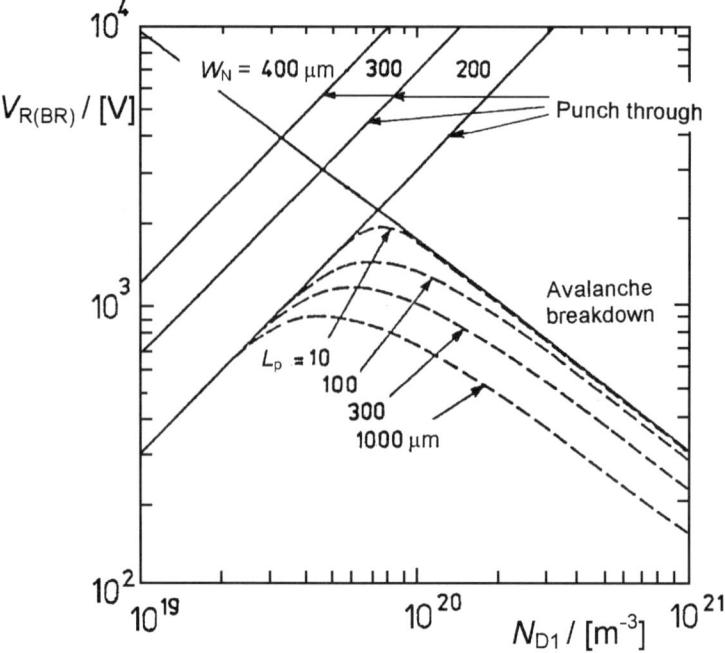

**Figure 7.2** The influence of the impurity concentration and thickness of the layer $N_1$ on the maximum reverse blocking voltage

### 7.1.3 Surface Profiles for High Breakdown Voltages

The absolute maximum reverse voltage $V_{RRM}$ and the absolute maximum forward blocking voltage $V_{DRM}$ that may be repeatedly applied at any permitted working temperature are normally specified for any given thyristor. These limiting voltages are established by leaving an adequate safety margin below the expected breakdown voltage. Thyristors are divided into groups according to the $V_{DRM}$ specification, some of which are set to match the standard line voltages used in different applications. Higher maximum voltages for non-repetitive surge conditions are also usually specified. They are designated $V_{RSM}$ and $V_{DSM}$, respectively, and may not be exceeded in any circumstances.

In order to obtain high values of forward and reverse blocking voltages, careful surface termination of the p–n junctions is essential, as described in Section 3.2.4. The high voltage junctions $J_1$ and $J_2$ are usually formed more than 40 $\mu$m below the silicon wafer surface, so that defects caused during the mechanical finishing of the wafers do not degrade the performance. Surface contouring of the wafer edge, using either positive or negative bevelling, is normally used to ensure high breakdown voltages for each high voltage junction and may be obtained either mechanically or by chemical etching.

Different combinations of positive and negative surface bevelling are used for different types of thyristor. A few examples are shown in Figure 7.3. For small-area

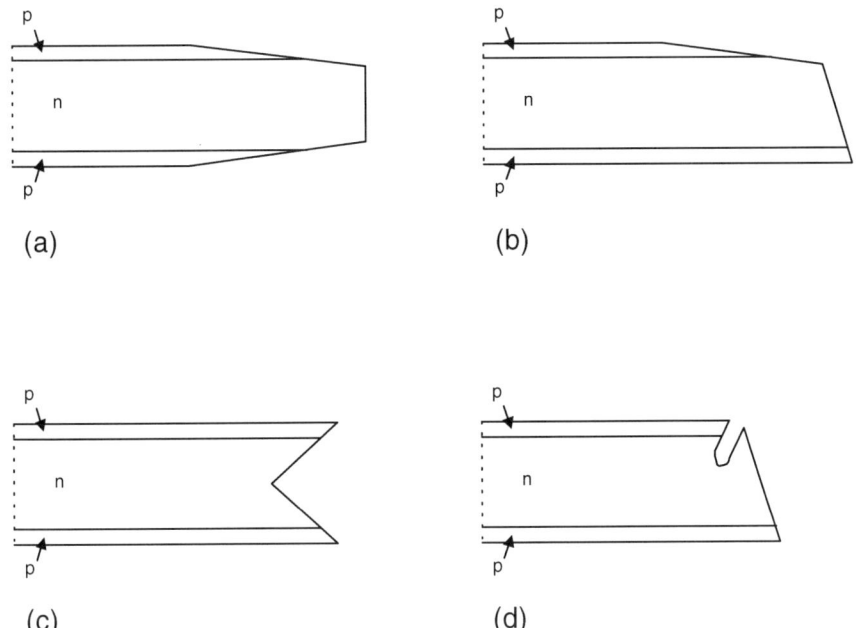

**Figure 7.3** Bevelled structures for high voltage thyristors: (a) double negative bevelling; (b) a combination of positive and negative bevelling; (c) double positive bevelling; (d) an alternative way to obtain a double positive bevel

thyristors with $V_{DRM} < 2000$ V, double negative bevelling, as shown in Figure 7.3(a), is often used. This can be also formed as a double mesa using chemical etching. Mechanical bevelling is used for circular thyristors, chemical etching for devices of rectangular cross-section. For devices whose diameter exceeds 10 mm and with $V_{DRM} < 3$ kV, a combination of positive and negative bevelling is often used, as shown in Figure 7.3(b). First, a high angle, positive bevel is formed (usually at $J_1$) and then a low angle ($<10°$), negative bevel is made (usually at $J_2$). A disadvantage of these configurations is the loss of active device area.

For high voltage thyristors, with $V_{DRM} > 3$ kV, the active area lost by the use of low angle, positive bevelling would be too great. Furthermore, it is difficult to reach a reverse breakdown voltage exceeding 3500 V using negative bevelling. As a result, high voltage, high current thyristors have both junctions $J_1$ and $J_2$ bevelled positively. Examples of such surface contouring are shown in Figure 7.3(c) and (d). The bevelled surface is usually etched and passivated by a silicone rubber or polyimide. In the case of smaller devices, high voltage planar technology is normally used, with sets of guard or field rings, or a combination of both, and with passivation by thermally grown oxide glass or else a SIPOS layer. These techniques are described in Section 3.2.4.

## 7.1.4 The Forward Conducting State

It will be shown in the next section that, for the thyristor to switch to the on-state, a large concentration of excess carriers must be injected, at least locally, into the vicinity of junction $J_2$. Once in the on-state, thyristors usually operate at quite high levels of current density, typically of order $1 \, \text{A/mm}^2$. This implies that the conditions are those of high injection and that the concentration of injected electrons and holes is much greater than the equilibrium concentrations of majority carriers in the inner layers of the thyristor structure ($N_1$ and $P_2$), as shown in Figure 7.4.

In the on-state, at these current densities, the thyristor behaves like a p–i–n structure, similar to a power diode. Consequently, the results obtained in Section 5.1 for the forward characteristics of the p–i–n diode can be used to model the on-state characteristics of thyristors. Equations (5.12) and (5.18) can be expected to apply equally well to the thyristor structure. According to the theory of Section 5.1, the forward on-state voltage $V_T$ depends strongly on the distance $w$ between the $n^+$- and $p^+$-layers. But, as shown in Figure 7.2, $w$ is determined by the required forward and reverse blocking voltages. The increase in $w$ needed to raise the blocking voltages increases $V_T$. This effect can be offset by increasing the carrier lifetime, but then the dynamic characteristics of the thyristor are degraded, as discussed in Section 7.4. The expected variation of $V_T$ with the overall thickness of the two base regions and the excess carrier lifetime $\tau$ is shown in Figure 5.2. Experimental measurements of the distribution of potential across the device are in good agreement with this theory, as shown in Figure 7.5. The theory assumes a high injection efficiency from both $n^+$- and $p^+$-emitters ($N_2$ and $P_1$) into their respective base layers ($P_2$ and $N_1$). This in turn requires that the impurity concentrations in the emitter layers are high. When this is not valid, $V_T$ rises rapidly at higher current densities, as analysed in Section 5.1.

Theoretical modelling of the thyristor on-state is complicated at lower levels of forward current. First, high injection conditions cannot be assumed to apply in the $P_2$-layer with current densities of less than $0.1 \, \text{A/mm}^2$. Secondly, in large-area thyristors it is found that only part of the device area is turned on [7.1]. The turned-on areas are separated from the non-conducting regions by a transition zone whose thickness is comparable to the carrier diffusion length. This is illustrated in Figure 7.6(a). The potential distributions across the on- and off-regions of the thyristor are shown in Figure 7.6(b). If the on-state current is increased slowly, the turned-on area spreads so that the current density in the turned-on region remains nearly constant, as does $V_T$.

Many aspects of thyristor behaviour can be understood using an equivalent circuit that models the device as two interconnected transistors, as discussed in the next section and illustrated in Figure 7.7. This shows that the thyristor can be considered to combine a pnp transistor and an npn transistor, with the collector of each attached to the base of the other. In this context a minimum on-state current density $J_M$ is needed to keep junction $J_2$ flooded with excess carriers and so maintain the two transistors in saturation. The value of $J_M$ increases with the total width of the two base regions and with the carrier recombination rate. That is, it is inversely

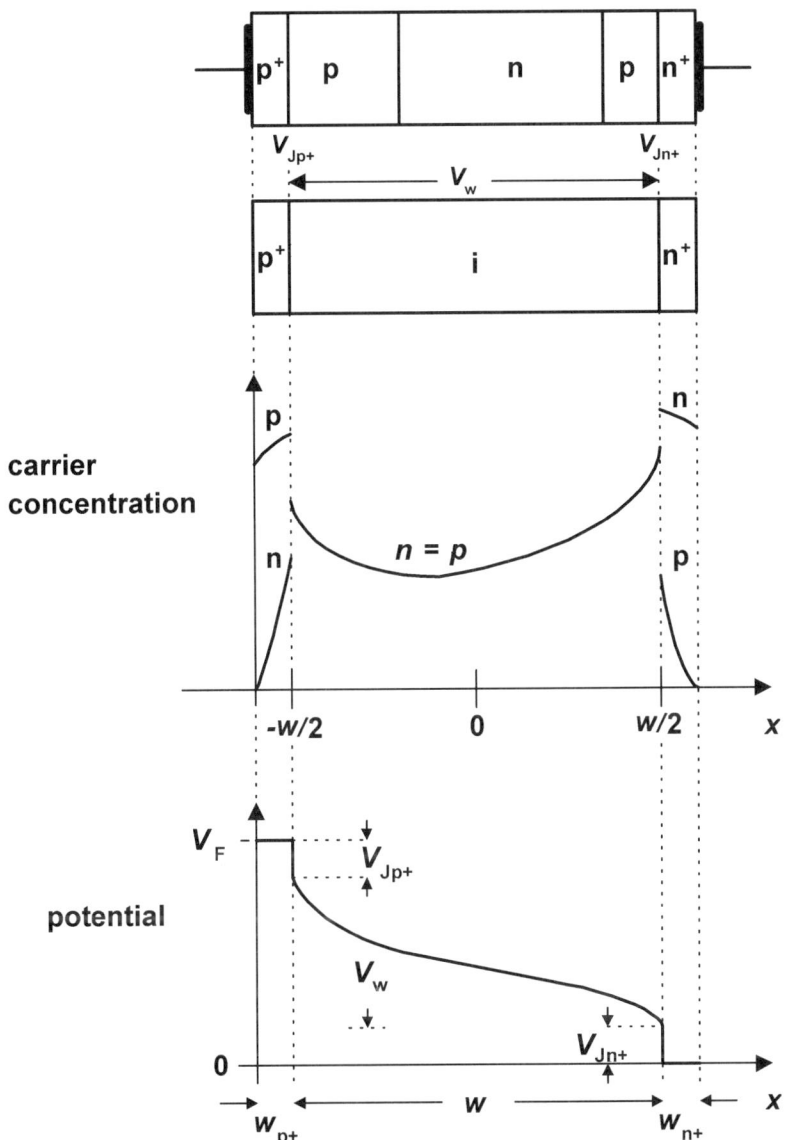

**Figure 7.4** Carrier concentration distributions in a thyristor in the on-state, showing that it can be modelled as a p–i–n diode structure

proportional to the carrier lifetime. It also increases when emitter shorts are used, as described in Sections 7.2 and 7.3.3.

During a slow decrease in the on-state current, the turned-on area contracts. Eventually, the thyristor turns off at a current called the holding current $I_H$. The on-state condition then exists only in a narrow filament whose cross-sectional area is typically about $10^{-2}$ mm$^2$. This is discussed in more detail in Section 7.4.2.

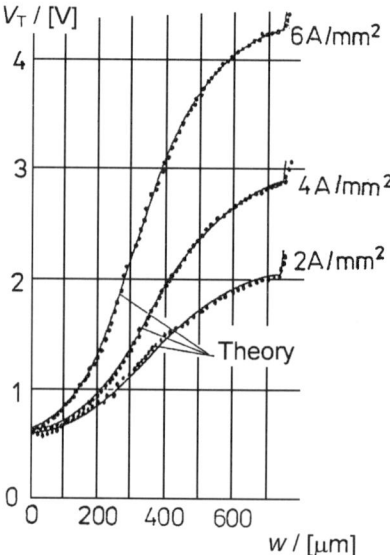

**Figure 7.5** A comparison of the measured potential distribution across a thyristor in the on-state with the predictions of the p–i–n diode theory. From a Technical Report by J. Homola. Reproduced by permission of Polovodiče, a. s., Prague

## 7.2 The Two-Transistor Model for Thyristor Switching

There are several mechanisms by which a thyristor can be turned on. When a thyristor is forward biased, with the gate terminal open circuited ($I_G=0$) and the applied forward voltage less than $V_{D(BO)}$, the thyristor is said to be in the forward blocking state. If the applied voltage is increased above $V_{D(BO)}$, the thyristor reverts to a low impedance state; it turns on. The mechanism by which this transition occurs can be modelled using the two-transistor equivalent circuit shown in Figure 7.7. When $I_G=0$ the anode current $I_A$ is given by

$$I_A = I_K = I_{E1} = I_{E2} = I_{C1} + I_{C2} \tag{7.5}$$

The currents are defined in Figure 7.7(c).

When turn-on occurs as a result of the breakdown voltage of junction $J_2$ being exceeded, carrier multiplication at the junction has to be taken into account. If the common-base current gain of the pnp transistor is $\alpha_1$ and that of the npn transistor is $\alpha_2$, the collector currents of the partial transistors can then be expressed as

$$I_{C1} = M(\alpha_1 I_{E1} + I_{C01}) \tag{7.6}$$

and

$$I_{C2} = M(\alpha_2 I_{E2} + I_{C02}) \tag{7.7}$$

## 7.2 THE TWO-TRANSISTOR MODEL FOR THYRISTOR SWITCHING

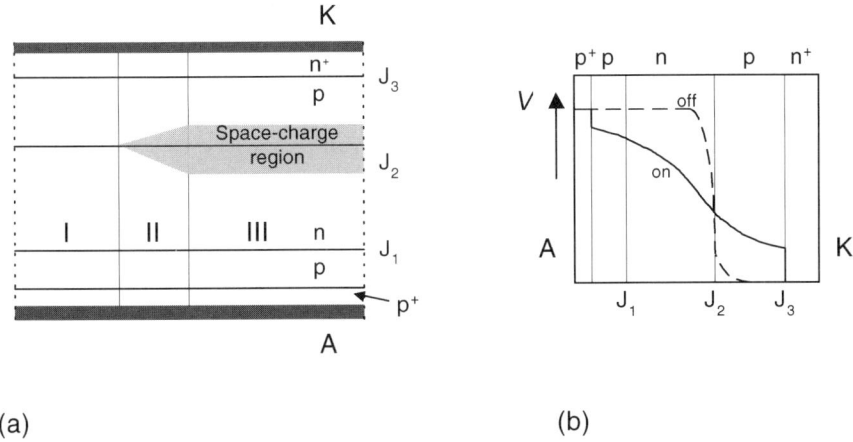

**Figure 7.6** (a) An illustration of the on- and off-state regions that occur in a thyristor at low forward current levels: (I) on-state region; (II) transition region; (III) off-state region. (b) The potential distributions across the on- and off-state regions

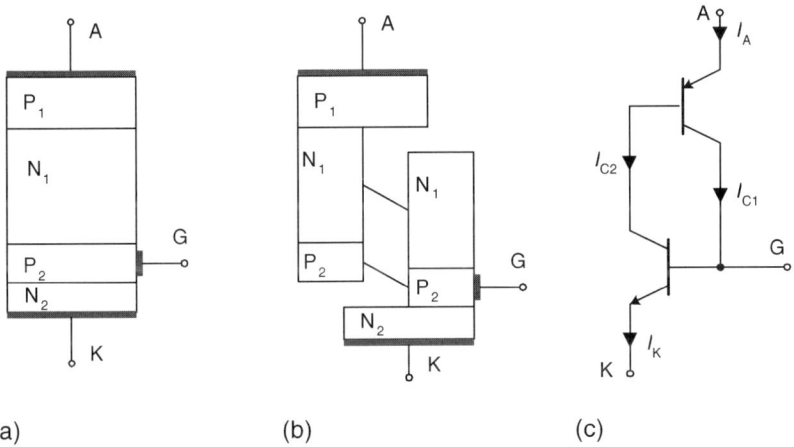

**Figure 7.7** The two-transistor model of the thyristor: (a) the basic thyristor structure; (b) showing the thyristor as two inter-connected transistors; (c) equivalent circuit

where $I_{C01}$ and $I_{C02}$ are the leakage currents of the partial transistors. The multiplication factor $M$ is given by (2.44). Note that the current gain coefficient $\alpha_1$ of the pnp transistor increases with the anode voltage because the expansion of the space charge region at junction $J_2$ into the lightly doped $N_1$-layer decreases the effective base width.

Substituting (7.6) and (7.7) into (7.5) shows the thyristor anode current to be

$$I_A = \frac{MI_{C0}}{1 - M(\alpha_1 + \alpha_2)} \tag{7.8}$$

where $I_{C0}=I_{C01}+I_{C02}$ is the total leakage current of the thyristor structure. Note that $I_{C0}$ is carried by majority carriers. As explained in Section 6.1, the current gain coefficients $\alpha_1$ and $\alpha_2$ are increasing functions of both the current flowing and the voltage applied. The anode current itself increases as $M(\alpha_1+\alpha_2)$ increases and goes to infinity when

$$M(\alpha_1 + \alpha_2) = 1 \tag{7.9}$$

The thyristor then switches into its low impedance state; it turns on.

If the gate is biased so that $I_G<0$, the injection efficiencies of junctions $J_1$ and $J_2$ are low. As a result, the current gains $\alpha_1$ and $\alpha_2$ are also low and (7.9) is fulfilled only when the multiplication coefficient $M$ is large. From (2.44) it follows that the breakover voltage $V_{D(BO)}$ at which turn-on occurs is then very close to the breakdown voltage $V_{BR}$ of the isolated $J_2$ junction.

When the gate is made positive with respect to the cathode, a positive gate current $I_G>0$ flows through the junction $J_3$ and the resulting increase of the emitter current causes the current gain coefficient $\alpha_2$ to increase. The thyristor can now turn on at a lower anode voltage, where we can assume $M=1$. The anode current is then

$$I_A = \frac{\alpha_2 I_G + I_{C0}}{1 - (\alpha_1 + \alpha_2)} \tag{7.10}$$

The current gain coefficients $\alpha_1$ and $\alpha_2$ increase with increasing current density. So starting in the absence of gate current, with $(\alpha_1+\alpha_2)<1$, $I_G$ can be increased to raise $(\alpha_1+\alpha_2)$ until it equals 1. The thyristor is then triggered into the low impedance state, where it remains even after the gate current is removed, provided the anode current has risen above a level known as the latching current $I_L$. The threshold gate current at which this occurs is designated $I_{GT}$. The latching current is somewhat higher than the holding current defined in Section 7.1.4.

The triggering condition can be derived more generally from the two-transistor model. Using (7.5) to (7.7), with the gate current added,

$$I_A = M(\alpha_1 I_A + I_{C01} + \alpha_2 I_A + \alpha_2 I_G + I_{C02}) \tag{7.11}$$

Turn-on occurs when an increasing gate current causes the anode current to rise uncontrollably, i.e. $dI_A/dI_G \to \infty$. Taking the derivative of (7.11) with respect to $I_A$ and setting $dI_G/dI_A=0$, the triggering condition is seen to be given by

$$\alpha_1 + \alpha_2 + I_A \frac{d\alpha_1}{dI_A} + (I_A + I_{GT})\frac{d\alpha_2}{dI_A} = \frac{1}{M} \tag{7.12}$$

If a positive gate current is supplied while the thyristor is reverse biased, a current of electrons $I_n$ is injected from the $n^+$-emitter. They diffuse through the $P_2$-layer to junction $J_2$, where they are collected by the electric field of the space charge layer. As $J_2$ is forward biased, this approximates to the built-in space charge region and the current $\alpha_2 I_n$ is essentially the reverse bias current crossing $J_1$. To balance this negative charge, holes are injected into $N_1$ from the anode layer $P_1$, and the hole current $I_p \approx \alpha_2 I_n$ is the emitter current of the pnp transistor. As a result, a gate current $I_G$ causes an increment in the anode current $\Delta I_A$, given by

## 7.2 THE TWO-TRANSISTOR MODEL FOR THYRISTOR SWITCHING

$$\Delta I_A = \frac{\alpha_1 \alpha_2}{1 + \alpha_1 \alpha_2} I_G \tag{7.13}$$

This can considerably increase the dissipation in the region surrounding the gate and it can lead to thermal breakdown.

The current gains $\alpha_1$ and $\alpha_2$ are increasing functions of temperature. This follows from (6.6) because both the injection efficiency and the diffusion length increase with temperature. A consequence is that the breakover voltage $V_{D(BO)}$ decreases with increasing temperature. In order to minimise this effect, the injection efficiency of junction $J_3$ can be reduced by connecting a resistance $R_{SH}$ between cathode and gate, as shown in Figure 7.8(a). The current flowing from the $P_2$-base region is divided into a current $I_{SH}$ flowing through the shunt resistance and a current $I_{J3}$ flowing through the p–n junction $J_3$. In practice $R_{SH}$ takes the form of some direct connections of the $P_2$-base region to the cathode contact, as shown in Figure 7.8(b). These are known as cathode shorts. As long as the voltage dropped across $R_{SH}$, and hence $J_3$, is less than about 0.5 V, the gate current flows directly through the $P_2$-base to the cathode without causing any injection of electrons from the $n^+$-emitter region across $J_3$. Above this, $I_{J3}$ rises more rapidly than $I_{SH}$ and the increased electron injection causes the current gain coefficient $\alpha_2$ to rise.

The triggering condition can now be expressed as

$$\alpha_1 + \alpha_{2\text{eff}} \geqslant 1 \tag{7.14}$$

where we have defined an effective current gain coefficient for the npn transistor as

$$\alpha_{2\text{eff}} = \frac{\alpha_2}{1 + I_{SH}/I_{J3}} \tag{7.15}$$

Since $V_{J3} = I_{SH} R_{SH}$ this becomes

$$\alpha_{2\text{eff}} = \alpha_2 \left[ 1 + \frac{V_{J3}}{R_{SH} I_{J30} \{\exp(eV_{J3}/kT) - 1\}} \right]^{-1} \tag{7.16}$$

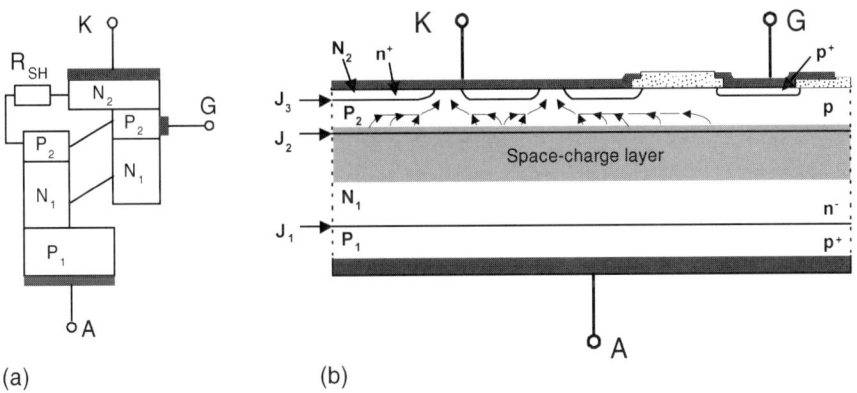

**Figure 7.8** An illustration of the use of cathode emitter shorts: (a) equivalent circuit in the two transistor model; (b) device cross-section

Figure 7.9 shows the reduced temperature dependence of the breakover voltage $V_{D(BO)}$ for a thyristor with cathode shorts, compared with one of similar construction made without shorts. The predicted breakdown voltage of the isolated junction $J_2$ is also given for comparison.

## 7.3 Transient Processes during Turn-On

The transition of a thyristor from the forward blocking off-state to the forward conducting on-state was initially analysed in terms of the two-transistor model, as described in the last section. However, the detailed physical dynamics of the turn-on and turn-off phases are much more difficult to analyse. The three steady-state conditions described in Section 7.1 can be modelled quite well using what is essentially a one-dimensional analysis. But the transient processes of turn-on and turn-off, as well as being inherently time-varying, also involve all three spatial dimensions. The discussions to be found in the literature are usually based on the two-transistor, charge-control approach and as a result are greatly oversimplified.

What distinguishes thyristor turn-on is that, once the anode current exceeds the latching current, the gate loses control of the turn-on process. The subsequent rise of the anode current and the fall of the anode voltage are determined by the physical processes inside the device and the constraints of the external circuit. The circuit configuration and the initial conditions set the boundary conditions that determine the way these physical processes evolve.

At its simplest, the external circuit can be approximated by the equivalent circuit shown in Figure 7.10. $V_D$ is the anode–cathode voltage across the device; $Ri_F$ is the voltage across any circuit resistance; and $L di_F/dt$ is the voltage across any circuit inductance. At any instant their sum has to equal the supply voltage $V_0$:

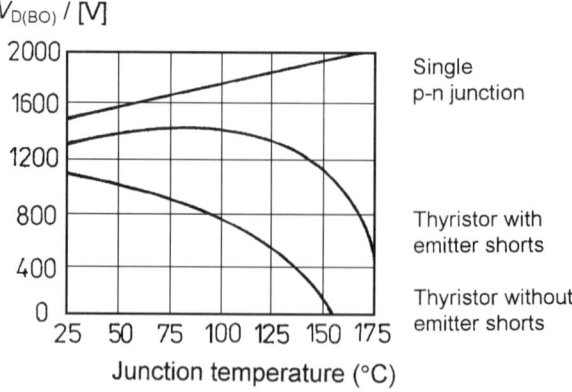

**Figure 7.9** The influence of emitter shorts on the temperature dependence of the forward blocking voltage. Taken from R. A. Kokosa, 'A High-Voltage, High-Temperature Reverse Conducting Thyristor', *IEEE Trans. on Electron Devices*, **ED-17**, 667–7 (1970). Reproduced by permission of The Institution of Electrical and Electronic Engineers

## 7.3 TRANSIENT PROCESSES DURING TURN-ON

$$V_0 = V_D + Ri_F + L\frac{di_F}{dt} \qquad (7.17)$$

Before the thyristor current can start to rise, $V_D$ has to fall. This requires the self-capacitance of the device to be discharged. The capacitance mainly arises in the space charge region on either side of $J_2$. The whole of the cross-sectional area of the junction has to be discharged, the non-conducting regions as well as those sections where conduction is initiated. The situation is illustrated in Figure 7.11 for a thyristor triggered in the usual way by a pulse of gate current.

A current of electrons is needed to discharge the region of positive space charge in the $N_1$-layer. They cannot be supplied from the anode, across $J_1$, but have to be injected from the cathode–emitter layer over $J_3$. They then diffuse across the $P_2$-layer and pass through the space charge region into the $N_1$-layer. Likewise, the holes required to discharge the negatively charged space charge region in layer $P_2$ have to be injected over $J_1$, from where they diffuse through layer $N_1$ and pass through the space charge region into layer $P_2$.

The current required to discharge the junction capacitance $C_J$ is additional to any current that flows in the external circuit during the turn-on period. The capacitive energy stored at the junction is dissipated in the conducting channels. It is given by

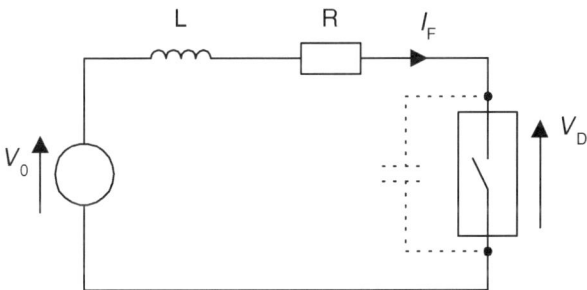

**Figure 7.10** A simple equivalent circuit determining the relationship between the anode current and voltage during turn-on

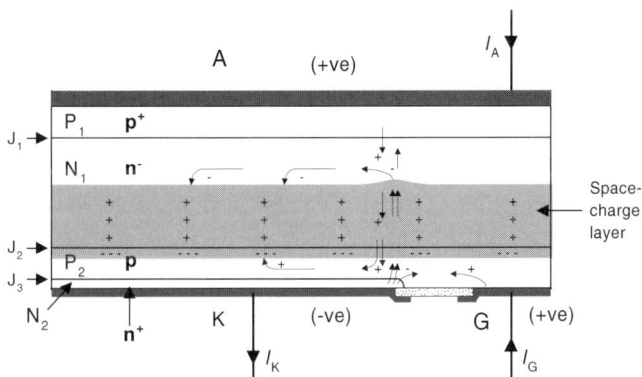

**Figure 7.11** Carrier flow during gate-triggered turn-on

$$C_J V_0 = A\sqrt{e\varepsilon_r\varepsilon_0 N_D V_0} \qquad (7.18)$$

where $A$ is the junction area and where the total junction capacitance $C_J$ is twice the differential capacitance given by (2.34). We have assumed that $N_D$, the donor concentration in layer $N_1$, is much less than $N_A$, the acceptor concentration in layer $P_2$.

Increasing the device area increases the current rating but also increases the stored capacitive energy. In a high voltage, large-area device this can cause problems if the device is turned on by a locally initiated breakdown and the energy is discharged through a very narrow channel. Breakover, through overvoltage or a rise in temperature, may occur at a local inhomogeneity or result from the carrier generation caused by a high energy particle such as a cosmic ray. If the stored energy exceeds a critical value, a microplasma may form and the localised dissipation can cause instantaneous device failure. This effect can limit the practical cross-sectional area of high voltage, high current thyristors.

If the circuit current is allowed to rise significantly while $V_D$ is falling, the dissipation in the conducting channels is greatly increased. For these reasons it is desirable that they occupy as large an area as possible during the period of peak dissipation. Making the circuit time constant $L/R$ significantly longer than the anode voltage fall time limits the rise rate of the anode current.

In the following three subsections we discuss first the normal method of turning on a thyristor by means of a gate current pulse, then the limitation that has to be imposed on the rise rate of the anode current and finally the risk of triggering a thyristor inadvertently as a result of an excessively high rise rate of the anode voltage. Light-triggered thyristors and breakover diodes, which are ungated, are considered in Sections 8.5 and 8.6.

### 7.3.1 Gate Turn-On

The application of a gate voltage, positive with respect to the cathode, causes a flow of gate current across the forward-biased gate–emitter junction ($J_3$). In normal gate triggering, a short voltage pulse is applied through a resistor. Sometime after the start of the resulting gate current pulse, the thyristor starts to conduct: the anode current rises, eventually reaching its full value, as determined by the circuit impedance. The required gate triggering current $I_{GT}$ is an important thyristor parameter. Together with the gate voltage $V_{GT}$ needed to turn on the thyristor, it is specified in data sheets over the whole range of working temperatures, i.e. up to $T_{Jmax}$. Likewise, the highest gate current and voltage, $I_{GD}$ and $V_{GD}$, that will *not* cause the thyristor to turn on are also specified. These parameters are important since they determine the maximum noise level permissible in the gate circuit. Figure 7.12 shows an example of a set of gate–cathode characteristics with the areas of safe triggering and maximum allowable power dissipation defined. The use of emitter shorts causes an increase in $I_{GT}$.

Figure 7.13 shows typical turn-on waveforms. The turn-on period may be divided into three phases: the initial delay time $t_d$, the anode voltage fall time $t_f$, and the final rise of the anode current. The duration of this last period is normally determined by the inductance of the circuit.

The physical processes during gate-triggered thyristor turn-on can be explained as follows. The application of a positive gate voltage to a thyristor in the forward blocking state causes electrons to be injected from the $n^+$-cathode layer (the $N_2$-emitter) into the gate layer (the $P_2$-base). The electron injection takes place along the edge of the gate contact, in a similar manner to that described in Section 6.2 for the case of a transistor. The injected electrons diffuse across the $P_2$-layer towards $J_2$. The time required for them to reach the edge of the $J_2$ space charge region is given by (6.45) as $w_p^2/2D_n$. On reaching the space charge region, they are accelerated into the $N_1$-base layer by the built-in electric field. The negative charge of these electrons lowers the potential of the $N_1$-layer and increases the forward bias across junction $J_1$. As a result, holes are injected across junction $J_1$, from the $P_1$-layer into the $N_1$-layer. During a transit time given by $w_n^2/2D_p$, some diffuse to the $N_1$-edge of the $J_2$ space charge layer. They are then accelerated across $J_2$ and into the $P_2$-region, which they cause to become more positively biased. This, in turn, increases the electron injection from the $N_2$-layer across $J_3$ and so gives rise to positive feedback.

When the increase of electron injection resulting from these processes exceeds the recombination rate, there is net regeneration and ever increasing concentrations of excess carriers (electrons and holes, respectively) are injected over junctions $J_3$ and $J_1$. Some of them discharge the self-capacitance of the space charge region of $J_2$, while the rest flow out into the external circuit as the thyristor

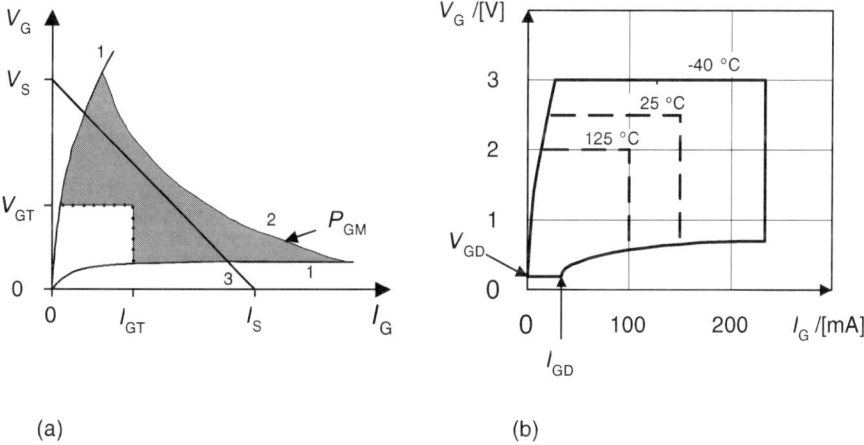

**Figure 7.12** Gate operating limits: (a) Curves (1) represent the extreme characteristics for all devices at any allowed operating temperature. Curve (2) represents the maximum permitted gate power dissipation. The lines $V_{GT}$ and $I_{GT}$ indicate the gate voltage and gate current that are guaranteed to turn-on the thyristor under any circumstances. Thus the shaded area is the effective working area for gate triggering. Any gate circuit load line, such as (3) should therefore lie entirely within the shaded area. (b) The $I_{GT}$ and $V_{GT}$ limits are shown for three different temperatures. $I_{GD}$ and $V_{GD}$ are the limiting current and voltage that are guaranteed not to turn-on the thyristor

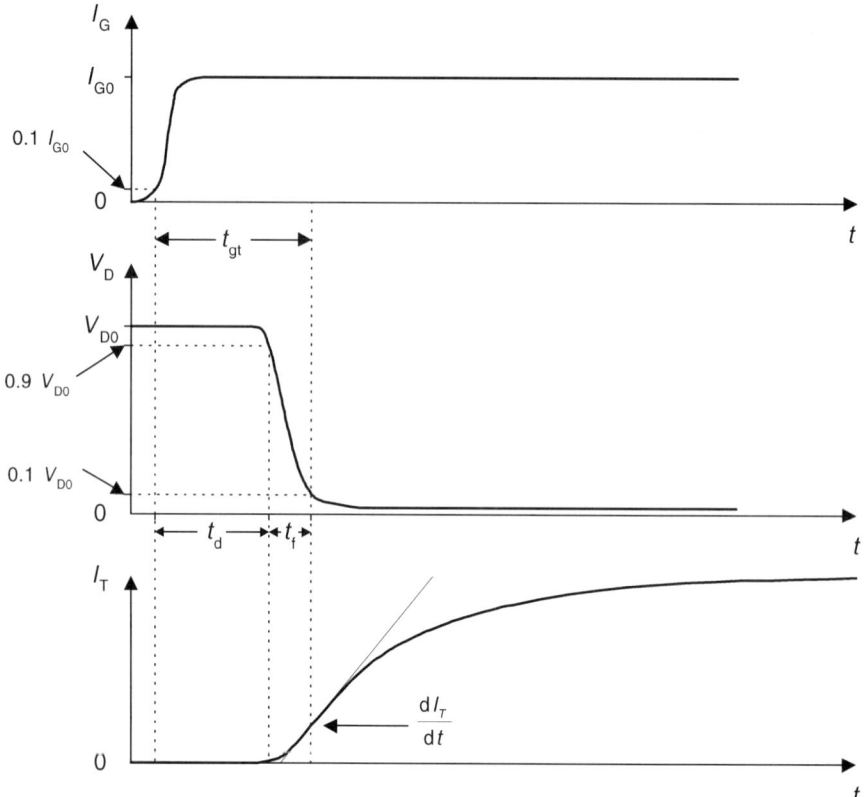

**Figure 7.13** Typical current and voltage waveforms following gate-triggered turn-on

switches into its low impedance state. In the two-transistor analogy, the process of flooding junction $J_2$ with excess carriers can be identified with the switching of at least one of the two transistors into the saturation mode. This requires a sufficient excess carrier concentration at junction $J_2$ to inhibit the formation of a space charge layer.

As we found for a bipolar junction transistor, the current density is initially highest along the edge of the $n^+$-emitter area next to the gate contact, and this is the first region of the thyristor structure to turn on. Increasing the amplitude and the rate of rise of the gate current increases the initially turned-on area. In order to maintain the turned-on state after the gate signal is removed, the dimensions of the turned-on area have to exceed the carrier diffusion length. This is typically of the order of tens of microns. The thyristor is able to remain in the on-state, without a gate current, only if the anode current $I_A$ is greater than the latching current $I_L$ shown in Figure 3.4(c).

The delay time $t_d$ is normally defined as the time interval from the moment the gate current reaches 10% of its final value to the time when the anode voltage falls to 90% of its initial value. It represents the time taken for the $J_2$ space charge region to

## 7.3 TRANSIENT PROCESSES DURING TURN-ON

start to become flooded with carriers, so that the voltage across it starts to decrease. It is thus somewhat greater than the sum of the transit times, by diffusion, of the carriers crossing the neutral regions of the base layers:

$$t_d > \frac{w_n^2}{2D_p} + \frac{w_p^2}{2D_n} \tag{7.19}$$

Initially, the distances $w_n$ and $w_p$ are much less than the full widths of the base regions because of the extent of the space charge layer. A complicating factor is that the space charge of the free carriers, initially the electrons, may be sufficient to disturb the space charge layer by increasing its thickness in the p-base and reducing it in the n-base.

In practice the delay time depends on the amplitude and the rise rate of the gate current and on the initial anode voltage. A typical value for a 4 kV thyristor is about 2 μs, whereas in lower voltage devices it usually about 0.5 μs. The longest delay time is for the minimum gate triggering current $I_{GT}$, defined in Section 7.2. Increasing the gate current, $I_G > I_{GT}$, decreases the delay time. The gate current introduce excess carriers into the inner layers of the thyristor. There is a minimum excess carrier charge $Q_{GT}$ needed to initiate positive feedback and cause turn-on. With the simplified assumption that $Q(t_d) = Q_{GT}$,

$$t_d = \tau \ln\left[\frac{I_G}{I_G - I_{GT}}\right] \tag{7.20}$$

The thyristor can turn on if the amplitude and duration of the gate current pulse satisfy the condition $I_G t_{Gp} > Q_{GT}$ when the current is constant, or

$$\int_0^{t_{Gp}} I_G(t) dt \geq Q_{GT} \tag{7.21}$$

otherwise. The excess carriers injected into the $P_2$-layer by an impulse of gate current can induce an increase of anode current that fulfils the turn-on conditions even after the gate pulse is finished.

During the next phase of the turn-on process, a section of the $J_2$ space charge region is rapidly flooded with excess carriers, its thickness decreases and the voltage across it falls, leading to a reduction of the anode voltage and the rise of the anode current. The time needed to achieve this is normally defined from the time $t_f$ during which the blocking voltage $V_D$ decreases from 90% to 10% of its initial value. The overall gate triggering turn-on time is then $t_{gt} = t_d + t_f$.

The gate triggering current has to flow laterally through the $P_2$-layer, as shown in Figure 7.11. This requires a lateral potential gradient, which causes the forward bias of junction $J_3$ to be highest along the edge nearest to the gate contact. It is here that turn-on originates, in a strip about a diffusion length wide (say about 100 μm) along the periphery of the cathode–gate junction. The current density and the excess carrier concentration in this channel rapidly become very high. There is a high lateral gradient of the excess carrier concentration which promotes the lateral diffusion of the excess carriers.

Lateral fields are also set up in the $N_1$-base. These promote the lateral flow of the electrons that have come through the space charge layer, so they are able to discharge the remoter regions of the space charge layer. This helps to expand the area of junction $J_1$ that is forward biased, injecting holes in the reverse direction. On their arrival back in the $P_2$-base layer, some of these holes must also flow laterally, in order to discharge the space charge layer on the cathode side of junction $J_2$.

Under the influence of the lateral diffusion and the lateral electric fields that persist after the collapse of the anode voltage and the discharge of the space charge layer, the boundary of the turned-on area spreads out from the edge of the cathode emitter. It travels at a speed $v_s$ called the spreading velocity, that varies sublinearly with the current density. Various analytical expressions have been proposed [7.2, 7.3], but as the current density itself is increased by the rising current and diminished as the area in conduction expands, they are of limited value. The turned-on area reaches its maximum extent ($v_s \to 0$) when the current density falls to the minimum value $J_M$ needed to keep the partial transistor structures in saturation. The spreading velocity increases with increasing temperature, but decreases with increasing base width, with a higher density of cathode shorts and with a decrease in the carrier lifetime.

An example of the experimentally observed dependence of the spreading velocity on the current density with different patterns of emitter shorts is shown in Figure 7.14. The turn-on time $t_{gt}$ depends on the rise rate of the on-state current $di_T/dt$. When this is high, the on-state voltage is increased and fall time $t_r$ of the anode voltage is longer. When the turn-on time $t_{gt}$ is used to compare different devices, it must be measured with parameters such as $V_D$, $di_T/dt$ and the gate current waveform specified.

The spreading of the turned-on area influences the dynamic turn-on characteristics. At high $di_T/dt$ the current density in the turned-on area may be much greater than $J_M$ because the device area available for conduction at any given time is limited by the spreading velocity. The instantaneous value of the on-state voltage thus depends on the rise rate of the current and is higher than under static conditions. A comparison of typical static and dynamic on-state current–voltage characteristics is given in Figure 7.15. It is clear that phenomena associated with the spreading of the turned-on area lead to higher power dissipation than under steady-state conditions. The thermal management of the device must take account of this, and suitably robust gating waveforms must be provided to ensure the area of the device that takes part in the initial phase of conduction is as large as possible.

### 7.3.2 Critical di/dt

There is an upper limit to the permissible rise rate of the on-state current if irreversible damage to the electrical characteristics of the thyristor is to be avoided. The parameter $(di_T/dt)_{crit}$, the critical rise rate of the on-state current, is a direct function of the area initially turned-on and the spreading velocity. We have seen that the current density for the turned-on area may be much higher than for the steady on-state. It increases with increasing $di_T/dt$ and can be as high as 10–1000 A/mm$^2$ ($10^3$ to $10^5$ A/cm$^2$). As a result, the voltage across the thyristor can remain high

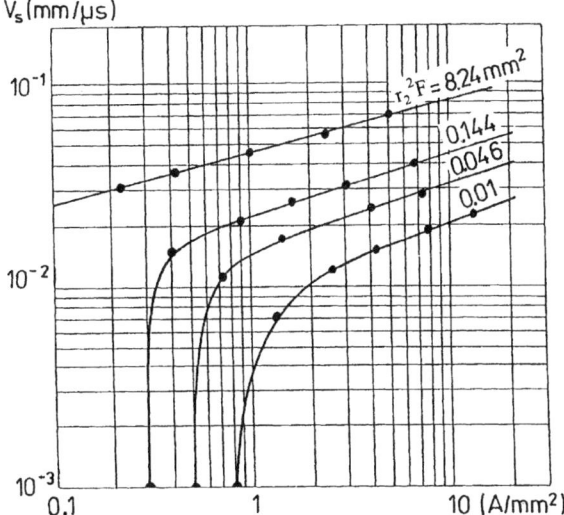

**Figure 7.14** The variation of the spreading velocity with the on-state current density, showing the effect of the density of cathode shorts. The parameter, $(r_2^2 F)$ is defined in (7.27). From a Technical Report by B. Pína. Reproduced by permission of Polovodiče, a. s., Prague

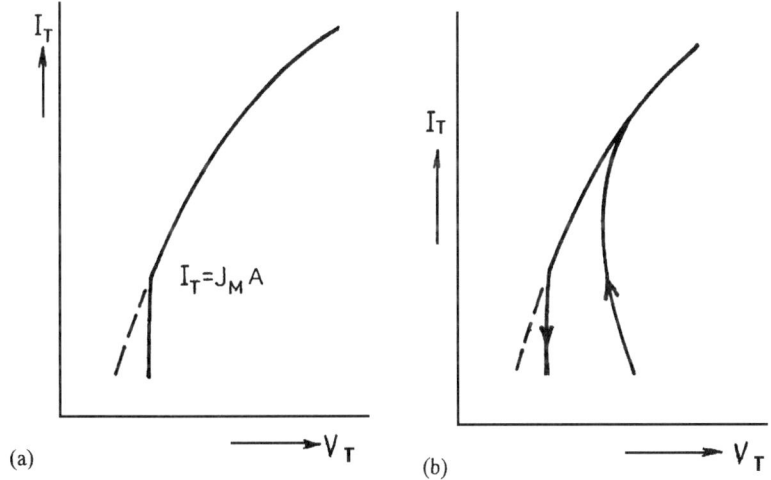

**Figure 7.15** On-state characteristics of a thyristor, showing the effect of the spreading resistance: (a) static characteristic; (b) dynamic characteristics, showing the higher voltage drop that occurs during turn on

during the rise of current, typically in the range 10–100 V. It is an increasing function of $di_T/dt$.

Examples of current, voltage and power waveforms are given in Figure 7.16. With $di_T/dt$ of order 100 A/μs, the instantaneous power can reach tens of kilowatts. All this energy is concentrated in a very small volume within the thyristor structure. The energy density is given by

$$W_V = \frac{1}{A_0 d} \int_0^t P(t) \frac{A_0}{A(t)} dt \tag{7.22}$$

where $A_0$ is the initial turn-on area, $d$ is the overall thickness of the silicon wafer, $A(t)$ is the turned-on area and $P(t)$ is the instantaneous power dissipation.

When the rise rate of the current exceeds $(di_T/dt)_{crit}$, the local overheating of the turned-on region is likely to cause a significant deterioration of the thyristor characteristics. The device may even be destroyed in a single shot. There are two possible causes. The temperature may rise sufficiently to exceed the melting point at the metal–semiconductor contact. Alternatively, the internal mechanical stress caused by the local overheating can cause the silicon crystal to fracture, especially under repetitive operation.

The critical rate of rise of the current can be increased by increasing the cross-sectional area of the thyristor that is initially turned on. The application of a gate current pulse that rises rapidly to a high value enables the condition for triggering to be reached all along the periphery of the cathode emitter surrounding the gate contact.

The gate–cathode structure known as the amplifying gate is effective in increasing $(di_T/dt)_{crit}$. Its structure is shown in cross-section in Figure 7.17(a) and in plan in Figure 7.17(b). It is an integrated structure that can be represented by the equivalent

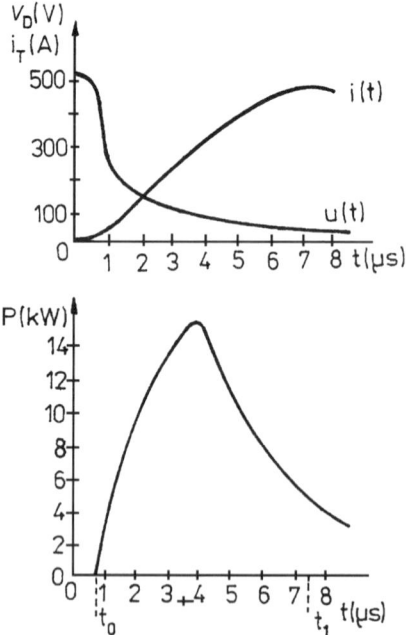

**Figure 7.16** Measured variation of voltage, current and dissipated power during thyristor turn-on at high $di_T/dt$

circuit of Figure 7.17(c). Effectively, part of the rising anode current is used to provide a large gate current impulse. Initially the auxiliary thyristor turns on (1) and its anode current flows out through the auxiliary contact to become the gate current of the main device (2). In this way, almost all the periphery of the cathode emitter adjacent to the auxiliary contact is triggered. Values of $(di_T/dt)_{crit}$ in repetitive operation typically exceed 200 A/μs and in some special types of thyristor can exceed 1000 A/μs. Non-repetitive values can be some two to three times higher.

The amplifying gate depends on a high rise rate of the current in the main thyristor to generate a large initial turned-on area. Paradoxically, it can fail when the applied gate current pulse is too high. This may cause the main thyristor to be triggered locally at the same time as the auxiliary thyristor.

The interdigitated gate–cathode geometry, shown in Figure 7.17(b), provides a long gate–cathode periphery. As long as the gate current is large enough to turn on the whole length, the initial turned-on area can be large and the turn-on losses can be minimised. For this reason the amplifying gate, with an interdigitated gate–cathode structure, is often used in thyristors designed to work at higher frequencies.

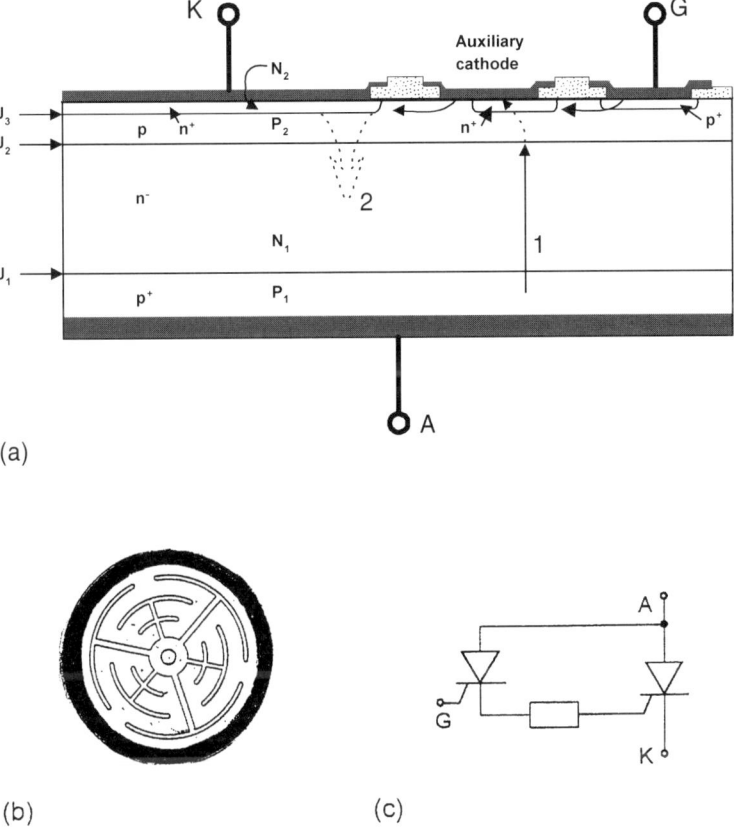

**Figure 7.17** The amplifying gate structure: (a) cross-section through the active region; (b) plan view; (c) the equivalent circuit. Part (b) reproduced by permission of Westcode Semiconductors Limited, Chippenham, England

An alternative layout is the field-initiated gate structure shown in Figure 7.18. When a gate triggering pulse is applied, the area turned on initially is located at the edge of the $n^+$-emitter. The anode current flows laterally through the $n^+$-emitter to the auxiliary contact and then through the non-contacted part of the emitter to the main cathode contact. This sets up a lateral electric field in the $P_2$-layer under the non-contacted part, which accelerates the initial spreading of the turned-on area.

### 7.3.3 Critical $dV/dt$

Another dynamic parameter connected with the turn-on process is the critical rate of rise of the blocking voltage $(dV_D/dt)_{crit}$. This is the maximum rise rate for the forward voltage that can be sustained by the thyristor in the blocking state under open gate conditions. The displacement current density needed to charge the differential capacitance $C_j$ of the space charge layer at junction $J_2$ is

$$J_q = \frac{d}{dt}(C_j V_D) = \left(C_j + V_D \frac{dC_j}{dV_D}\right) \frac{dV_D}{dt} \tag{7.23}$$

In the $P_2$-layer this current density represents the current of holes that are displaced by the expansion of the space charge region towards $J_3$. It has an effect similar to a positive gate current. It is possible for $J_q$ to fulfil the triggering condition (7.14). The parameter $(dV_D/dt)_{crit}$ is the highest value of $dV_D/dt$ which does not cause the thyristor to trigger in this way. It can be increased considerably by the use of cathode shorts, as described in Section 7.2. They enable the hole displacement current to be conducted to the cathode without the injection of electrons from the cathode emitter.

The shorts normally take the form of circular gaps in the emitter diffusion. For the most effective pattern of shorts, their centres should lie on the vertices of a network of equilateral triangles, as shown in Figure 7.19. Their effectiveness depends on the conductivity of the $P_2$-layer and on the ratio of their radius $r_2$ to their separation $2r_1$.

In this configuration the axial displacement current density $J_q$ has to flow radially in the $P_2$-layer into the shorts through a cylindrical annulus of internal diameter $2r_2$,

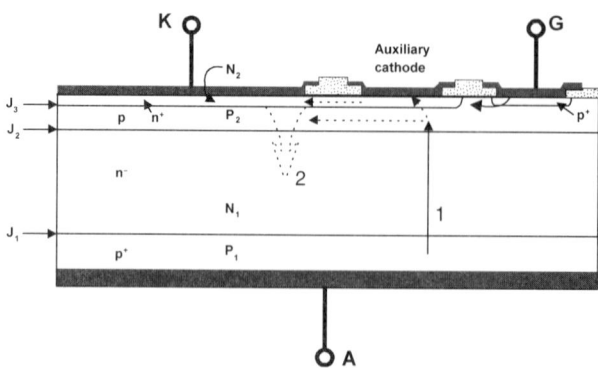

**Figure 7.18** The field-initiated gate structure

## 7.3 TRANSIENT PROCESSES DURING TURN-ON

external diameter $2r_1$, thickness $w_P$ and resistivity $\rho$. The maximum lateral voltage $V_E$ that can be allowed is limited to about 0.5 V by the need to avoid the direct injection of electrons over $J_3$. The lateral current $I(r)$ flowing into a given short at a distance $r$ from its centre is

$$I(r) = J_q \int_r^{r_1} 2\pi r \, dr = J_q \pi (r_1^2 - r^2) \tag{7.24}$$

The radial potential gradient is

$$\frac{dV}{dr} = \frac{I(r)\rho}{2\pi r w_P} \tag{7.25}$$

The voltage between $r_1$ and $r_2$, can be expressed as

$$V_{SH} = \int_{r_1}^{r_2} \frac{dV}{dr} dr = \frac{\rho J_q r_2^2}{2 w_P} F\left(\frac{r_1}{r_2}\right) \leqslant V_E \tag{7.26}$$

where $F$ is a geometrical factor that takes account of the pattern of shorts. In the case of the triangular pattern, the outer limit of the integration is not a circle of radius $r_2$, as indicated in (7.26), but the hexagon shown in Figure 7.19. Then it can be shown that

$$F = \frac{1}{4}\left[1 - \frac{r_1^2}{r_2^2}\left(1 - 2\ln\frac{r_1}{r_2}\right)\right] \tag{7.27}$$

which is plotted in Figure 7.20.

If the thyristor triggers when $V_{SH} > V_E$, the critical displacement current density is

$$J_{crit} = \frac{2 w_P V_E}{\rho r_2^2 F} \tag{7.28}$$

The critical current density $J_{crit}$ increases with the thickness of the $P_2$-layer and decreases with its resistivity and with the geometrical factor $F$. Often the resistivity of the p-base is reduced immediately under the $n^+$-emitter and in the surface region by a shallow diffusion of acceptors (often gallium).

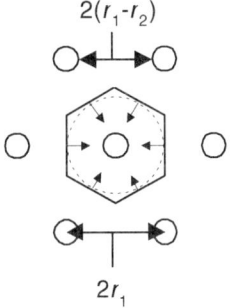

**Figure 7.19** Plan view of cathode shorts forming a pattern of equilateral triangles

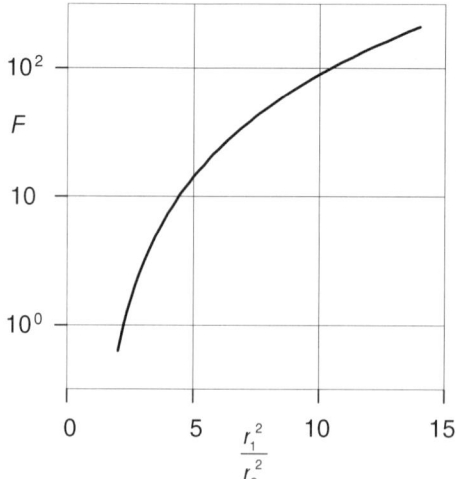

**Figure 7.20** Plan view of the cathode-short geometric factor, $F$, with the ratio, $(r_1/r_2)$, for a pattern of equilateral triangles

A special cathode short normally surrounds the whole device perimeter when a double-bevel, high voltage termination is used. This diverts to the cathode any displacement current flowing in the peripheral region.

Values of $(dV_D/dt)_{crit}$ exceeding $1000\,V/\mu s$ can be obtained with optimised patterns of cathode shorts. Without cathode shorts, this limit reduces to a few volts per microsecond. The thermal stability of the breakover voltage is also improved. However, $V_E$ decreases with temperature and $\rho$ increases with temperature. Consequently, $J_{crit}$ and $(dV_D/dt)_{crit}$ decrease with temperature. Any contact resistance between the cathode metallisation and the $P_2$-layer reduces the effectiveness of the cathode shorts.

Cathode shorts reduce the spreading velocity following turn-on and so may increase the turn-on losses. They are also likely to make a slight increase in the static on-state voltage. An example of a short pattern used on a 1 kHz thyristor with an amplifying gate structure and high $dV_D/dt$ rating is shown in Figure 7.21.

In GTO thyristors, where cathode shorts cannot be used, a considerable improvement in $(dV_D/dt)_{crit}$ can be obtained by applying a negative gate voltage.

## 7.4 Transient Processes during Turn-Off

Turn-off is the name given to the transition from the forward conducting on-state to the forward blocking state. In the on-state the inner layers of the thyristor structure are flooded with excess carriers which create an electron–hole plasma. Before the blocking state of the thyristor can be restored, the carrier concentrations must reduce to values near to those of thermodynamic equilibrium so that the space charge region at junction $J_2$ can be re-established. This can come about in one of several ways,

## 7.4 TRANSIENT PROCESSES DURING TURN-OFF

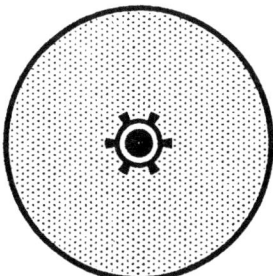

**Figure 7.21** A plan view of the pattern of cathode shorts in a thyristor with an amplifying-gate structure. Reproduced by permission of Polovodiče, a. s., Prague

depending on the circuit in which the device is being operated. There are normally three possibilities:

1. Turn-off by circuit commutation
2. Turn-off when the on-state current decreases below the holding current
3. Turn-off brought about by the application of a negative gate voltage

The last of these is the basis of the gate turn-off thyristor, one of the most important power electronic devices.

### 7.4.1 Turn-Off using Circuit Commutation

When a thyristor is operating in an ac circuit, the reverse recovery phase that follows the reversal of polarity in the circuit is similar to that of a power diode. Provided that the supply frequency is not too high, the excess carrier concentration decreases to zero during the reverse-biased period and the blocking state is restored when the anode voltage returns to a positive value. With a dc supply, turn-off requires an auxiliary source that is able to commutate the anode current when it is applied to the thyristor for a short time. Anode current commutation is usually obtained by connecting a capacitance through an auxiliary thyristor, as shown in Figure 7.22.

As explained in Section 7.1.4, the excess carrier distribution in a thyristor in the on-state is similar to that in a forward-biased $p^+$–n–$n^+$ diode. When the polarity of the anode voltage is reversed, the recovery process is similar to that of a diode and the analysis developed in Section 5.3.2 can be applied directly. Typical voltage and current waveforms during the turn-off period are shown in Figure 7.23. The turn-off time is the interval from the instant the falling anode current crosses zero until the instant the anode voltage again becomes positive. It is shown in Figure 7.23 as $t_q$. The thyristor turns off successfully only if the excess free charge remaining in the device after $t_q$ is less than some critical value $Q_{cr}$.

It is conventional here to use the instant of anode current reversal as time zero. With this modification the turn-off condition is derived from (5.29) as

$$Q(t_q) = Q_{cr} = Q_1 \exp\left(-\frac{t_q - t_s}{\tau_{eff}}\right) \qquad (7.29)$$

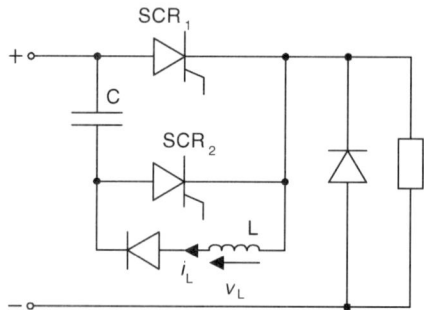

**Figure 7.22** A circuit designed to provide forced commutation at turn-off (auxiliary resonance turn-off)

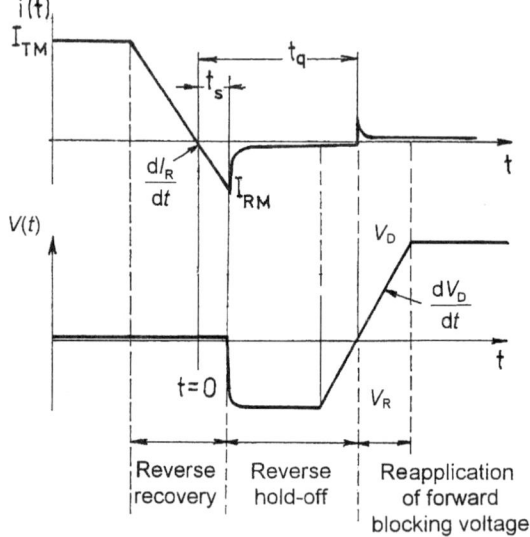

**Figure 7.23** Current and voltage waveforms in a thyristor turning-off by forced commutation

where $Q_1$ is the free charge present at time $t_s$, the instant at which a space charge layer starts to form at junction $J_1$. Thus

$$t_q = t_s + \tau_{\text{eff}} \ln\left(\frac{Q_1}{Q_{\text{cr}}}\right) \approx \tau_{\text{eff}} \ln\left(\frac{Q_1}{Q_{\text{cr}}}\right) \qquad (7.30)$$

because normally $t_s \ll t_q$. As in the case of the p–i–n diode, the charge $Q_1$ depends on the circuit commutation conditions ($di_R/dt$ and $V_R$). The critical charge $Q_{\text{crit}}$ decreases with increasing rate of rise of the anode voltage at recovery.

In many applications, e.g. for high frequency operation, the turn-off time is required to be as short as possible. This also means that a small capacitor can be used for commutation. One way of achieving this is to reduce the effective carrier lifetime $\tau_{\text{eff}}$ using gold diffusion or electron irradiation to increase the rate of recombination. In addition, the thickness of the base layers, especially the n-base, should be minimised. This helps to minimise the on-state voltage, which is increased by shortening the carrier lifetime. Gold doping also increases the leakage currents under blocking conditions.

The use of cathode shorts increases $Q_{crit}$ and, with an efficient pattern, it is possible to reduce the turn-off time to about $2\tau_{eff}$. The danger here is that the turn-on characteristics are poorer and high frequency operation is therefore limited.

Almost all the changes of thyristor design and technology made to reduce the turn-off time cause a deterioration in other important device parameters: $V_T$ is increased, whereas $V_{DRM}$, $I_{AV}$ and $(di_T/dt)_{crit}$, are decreased; the only parameter to be improved is $(dV_D/dt)_{crit}$, which is increased. The circuit operating conditions also influence $t_q$, as shown in Figures 7.24 and 7.25. Several special types of thyristor have been proposed with the aim of achieving the best compromise among these conflicting requirements. They are described in Chapter 8.

### 7.4.2 Turn-Off by a Decrease of Forward Current

As is discussed in Section 7.2, a relatively high concentration of excess carriers is needed in the region of junction $J_2$ in order to keep the thyristor in the on-state. In the two-transistor model, at least one of the partial transistor structures must be saturated. This can be expressed by the condition

$$\frac{\partial \alpha_1}{\partial J_T} + \frac{\partial \alpha_2}{\partial J_T} < 0 \qquad (7.31)$$

The minimum on-state current density for which (7.31) is satisfied is defined in Section 7.1.4 as $J_M$. If the on-state current decreases, the turned-on area decreases, so the current density remains at $J_M$. Below the holding current $I_H$, the lateral diffusion

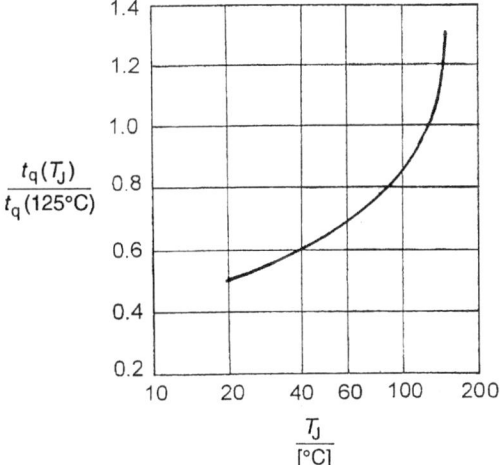

**Figure 7.24** An example of the temperature dependence of the thyristor turn-off time

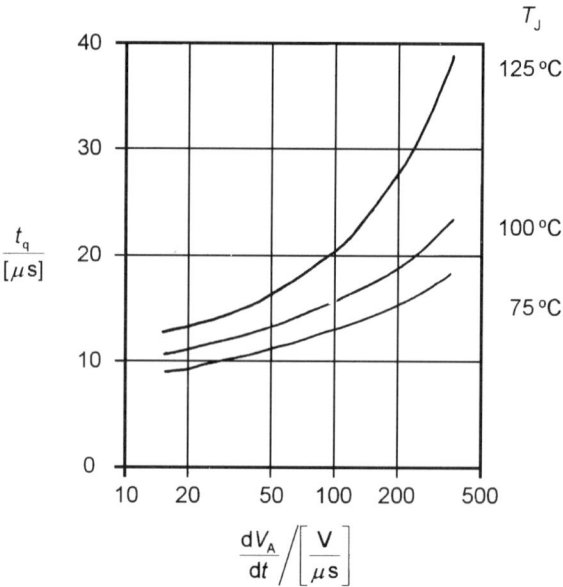

**Figure 7.25** An example of the dependence of the thyristor turn-off time on the rate of rise of the re-applied anode voltage

of carriers out of the constricted conducting filament causes the carrier concentration to fall, so a space charge layer forms at $J_2$ and the thyristor turns off. In a transient situation the holding current $I_H$ depends on the rate of decrease of the on-state current. Its smallest value $I_{HO}$ occurs when the decrease of current is very slow. It is the lowest current at which the thyristor can be maintained in the on-state when the gate is open circuit.

The holding current depends on several thyristor parameters, including the ratio of base width to diffusion length and the density of any cathode shorts. This is shown by the experimentally measured static characteristics in Figure 7.26. When the on-state current falls rapidly, the thyristor turns off at a current much greater than $I_{HO}$; sometimes several amperes are still flowing [7.5].

It can be seen in Figure 7.26 that the minimum on-state current density is much higher in a thyristor with cathode shorts. If the geometric factor $F$ can be reduced rapidly, by making cathode shorts electronically, it is possible to turn off the thyristor. This principle is used in the turn-off process of the MOS-controlled thyristor discussed in Section 11.2.

### 7.4.3 Gate Turn-Off

Provided there are no cathode shorts, it is possible for a thyristor in the on-state to be turned off by applying a negative voltage to the gate. To make gate turn-off feasible, a special structure is required. An example is shown in Figure 7.27. The strip cathode takes the form of a mesa surrounded by the gate contact. In the on-state the inner layers of the thyristor structure are flooded with excess carriers

## 7.4 TRANSIENT PROCESSES DURING TURN-OFF

injected from both anode and cathode emitters. In the two-transistor model, the partial transistors are in saturation, with

$$\alpha_1 + \alpha_2 \geqslant 1 \tag{7.32}$$

A negative gate voltage draws the hole current away from the cathode emitter junction $J_3$, and so decreases electron injection into the gate layer from the $n^+$-emitter, as shown in Figure 7.28. Initially, the turned-on area is squeezed in from the gate contact. This reduces $\alpha_2$ so that, with a sufficiently negative gate voltage, condition (7.32) is no longer fulfilled, the electron–hole plasma decays away and the thyristor turns off.

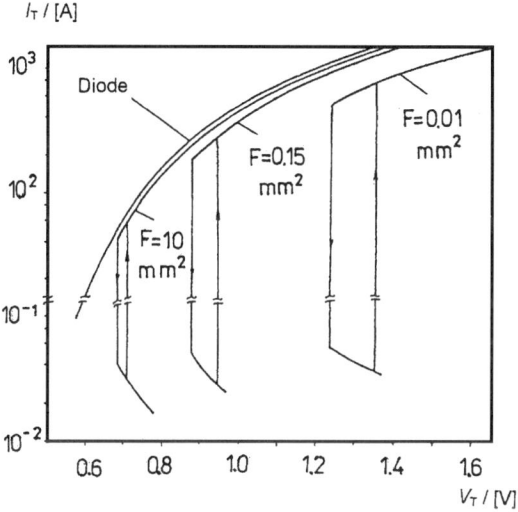

**Figure 7.26** The effect of the cathode shunt density on the static current/voltage characteristics. Adapted from Reference [7.4]

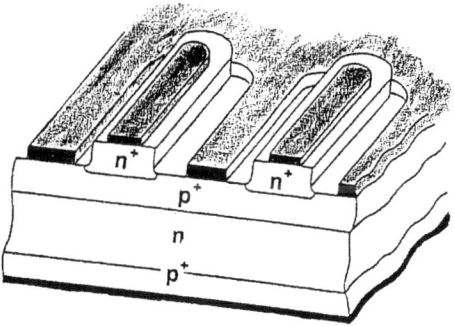

**Figure 7.27** A perspective view of the gate-turn-off (GTO) thyristor structure

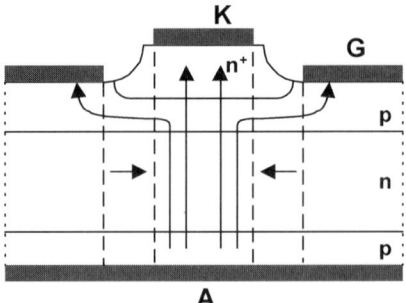

**Figure 7.28** An illustration of the squeezing of the turned-on region during gate turn-off

In the two-transistor analogy, the anode current can be expressed as

$$I_A = I_A \alpha_1 + I_K \alpha_2 = I_A \alpha_1 + (I_A - I_G)\alpha_2 \tag{7.33}$$

If the smallest negative gate current needed to turn off a given steady anode current is $-I_{GQ}$, the maximum available turn-off gain is

$$G_{offM} = \frac{I_A}{I_{GQ}} = \frac{\alpha_2}{\alpha_1 + \alpha_2 - 1} \tag{7.34}$$

It follows that $G_{offM}$ is biggest when $\alpha_2$ is high and $\alpha_1$ is low. There are no cathode shorts to decrease $\alpha_2$, but $\alpha_1$ can be decreased by the use of anode shorts, as shown in Figure 7.29. These reduce the $J_1$ injection efficiency. A large $N_1$-base thickness and a low carrier lifetime also help to keep $\alpha_1$ low. When anode shorts are created, junction $J_1$ has no blocking capability and the reverse blocking voltage is governed by the breakdown voltage of junction $J_3$. The presence of anode shorts complicates the fabrication process, because they need to be aligned with the cathode strips. However, it is better to reduce $\alpha_1$ in this way rather than by increasing $w_N/L_p$, which would increase the on-state voltage.

The maximum allowable negative gate current is limited by $V_{G(BR)}$, which is the breakdown voltage of junction $J_3$, and by the lateral resistance of the p-base layer. If $R_G$ is the resistance between any two gate contacts on opposite sides of an emitter strip, the gate current is limited to

$$I_{GM} = \frac{4V_{G(BR)}}{R_G} \tag{7.35}$$

This is because the current is assumed to divide into two equal parts, each of which has to pass through a resistance of $\frac{1}{2}R_G$. The maximum anode current that can be turned off is then given by

$$I_{TGQM} = \frac{4G_{offM}V_{G(BR)}}{R_G} = \frac{4\alpha_2 V_{G(BR)}}{(\alpha_1 + \alpha_2 - 1)R_G} \tag{7.36}$$

Although the breakdown voltage $V_{G(BR)}$ is increased if the $P_2$-base resistivity is increased, this does not increase $I_{TGQM}$ because of the simultaneous increase in $R_G$.

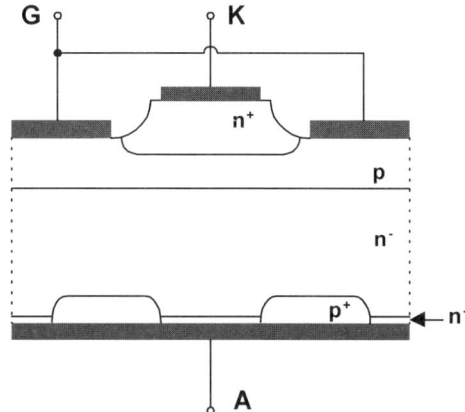

**Figure 7.29** The use of anode shorts to reduce the current gain of the pnp transistor, forming a reverse conducting GTO structure

With a normal doping profile, $V_{G(BR)}$ is about 20 V. Reducing the width of the cathode emitter strip decreases $R_G$ and so increases $I_{TGQM}$. This is shown in Figure 7.30. Cathode strips are typically 200–400 μm wide, although they can be as narrow as 100 μm.

The gate turn-off process can be divided into three phases. Initially, the excess charge of free carriers stored in the $P_2$-base layer is extracted into the gate. This continues until a space charge layer can form at junction $J_2$, which then becomes reverse biased. Typical anode and gate waveforms are shown in Figure 7.31. A one-dimensional charge-control analysis gives an approximate estimate of the storage time $t_{gs}$, which is defined in the figure. If, during this period, $I_G$ is assumed to be constant, the stored charge removed from the p-base is

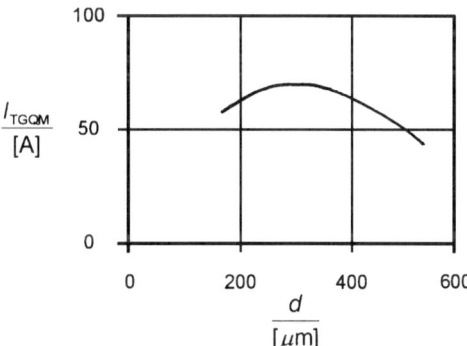

**Figure 7.30** Influence of the GTO cathode segment width on the maximum gate turn-off current, switching into a clamped 400 V supply. The segment length was 4 mm. Data taken from Nagano, T., Fukui, H., Yatsuto, T. and Okamura, M. A snubberless GTO, *Record of IEEE Power Electronics Specialist Conference (PESC 82)* (1982)

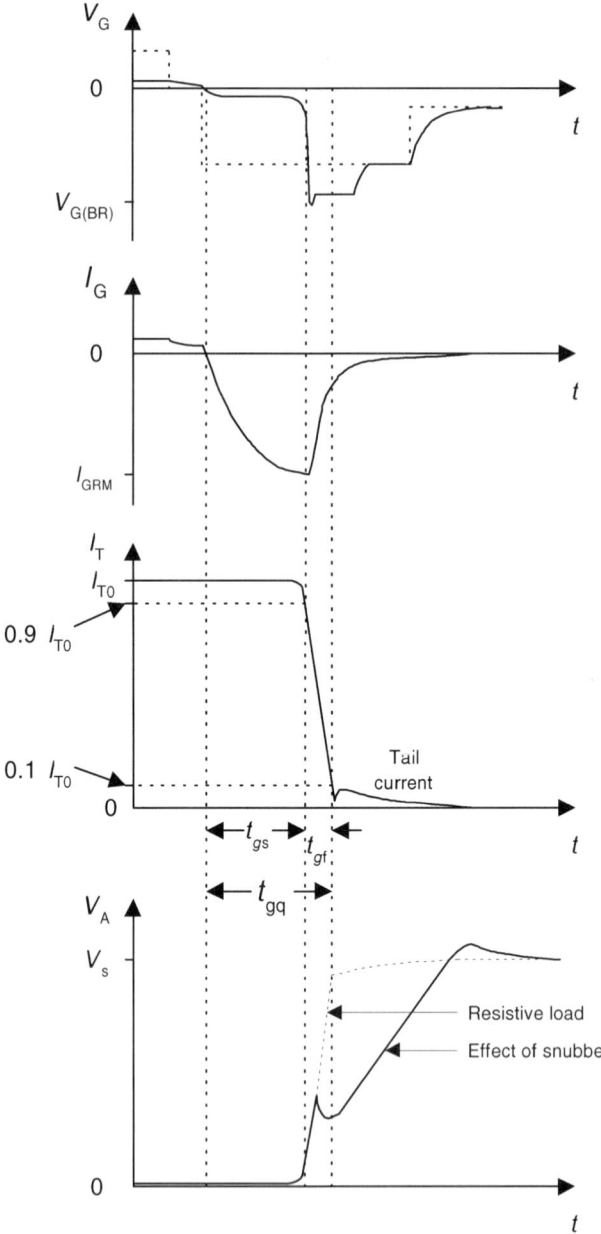

**Figure 7.31** Current and voltage waveforms during GTO switching. Detailed features, such as the effects of a snubber circuit, are discussed in Section 8.3

$$Q_P = I_G t_{gs} = e w_p n_{av} A \tag{7.37}$$

where $A$ is the thyristor area and $n_{av}$ is the average excess carrier concentration. Expressing this as

## 7.4 TRANSIENT PROCESSES DURING TURN-OFF

$$n_{av} = \frac{Q_{on}}{(w_P + w_N)A} = \frac{I_A \tau_H}{(w_P + w_N)A} \qquad (7.38)$$

it can be seen that the storage time is given by

$$t_{gs} = \frac{Q_P}{I_G} = \tau_H \frac{I_A}{I_G} \frac{w_P}{(w_P + w_N)} = \tau_H G_{off} \frac{w_P}{(w_P + w_N)} \qquad (7.39)$$

where we have written $G_{off} = I_A/I_G < G_{offM}$ as the turn-off gain used in practice. From this equation it follows that the storage time can be reduced by reducing the carrier lifetime and increasing the negative gate current. This means operating the thyristor at a lower turn-off gain $G_{off}$. To increase the thickness of the inner layers $(w_P + w_N)$ would increase the on-state power losses. These have to be limited if the device is to remain within its allowed temperature range. It is seen in (5.12) that dissipation depends on the ratio $(w_P + w_N)/L_a$, where $L_a$ is the ambipolar diffusion coefficient. From this it follows that, for thick devices, we require $\tau_H \propto (w_P + w_N)^2$. Shortening the carrier lifetime $\tau_H$ also increases the on-state losses, so a careful trade-off is necessary.

A space charge first forms in those parts of junction $J_3$ under the gate contacts. The lateral electric fields that develop in the n-base tend to pinch the electron–hole plasma into a narrow channel of quite high current density. At time $t_{gs}$ the space charge and the electric field in the conducting channel modify the conditions so much that the channel pinches off completely and a space charge layer develops across the whole device cross-section.

An early theory by Wolley [7.6] describes the rate of pinch-off when the lateral flow is dominated by diffusion. Consider the section shown in Figure 7.28 to represent a segment running along a length $z$ of a cathode strip of width $d$. With the x-coordinate at right angles to the strip and its origin in the centre, we can assume that the portion between $x=0$ and $x=x_0$ ($0 < x_0 < d/2$) is in the on-state, whereas that between $x_0$ and $d/2$ has been turned off by the extraction of excess carriers by the applied negative gate voltage.

The excess carrier concentration can be written as

$$n(x, y) = n(y)\,f(x) \qquad (7.40)$$

where, for $0 \leqslant x < x_0$, $f(x) = 1$ and for $x_0 \leqslant x < d/2$, as a result of lateral diffusion,

$$f(x) = \exp\left(\frac{x_0 - x}{L}\right) \qquad (7.41)$$

where $L$ is the bipolar diffusion length defined as $\sqrt{D\tau}$ using (1.74) and (1.76).

The charge in the thin transition region $dx_0$, between the turned-on and turned-off regions, can be written as

$$dQ = en^* w_P z\, dx_0 \qquad (7.42)$$

where $n^*$ is the average concentration of excess carriers in the p-base.

A lateral current flows by diffusion from the on-region to the turned-off portion of the segment. From (7.41) and (7.42) the lateral current density for $x \geqslant x_0$ is given by

$$J_x = \frac{eD}{L}n(y)\exp\left(\frac{x_0 - x}{L}\right) \qquad (7.43)$$

The lateral current in the p-base entering through the boundary between the on and off portions at $x_0$ is given by

$$I(x_0) = z\int_0^{w_P} J_x(x_0)dy = \frac{eD}{L}zw_P n^* \qquad (7.44)$$

The current flows from the transition region to the gate electrode. Neglecting the recombination in the transition layer, the rate of change of excess charge is given by

$$\frac{dQ}{dt} = I(x_0) - \frac{I_G}{2} = \frac{eD}{L}zw_P n^* - \frac{I_G}{2} \qquad (7.45)$$

Differentiating (7.42) with respect to time and using (7.45), it follows that

$$en^* w_P z \frac{dx_0}{dt} = \frac{eD}{L}w_P z n^* - \frac{I_G}{2} \qquad (7.46)$$

where $dx_0/dt$ represents the speed of the moving boundary between the on- and off-regions. Note that the excess carrier concentration $n^*$ depends on the cathode current density $J_K$. If the gate current is constant, i.e. a rectangular gate current pulse is applied, then

$$\frac{dx_0}{dt} = \frac{D}{L} - \frac{I_G}{2en^* w_P z} \qquad (7.47)$$

The average excess carrier concentration in the p-base is proportional to the average excess carrier concentration $n_{av}$ in the thyristor structure, $n^* = \zeta n_{av}$. With

$$J_K = \frac{en_{av}(w_P + w_N)}{\tau_H} \qquad (7.48)$$

flowing through the turned-on part, which we assume to extend over $0 \leqslant x < x_0 + L$,

$$I_K = 2J_K z(x_0 + L) \qquad (7.49)$$

and

$$\frac{dx_0}{dt} = \frac{D}{L} - \frac{I_G}{I_K}\frac{(w_P + w_N)}{\zeta w_P \tau_H}(x_0 + L) \qquad (7.50)$$

The cathode current can be expected to remain constant until the width of the conducting region is reduced to a value $L^*$ in the centre of the $n^+$-emitter. It then starts to decrease. The storage time $t_{gs}$ can be calculated from (7.50), which for a rectangular pulse of negative gate current can be integrated to give

$$t_{gs} = \int_{d/2}^{L^*}\left[\frac{D}{L} - \frac{(w_P + w_N)(x_0 + L)}{(G_{off} - 1)\zeta\tau_H w_P}\right]^{-1} dx_0 = (G_{off} - 1)\frac{\tau_H w_P}{(w_P + w_N)}\ln(M) \qquad (7.51)$$

## 7.4 TRANSIENT PROCESSES DURING TURN-OFF

We have substituted $(I_K/I_G) = (G_{off} - 1)$ and introduced the parameter $M$, which is a function of the dimensions of the structure and the carrier lifetime. Comparing this result with (7.39), shows that the two-dimensional analysis makes $t_{gs}$ proportional to $(G_{off} - 1)$ rather than $G_{off}$. In both cases it increases with increasing carrier lifetime and layer thickness.

The GTO gate drive can usually be represented as a voltage source in series with an inductance, so the negative gate current increases linearly with time during the storage phase. Then

$$\frac{dx_0}{dt} = \frac{D}{L} - \frac{(w_P + w_N)(x_0 + L)}{I_K \zeta \tau_H w_P} \frac{dI_G}{dt} t \qquad (7.52)$$

and

$$t_{gs} \propto \frac{I_K}{(dI_G/dt)} \qquad (7.53)$$

During the storage phase, the electron–hole plasma in the on-state region is pinched into a very narrow channel of quite high current density. This gives rise to an increase in the anode voltage. At the end of the storage interval, the junction $J_3$ is reverse biased and at the same time junction $J_2$ also starts to recover its blocking ability. This requires a space charge region to form at the junction and leads to a rapid decrease of anode current and a rapid increase of anode voltage. The storage phase is regarded as complete when the on-state current has decreased to 90% of its initial value. During the next phase, the anode current is determined by the current of carriers that has to be swept out during the formation of the space charge region at $J_2$. If $y(t)$ is the width of this space charge region and $V_A(t)$ is the instantaneous anode voltage,

$$y(t) = \sqrt{\frac{2\varepsilon_r \varepsilon_0 V_A(t)}{e N_D}} \qquad (7.54)$$

To a first approximation the anode current represents the rate at which carriers are swept into the space charge region through the moving boundary, so that

$$I_A(t) = e A n_a \frac{dy(t)}{dt} \qquad (7.55)$$

where $n_a$ is the average concentration of excess carriers in the still flooded part of the n-base and $A$ is the cross-sectional area. Substituting $y(t)$ from (7.54) into (7.55), the anode current becomes

$$I_A(t) = A n_a \sqrt{\frac{e \varepsilon_r \varepsilon_0}{2 N_D V_A(t)}} \frac{dV_A}{dt} \qquad (7.56)$$

The thyristor is assumed to be part of an external circuit comprising a voltage source $V_s$ and an impedance $R_L$, so the voltage across the thyristor is

$$V_A(t) = V_s - V_{RL}(t) \qquad (7.57)$$

because the voltage $V_{RL}$ across the load varies with the anode current. Thus, $I_A(t)$ and $V_A(t)$ depend on the load.

The fall time $t_{gf}$ is defined as the time interval during which the anode current decreases from 90% to 10% of its initial value. Equation (7.56) shows that it can be approximated for a resistive load by

$$t_{gf} = \frac{1.82\tau_H}{(w_P + w_N)}\sqrt{\frac{2\varepsilon_0\varepsilon_r V_s}{eN_D}} \qquad (7.58)$$

At the end of this phase of the turn-off process, the space charge layer at junction $J_2$ has reached its maximum value as determined by the supply voltage. However, some excess carrier charge is still stored at junction $J_1$. This decays partly by recombination and partly by extraction through the space charge region of the blocking junction. It causes a current tail that decreases approximately exponentially with a time constant that is the effective carrier lifetime. Because the blocking voltage is high at this time, the tail current results in a high power dissipation. A small and rapidly decreasing tail current is therefore desirable. Most commonly, this is achieved by the use of anode shorts and by reducing the carrier lifetime in the region close to junction $J_1$ using, for example, proton irradiation.

Examples of devices designed to provide gate turn-off and gate-assisted turn-off are discussed in Chapter 8, along with other important types of thyristor.

## Summary

Thyristors are inherently four-layer pnpn devices. They are made in many different types and sizes and constitute one of the most important groups of power electronic devices. All types have the property that they can be switched from a forward blocking state to a forward conducting state, by several possible means. Some have reverse blocking characteristics, some do not. The GTO thyristor can be turned off by applying a negative signal to the gate electrode; the other types only recover their blocking characteristic once the anode current has fallen below a certain critical level for a period.

Many aspects of thyristor behaviour can be modelled in terms of two interconnected transistors. The on-state characteristics are similar to those of a power diode. However, the turn-on and turn-off transients involve complex three-dimensional processes in which very high power densities may be generated. During turn-on the rise rate of the anode current must be limited to allow time for the area of the device in conduction to spread, so the current density does not become too great while a significant voltage remains across the device. During turn-off the rate of rise of the anode voltage may trigger the device back into conduction.

The turn-off time of a standard thyristor and the current tail of a GTO can both be reduced by reducing the carrier lifetime. However, this has an adverse effect on other operating parameters, especially the on-state volt-drop. Other than for GTOs, thyristor characteristics can be improved by an optimum density of cathode shorts. When a reverse blocking capability is not required, the turn-off characteristics of GTOs are improved by the use of anode shorts.

## References

[7.1] Azuma, M. and Takigami, K. Anode Current Limiting Effect of High Power GTOs, *IEEE Electron Device Letters*, **EDL-1**, 203–5 (1980).

[7.2] Somos, I. and Piccome, D. E. Plasma Spread in High-Power Thyristors Under Dynamic and Static Conditions, *IEEE Trans. on Electron Devices*, **ED-17**, 680–7 (1970).

[7.3] Ruhl, H. J. Spreading Velocity of the Active Area Boundary in a Thyristor, *IEEE Trans. on Electron Devices*, **ED-17**, 672–80 (1970).

[7.4] Pína, B. Plasma in Four-Layer Semiconductor Structures (in Czech), *Electronic Horizons* (*Slaboproudý obzor*), **48**, 215–19 (1987).

[7.5] Garrett, J. M. The Study of an Alternative Blocking Recovery Mode in Silicon Thyristors, MSc Thesis (Brunel University, Uxbridge, UK, 1980).

[7.6] Wolley, D. E. Gate Turn-off in pnpn Devices, *IEEE Trans. on Electron Devices*, **ED-13**, 590–7 (1966).

# 8
# THYRISTOR TYPES AND APPLICATIONS

The silicon controlled rectifier (SCR) is the preferred term in the United States for what in Europe is called a thyristor. In its basic form it has approximately equal forward and reverse blocking voltages. However, many variations on the thyristor structure have been developed in an attempt to match the characteristics of the device to the demands of different applications. In general, they can be divided into low frequency thyristors and high frequency thyristors. In this chapter we detail the various types that are turned on by the injection of gate current. The aim is to show how specific structures can be adapted to meet specific requirements. We also include brief descriptions of breakover diodes, which are untriggered devices, and light-triggered thyristors. Voltage-controlled devices, some of which have thyristor-like structures, are discussed in Chapters 9 to 11.

## 8.1 Phase-Control Thyristors

The phase-control thyristor is designed for power frequency applications, such as low frequency converters or phase-control rectifiers and is required to have as low an on-state volt-drop as possible and a blocking voltage designed for the particular application. The main applications are for dc drives, soft starts of ac motors, resistance welding and power transmission applications, including high voltage dc systems (HVDC).

A key feature of thyristor design is the trade-off between the blocking voltage and the on-state current rating. The limit for the blocking voltage is related to the thickness and resistivity of the $N_1$-layer, as discussed in Section 7.1.1. For both forward and reverse blocking voltages to be high, layers $P_1$ and $P_2$ should be relatively thick, typically 70–100 $\mu$m. They are normally formed by deep diffusions. Increasing the thickness of the $N_1$-layer and the distance between the $n^+$- and $p^+$-emitters increases the on-state voltage drop (Section 7.1.4) and consequently increases the power dissipated at a given current density. As the maximum junction temperature is limited by $T_{Jmax}$ and the extraction of the heat generated is limited by the device diameter, the construction of the case and heat sink becomes important, as

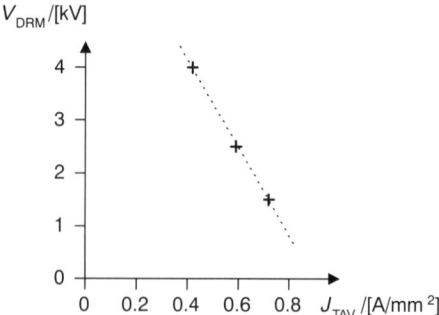

**Figure 8.1** The reduction in the current density at the rated mean current with increasing thyristor blocking voltage

discussed in Chapter 13. Figure 8.1 illustrates the maximum usable current density in devices made to have a higher blocking voltage. To obtain a low on-state loss, long carrier lifetime is necessary and various gettering techniques are commonly used during device fabrication.

Typical repetitive requirements for $(dI_T/dt)_{crit}$ and $(dV_D/dt)_{crit}$ are in the region of 100 A/µs and 200 V/µs, respectively. These ratings can be obtained by optimising the cathode and gate geometry. In particular, the $(dV_D/dt)_{crit}$ requirement can be met using an efficient pattern of cathode shorts and a $P_2$-layer with a relatively low sheet resistance, as described in Section 7.3.3. However, the pattern density of the shorts should not be so great that it inhibits plasma spreading during turn-on. In large devices the size and density of the shorts can be varied in different regions in order to optimise the spread of the conduction region.

The value of $(dI_T/dt)_{crit}$ is heavily dependent on the gate geometry. Using either the amplifying gate structure or the field-enhanced gate structure, described in Section 7.3.2, a long length of the gate perimeter can be turned on in the early stages of conduction and high $(dI_T/dt)$ ratings can be obtained. In large-area, phase-control devices an interdigitated amplifying gate, having just a few fingers, is often used to shorten the time needed to spread the on-state over all the active device area. The $(dI_T/dt)$ rating is also influenced by the overall wafer thickness. The higher thermal impedance of devices fabricated from thicker wafers causes the instantaneous temperature to be higher than in a thinner device. When the initially turned-on area is small and the current density very high, Auger recombination further increases the transient losses, especially in a thicker structure. The standard $(dI_T/dt)$ test requires the thyristor to be switched on from 66% of $V_{DRM}$ and so results in higher turn-on losses for high voltage devices. Thus, a high voltage thyristor normally has a lower $(dI_T/dt)$ rating than a thinner, low voltage device of otherwise similar gate geometry.

The operating frequency of converter type thyristors (SCRs) is normally less than 400 Hz, so the turn-off time can be relatively long. However, the reverse recovery charge $Q_{rr}$ should be minimised when devices are operated in series, in order to reduce the size of the capacitance divider, as discussed in Chapter 13. A low dose of electron irradiation can be used to decrease $Q_{rr}$ to an acceptable level without significantly increasing the on-state voltage drop. This has the additional benefit of reducing the carrier lifetime and making it more uniform across the device area.

The surge characteristics of high power thyristors are very important. The most typical overload situation occurs during a load short circuit when the device is protected by a fuse which may take several milliseconds to operate. The non-repetitive surge current rating $I_{TSM}$ is commonly required to be at least 10 times higher than the mean on-state current rating. In 50 Hz circuits the thyristor should be able to carry the surge current for up to 10 ms. Under such conditions the junction temperature may rise transiently to about 300 °C, which is above $T_{Jmax}$. At such temperatures, neither forward nor reverse blocking voltage can be reapplied without the danger of thermal breakdown and such overloading must only be allowed to occur infrequently. A high doping concentration in both $n^+$- and $p^+$-emitters is needed if the on-state voltage drop at the high current densities experienced during surge conditions is to be kept to an acceptable level.

Phase-control thyristors have a very broad range of current ratings, which are normally specified in terms of the average on-state current $I_{TAV}$. In order to minimise the need for the series and parallel connection of devices in very high power applications, high blocking voltage and high current ratings are desirable. Thyristors have been developed that can support in excess of 6 kV and conduct mean currents greater than 3 kA. Such devices use silicon wafer diameters of 100 mm and above and provide an active area greater than 75 cm$^2$.

## 8.2 Thyristors for High Speed Applications

For inverter applications, thyristors need a short turn-off time. This parameter sets the upper limit to the working frequency and determines the capacitance required in the auxiliary commutation circuit shown in Figúre 7.22.

From the analysis given in Section 7.4.1, it follows that the device parameters available for controlling the turn-off time $t_q$ are the lifetime and the parameters influencing the critical charge $Q_{cr}$, namely the emitter efficiency, the base widths and the cathode short density. The turn-off time can be decreased by reducing the carrier lifetime and by increasing the density of the cathode shorts. On the other hand, the turn-off time increases with the charge $Q_1$ that remains in the inner layers of the device after junction $J_1$ has recovered. Thus, the turn-off time increases with the thickness of the $N_1$-layer. It follows from (5.33) that the effective carrier lifetime also increases with the $N_1$-layer width and this results in a conflict between the requirements for short turn-off times and high blocking voltages.

The parameters that reduce the turn-off time, namely a short carrier lifetime and a high density of cathode shorts, adversely affect the on-state voltage drop, the turn-on time and the plasma spreading velocity. Therefore, designing for a short turn-off time is always in conflict with the need for low conduction and turn-on losses. In reverse blocking thyristors ($V_{DRM} = V_{RRM}$), this results in a trade-off between the blocking voltage and the turn-off time, as shown in Figure 8.2.

The average power loss, which may again be expressed by (6.56), also limits high frequency operation. As the operating frequency rises, the turn-on and turn-off losses make up an increasing proportion of the total power loss. The turn-on losses normally dominate because the $dI_T/dt$ requirements are high. Complete current spreading is achieved in a shorter time when turn-on starts from a larger initial area,

# 8 THYRISTOR TYPES AND APPLICATIONS

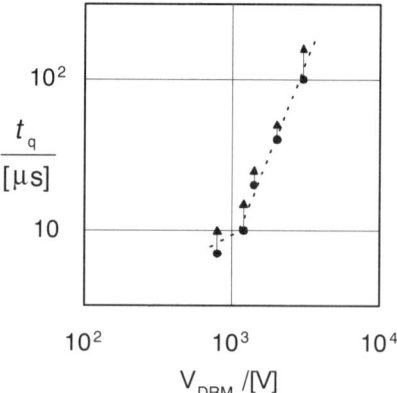

**Figure 8.2** The variation of turn-off time with blocking voltage

so the gate geometry must ensure that a large area of the device is turned on in the early stages of conduction. The simple ring amplifying gate gives an adequate performance for many inverter applications. Otherwise, an amplifying gate with an interdigitated auxiliary contact is used. The repetitive $(dI_T/dt)_{crit}$ parameter usually exceeds $200\,A/\mu s$. The geometry of a 10 kHz, 1200 V, 400 A thyristor is shown in Figure 8.3.

The special types of thyristor structure discussed in the following sections have been developed for inverter applications.

## 8.2.1 The Asymmetric Thyristor (ASCR)

In order to use thyristors at higher operating frequencies it is necessary to decrease the turn-on losses. This requires a high spreading velocity and a low voltage drop during turn-on. The ratio $w/L$ should be minimised; that is, the distance between the

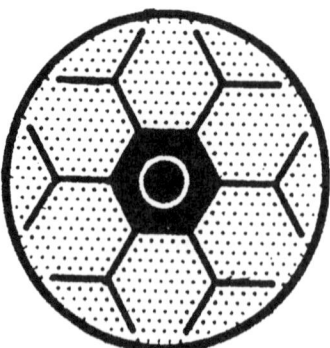

**Figure 8.3** An example of an inverter-grade thyristor. Reproduced by permission of Polovodiče, a. s., Prague

p+- and n+-emitters should be small compared to the diffusion length. Unfortunately, a thick $N_1$-layer is necessary if high values of both reverse and forward blocking voltage are to be obtained. A high forward blocking voltage $V_{DRM}$, a low forward voltage drop and good dynamic characteristics can all be obtained simultaneously by making design changes that decrease the reverse blocking voltage $V_{RRM}$. Devices in which $V_{RRM} < V_{DRM}$ are known as asymmetric thyristors (ASCRs). They are widely used in inverter circuits, in combination with an antiparallel diode. A series diode can be used to provide a reverse blocking capability, when this is needed.

One way to achieve this combination of characteristics is to reduce the thickness of the $P_1$-layer and shorten the carrier lifetime. The n+pnpp+ structure of a conventional thyristor becomes n+pnp+. This enables the turn-off time to be reduced, without degrading the on-state characteristics. A more effective method is to use the compressed-field structure (n+pvnp+) shown in Figure 8.4.

In a normal reverse blocking thyristor with an n+pnpp+ structure, the space charge thickness $w_N$ at junctions $J_1$ and $J_2$ is less than the thickness of the $N_1$-layer and the maximum blocking voltage is given by

$$V_{BO} = \tfrac{1}{2} E_{BR} w_N \qquad (8.1)$$

where the critical electric field $E_{BR}$ is related to the donor concentration in the $N_1$-layer by (2.38). Increasing the resistivity of the $N_1$-layer, reducing its thickness to $w_v$ and including a further n-type layer, with a higher donor concentration between it and the p+-layer, produces a compressed-field structure similar to the diode and transistor structures described in Sections 5.2 and 6.2 [8.1]. The extra layer is known as a stop or buffer layer. It leads to a forward breakover voltage given by

$$V_{D(BO)} = E_{BR} w_v - \frac{e N_D w_v^2}{2 \varepsilon_r \varepsilon_0} \qquad (8.2)$$

With this structure, a higher breakover voltage $V_{D(BO)}$ and a higher repetitive peak blocking voltage $V_{DRM}$ can be obtained for a given wafer thickness. A comparison of the concentration profiles and the electric field distribution under conditions of forward hold-off for symmetrical and asymmetrical thyristors is shown in Figure 8.4. The variation of the maximum breakover voltage with $N_1$-layer thickness and donor concentration is shown in Figure 8.5 for the two structures.

The reduction of the distance $w$ between the n+- and p+-emitters that is possible in the n+pvnp+ structure has beneficial effects on the on-state current–voltage characteristics. It increases the spreading velocity during turn-on and therefore decreases the turn-on losses. And it enables the carrier lifetime to be reduced, in order to reduce the turn-off time, without degrading the surge current characteristics. However, because the donor concentration at junction $J_1$ is considerably greater than at junction $J_2$, $V_{RRM}$ is normally reduced to something in the region of 20–50 V. The thermal stability of the forward blocking voltage is improved by creating shunts across junction $J_1$, when the device effectively becomes the reverse conducting thyristor described in Section 8.2.2.

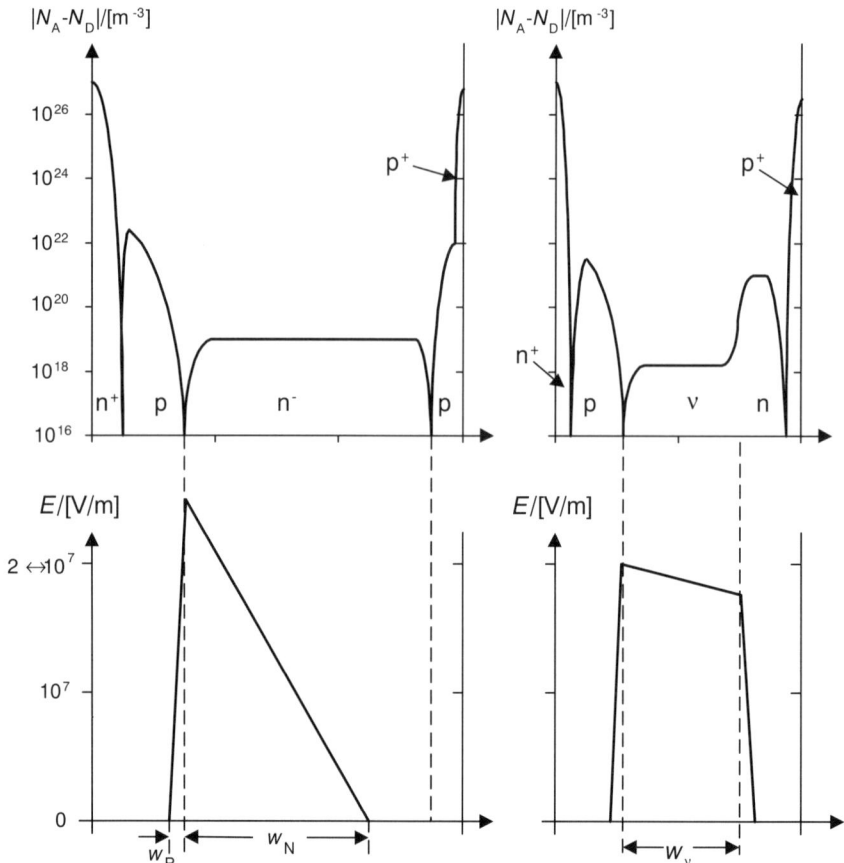

**Figure 8.4** A comparison of the symmetric and asymmetric thyristor structures

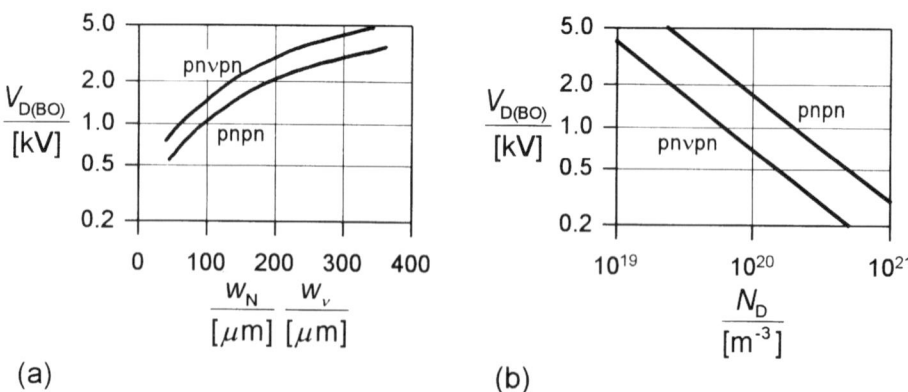

**Figure 8.5** The variation of the breakdown voltage of symmetric and asymmetric thyristor structures as a function of: (a) the $N_1$-layer thickness, with optimum donor concentration; (b) the donor concentration in an $N_1$ layer of minimum thickness. Reproduced from Reference [8.1] by permission of The General Electric Company, plc

## 8.2 THYRISTORS FOR HIGH SPEED APPLICATIONS

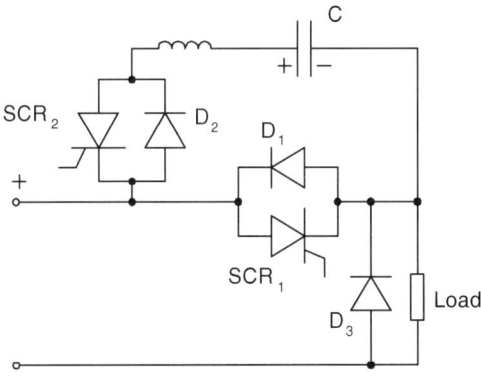

**Figure 8.6** An inverter circuit, showing the use of thyristors with anti-parallel diodes

In chopper and inverter circuits, asymmetric thyristors are often used in conjunction with an antiparallel diode. An example of such a circuit is given in Figure 8.6. In this case the reverse voltage during the turn-off process is determined by the forward voltage drop across the antiparallel diode. As this is very low, the turn-off losses are considerably reduced, although the turn-off time is increased slightly.

### 8.2.2 The Reverse Conducting Thyristor (RCT)

The antiparallel thyristor–diode combination causes a problem in inverter circuits because inductance in the interconnection adversely affects the thyristor turn-off. Current is required to commutate rapidly into the freewheeling diode (D3 in Figure 8.6), when the main thyristor (SCR1) turns off. Parasitic inductance in the lead of D1 makes it unable to clamp the voltage across SCR1 initially and, with an ASCR, the reverse breakdown voltage may be exceeded. The structure shown in Figure 8.7 minimises these problems by integrating an antiparallel diode with an asymmetric thyristor. It is known as the reverse conducting thyristor (RCT). The inner pνn

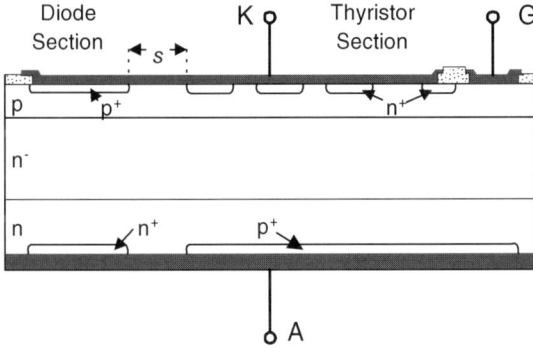

**Figure 8.7** The structure of the reverse conducting thyristor

structure has $p^+$- and $n^+$-regions selectively diffused into different regions. In the centre of the wafer the $p^+n\nu pn^+$ structure of an asymmetrical thyristor is formed; whereas at the periphery, the $p^+p\nu nn^+$ structure of an antiparallel diode is created. This simplifies the bevelling required.

During turn-off, sometime after the commutation of the anode current, the thyristor part of the RCT recovers and becomes reverse biased. This biases the diode part into forward conduction, so the reverse voltage on the thyristor part is limited to the forward voltage drop of the diode part. The thyristor and diode parts are normally isolated from one another by forming a ring-shaped short at the diode edge of the $n^+$-emitter. Excess carriers in one part do not then adversely affect the operation of the other.

### 8.2.3 The Gate-Assisted Turn-Off Thyristor (GATT)

It follows from (7.30) that it is possible to shorten the turn-off time $t_q$ by an increase of the critical charge $Q_{cr}$ which can still remain in the device without causing it to conduct on reapplication of a forward voltage. A considerable increase can be obtained with a properly constructed cathode–gate layout by extracting negative gate current before the rise of anode voltage. Typical waveforms are shown in Figure 8.8. Thyristors designed for this mode of operation are known as gate-assisted turn-off thyristors (GATTs). A suitable cathode–gate geometry takes the form of cathode strips surrounded by the gate contact. Figure 8.9 illustrates the structure with and without a strip cathode short.

During the increase of forward voltage, the total current flowing in the device consists of the sum of the displacement current and the current due to the remaining excess carriers. These two components give rise to a total current density $J_{tot}$ entering layer $P_2$ through junction $J_2$. The thyristor turns on when $J_{tot}$ exceeds the critical level

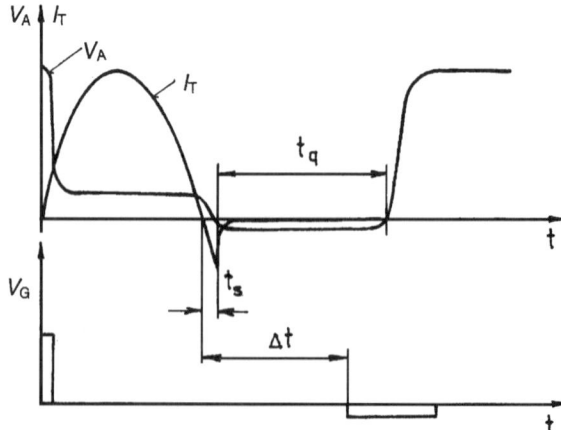

**Figure 8.8** Current and voltage waveforms for a gate-assisted turn-off thyristor during the turn off period

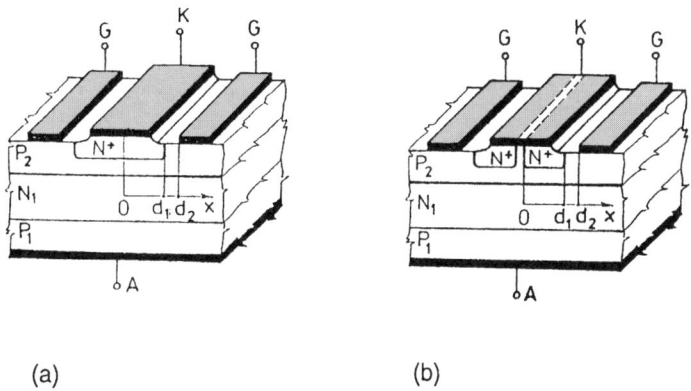

**Figure 8.9** Examples of the gate-assisted turn-off thyristor structure: (a) without a cathode short; (b) with a cathode short

$J_{cr}$. Electrons are injected from the cathode emitter if the lateral current flowing in layer $P_2$ sets up a voltage $V_E$ across $J_1$. The condition for turn-off can be expressed as

$$\max(J_{tot}) \leqslant J_{cr} = \frac{V_E}{K} \qquad (8.3)$$

where $K$ is a factor that depends on the gate-cathode layout and the concentration profiles in the $N_2$- and $P_2$-layers [8.2].

After the application of the negative gate voltage, current flows laterally through the $P_2$-layer under the $n^+$-emitter. The lateral current density is

$$J_{lat}(x) = \frac{1}{g} \int_0^x J_{tot} \, dx \qquad (8.4)$$

where $g$ is the thickness of the $P_2$-layer. The lateral current induces a lateral voltage along junction $J_2$. Without cathode shorts and assuming uniform resistivity $\rho$ of the $P_2$-layer, the voltage between the centre of the $n^+$-emitter strip and the gate contact is

$$V_{lat} = \rho \int J_{lat}(x) \, dx \qquad (8.5)$$

The turn-on condition (8.3) is met when $V_{lat}(J_{cr}) = V_G + V_E$. Using a simple approximation [8.3] in (7.30) gives

$$t_q = \tau_{eff} \ln\left(\frac{K\rho d^2}{8 V_G}\right) \qquad (8.6)$$

where $d = 2d_1$ is the width of the cathode emitter strip.

It is evident from (8.6) that the turn-off time can be influenced by the effective carrier lifetime, the resistivity of the $P_2$-layer, the width of the $N_1$-layer and the applied negative gate voltage. The maximum applied voltage must be less than the

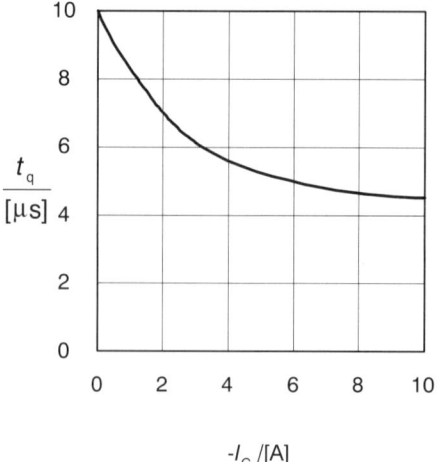

**Figure 8.10** An example of the variation of the turn-off time with the reverse gate current for a gate-assisted turn-off thyristor

breakdown voltage of $J_3$. A typical example of the variation of the turn-off time with the negative gate current is shown in Figure 8.10.

Examples of typical emitter–gate layout patterns for gate-assisted turn-off thyristors are shown in plan view in Figure 8.11. The long length of the emitter–gate periphery makes it possible for a large area to be turned on initially and so lowers the power dissipated during turn-on. Therefore, the gate-assisted turn-off thyristor can be used at frequencies above 10 kHz. In particular, the structure shown in Figure 8.11(b), with a very narrow emitter strip, has been called the zero turn-off

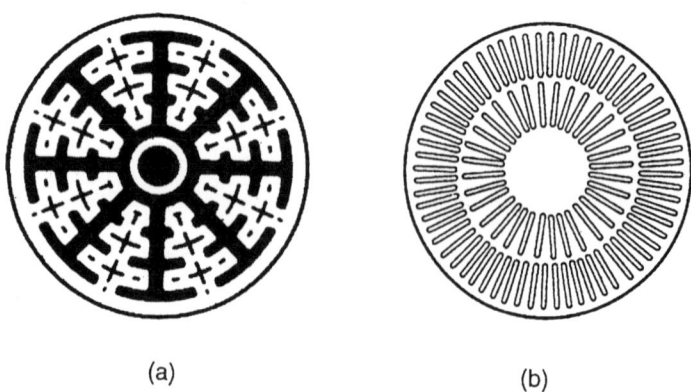

**Figure 8.11** Examples of cathode-emitter geometries for gate-assisted turn-off thyristors: (a) with an amplifying gate; (b) zero-turn-off structure. Part (a) is taken from E. S. Schlegel, 'A Technique for Optimising the Design of Power Semiconductor Devices', *IEEE Trans. on Electron Devices*, **ED-23**, 924–7, (1976) and is reproduced by permission of The Institution of Electrical and Electronic Engineers.

(ZTO) thyristor and can operate at 50 kHz with a 550 A sine wave and $V_{DRM} = 1100$ V.

## 8.3 The Gate Turn-Off Thyristor (GTO)

For many years, gate turn-off thyristors have dominated high power applications, typically for voltages over 2500 V and currents over 400 A, where the current and voltage handling capabilities of other devices are insufficient to satisfy the requirements. Several varieties of GTO have been developed to meet the needs of different applications. Their supremacy in this area is increasingly being challenged by IGBTs, whose voltage and current handling capabilities are constantly improving with the advance of fabrication technology.

In Section 7.4.3, where the principle of gate turn-off is described, it is shown that this technique is effective only when the cathode region takes the form of a narrow $n^+$-emitter strip surrounded by the gate contact. High power operation is obtained by the interconnection of many hundreds of such elementary GTOs in parallel on the wafer. It is important that all the segments turn off simultaneously, otherwise the current is concentrated into a few segments which become overloaded and are likely to be destroyed.

It is shown in Section 7.4.3 that a high turn-off gain $G_{off}$ can be obtained by decreasing the current gain $\alpha_1$ of the $P_1N_1P_2$ part of the thyristor structure. Two basic types of GTO use this principle. The first is a symmetrical (i.e. reverse blocking) structure, where the reduction of $\alpha_1$ depends on having as thick an $N_1$-base and as short a carrier lifetime as is consistent with the requirements for the on-state voltage drop. This is very difficult to achieve. Such devices are useful in converters for traction motors when disconnection of the source can occur. Trolley-bus drives are an example.

In other inverter and pulse converter applications, GTOs are normally used together with an antiparallel diode. A reverse blocking capability is then not necessary and $\alpha_1$ can be decreased by creating microshorts across junction $J_1$. The use of anode emitter shorts and a stop layer between $N_1$ and $P_1$ allows the distance between the $n^+$- and $p^+$-emitters to be minimised, and a given on-state characteristic can be realised with a shorter carrier lifetime and hence a reduced tail current. Although fabrication is complicated by the need to align masks on opposite sides of the wafer, this structure is so widely used that a GTO is normally assumed to have an asymmetric structure with anode emitter shorts, as shown in Figure 8.12. The $n^+p\nu np^+$ compressed-field structure is sometimes known as a punch-through GTO (PT-GTO), and the conventional $n^+pnp^+$ structure, where the space charge layer does not fill the $N_1$-layer, as a non-punch-through GTO (NPT-GTO). The tail current can be further decreased by proton irradiation to reduce the carrier lifetime in the region of $J_1$.

The layout of the individual segments must use the silicon wafer area in an effective way in order to obtain a large active cathode area and at the same time ensure identical gating conditions for all the individual GTO segments. Each should have approximately the same gate–cathode breakdown voltage $V_{G(BR)}$. Some typical layouts are shown in Figure 8.13. The one shown in Figure 8.13(a) is the more

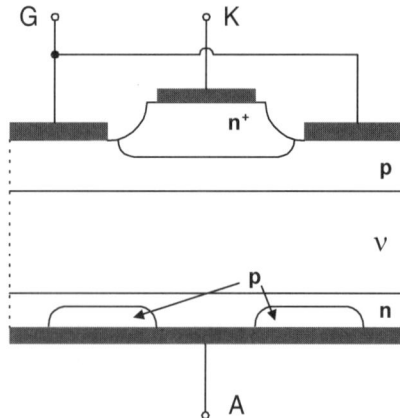

**Figure 8.12** Schematic cross-section through the active region of a typical GTO thyristor

efficient in the use of cathode active area and gate current density for smaller thyristors, say for wafer diameters up to 35 mm, and $I_{TGQM}$ up to 600 A. For larger GTO thyristors, with $I_{TGQM} > 600$ A, the layout shown in Figure 8.13(b) is often used, usually in the form of several concentric rings of elementary strip segments. The thickness of the aluminium layer forming the gate contact is normally limited to

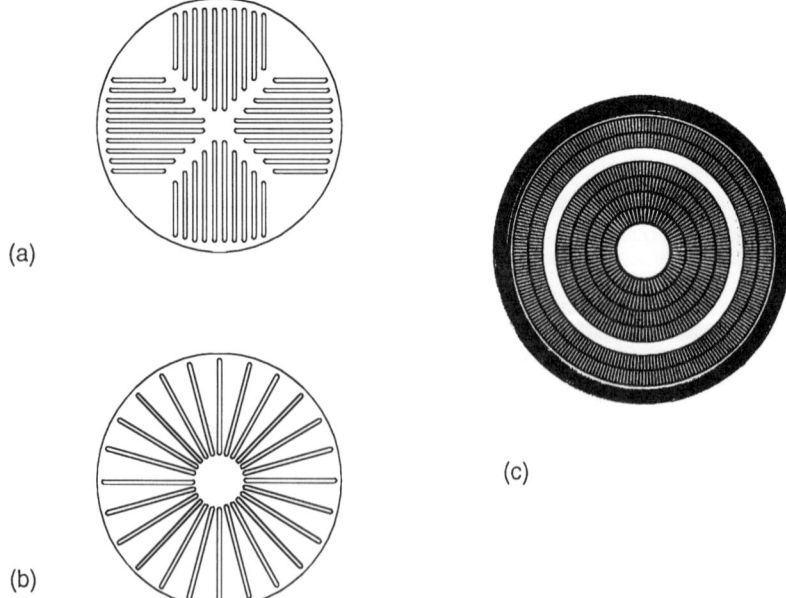

**Figure 8.13** Plan views of typical arrangements for the cathode-gate geometry of a GTO thyristor: (a) a geometry suitable for smaller devices; (b) a geometry used for a medium-sized thyristor: (c) the layout for a large device. Part (c) is reproduced by permission of Polovodiče, a.s., Prague

about 10 μm. For the largest devices, say $I_{TGQM} > 1000$ A, the effect of the gate layer resistance is minimised by distributing the gate signal both from the centre of the wafer and from a ring contact located among the concentric zones. This helps to ensure the same gating conditions for all the segments in a large-area device. An example of this type of layout is shown in Figure 8.13(c). The maximum gate turn-off current $I_{TGQM}$ has been shown to be proportional to the square root of the number of GTO segments. The cathode segments are united by a molybdenum plate pressing on the aluminium metallisation on top of the cathode islands.

The individual segments of the integrated structure of the power GTO must have nearly the same current–voltage characteristics in order to ensure a uniform distribution of the on-state current over all the GTO area. A segment with a higher carrier lifetime carries a higher on-state current density. At the same time, the gate storage time $t_{gs}$ increases with the carrier lifetime and all the anode current may be concentrated in the few segments with the longest carrier lifetimes, at the end of the gate turn-off period. This can result in enormously high power loss in a very small volume of the device and consequently in device failure. For this reason, a very high level of homogeneity of all the layers, both in their concentration profiles and their carrier lifetimes, is necessary. Several methods are used to study the homogeneity of large-area GTOs and the development of simple and effective methods of in-process monitoring is particularly important [8.4].

It is equally important that all the segments of the GTO are turned on simultaneously by the gate pulse. For this reason, the gate current should rise rapidly to quite a large amplitude: $di_G/dt > 10$ A/μs, $I_{GM} > 10$ A. If the rate of rise or the amplitude of the gate current is too low, not all of the segments are turned on, which results in a deterioration of the characteristics. GTO thyristors usually have a small $\alpha_1$ and a high holding current density. For this reason, only some of the elementary GTO segments may be turned on at low values of anode current $I_T$. Also it can happen that a sudden decrease in the on-state current can cause some of the segments to turn off. Then, during a subsequent increase in the current, the turned-on segments become overloaded, locally overheat and suffer a worsening of their gate turn-off ability. To avoid these effects, the turn-on state of all the segments is maintained by the application of a relatively small gate current during the whole time the GTO is in the on-state. This current is called the *backporch* current and can be seen in Figure 7.31, on the left-hand side of the gate current waveform.

The $di_T/dt$ capability of GTO thyristors is very high because of the high level of cathode–gate interdigitation. For high $di_T/dt$ operation, say of the order of 1000 A/μs, special gate drive circuits, in which $I_{GM} \approx 100$ A and $dI_G/dt \geqslant 100$ A/μs, are required.

In order to turn off the GTO, the negative gate pulse has to attain sufficient magnitude to satisfy the maximum gain limitation $G_{offM}$. But the rise rate of the negative gate current $-di_G/dt$ must also be optimised. Making it high decreases the turn-off time but also decreases $G_{off}$. In practice a typical maximum value of the turn-off gain coefficient $G_{off}$ is about 5. When $-di_G/dt$ is low, the turn-off time increases and there is increased probability of local instabilities in the structure. Limits for the turn-off gate signal are specified in device data sheets.

There is considerable power dissipation during the GTO turn-off process. As can be seen clearly in Figure 8.14, the anode blocking voltage starts to increase before the

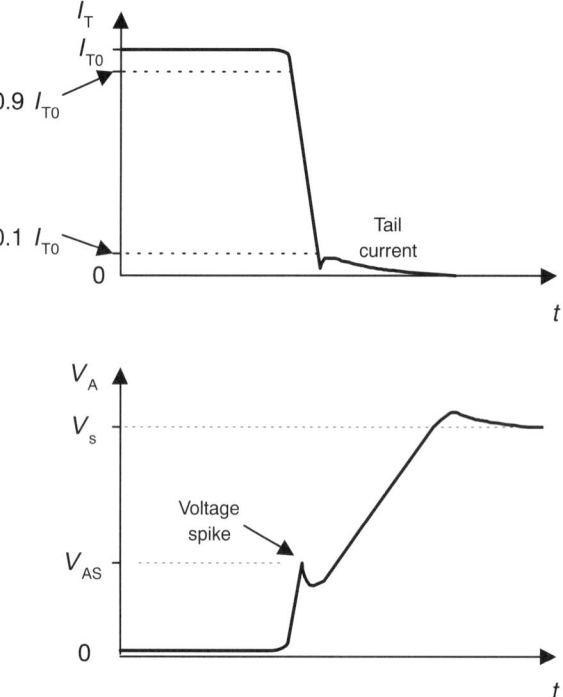

**Figure 8.14** GTO turn-off waveforms

constriction of the conducting channel is complete and the power dissipation increases with both the blocking voltage and its rate of rise. Local overheating can occur and, as with the power transistor at very high current density in the quasi-active region, a redistribution of the electric field can induce avalanche injection. Either effect can cause second breakdown in the thyristor structure. The risk of this increases with the width of the $n^+$-emitter segments.

GTO turn-off is dominated by the requirement to restrict the rise rate of the anode voltage to a safe value, hence a snubber is mandatory. The switching trajectory of the GTO is therefore never permitted to enter the square area delimited by the maximum turn-off current and the full voltage rating. A typical snubber arrangement is shown in Figure 8.15(a). Its effectiveness is limited by the stray inductance of the capacitor and the leads, which are shown explicitly in Figure 8.15(b). They induce the voltage spike shown in Figure 8.14, which arises because the parallel capacitor is unable to reduce $dV_A/dt$ before the blocking voltage reaches $V_{AS}$. The height of the spike depends on the diode characteristics as well as the parasitic inductance in the snubber circuit. It limits the maximum on-state current that can be turned off and influences the power dissipation. For a snubber to be effective, the parasitic inductance of the thyristor/diode/capacitor circuit must be very low ($<1\,\mu H$). This requires special types of diode and capacitor.

Non-uniform distribution of the current density during the turn-on and turn-off transients causes a non-uniform distribution of power dissipation that can induce a

## 8.3 THE GATE TURN-OFF THYRISTOR (GTO)

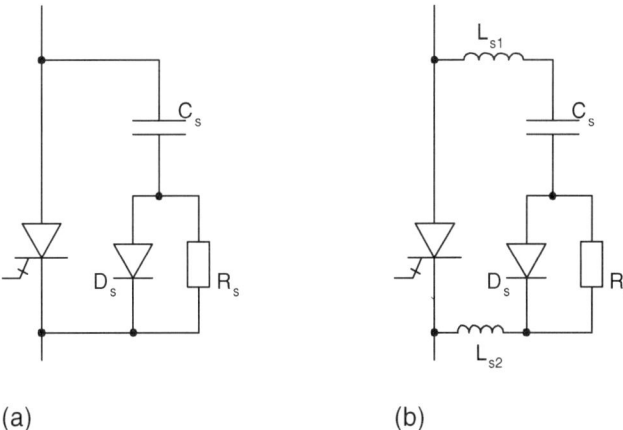

**Figure 8.15** A GTO snubber circuit: (a) the basic snubber circuit; (b) showing parasitic inductance

local increase in temperature. Therefore, a time interval between turn-on and turn-off and vice versa is needed to obtain a more uniform temperature distribution and prevent hot spots occurring, which might induce second breakdown. This interval depends on the type of device and is usually 20–150 $\mu$s.

In the case of GTO thyristors, the maximum gate turn-off current $I_{TGQM}$ is normally given as a basic device parameter, whereas for other types of thyristor it is usual to specify the average on-state current $I_{TAV}$. This is usually about 20% of $I_{TGQM}$. Thus, in a GTO for which $V_{DRM} = 4500$ V and $I_{TGQM} = 3000$ A, the maximum average on-state current is $I_{TAV} = 600$ A. The high power dissipation generated by the gate turn-off process causes a decrease of useable $I_{TAV}$ at higher operating frequencies.

For operating frequencies above 1 kHz, special devices with more finely structured cathode segments have been developed. These are designed for blocking voltages up to 3000 V and can be operated in gate turn-off mode at frequencies of several kilohertz. GTO thyristors can also operate in the GATT regime. The power dissipation during turn-off is then much lower and consequently the maximum operating frequency can be much higher.

GTO thyristors are used especially in inverter circuits requiring a blocking voltage $V_{DRM} > 2000$ V and an on-state current $I_{TAV} > 600$ A. In such applications the asymmetric GTO, together with an antiparallel diode and a snubber, is often preferred, as in the circuit of Figure 4.11. To eliminate parasitic inductance, an integrated connection of GTO and antiparallel diode can be made, similar to the arrangement shown in Figure 8.7. This is called a reverse conducting GTO (RC-GTO).

In devices designed to have a blocking voltage higher than 5 kV, a long carrier lifetime is necessary to keep the on-state voltage to an acceptable level. The gate storage time $t_{gs}$ increases, as do the differences between segments. One solution is to decrease the width of the cathode strips to about 100 $\mu$m and use proton irradiation to create an appropriate carrier lifetime gradient around $J_1$. A GTO with $V_{DRM} = 6$ kV and $I_{TGQM} = 6$ kA has been made in this way [8.5].

During gate turn-off the anode emitter continues to inject holes into the inner layers until the cathode emitter becomes reverse biased. This relatively large charge of excess carriers extends the gate storage time period and gives rise to a high tail current, both of which increase the turn-off losses. To decrease the tail current period during the turn-off process in high voltage applications, the double-gate GTO (DGTO) has been developed [8.6]. The structure is shown schematically in Figure 8.16. When gate $G_2$ is positively biased with respect to the anode, anode emitter injection is decreased and the excess charge stored in the $N_1$-base can be extracted via $G_2$. A short time after applying the positive turn-off pulse to $G_2$, a negative voltage with respect to the cathode is applied to $G_1$ and the DGTO turns off. Under these conditions the excess carriers in the inner layers of the GTO structure at turn-off are considerably reduced. This reduces the tail current and decreases the turn-off losses to less than 20% of those for a conventional GTO of similar voltage and current rating. It allows the working frequency of a 6 kV DGTO to be increased to 1 kHz. The disadvantage of the double-gate arrangement is the need to provide two isolated gate drivers, one necessarily on the high voltage side.

In a normal single-gate GTO, turn-off is achieved by applying a negative voltage $-V_{GN}$ directly to the gate. The source is usually a parallel bank of precharged electrolytic capacitors and the switch connecting them to the gate usually comprises several parallel-connected MOSFETs. Since they are only required to carry short pulses of current, full use is made of the MOSFET pulse current rating. The resistance of the gate circuit is made very low, so the waveform of the gate current when the MOSFETs turn on is determined chiefly by the parasitic inductance $L_G$ in the gate circuit. The resulting negative gate current waveform therefore approximates to a ramp of slope $(-V_{GN}/L_G)$. When $L_G$ is large, the current reaches only a relatively low value by the time that sufficient charge has been removed from the thyristor to turn it off. The turn-off gain is high but the storage time is long and the dissipation high. On the other hand, if the gate circuit inductance is low, the negative gate current reaches a relatively high value by the time sufficient charge has been

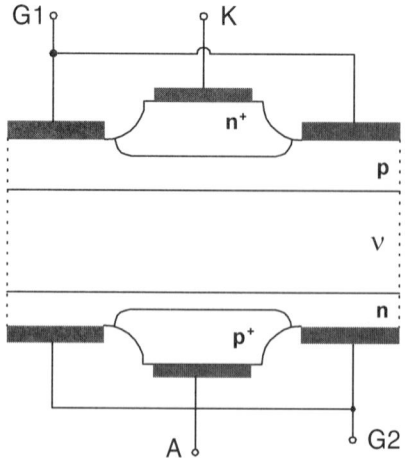

**Figure 8.16** A double-gate GTO

removed to achieve turn-off. The turn-off gain is low but the charge storage time and the turn-off losses are reduced. In addition, the difference between the instants of turn-off at the fastest and slowest switching cathode fingers is minimised, reducing the possibility of hot spots developing.

One means by which fast turn-off can be obtained is to use a GTO in conjunction with a MOSFET in the cascode configuration. The low voltage MOSFET is connected in series with the GTO in the cathode lead. The gate of the GTO is connected to a low impedance voltage source. Turning off the MOSFET forces a voltage to be injected into the gate–cathode circuit, due to the rising resistance of the MOSFET. This voltage is effectively applied to the parasitic capacitance in the gate–cathode circuit and results in a rapid rise of negative gate current and fast turn-off of the GTO. Although this arrangement can be effective at generating a rapid rise of current in the gate circuit at turn-off, it suffers from the serious disadvantage that the MOSFET is in the path of the main current when the GTO is conducting in the on-state, thereby increasing the forward voltage drop and the losses associated with the device. It is clearly much more efficient to inject the negative voltage required at turn-off from an external source. However, to reach the values of $di/dt$ achieved in cascode switching, it is necessary to use gate drive circuits with very low parasitic inductance. This is achieved by the close integration of the GTO and the gate circuit. Such thyristors are known as gate-commutated turn-off GTOs (GCTs) [8.7] and sometimes, when made as a single unit with a closely coupled gate drive, as integrated gate-commutated turn-off GTOs (IGCTs) [8.8, 8.9]. They can operate with a very small snubber capacitance and in many instances without any snubber at all.

The safe operation of normal GTOs requires a snubber capacitance of a few microfarads, with a high voltage and very low inductance construction. With the GCT this capacitance can be reduced in size, or even eliminated. The large negative gate current pulse with its short rise time sweeps out the excess carriers from the $P_2$-layer in a very short time and reduces the differences in the storage times of the individual GTO segments. In effect, practically all the anode current flows through the gate, giving unity turn-off gain. In a conventional GTO this mode of turn-off is impossible to achieve practically. The gate supply voltage needs to be below 20 V (it is typically $-15$ V with respect to the cathode), so the total inductance of the gate drive circuit, including the connections internal to the GTO, limits the maximum rate of rise of the negative gate current. In a conventional GTO the internal inductance is typically about 30 nH, whereas the inductance of the gate drive circuit, including the coaxial connecting cable, is usually about 300 nH. As a result, the rate of rise is limited to something less than 50 A/$\mu$s.

To improve on this requires a special construction for both the gate drive circuit and the internal thyristor geometry. With flatband cable and using a special gate drive circuit construction, it is possible to reduce the overall inductance below 100 nH and achieve what are known as hard drive conditions [8.9]. The GCT uses a coaxial thyristor housing in which the internal inductance is less than 5 nH. Integrating this with an ultralow inductance gate drive unit in a special construction enables the overall inductance to be reduced to less than 20 nH and $(-dI_G/dt)$ to exceed 1 kA/$\mu$s.

In these devices the gate connection is made to the circumference of the GTO wafer, which provides excellent low inductance access to all the cathode islands. The gate driver circuit is then built around the GTO in such a way as to minimise inductance. Using this type of construction, it has proved possible to dispense with the snubber. Turn-off times are improved and turn-off losses reduced. Devices have been developed with a voltage rating of 4.5 kV and a turn-off current capability of 4 kA. They are expected to find application in electrical power systems, perhaps in static VAR generators, and in high power, variable-speed drives.

In inverter applications, where a freewheeling diode is used in antiparallel connection with the semiconductor switches in each leg, parasitic inductance in the diode connection can limit its ability to clamp the reverse voltage applied to the switch under conditions of very high $di/dt$. The inductance in the diode path can be reduced to negligible levels by the monolithic integration of the diode with the switch, forming a reverse conducting GCT [8.10] similar to the conventional RCT shown in Figure 8.7.

## 8.4 The Triac

The triac (triode ac switch) is a bidirectional switch controlling the flow of load current in either direction. The triac structure can be thought of as a pair of integrated antiparallel thyristors with a special gate layout that enables the device to be turned on with either a positive or a negative gate voltage. The triac structure and its basic characteristics are shown in Figure 3.5. It turns off when the main current flowing between electrodes $MT_1$ and $MT_2$ decreases below the holding current $I_H$. The triac replaces two thyristors in antiparallel connection and requires only a single gate circuit for ac power regulation.

There are four combinations of polarity of the main voltage (between electrodes $MT_1$ and $MT_2$) and the gate voltage (between electrodes G and $MT_1$):

1.  Terminals $MT_2$ and G are both positive with respect to $MT_1$ (quadrant I). In this case the triac turns on in the same way as an ordinary thyristor, as demonstrated in Figure 8.17(a).
2.  Terminal $MT_2$ is positive with respect to $MT_1$ but the gate G is negative with respect to the terminal $MT_1$ (quadrant II), as shown in Figure 8.17(b). In this case a space charge region forms at junction $J_2$. When a negative gate voltage is applied, electrons are injected into layer $P_2$ from the $n^+$-region of the gate. The excess electrons diffuse to junction $J_2$ and, on entering layer $N_1$, they induce the injection of holes from layer $P_1$ into $N_1$. When this is high enough, the electron injection from the $n^+$-region flows through to terminal $MT_2$ and the thyristor structure $P_1N_1P_2N_2$ turns on.
3.  The electrode $MT_2$ and the gate G are negative with respect to $MT_1$ (quadrant III), as shown in Figure 8.17(c). Now there is a space charge region at junction $J_1$. When the gate is made negative with respect to $MT_1$, electrons are injected from the $n^+$-region into layer $P_2$. Even if junction $J_1$ is not reverse biased, a built-in space charge region exists at $J_2$ and some excess electrons enter layer $N_1$, as expressed by (7.13). These make $N_1$ negative with respect to $P_2$, and junction $J_2$ becomes forward biased. As a result, holes are injected into $N_1$ and diffuse towards the reverse-biased junction $J_1$, where

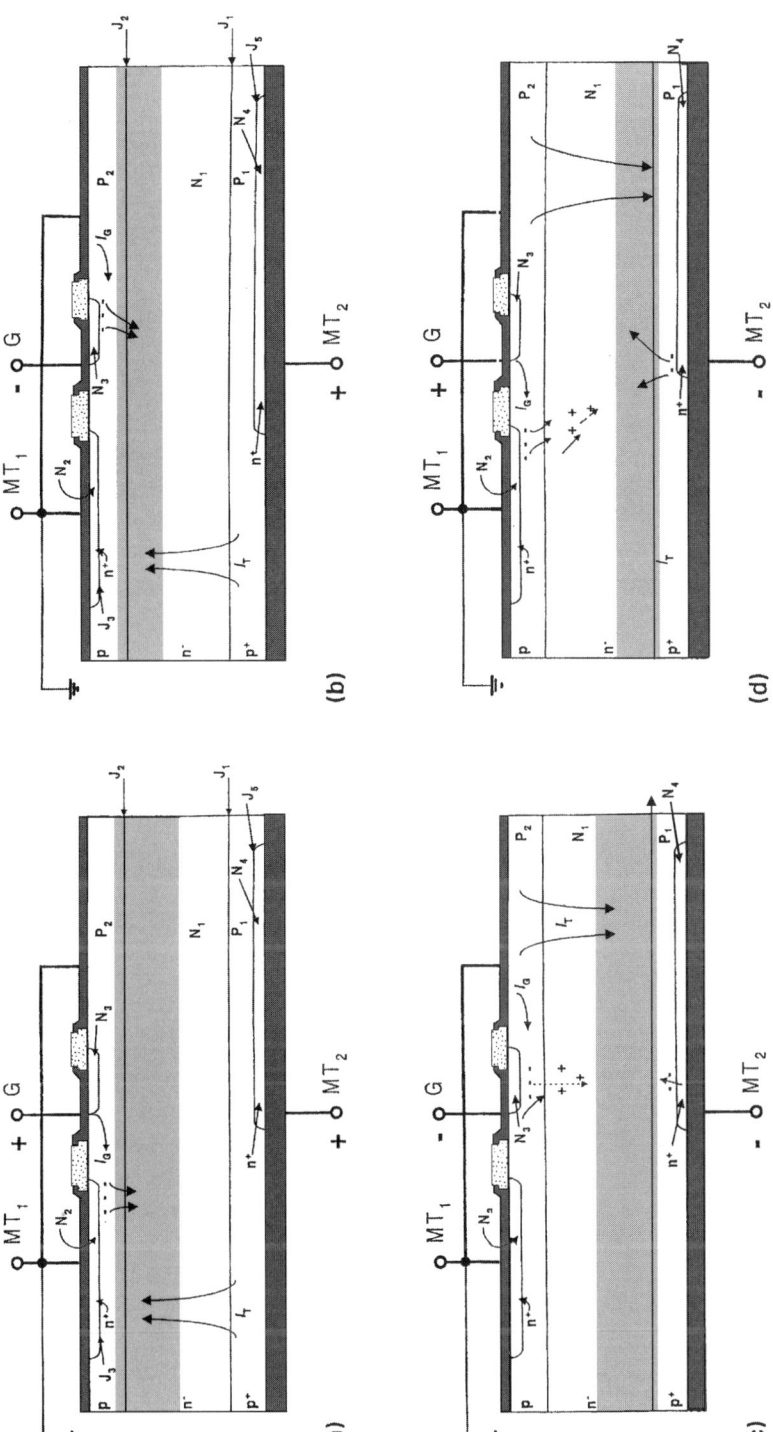

**Figure 8.17** The four possible triac operating conditions: (a) quadrant I; (b) quadrant II; (c) quadrant III; (d) quadrant IV

they are transported into region $P_1$ by the electric field of the space charge region. Layer $P_1$ becomes positively charged and this induces electron injection from the $n^+$-layer connected to contact $MT_2$. As a result of this positive feedback, the thyristor structure $N_4P_1N_1P_2$ turns on.

4. The electrode $MT_2$ is negative and the gate G is positive with respect to $MT_1$ (quadrant IV), as shown in Figure 8.17(d). In this case the mechanism of the remote gate triggering is the same as in item 3; the only difference is that electrons are injected into layer $P_2$ from the $n^+$-emitter region connected to electrode $MT_1$.

To obtain sufficiently low gate triggering currents, a complicated arrangement of layers $N_2$, $P_1$, $N_4$ and $P_2$ must be used. In consequence, the thyristor structures are not fully separated. At turn-off a charge of unrecombined excess carriers still remains in one of these antiparallel structures when the blocking voltage starts to rise. This can cause the triac structure to turn on again at quite a low rise rate of the voltage, because the emitters cannot be shorted effectively in the gate region.

Some problems can occur when triacs are used in circuits with an inductive load. The current and voltage waveforms are shown in Figure 8.18. The load current is delayed with respect to the supply voltage and passes through zero while the supply voltage is high. The triac turns off when the current decreases below the holding current, which is close to zero. We assume that the rate of fall of current is quite low. At the moment of turn-off, the forward blocking voltage $V_D$ rises. If its rise rate exceeds a certain value, $(dV_D/dt)_{com}$, the triac turns on. This value is much lower than $(dV_D/dt)_{crit}$, the rate that turns on the device in the absence of a preceding on-state current. Therefore, the use of triacs in circuits with inductive loads is relatively limited and two ordinary thyristors, connected in antiparallel, are usually employed. Triacs primarily find application for power regulation in low power circuits ($I_{RSM} < 25$ A).

## 8.5 Light-Triggered Thyristors (LTTs)

Light-triggered thyristors – sometimes called photothyristors, optothyristors or light-activated SCRs (LASCRs) – are four-layer structures of the thyristor type specially constructed to be triggered by the optical generation of excess carriers.

When a piece of semiconductor is illuminated by light whose photon energy is greater than the bandgap energy ($hv > W_g$), optical generation of electron–hole pairs occurs, as described in Section 1.2.5. Light so absorbed in the $P_2$-layer of a thyristor in the blocking state, generates electrons that are collected by the space charge region of junction $J_2$ and transported into layer $N_1$. This leads to the injection of excess holes into $N_1$ from layer $P_1$. The effect of the optical carrier generation is similar to that of the electron injection from the $n^+$-emitter following the application of a positive gate voltage. If enough carriers are generated, the condition for turn-on (7.14) is fulfilled and the thyristor turns on through the normal positive feedback mechanism.

The basic problem of the construction of light-triggered thyristors is solving the contradictory requirements of the thyristor structure. On one hand, it is necessary to

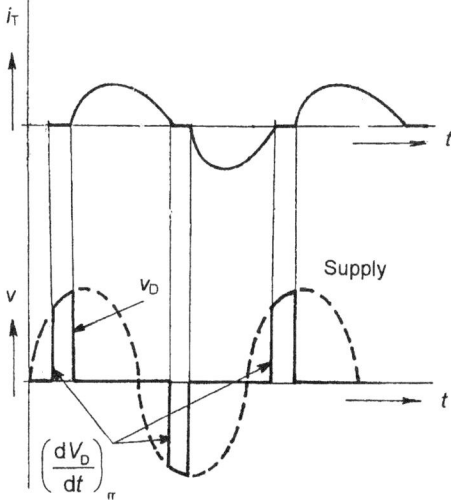

**Figure 8.18** Triac current and voltage waveforms in a circuit with an inductive load

obtain satisfactory turn-on with a reasonably low optical power; on the other hand, high values of the parameters $(dV_D/dt)_{crit}$ and $(dI_T/dt)_{crit}$ are required. The structure and layout of the optical gate region is very important in determining these parameters. A typical example is shown in Figure 8.19. The optical signal is supplied through a fibre-optic light-guide to the thyristor surface. The wavelength should be in the region 0.85–0.95 µm so that the absorption depth is about 50 µm, as shown in Figure 8.20. It is essential to have optical sources of suitable power (10 mW and more) and a reliability comparable to that of power thyristors.

The incident light generates excess carriers in proportion to the optical power density. Where this is high enough to initiate the turn-on process locally, the anode current flowing to the gate via the auxiliary contact gives the effect of a large gate current and, as in the case of the amplifying gate thyristor, creates the larger turned-on region that is necessary for a good $di_T/dt$ capability.

It is desirable that the radius of the illuminated region should match the radius $R$ of the auxiliary emitter region if good triggering sensitivity is to be obtained. However, this is normally too small to make $(dV_D/dt)_{crit}$ adequately high or to provide a sufficiently large initial turned-on region in the main part of the cathode emitter of a large-area thyristor. Therefore, the optical gate of a high power, light-triggered thyristor is often arranged as a cascade of auxiliary thyristors, as shown in Figure 8.21. The configuration of the auxiliary cathode AK2 determines the gate area of the main part of the thyristor cathode and can be arranged as for a high frequency thyristor, to guarantee a large initial turn-on area. Optically triggered triacs have also been developed.

Optical triggering is desirable in high voltage applications, especially where several thyristors are connected in series and the insulation of gate and anode circuits is required to be higher that 10 kV. A typical light-triggered thyristor suitable for HVDC applications might have ratings of $V_{DRM} = 8$ kV and $I_{TAV} = 3$ kA, or $V_{DRM} = 6$ kV and $I_{TAV} = 5.5$ kA. Developments continue [8.11].

268  8 THYRISTOR TYPES AND APPLICATIONS

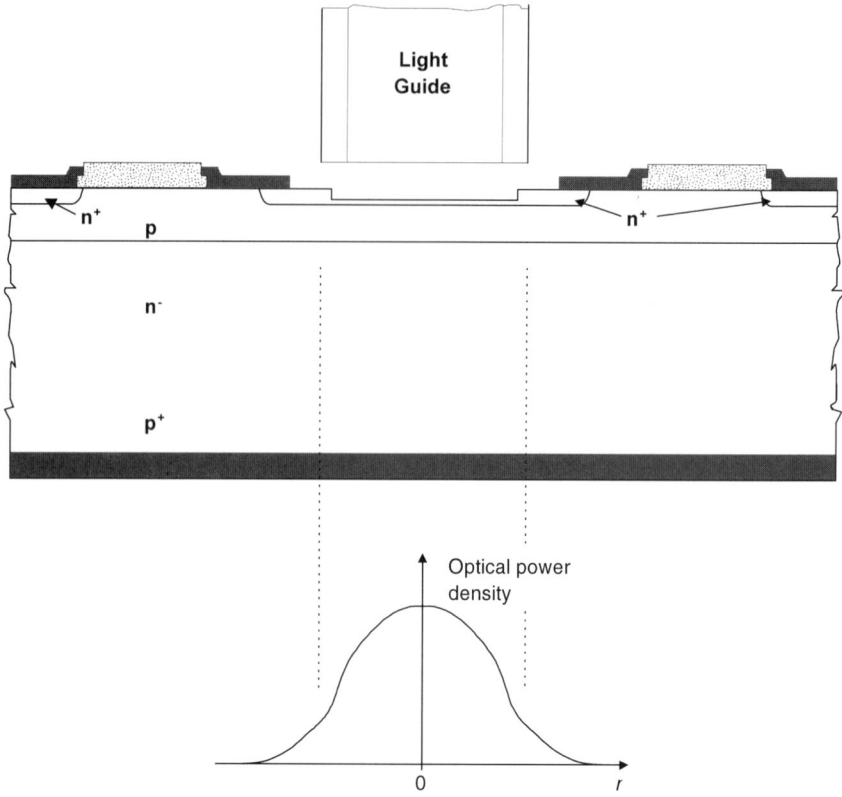

**Figure 8.19** The basic arrangement of an optically triggered thyristor showing the optical power distribution in the gate area

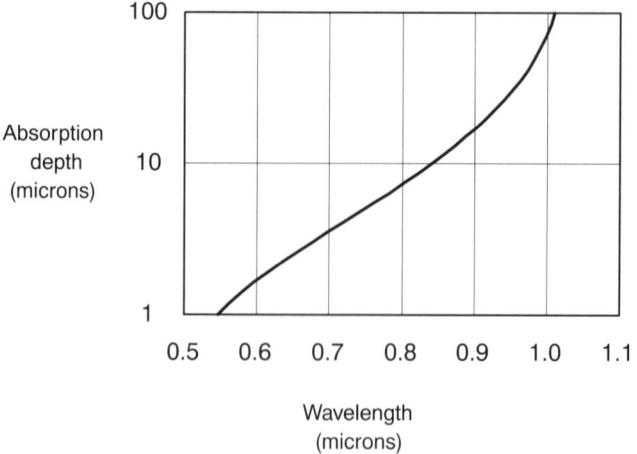

**Figure 8.20** The optical absorption depth in silicon as a function of wavelength

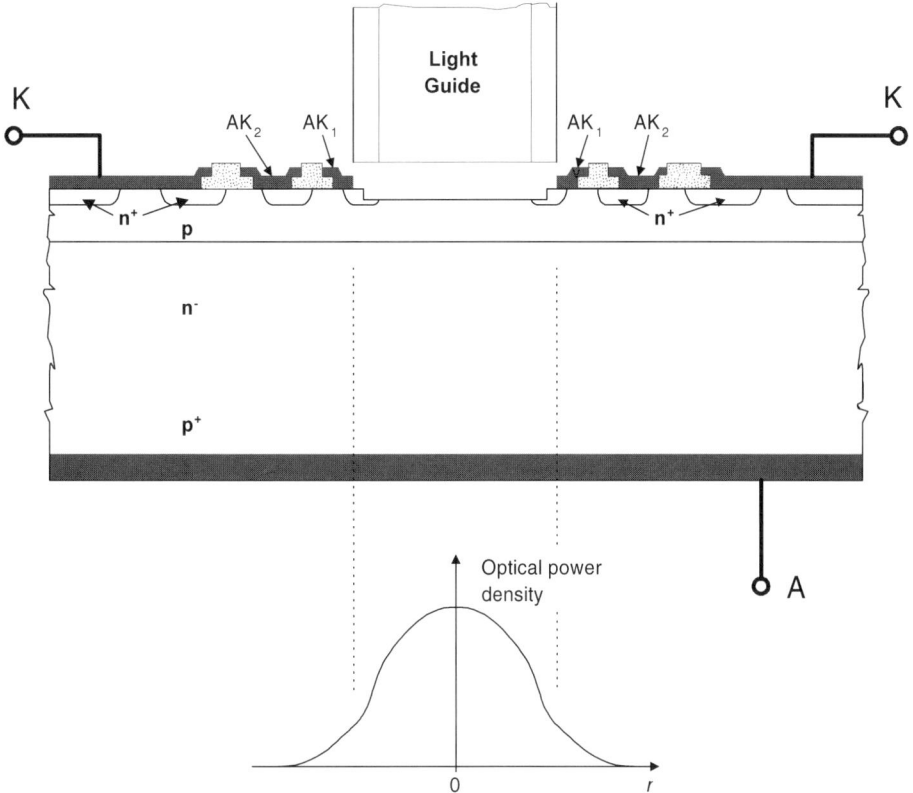

**Figure 8.21** The cascaded gate of an optically triggered thyristor

As an alternative to direct optical triggering, opto-isolators can be used to drive a low power device in the gate circuit of an otherwise conventional high power thyristor, as shown in Figure 8.22. In this arrangement the conflicting requirements can be more easily accommodated.

## 8.6 Breakover Diodes (BODs)

Breakover diodes (BODs) are two-terminal, ungated pnpn structures designed to turn on at a specified blocking voltage or a specified rate of change of anode voltage. They are sometimes called Shockley diodes or dinistors.

The switching voltage is determined by (7.4). The structure can be designed to use either the avalanche breakdown voltage of $J_2$ or the punch-through voltage. To obtain good thermal stability of the breakover voltage, the $n^+$-emitter is usually shorted. The short pattern can be designed so that the structure turns on at a predetermined $dV/dt$. The BOD can be constructed as a reverse conducting device.

**270**  8 THYRISTOR TYPES AND APPLICATIONS

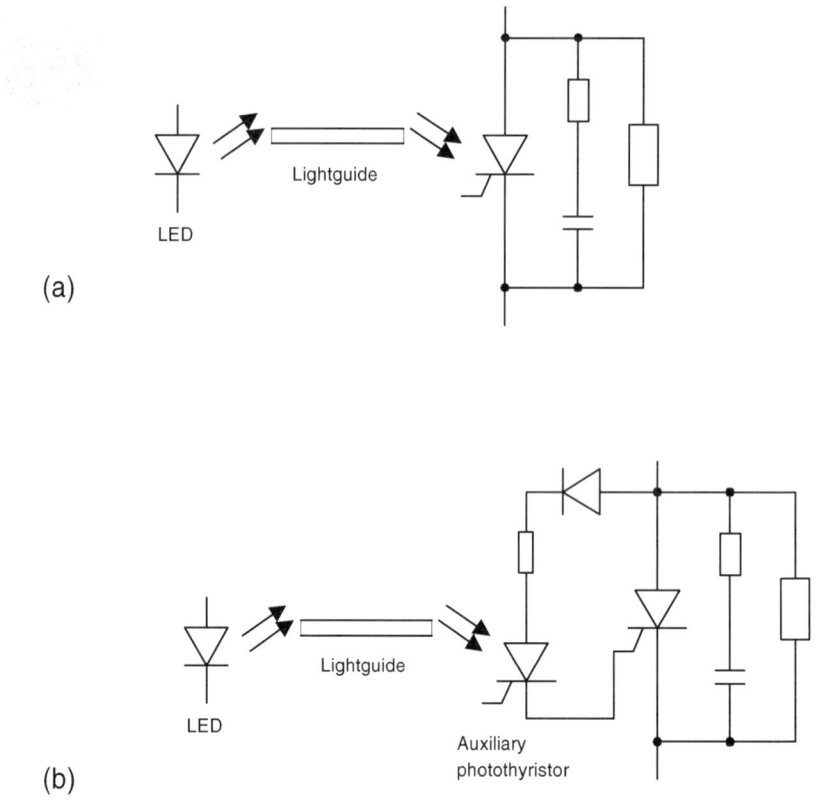

**Figure 8.22** The application of opto-isolators in thyristor gate-drive circuits: (a) directly triggered; (b) using an auxiliary thyristor in the gate circuit

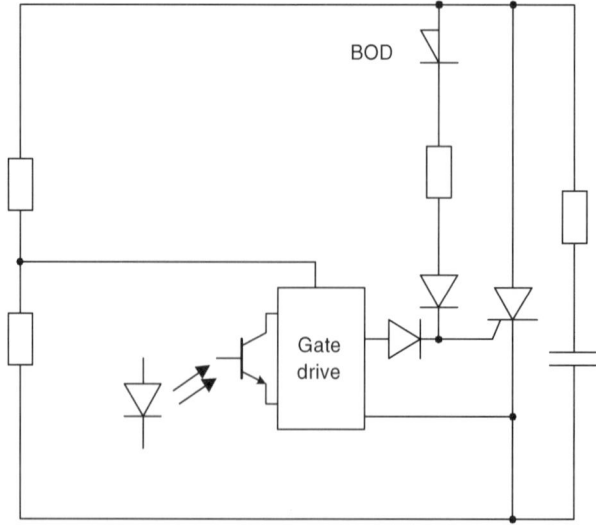

**Figure 8.23** Use of a breakover diode to protect a high current thyristor

Breakover diodes are used to provide overvoltage protection for large-area thyristors in which breakover switching would be potentially destructive. An arrangement that also takes advantage of optical triggering is shown in Figure 8.23.

## Summary

The two main categories of SCR are the phase-control thyristors, designed for power frequency applications and the high frequency thyristors used mainly in inverter applications.

With phase-control thyristors, turn-off time is not normally critical and long carrier lifetimes are necessary. There is a trade-off between the blocking voltages and the surge current capability.

With inverter-grade devices, optimisation for a given application is much more complex, as a trade-off has to be reached between the turn-off time on the one hand and the on-state voltage and the turn-on behaviour on the other. Turn-off time can be reduced by cathode shorts and by reducing the carrier lifetime. However, this increases the on-state voltage, reduces the surge current capability and limits the maximum allowable rate of rise of the current. The reach-through design with a stop layer, enhances these parameters with the loss of the reverse blocking capability, the ASCR. A highly interdigitated gate–cathode structure improves both the turn-on and the turn-off characteristics. Extracting negative gate current before turn-off reduces the turn-off time (GATT).

One of the most important power semiconductor devices is the GTO, which also requires a highly interdigitated gate–cathode arrangement. This can be switched off (with a delay and a 'tail' of anode current) by a sufficiently large negative gate current. Cathode shorts cannot be used, but anode shorts improve thermal stability and aid the gate turn-off process.

Important variants on the basic thyristor include the triac two-way switch, the light-triggered thyristor and the breakover diode.

## References

[8.1] Cordingley, B. V. and Chamund, D. J. Design and Performance of Medium Power Asymmetrical Thyristors, *GEC Journal of Science & Technology*, **46**, 67–72 (1980).

[8.2] Benda, V. The turn-off process of the thyristor and the measurement of the turn-off time, *Acta Polytechnica*, **9 (III-2)**, 79–96 (1980).

[8.3] Shimizu, J., Oka, H., Funakawa, S., Samo, H., Iido, T. and Kawakami, A. High Voltage High Power Gate-Assisted Turn-off Thyristor for High Frequency Use, *IEEE Trans. on Electron Devices*, **ED-23**, 883–7 (1976).

[8.4] Benda, V. and Spur, P. A Simple Method of In-Process Checking of the Large-Area GTO Homogeneity, *Microelectronics and Reliability*, **33**, 1441–5 (1993).

[8.5] Nakagawa, T., Tokunoh, F., Yamamoto, M. and Koga, S. A New High Power Low Loss GTO, *Proc. ISPSD'95*, 84–8 (Yokohama, 1995).

[8.6] Ogura, T., Nakagawa, A., Atusta, M., Kamei, Y. and Takagami, K. High Frequency 6000-V Double-Gate GTOs, *IEEE Trans. on Electron Devices*, **ED-40**, 628–33 (1993).

[8.7] Satoh, K., Yamamoto, M., Nakagawa, T. and Kawakami, A. A New High Power Device, GCT (Gate Commutated Turn-off) Thyristor, *Proc. EPE'97*, vol. 2, 70–5 (Trondheim, 1997).

[8.8] Steimer, P. K., Grüning, H. E. G. and Werninger, J. The IGCT – The Key Technology for Low Cost, High Reliable High Power Converters with Series Connected Turn-off Devices, *Proc. EPE'97*, vol. 1, 384–9 (Trondheim, 1997).

[8.9] Grüning, H. E. and Ødegard, B. High Performance Low Cost MVA Inverters Realised with Integrated Gate Commutated Thyristors (IGCT), *Proc. EPE'97*, vol. 2, 60–5 (Trondheim, 1997).

[8.10] Linder, S., Klaka, S., Frecker, M., Carroll, E. and Zeller, H. A New Range of Reverse Conducting Gate-Commutated Thyristors for High-Voltage, Medium Power Applications, *Proc. EPE'97*, vol. 1, 117–24 (Trondheim, 1997).

[8.11] Katoh, S., Choi, J. H., Yokota, T., Watanabe, A., Yamaguchi, T. and Saito, K. 6-kV, 5.5-kA Light-Triggered Thyristor, *Proc. ISPSD'97*, 73–6 (1997).

# 9

# STATIC INDUCTION POWER DEVICES

With the bipolar devices described in Chapters 5 to 8, conductivity modulation is obtained by the injection of excess minority carriers across forward-biased p–n junctions. Since 1952, other structures have been in use, in which the conductance is based on majority carriers in a region whose dimensions are varied by varying an applied voltage and so expanding or contracting a space charge region. In the case of the junction field effect transistor (JFET) this is at a p–n junction. In the case of the metal–semiconductor field effect transistor (MESFET) it is at a rectifying metal–semiconductor contact.

For power devices, a high blocking voltage in the off-state and high conductance in the on-state are required. A structure which achieves this is the static induction transistor (SIT) [9.1], which is effectively a high power, short-channel JFET. The development of the power SIT started in 1970 and devices of this family remain the subject of research and development. They differ from most other three-terminal devices in that they are normally on and require the application of a gate voltage (normally negative) to turn them off.

A related device, in which an extra injecting layer is added to the conduction path so that the conductivity can be modulated, is known variously as the static induction thyristor (SITh), the field-controlled thyristor (FCT) and the field-controlled diode (FCD); FCD is arguably the most appropriate name. At present, static induction transistors and static induction thyristors have limited practical significance.

## 9.1 The Static Induction Transistor

The basic structure of a static induction transistor (SIT) with a buried gate is shown in Figure 3.7. Two alternative designs are shown in Figure 9.1. When a reverse voltage is applied between the source terminal S and the gate G (G is negative), the space charge region at the $p^+$–n junction restricts the area of the conducting channel between the source and the drain and so increases its resistance. When the applied negative gate voltage $-V_{GS}$ is high enough, the space charge region fills all the space between the gate regions, thus closing the channel completely, as shown by the

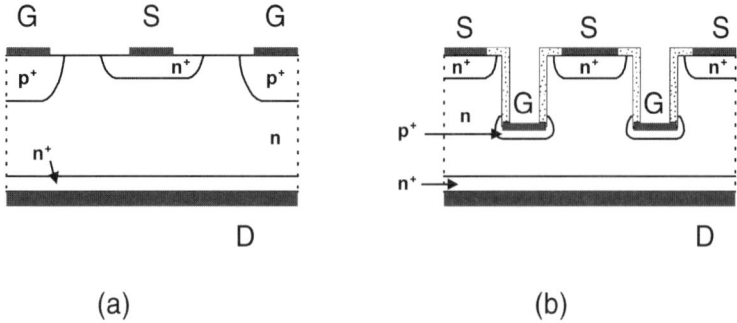

**Figure 9.1** Alternative designs of static induction transistor: (a) planar structure; (b) trench structure

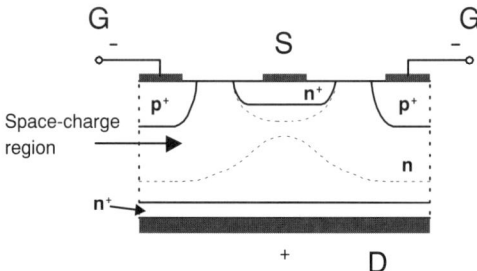

**Figure 9.2** The SIT structure with the conduction channel pinched off

dashed lines in Figure 9.2. Decreasing $V_{GS}$ or changing its polarity causes the thickness of the space charge region to decrease so that there is again a conducting channel between the source and drain contacts. The conductance of this channel depends on the drain–source voltage $V_{DS}$, the gate–source voltage $V_{GS}$, and the dimensions of the structure.

In a conventional JFET the channel length $l$ is much greater than its width $w_{ch}$. Figure 9.3 illustrates how this might be achieved in a vertical channel device. When a voltage $V_{DS}$ is applied between the drain terminal D and the source, the thickness of the space charge region at the $p^+$–n junction is a function of the voltage $V_{GS}$ and the distance from the source. Near the source it is

$$d_S = \sqrt{\frac{2\varepsilon_r \varepsilon_0}{eN_d}(V_{GS} + V_{diff})} \qquad (9.1)$$

while at the drain end it becomes

$$d_D = \sqrt{\frac{2\varepsilon_r \varepsilon_0}{eN_d}(V_{GS} + V_{DS} + V_{diff})} \qquad (9.2)$$

## 9.1 THE STATIC INDUCTION TRANSISTOR

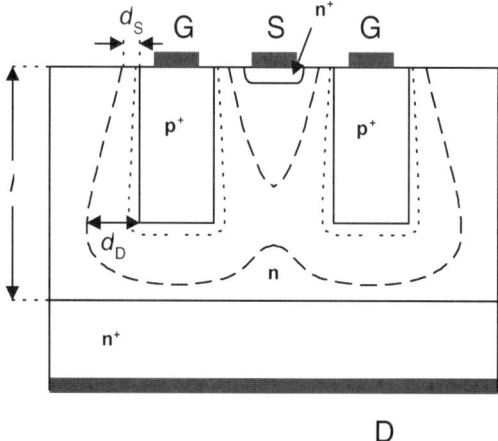

**Figure 9.3** The structure of a long channel JFET. The dotted line shows the extent of the space charge region at a low negative gate voltage; the dashed line represents the boundary of the space charge region when the negative gate voltage is increased to pinch off the channel

The donor concentration $N_d$ in the n-region is assumed to be uniform and the diffusion voltage $V_{diff}$ is given by (2.1).

Assume that the gate regions are separated by a distance $w_{ch}$, the space between them is rectangular with a length $z$, and the distance between the source and drain electrodes is $l$. Then the drain current $I_D$ can be expressed as

$$I_D = 2z\left[\frac{w_{ch}}{2} - d(x)\right] e\mu N_d \frac{dV}{dx} \quad (9.3)$$

where

$$d(x) = \sqrt{\frac{2\varepsilon_r \varepsilon_0}{eN_d}(V_{GS} + V_{diff} + V(x))} \quad (9.4)$$

By integrating (9.3) the current–voltage characteristic can be derived as

$$I_D = e\mu N_d z \frac{w_{ch}}{l} \left\{ V_{DS} - \frac{4}{3w_{ch}} \left(\frac{2\varepsilon_r \varepsilon_0}{eN_d}\right)^{1/2} [(V_{GS} + V_{DS} + V_{diff})^{3/2} - (V_{GS} + V_{diff})^{3/2}] \right\} \quad (9.5)$$

The conducting channel narrows towards the drain end and when

$$(V_{GS} + V_{DS} + V_{diff}) = \frac{eN_d w_{ch}^2}{8\varepsilon_r \varepsilon_0} = V_P \quad (9.6)$$

it is pinched off, as shown by the dashed line in Figure 9.3. For larger values of $V_{DS}$, less than the breakdown voltage, the pinch-off point moves towards the source,

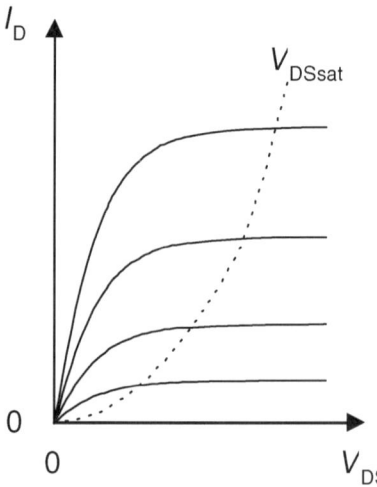

**Figure 9.4** Typical current–voltage characteristics of a JFET

leaving the drain current essentially unchanged. It is said to have saturated. A set of JFET characteristics is shown schematically in Figure 9.4.

The transconductance $g_m$ is an important parameter which is maximum in saturation, i.e. when $d_D = w_{ch}/2$:

$$g_m = \left.\frac{\partial I_D}{\partial V_{GS}}\right|_{V_{DS}} = e\mu N_D \frac{w_{ch} z}{l}(d_D - d_S) \qquad (9.7)$$

Another important parameter is the channel (or drain) conductance:

$$g_D = \left.\frac{\partial I_D}{\partial V_D}\right|_{V_G} = e\mu N_D \frac{z}{l}(w_{ch} - 2d_D) \qquad (9.8)$$

This is maximum at low drain voltage and zero in the saturation region.

For a more exact solution it is necessary to take account of the dependence of the mobility on the electric field (Section 1.3.2). However, long-channel JFET structures are not usually suitable for power semiconductor devices.

In the case of a short channel, the space charge region can be considered to form a potential barrier to the flow of electrons between source and drain, as shown in Figure 9.5. To a first approximation, the height $V_B$ and width of the barrier depend linearly on the voltages $V_{GS}$ and $V_{DS}$. Thus:

$$V_B = \alpha V_{GS} - \beta V_{DS} \qquad (9.9)$$

where $\alpha$ and $\beta$ are constants which depend on the geometric configuration and the concentration profiles. The current is determined by those electrons that have sufficient kinetic energy to surmount the barrier. The current density is therefore

$$J_D = J_0 \exp\left(-\frac{eV_B}{kT}\right) \qquad (9.10)$$

## 9.1 THE STATIC INDUCTION TRANSISTOR

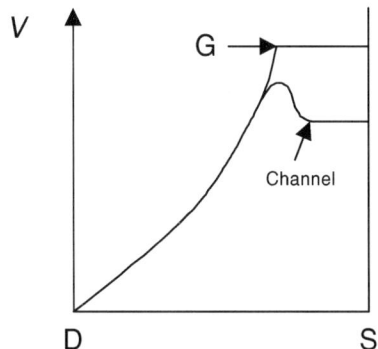

**Figure 9.5** The potential distribution in the short-channel SIT structure

Electrons cross the space charge region under the influence of both diffusion and drift, so that

$$J_D = en\mu_n E(x) + eD_n \frac{dn}{dx} = eD_n \left( n \frac{kT}{e} \frac{dV}{dx} + \frac{dn}{dx} \right) \tag{9.11}$$

Using the Einstein relation (1.70) and multiplying both sides of (9.11) by $\exp(-eV(x)/kT)$, we obtain

$$J_D \int \exp\left(\frac{-eV(x)}{kT}\right) dx = eD_n N_D \tag{9.12}$$

When the space charge density is uniform, the voltage in the region varies as

$$V(x) = \frac{4V_B}{l^2}(x^2 - xl) \tag{9.13}$$

where $l$ is the width of the potential barrier, i.e. the length of the channel. The drain current density $J_D$ then varies with the height of the potential barrier as

$$J_D = \frac{2eD_n N_D}{l} \sqrt{\frac{eV_B}{\pi kT}} \exp\left(-\frac{eV_B}{kT}\right) \tag{9.14}$$

Using (9.9) this can be expressed in terms of $V_{DS}$ and $V_{GS}$ as

$$J_D = \frac{2eD_n N_D}{l} \sqrt{\frac{e}{\pi kT}(\alpha V_{GS} - \beta V_{DS})} \exp\left[\frac{-e}{kT}(\alpha V_{GS} - \beta V_{DS})\right] \tag{9.15}$$

A comparison of (9.5) and (9.15) shows there is a considerable difference between the characteristics of the JFET (Figure 9.4) and the characteristics of the SIT structure (Figure 9.6).

Often the SIT structure is made with a very low donor concentration and a very short distance between the gate junctions. A potential barrier can then form even

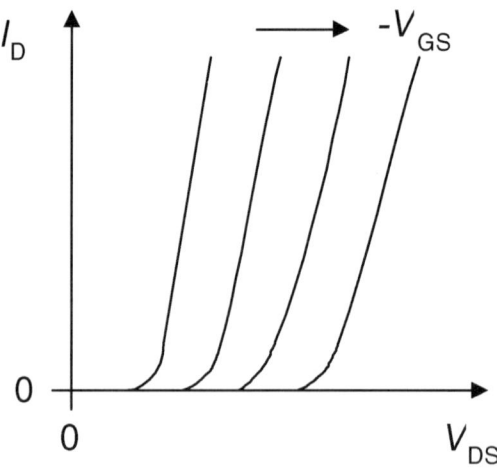

**Figure 9.6** Typical current–voltage characteristics of an SIT

without an applied negative gate voltage, simply as a result of the diffusion potential. In this case, channel pinch-off occurs when

$$w_{ch}^2 \leqslant \frac{3\varepsilon_r \varepsilon_0}{l} \frac{V_{diff}}{eN_D} \qquad (9.16)$$

Figures 3.8 and 9.1 show how the n⁺-source and p⁺-gate regions can be formed in different ways. Planar technology can be use to form the SIT structure shown in Figure 9.1(a), and the channelled device shown in Figure 9.1(b) is known as the USIT structure. In all cases the distance between p⁺-regions has to be in the order of 1–10 μm if condition (9.16) is to be fulfilled. The high voltage blocking capability of the structure can be obtained only when a sufficiently large negative gate impulse is applied. It is characterised by the voltage gain coefficient $G_D$:

$$G_D = \frac{\partial V_D}{\partial V_G} = \frac{\alpha}{\beta} \qquad (9.17)$$

A high value of $G_D$ requires the channel length and width to be of comparable dimensions.

Current is carried through the SIT by the majority carriers, so the frequency dependence of parameters such as the gain coefficient is determined by the capacitance of the p⁺–n junction and the resistance of individual layers rather than by the carrier transit time. An equivalent circuit for the SIT transistor is shown in Figure 9.7.

The cut-off frequency is determined mainly by the gate–source capacitance $C_{GS}$, and the resistance $R_S$:

$$f_m = \frac{1}{2\pi C_{GS} R_S} \qquad (9.18)$$

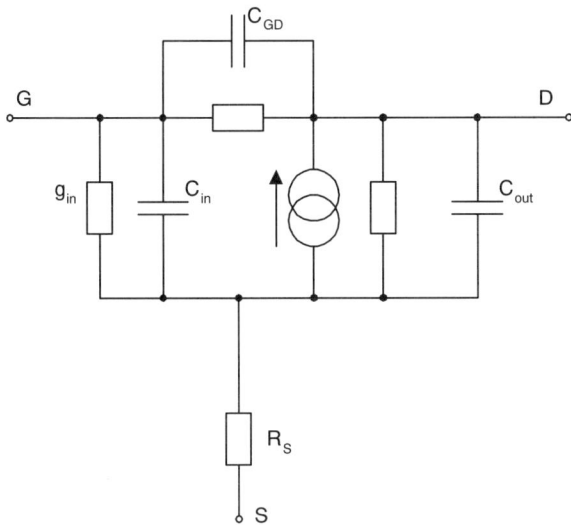

**Figure 9.7** An equivalent circuit for the SIT

Both $C_{GS}$ and $R_S$ depend on the gate voltage. The capacitance decreases and the resistance increases as $V_G$ increases. Generally, the cut-off frequency of SIT transistors is high in comparison with that of bipolar devices; the predicted [9.2] theoretical limit for low voltage devices is in excess of 100 GHz. A 100 W SIT has been reported [9.3] as operating at 1 GHz ($V_{GDM} = 90$ V). A 1 kW device ($V_{GDM} \leqslant 1500$ V, $I_{DM} \leqslant 60$ A) with a maximum operating frequency of 10 MHz and a 3 kW device ($V_{GDM} \leqslant 1500$ V, $I_{DM} \leqslant 180$ A) with a maximum operating frequency of 7 MHz have been described.

The resistance of the conducting channel, $R_{on}$, increases with temperature (as $T^{5/2}$). When SITs are connected in parallel, the structure with the lowest $R_{on}$ attracts the highest current density and hence the highest power dissipation. Because the associated temperature rise increases $R_{on}$, the effect is self-compensating, provided the temperature remains below the intrinsic temperature $T_i$. If $T_i$ is exceeded, the resulting decrease in resistivity induces an increase in current density and thermal breakdown follows.

Although the use of majority carriers for the transport of charge enables the SIT to have a very good high frequency performance, high voltage devices require a wide and lightly doped n$^-$-region, which makes $R_{on}$ high and limits current-carrying capacity.

It is possible to decrease $R_{on}$ by allowing the gate–source junction to become forward biased. In this case, excess carriers are injected into the n$^-$-layer from the p$^+$- and n$^+$-regions and conditions are similar to those in a p–i–n diode, as described in Section 5.1. The excess carriers modulate the conductivity of the region between source and drain, and $R_{on}$ is decreased by several orders of magnitude. It can be shown [9.4] that $R_{on} \propto I_{GS}^{-0.5}$. However, the excess carrier injection limits high frequency operation, which becomes dependent on the carrier lifetime. SITs especially constructed to operate with a forward-biased gate–source junction are

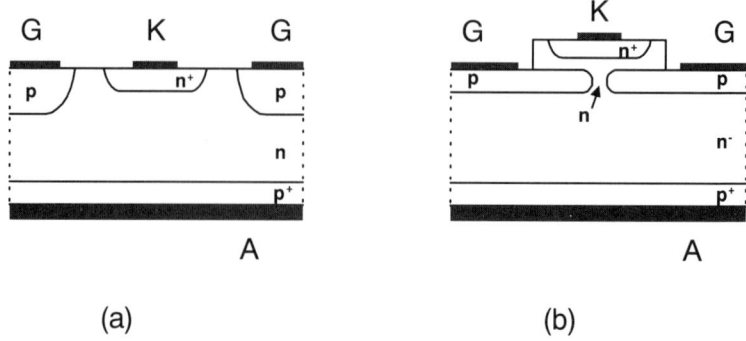

**Figure 9.8** The basic construction of the Static Induction Thyristor: (a) a planar geometry device; (b) a device with mesa geometry

called bipolar SIT (BSIT) or bipolar-mode field-effect transistor (BMFET) devices. They remain the subject of research.

## 9.2 The Field-Controlled Diode or Static Induction Thyristor

The static induction thyristor (SITh) is a three-terminal device derived from the SIT. It is also known as a field-controlled diode (FCD) and a field-controlled thyristor (FCT). As discussed in Section 9.1, the on-state resistance of the SIT structure can be decreased by forward biasing the gate junction and operating in a bipolar mode. Replacing the $n^+nn^+$ structure of the SIT by an $n^+np^+$ structure greatly increases this effect because of the bipolar injection that results. The basic FCD structure is shown in Figure 3.9 and alternative designs are shown in Figure 9.8.

A negative gate–cathode voltage creates a potential barrier which shuts off the conducting channel at the $n^+$–n junction. Making $V_{GK}$ zero, or positive, removes this potential barrier and opens a conducting channel between the $n^+$- and n-layers. The FCD then behaves like a $p^+nn^+$ diode. Consequently, the on-state current–voltage characteristics of the FCD can be approximated by the forward characteristics of an equivalent diode structure.

As with the SIT, the anode current takes the form

$$I_A = I_0 \exp\left(-\frac{eV_B}{kT}\right) \tag{9.19}$$

with

$$V_B = \alpha(V_{GK} + V_{\text{diff}}) - \beta V_A \tag{9.20}$$

where $\alpha$ and $\beta$ are constants. Substituting (9.20) into (9.19), it follows that

$$I_A = I_0 \exp\left\{\frac{e}{kT}[\beta V_A - \alpha(V_G + V_{\text{diff}})]\right\} \tag{9.21}$$

## 9.2 THE FIELD-CONTROLLED DIODE OR STATIC INDUCTION THYRISTOR

In the blocking state, the hole current entering the space charge region at the gate junction flows through the p-type gate layer to the gate contact. If the resistance of the path through the gate layer is $R_G$ and the gate contact is biased to $-V_{G0}$, the voltage at the potential barrier is

$$V_G = V_{G0} - \gamma I_A R_G \tag{9.22}$$

where $\gamma$ is a constant.

With increasing anode voltage, the gate voltage decreases and a region of negative differential resistance appears on the current–voltage characteristic at the breakover voltage $V_{(BO)}$. This occurs when $dI_A/dV_A = 0$; that is

$$V_{BO} = \frac{kT}{e}\frac{1}{\beta} \ln\left(\frac{kT}{e}\frac{1}{2\gamma R_G I_0}\right) + \frac{\alpha}{\beta} V_{G0} - 1 \tag{9.23}$$

As with the static induction transistor, the blocking capability of the static induction thyristor can be characterised by a blocking-state voltage amplifying coefficient

$$G_D = \left.\frac{\partial V_A}{\partial V_G}\right|_{I_A} = \frac{\alpha}{\beta} \tag{9.24}$$

For successful operation of a power FCD, $G_D \gg 10$. The parameters $\alpha$ and $\beta$ depend on the impurity concentration profiles and on the geometry of the conducting channel.

An FCD can be constructed in such a way that the space charge region at the gate junction cuts off the channel at $V_G = 0$. In this case a very fine structure is required in the gate–cathode region and fabrication is very complicated. Samples of static induction thyristors, with $V_{(BO)} > 500$ V at $V_G = 0$, i.e. $G_D > 1000$, have been fabricated [9.5].

The behaviour of the FCD can be also modelled using the two-transistor approximation. One transistor is a bipolar $p^+np^+$ transistor with a current gain coefficient $\alpha_1$ and the other is a SIT with a current gain coefficient $\alpha_2$. Following the derivation of Section 7.2, the anode current is given by

$$I_A = \frac{\alpha_2 I_G + I_0}{1 - (\alpha_1 + \alpha_2)} \tag{9.25}$$

Because the current gain of the SIT for $V_{GS} > 0$ is very high, the triggering condition $(\alpha_1 + \alpha_2 \geqslant 1)$ is reached at a relatively low gate current. The current turn-on coefficient, $G_{on} = I_A/I_G$, can be of the order $10^4$.

After the FCD has switched into the on-state, bipolar injection from both $n^+$- and $p^+$-emitters occurs and the space charge region cannot reform when the gate current is removed. The on-state current–voltage characteristics are very similar to those of a $p^+nn^+$ diode, as discussed in Section 5.1. They thus depend in the main on the distance between the cathode $n^+$-emitter and the anode $p^+$-emitter and on the excess carrier lifetime, as follows from (5.13) and (5.18). As with diode structures, it is also possible to decrease the on-state volt-drop while keeping the same blocking capability by reducing the distance between the emitters through the use of a 'stop

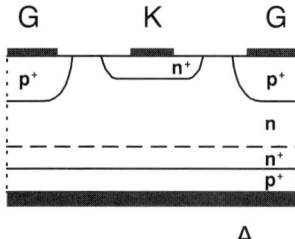

**Figure 9.9** The structure of the asymmetric SITh

layer'. This is a layer of more heavily doped n-type material between the high resistivity n-type layer and the p⁺-anode, as shown in Figure 9.9.

The reverse characteristics of the FCD depend on the breakdown voltage of the p⁺np⁺ structure between the gate and the anode regions. The reverse breakdown voltage (as for an ordinary thyristor) can be expressed by (7.1) and (7.2). When $V_{BR} = V_{(BO)}$, the device is called a symmetrical FCD. When a stop layer is included, $V_{BR} \ll V_{(BO)}$ and the structure is called an asymmetrical FCD, by analogy with the asymmetric thyristor discussed in Section 8.2.1.

In applications the switching parameters of static induction thyristors are very important. They are discussed in the following sections.

### 9.2.1 Gated Turn-Off

As with a conventional thyristor, in order to turn off an FCD it is necessary to interrupt the bipolar injection of carriers. This can be achieved either by circuit commutation or by the application of a negative gate signal. The required condition can be derived using the two-transistor model in the same way as for a GTO thyristor. (Section 7.4.3). A negative gate current $-I_G$ is required, where

$$I_G = \frac{(\alpha_1 + \alpha_2 - 1)I_A}{\alpha_2} \tag{9.26}$$

Although the underlying mechanism is different, an amplifying coefficient for turn-off, $G_{gq}$, can be defined in an identical way to the turn-off gain coefficient of a GTO:

$$G_{gq} = \frac{I_A}{I_G} = \frac{\alpha_2}{\alpha_1 + \alpha_2 - 1} \tag{9.27}$$

Because the value of $\alpha_2$ is much higher for an FCD than in a GTO thyristor, the turn-off gain is approximately double. However, it varies with time and is a function of the gate circuit resistance. Regarding the gate circuit as a voltage source $-V_{GK}$ in series with a resistance $R_G$,

$$G_{gq} = \frac{I_A R_G}{V_{GK}} \tag{9.28}$$

## 9.2 THE FIELD-CONTROLLED DIODE OR STATIC INDUCTION THYRISTOR

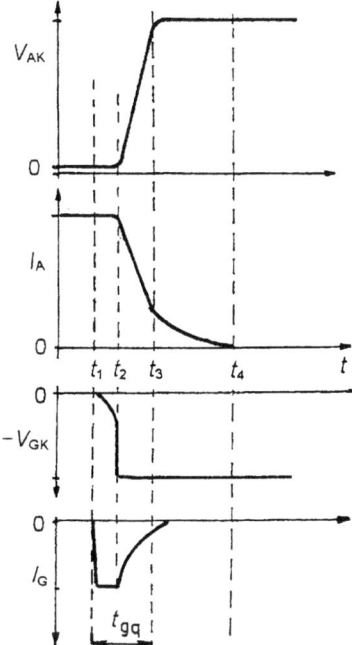

**Figure 9.10** Current and voltage waveforms during SITh turn-off

The turn-off time $t_{gq}$ can be considered to be approximately the time needed to remove the excess carriers from the $p^+$-gate region to permit a space charge layer to be re-established at the gate junction.

Current and voltage waveforms during the turn-off process are shown in Figure 9.10. Turn-off times are typically of the order of 1 µs, increasing with the reapplied blocking voltage because a thicker space charge region has to be re-established at the gate junction. Static induction thyristors can also be turned off using anode current commutation similar to the gate-assisted turn-off described in Section 8.2.3.

### 9.2.2 The d$V$/d$t$ Capability

As with thyristors, a high rise rate of the blocking voltage can cause the FCD to turn on because of the displacement current generated in the gate circuit through the charging of the junction space charge region. Approximately,

$$I_G = C_{GD} \frac{dV_{AK}}{dt} \tag{9.29}$$

The FCD turns on when the gate current reaches its triggering value, which occurs when the rise rate of the voltage is $(dV_{AK}/dt)_{crit}$. With a low gate resistance this can be of the order $10^4$ V/µs. This is much higher than the corresponding figure for GTOs and can be improved by reverse biasing the gate.

### 9.2.3 The Turn-On Process

To switch an FCD from the blocking state to the on-state, it is necessary to switch the bias of the gate–cathode junction from reverse to forward. During the gate turn-on process the capacitance of the gate–cathode junction has to be charged so that the injection of excess holes can accelerate the turn-on process. Therefore, the gate current used in the triggering gate pulse is significant, but not as significant as the current needed to turn on a GTO. The gate–cathode capacitance $C_S$ is charged through the resistance $R_G$ of the interdigitated gate contact. The turn-on time $t_{on}$ is determined by the time constant $R_G C_S$ and by the transit time of the carriers through the device by diffusion, which can be expressed by (6.47). Therefore

$$t_{on} = R_G C_S + \frac{w_N^2}{2D_n} \tag{9.30}$$

This is typically between 0.1 and 1 $\mu$s, considerably shorter than the turn-off times of GTO thyristors.

The gate electrode is normally interdigitated and has to be treated as a distributed circuit when an abrupt gate signal is applied. The potential barrier is not removed over the whole area of the structure simultaneously. Regions nearest the gate terminal are turned on before the rest of the structure. A locally turned-on area can overheat when $di_T/dt$ is high. This can degrade the device parameters or cause its destruction through damage to the fine structure of the contact metallisation. Therefore, thick metal contacts with a low resistivity are desirable and long gate–cathode segments are unsuitable. The maximum permissible value of $di_T/dt$ is about 200 A/$\mu$s.

Static induction thyristors are particularly suited to high frequency power converters. A 1200 V, 300 A device used in a 60 kHz, 100 kW inverter circuit has been described [9.6]. The relatively complicated fabrication technology of SIT and FCD devices has hindered their application. In addition, they require a negative gate voltage in order to achieve their full blocking capability. This is a serious disadvantage in many power electronics circuits.

Static induction thyristors have been the subject of research and development since the early 1970s and many papers have been published. However, in spite of the progress that has been made in improving the operating parameters of commercial devices, SIThs have been selected for relatively few applications.

## Summary

The static induction transistor is essentially a vertical-geometry, short-channel field-effect transistor. It requires a negative gate voltage to achieve its full blocking voltage. The on-state characteristic of the field-controlled diode is improved by conductivity modulation. Although these devices have the potential for high power operation at high frequency, they have not come into general use.

# References

[9.1] Nishizawa, J.-I. and Watanabe, Y. Japanese Patent 205068 (1950).
[9.2] Nishizawa, J.-I. Static induction transistor, in *Semiconductor Technologies 1982*, Ed. J.-I. Nishizawa (North-Holland, 1981).
[9.3] Shirahata, K. and Yukimuto, Y. Microwave static induction transistors, in *Semiconductor Technologies 1982*, Ed. J.-I. Nishizawa (North-Holland, 1981).
[9.4] Baliga, B. J. Bipolar Operation of Power Junction Field Effect Transistors, *Electronics Letters*, **16**, 300–1 (1980).
[9.5] Terasawa, Y., Miyata, K., Murakami, O., Nagano, T. and Okamura, M. High Power Static Induction Thyristor, *Digest of IEEE International Electron Devices Meeting (IEDM)*, 250–3 (1979).
[9.6] Muraoka, K., Kawamura, Y., Ohtsuho, Y., Sugawara, S., Tamamushi, T. and Nishizawa, J.-I. Characteristics of the High-Speed SI Thyristor and Its Application to the 60-kHz 100-kW High-Efficiency Inverter, *IEEE Trans. on Power Electronics*, **4**, 92–100 (1989).

# 10

# POWER METAL–OXIDE–SEMICONDUCTOR FIELD-EFFECT TRANSISTORS

Metal–oxide–semiconductor field-effect transistors (MOSFETs) are among the most commonly used electronic devices. Subminiature examples form the basic components of digital integrated circuits (VLSI, ULSI). They are voltage controlled and easy to drive. In addition to these devices, since 1974 several quite different types of MOS transistor have been developed that are able to switch relatively high currents and voltages and which therefore can be used in power electronic circuits. Power MOS brings to power electronic applications the following benefits:

- high input impedance
- high power gain
- voltage control
- thermal stability

## 10.1 Principles of MOS Transistor Operation

Chapter 9 describes the operating principles of the junction field-effect transistor. In that type of device the current is controlled by the voltage-dependent variation of the size of the conducting channel, which in turn is governed by the expansion and contraction of the space charge region at the gate–source p–n junction. The depletion mode MOSFET operates in a similar way, but the depletion layer is formed by applying a potential to a gate electrode that is separated from the semiconductor by a thin layer of gate oxide, as described in Section 3.1.5. Figure 2.18 shows how accumulation and inversion layers, as well as depletion layers, can form at a semiconductor surface when the surface potential is varied. These effects allow the field-controlled MOS transistor to operate in the enhancement mode as well as the depletion mode [10.1]. Enhancement mode devices are normally off and require the application of a gate voltage to turn them on. The higher electron mobility in silicon favours the use of n-channel rather than p-channel devices. In low to medium power applications, p-channel enhancement mode devices are sometimes used in order to simplify circuit design. Examples may be found in high-sided switches for automotive applications and in low voltage inverters in combination with an n-channel device in a complementary circuit. Otherwise, for the reasons stated, n-channel enhancement mode devices are normally used in power applications.

The structure of a basic, lateral, small-signal MOSFET is illustrated in Figure 10.1.

### 10.1.1  The On-State

When a voltage $V_{GS}$ is applied to make the gate G of an n-channel enhancement mode MOS transistor positive with respect to the source S, the potential at the semiconductor surface underneath the oxide layer is raised and the energy band diagram is altered in the way illustrated in Figure 10.2. When the applied voltage is low, a space charge region that is depleted of holes forms under the oxide. Increasing the applied gate voltage initially causes the space charge layer to expand but, at the threshold voltage $V_{GS(th)}$, a sufficient concentration of free electrons is attracted into the region immediately below the gate oxide to form a thin conductive layer. In the structure shown in Figure 10.1 this creates a conducting channel between the source and the drain D, the conductance of which can be controlled by the gate voltage.

The theory of this type of device has been presented in detail in several books [10.2, 10.3]. With $V_{GS} > V_{GS(th)}$ the free electron concentration in the inversion layer at the surface exceeds the hole concentration in the bulk p-type region, but it diminishes rapidly with the depth $x$ below the surface. In the surface layer, the concentration of scattering centres is higher than in the bulk semiconductor, so the carrier mobilities at the surface, $\mu_{ns}$ and $\mu_{ps}$, are lower than those of the bulk material. Their temperature variation is also different. The mobilities, too, vary with depth below the semiconductor surface. Taking this into account, we can express the variation of the lateral conductivity with depth as

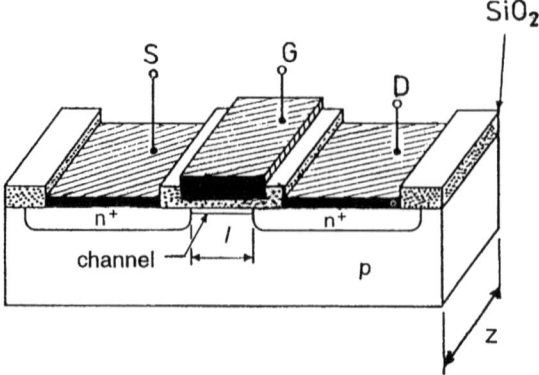

**Figure 10.1**  The structure of a typical lateral n-channel enhancement-mode MOSFET. Reproduced from *Physics of Semiconductor Devices* by S. M. Sze, by permission of John Wiley & Sons, Inc.

## 10.1 PRINCIPLES OF MOS TRANSISTOR OPERATION

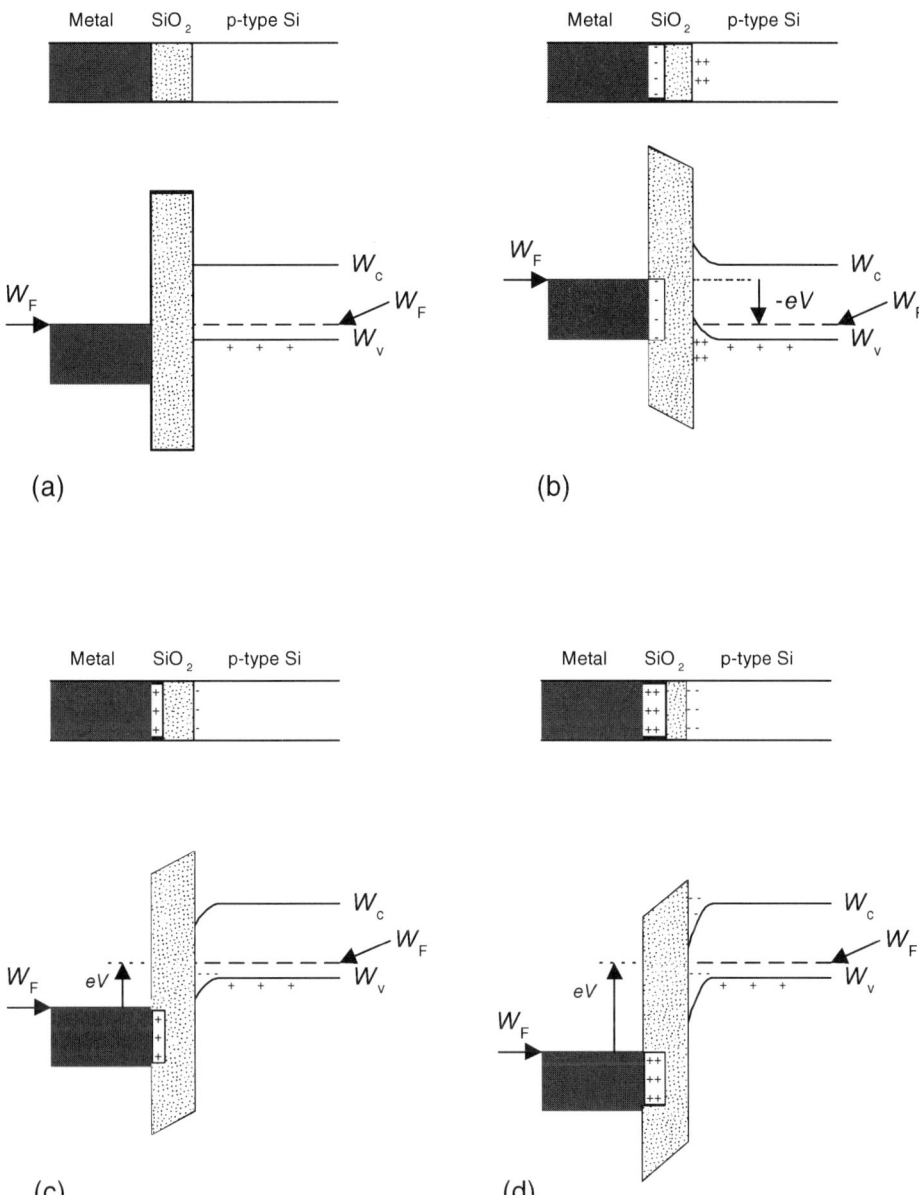

**Figure 10.2** Electron energy bands at an MOS interface with p-type semiconductor: (a) no interface charge or gate bias; (b) the formation of an accumulation layer with a negative gate voltage; (c) the formation of a depletion layer with a small positive gate voltage; (d) the formation of an inversion layer with a higher positive gate voltage

$$\sigma(x) = en(x)\mu_{ns}(x) \tag{10.1}$$

If the thickness of the inversion layer is $x_{inv}$, its width is $z$ and its length $l$, the resistance $R_{ch}$ of the channel it forms between the source and drain regions is given by

$$\frac{1}{R_{ch}} = \frac{z}{l}e\int_0^{x_{inv}} \mu_{ns}(x)n(x)\,dx = \frac{z}{l}\mu_{eff}Q_n \tag{10.2}$$

where

$$Q_n = e\int_0^{x_{inv}} n(x)\,dx \tag{10.3}$$

gives the free electron charge per unit area in the inversion layer and

$$\mu_{eff} = \frac{e}{Q_n}\int_0^{x_{inv}} \mu_{ns}(x)n(x)\,dx \tag{10.4}$$

gives the effective average mobility of these electrons.

The total charge per unit area on either side of the oxide layer, $Q_{tot}$, results from the local potential difference across the oxide, $V_{ox}$:

$$Q_{tot} = C_{ox}V_{ox} = \frac{\varepsilon_0 \varepsilon_{ox}}{d_{ox}} V_{ox} \tag{10.5}$$

where $C_{ox}$ is the capacitance per unit area of the oxide layer, whose thickness and relative permittivity are $d_{ox}$ and $\varepsilon_{ox}$, respectively. In general, $V_{ox}$ and $Q_{tot}$ are functions of position along the channel, represented by the coordinate $y$, whose origin is taken to be at the source–channel junction. The drain–channel junction is at $y=l$.

On the semiconductor side of the oxide, the total surface charge density at any point comprises the negative charge of the depleted region of the p-type semiconductor, $Q_D$, the free electrons in the inversion layer, $Q_n$, and the layer of positive ions that tend to be attracted to the semiconductor–oxide interface, $Q_{SS}$. Thus,

$$Q_{tot}(y) = Q_D(y) + Q_n(y) - Q_{SS} \tag{10.6}$$

The interface charge is partly formed from ions of alkali metal contaminants in the oxide, such as sodium and potassium. These tend to migrate to the semiconductor surface where they join the positive charge that results from the unfilled electron energy levels at the surface, the so-called dangling bonds. If the resulting $Q_{SS}$ is high, an inversion layer can form, even in the absence of a positive gate voltage. This means the transistor is normally on and requires a negative gate voltage to turn it off.

The voltage across the oxide at any point along the channel, which determines the value of $Q_{tot}(y)$, is made up from four components:

$$V_{ox}(y) = V_{GS} + \psi_{GB} - \varphi_{Bs} - \varphi_s(y) \tag{10.7}$$

## 10.1 PRINCIPLES OF MOS TRANSISTOR OPERATION

where $V_{GS}$ is the applied gate–source voltage; $\psi_{GB}$ is the contact potential between the gate electrode and the bulk p-type semiconductor; $\varphi_{Bs}$ is the internal potential difference between the interior of the p-type region and the semiconductor surface at the source end of the channel ($y=0$) that is needed before an inversion layer can form; $\varphi_s(y)$ is the potential difference along the channel resulting from any applied drain–source voltage.

The contact potential between a metal gate electrode and the bulk p-type semiconductor is defined in (2.66) as $V_S$. Thus, $\psi_{GB} = V_S$. It is more usual to form the gate structure from n-type polycrystalline silicon. Then the contact potential is identical to the diffusion potential that would occur if the gate and bulk semiconductor formed a p–n junction, as given by (2.1). Thus

$$\psi_{GB} = \frac{kT}{e} \ln\left(\frac{N_A N_D}{n_i^2}\right)$$

where $N_A$ and $N_D$ are the acceptor concentration in the bulk silicon and the donor concentration in the gate layer, respectively, and $n_i$ is the intrinsic carrier concentration. When the gate doping is sufficiently heavy to make the polycrystalline silicon degenerate, as described in Section 1.2, the Fermi level is situated close to the conduction band edge and the contact potential is then given by

$$e\psi_{GB} = \frac{W_g}{2} + kT \ln\left(\frac{N_A}{n_i}\right)$$

where $W_g$ is the bandgap energy. The use of a very heavily doped polycrystalline gate has the added benefit of reducing $Q_{SS}$ and so ensuring that a positive gate voltage is needed to form an inversion layer.

The definition of the inversion threshold is that the concentration of electrons at the semiconductor surface should equal the concentration of holes in the bulk material. This requires $W_i$ at the surface to be as far below the Fermi energy as it is above it in the interior of the semiconductor. Thus

$$\varphi_{Bs} = \frac{2kT}{e} \ln\left(\frac{N_A}{n_i}\right)$$

Once the inversion layer forms, $\varphi_{Bs}$ becomes 'pinned' at this value. Increasing the gate voltage above the value $V_{GS(th)}$, needed to form an inversion layer, increases the charge density in the inversion layer, but it does not significantly affect the potential distribution across the depletion region at the source end of the channel. Elsewhere along the channel, the depletion region is affected by any change in $\varphi_s(y)$.

Once the inversion layer has formed, the internal potential difference between the bulk p-type semiconductor and the surface is $\varphi_{Bs} + \varphi_s(y)$. From the analysis of Section 2.1.2, the charge per unit area in the resulting depletion layer is seen to be

$$Q_D(y) = \sqrt{2e\varepsilon_0 \varepsilon_{Si} N_A [\varphi_{Bs} + \varphi_s(y)]} \tag{10.8}$$

A rigorous and self-consistent derivation of the current–voltage characteristics of a MOS transistor from (10.5) to (10.8) is no easy task [10.3]. However, the threshold gate–source voltage needed to form an inversion channel is easily obtained by putting $\varphi_s = 0$ and $Q_n = 0$ in these equations. Thus

$$V_{GS(th)} = \varphi_{Bs} - \psi_{GB} + \frac{\sqrt{2e\varepsilon_0\varepsilon_{Si}N_A^2\varphi_{Bs}} - Q_{ss}}{C_{ox}} \quad (10.9)$$

When a current $I_D$ flows between drain and source, through the inversion layer, the voltage drop $dV$ across an element $dy$ is

$$dV = \frac{I_D \, dy}{z\mu_{eff}Q_n(y)} \quad (10.10)$$

where $Q_n(y)$ can be calculated from (10.5) to (10.9). Integrating (10.10) from $y = 0$ to $l$ enables a relationship for $I_D$ to be derived in terms of $V_{GS}$ and $V_{ch}$, the voltage between the source and drain regions. When $V_{ch}$ is small enough to be neglected,

$$Q_n = C_{ox}(V_{GS} - V_{GS(th)}) \quad (10.11)$$

and integrating (10.10) between $y = 0$ and $y = l$ gives

$$I_D = \frac{z}{l}\mu_{eff}C_{ox}(V_{GS} - V_{GS(th)})V_{ch} \quad (10.12)$$

Thus the resistance of the channel is given by

$$\frac{1}{R_{ch}} = \frac{I_D}{V_{ch}} = \frac{z}{l}\mu_{eff}C_{ox}(V_{GS} - V_{GS(th)}) \quad (10.13)$$

Increasing the drain current by increasing $V_{ch}$ causes the free electron charge density in the inversion layer to decrease along the channel for two reasons. First, the voltage $V_{ox}$ is reduced as $y$ increases and this reduces $Q_{tot}(y)$. Second, the voltage across the depleted region of the p-type semiconductor increases with $y$. Thus, the depletion layer expands towards the drain end of the channel and so contains more fixed negative charge, i.e. $Q_D(y)$ increases. As a result, the free electron charge in the inversion layer is reduced by an equal amount. For small values of $V_{ch}$, integration of (10.10) leads [10.3] to the following result:

$$I_D = \frac{z}{l}\mu_{eff}C_{ox}\left[(V_{GS} - V_{GS(th)})V_{ch} - \alpha\frac{V_{ch}^2}{2}\right] \quad (10.14)$$

where

$$\alpha = 1 + \frac{1}{C_{ox}}\left(\frac{e\varepsilon_0\varepsilon_{Si}N_A}{2\varphi_{Bs}}\right)^{1/2} \quad (10.15)$$

and $(e\varepsilon_0\varepsilon_{Si}N_A/2\varphi_{Bs})^{1/2}$ can be thought of as the capacitance of the depletion layer at the source end of the channel. In small-signal MOSFETs it is often assumed that $\alpha = 1$. But in power MOSFETs it is normal for there to be a much higher concentration of acceptors in the p-type semiconductor, so $\alpha = 4$ is a better approximation.

Note that the voltage drop $V_{\text{ch}}$ is a part of the total voltage $V_{\text{DS}}$ between the source and drain electrodes.

### 10.1.2 The Saturation Condition

Equation (10.14) can only be expected to apply for

$$\alpha V_{\text{ch}} \leqslant V_{\text{GS}} - V_{\text{GS(th)}} \tag{10.16}$$

at which point the drain current reaches its maximum value,

$$I_{\text{D(Sat)}} = \frac{1}{2\alpha} \frac{z}{l} \mu_{\text{eff}} C_{\text{ox}} \left( V_{\text{GS}} - V_{\text{GS(th)}} \right)^2 \tag{10.17}$$

and the surface layer is no longer inverted at the drain end of the channel. This is known as pinch-off. As with the JFET, discussed in Section 9.1, increasing $V_{\text{DS}}$ causes no further increase of $I_{\text{D}}$; it is said to saturate. More exact solutions of the approach to saturation require two-dimensional modelling [10.3, 10.4].

Typical characteristics for small-signal, n-channel, enhancement mode MOSFETs are shown in Figure 3.10(a).

## 10.2 Vertical Power MOSFET Designs

The fully turned-on condition represented by (10.12) is known, rather inappropriately, as the linear region of operation. Note that the resistance of the conducting channel is proportional to its length. For the drain junction to have a high breakdown voltage, the channel length must exceed the thickness of the space charge region on the channel side of the drain junction, in order to avoid punch-through. A high breakdown voltage requires a long channel, but that increases the channel resistance and reduces the current rating. The conflict between these requirements cannot be resolved satisfactorily using the simple lateral MOS transistor structure.

MOS transistors for power applications need a very short channel and, at the same time, a low doping level in the drain region so that the space charge layer at the reverse-biased drain–channel junction spreads into the n-type drain region rather than into the p-type channel. The base and collector regions of high voltage bipolar transistors have a similar requirement. Figures 3.11, 3.12 and 3.13 show some radically different configurations that overcome these problems. Very short channels are obtained by means of controlled diffusions. Acceptors are diffused into lightly doped n-type epitaxial material to create a p-type layer for the channel region. A further diffusion of donors forms the n$^+$-source regions.

For the devices shown in Figures 3.11 and 3.12, channels are etched into the diffused structure. In the V-MOS (or VVMOS) structure of Figure 3.11, which is named after the shape of the grooves and the vertical geometry, an anisotropic etch forms a V-groove. The growth of thermal oxide and the deposition of a metal gate contact inside the V-groove complete the device. The vertical sides of the trench structure shown in Figure 3.12 enable the packing density to be increased and the gate length per unit area to increase in proportion.

The configuration shown in Figure 3.13 is obtained by diffusing acceptors and donors successively through the same aperture in the oxide masking layer. This structure is called D-MOS (double-diffused MOS) and exploits all the advantages of diffusion technology. The channel region is shown in more detail in Figure 10.3. The first diffusion (boron) is usually from an implanted layer giving a junction depth of about 3 μm. The second diffusion (phosphorus) produces a junction 1–2 μm below the surface. A layer of high quality, thermally grown silicon dioxide forms the gate insulator. Its thickness is typically about 100 nm. The gate contact is normally heavily doped, n-type, polycrystalline silicon deposited onto the oxide. In special cases aluminium may be used. The gate contact is usually covered with an insulating layer of silica, put down by chemical vapour deposition.

The drain contact can be placed either on the same side of the silicon wafer as the source contact to create a lateral double-diffused MOS (LDMOS) structure, or on the opposite side to form the vertical double-diffused MOS (VDMOS) structure shown in Figure 3.13. In this case the maximum voltage is not limited by breakdown of the gate oxide, which is shielded by the JFET action that occurs between the cells. Thus, high voltage devices, up to 1000 V, can be made. The VDMOS structure tends to be used for higher voltages, and the trench structure is becoming popular for lower voltage devices.

### 10.2.1 Static Characteristics

Application of a gate voltage $V_{GS} > V_{GS(th)}$ allows current to flow from the source through the conducting inversion channel into the lightly doped n-type drain region. In this region, at the semiconductor surface immediately below the gate oxide, an accumulation layer of higher conductivity forms under the influence of the positive gate voltage. The drain current then passes out into this region and flows vertically through the bulk semiconductor of higher resistivity to the drain contact.

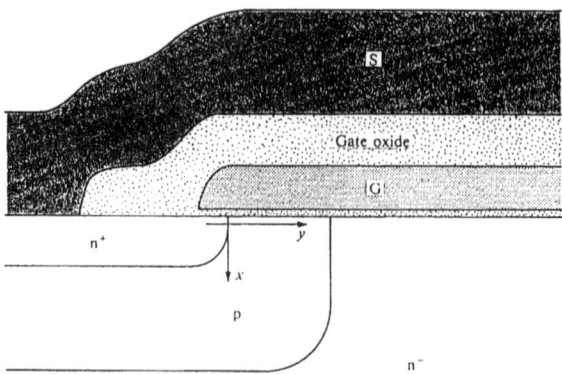

**Figure 10.3** The channel region of a double-diffused MOSFET. Reproduced from *Power MOSFETs Theory and Applications* by D. A. Grant & J. Gowar, by permission of John Wiley & Sons, Inc.

The total on-state resistance of the D-MOS transistor in the linear region, $R_{DS(on)}$, can be expressed as the sum of four terms: $R_{ch}$, the resistance of the inversion channel; $R_a$, the resistance of the accumulation region; $R_D$, the resistance of the bulk semiconductor drain region; and $R_{n+}$, the resistance of the n$^+$-substrate, the source regions and their contacts:

$$R_{DS(on)} = R_{ch} + R_a + R_D + R_{n+} \qquad (10.18)$$

The on-state voltage drop is given by

$$V_{DS} = R_{DS(on)} I_D \qquad (10.19)$$

and the on-state power dissipation by

$$P_D = I_D^2 R_{DS(on)} \qquad (10.20)$$

The maximum power dissipation is limited by the thermal impedance of the cooling path and is subject to the stability condition (1.84). Thus, the maximum on-state current $I_{DM}$ varies inversely as the square root of $R_{DS(on)}$ at the highest permitted operating temperature $T_{Jmax}$.

All the partial resistances in (10.18) increase with increasing temperature, and this facilitates the parallel connection of D-MOS transistors. Power VDMOS transistors are usually an integration of between a thousand and several million individual cells connected in parallel on a single chip and forming a regular pattern over the device area. This is illustrated in Figure 10.4. More detail can be found in the literature [10.3].

In order to achieve a high breakdown voltage together with planar technology, the active area is surrounded by several field rings, as described in Section 3.4. The maximum off-state voltage of D-MOS transistors, $V_{DSM}$, is limited by the basic n–p–n bipolar transistor structure of the device, as described in Section 6.2.1. The n$^+$-source region (the emitter of the parasitic bipolar transistor) is normally shorted to the p- region in which the channel is formed (the base of the parasitic bipolar). The breakdown voltage is therefore increased from $V_{CEO}$ to $V_{CES}$. However, it is also important that the electric field should be kept small at the interface between the gate oxide and the n-type semiconductor (the drain region), in order to avoid electrical breakdown across the oxide layer. The expansion of the space charge layer across the space between the cells at high drain voltages provides a JFET action that automatically screens the oxide.

The channel region is usually formed in a three-stage process. In the first step, a deep p$^+$-well is diffused through a window in the masking oxide. The p–n junction so formed can function as an avalanche diode and provide protection against overvoltage. It also serves as an integrated antiparallel diode that protects the device against breakdown of the n$^+$–p junction when the drain voltage becomes negative. Next an oxide mask is prepared for the double diffusion process and the main p- region with a lower surface concentration and a shallower junction is formed by diffusion. Finally, the n$^+$-source is diffused in through the same mask window.

The device parameters can be optimised by the careful control of the impurity concentration in the drain region under the gate oxide. An implant through the gate

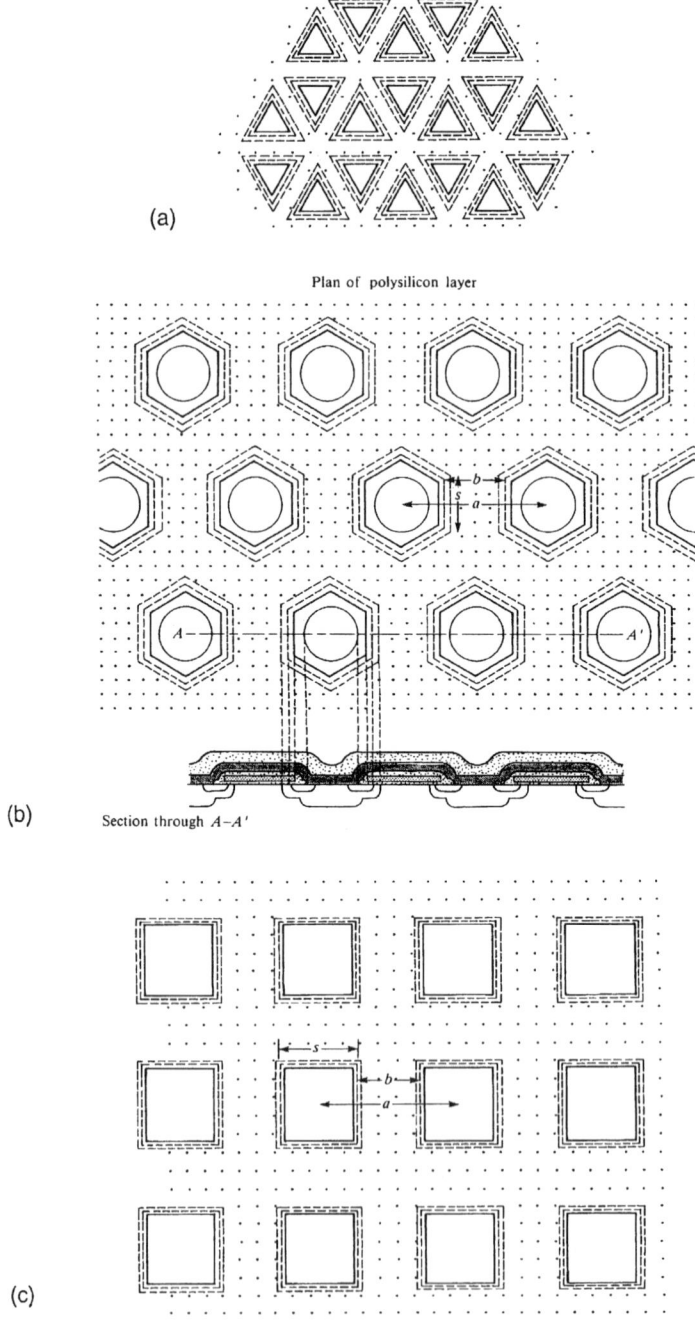

**Figure 10.4** Examples of the types of cell geometry used for power MOSFETs: (a) a triangular arrangement; (b) a hexagonal geometry; (c) a square geometry. Reproduced from *Power MOSFETs Theory and Applications* by D. A. Grant & J. Gowar, by permission of John Wiley & Sons, Inc.

oxide is normally used to give the typical concentration profile shown in Figure 10.5. In p-channel devices, dopants of the opposite polarity are diffused and implanted into p-type epitaxial material on a p$^+$-substrate. They have a higher $R_{DS(on)}$ because the hole mobility is lower, but may be used in circuits where they are complementary to n-channel devices and in applications where a negative drain voltage is desirable.

It can be seen in Figure 10.5(a) that the drain region below the gate oxide forms a throat whose resistance is a function of $V_{DS}$ because of the expansion and contraction of the space charge layer at the p–n junction. The region behaves like a parasitic JFET and causes the drain–source resistance to increase with increasing $V_{DS}$, as the throat becomes narrower. An equivalent circuit of the VDMOS transistor, including the various parasitic transistors, diodes and other components, is shown in Figure 10.6. The total on-state resistance of a VDMOS structure is given by (10.18), with $R_D$ made up of the basic drain resistance $R_{D0}$ and the extra resistance resulting from JFET action.

The values of the terms in (10.18) depend on the doping concentration profiles and the detailed device geometry. In particular, $R_{ch}$ is given by (10.13) and, by analogy, $R_a$ can be expressed as

$$R_a = \frac{K_a}{z} \frac{(l_G - x_p)}{\mu_{na} C_{ox} (V_{GS} - V_{GS(th)})} \tag{10.21}$$

where $z$ is the width of the gate, $l_G$ the width of the throat between the cells, $x_p$ the thickness of the p-layer, $\mu_{na}$ is the electron mobility in the accumulation layer and $(l_G - x_p)$ is its effective length. The coefficient $K_a$ takes account of the effect of the spreading of current from the thin surface layer into the volume of the device. Its value has been estimated to be approximately three.

The resistance $R_D$ of the lightly doped epitaxial layer of thickness $w_N$ and resistivity $\rho_N$ can be expressed as

$$R_D = \rho_N \frac{K_S w_N}{z l_G} \tag{10.22}$$

where the coefficient $K_S$ allows for the effect of the spreading out of the current from the throat of the parasitic JFET into the volume of the epitaxial layer by a distance $s$.

Following (9.8) the additional resistance resulting from the effect of the parasitic JFET can be expressed as

$$R_J = \rho_N \frac{x_p + d_D}{z(l_G - 2d_D)} \tag{10.23}$$

where $d_D$ is the thickness of the p–n junction space charge region, as given by (9.2). The resistance of the n$^+$-substrate is

$$R_{n+} = \rho_{n+} \frac{w_{n+}}{z(l_G + s)} \tag{10.24}$$

when its thickness is $w_{n+}$ and its resistivity is $\rho_{n+}$. The lateral resistance of the high conductivity source region and the contacts is usually less then 10% of $R_{n+}$ [10.3].

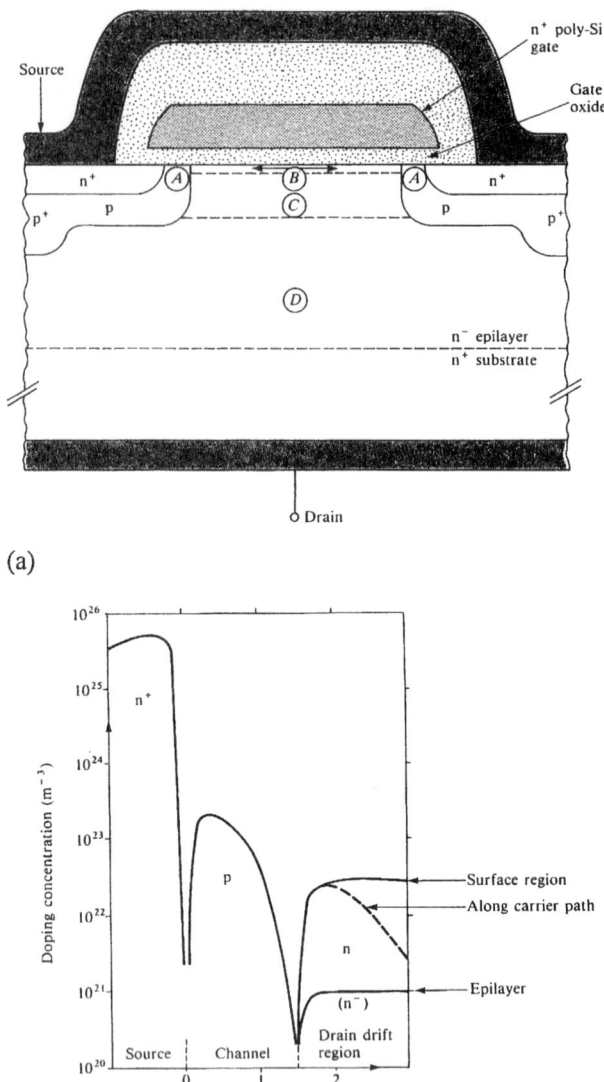

**Figure 10.5** The VDMOS transistor: (a) a section through a typical cell; (b) typical impurity concentration profiles. Reproduced from *Power MOSFETs Theory and Applications* by D. A. Grant & J. Gowar, by permission of John Wiley & Sons, Inc.

From this discussion it can be seen that the dimension $l_G$ of the gate region influences the individual components of $R_{DS(on)}$ in different ways. For example, the resistance of the accumulation layer increases with $l_G$, whereas $R_D$ and $R_J$ decrease. The optimum gate length has been estimated to be in the region of 1–2 μm for low voltage devices and rather longer when high voltages are required.

## 10.2 VERTICAL POWER MOSFET DESIGNS

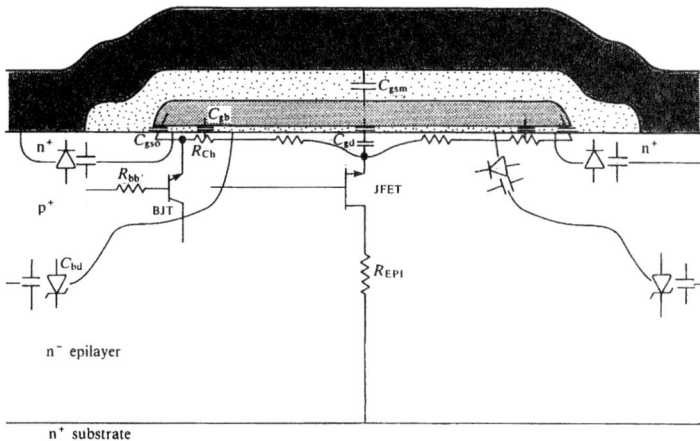

(a)

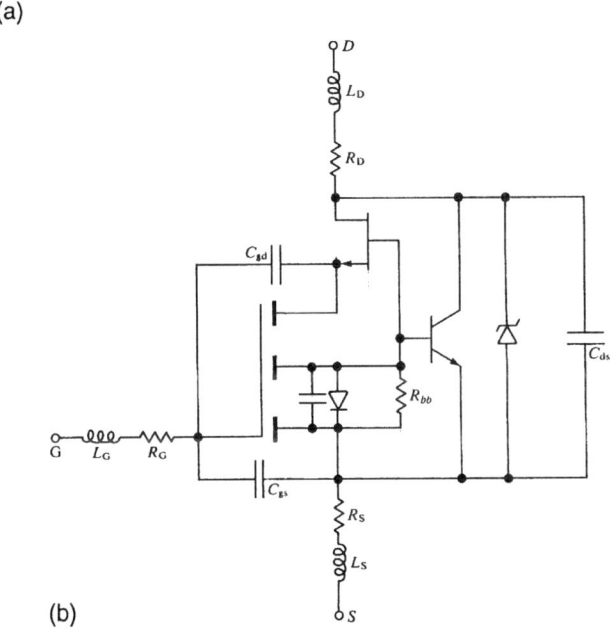

(b)

**Figure 10.6** The parasitic components associated with a VDMOS transistor: (a) section through a VDMOS cell showing the origin of the various parasitic elements; (b) an equivalent circuit. Reproduced from *Power MOSFETs Theory and Applications* by D. A. Grant & J. Gowar, by permission of John Wiley & Sons, Inc.

For low voltage devices ($V_{DS(BR)} < 100$ V) the resistivity and thickness of the epitaxial layer are lower, so that $R_D$ is low and $R_{ch} + R_a$ becomes a significant component of the on-state resistance. It can be reduced by making the gate oxide as thin as possible which increases $C_{ox}$ in (10.13) and (10.21). The limit is set by the need for the gate oxide breakdown voltage to be at least 100 V. The use of an oxynitride layer, which has a higher breakdown strength, may assist this. It is formed by the thermal oxidation of silicon in nitrous oxide ($N_2O$).

For high voltage devices ($V_{DS(BR)} > \approx 200$ V), the resistance of the epitaxial layer becomes the largest component of the on-state resistance. The thickness $w_N$ of the high resistivity epitaxial layer can be minimised if the space charge region reaches through to the n–n$^+$ junction. Detailed analysis [10.3] indicates that $R_D \propto V_{DS(BR)}^{2.6}$.

The transconductance $g_{fs}$ is an important MOS transistor parameter:

$$g_{fs} = \frac{\partial I_D}{\partial V_{GS}}\bigg|_{V_{DS}=\text{constant}} \quad (10.25)$$

It is influenced by the dependence of $R_{ch}$, $R_a$ and $R_J$ on $V_{GS}$. If these effects are neglected, and (10.14) and (10.17) apply in the linear and saturation regions, respectively, the corresponding expressions for the transconductance are

$$g_{fs} = \frac{z}{l}\mu_{eff}C_{ox}V_{ch} \quad (10.26)$$

and

$$g_{fs} = \frac{z}{l}\mu_{eff}C_{ox}(V_{GS} - V_{GS(th)}) \quad (10.27)$$

At higher values of $V_{DS}$, two effects cause a decrease in the transconductance. First, when the geometry of the throat region is such that $R_J$ increases significantly with $V_{DS}$, a quasi-saturation region appears in the device characteristics. Second, when the electric field along the channel exceeds $10^6$ V/m, the electron mobility decreases until, at $5 \times 10^6$ V/m, the saturation drift velocity of $9 \times 10^4$ m/s is reached. With a channel length of about 1 μm this occurs for channel voltages greater than 1 V. Once the saturation drift velocity is reached at the drain end of the channel, the drain current becomes independent of $V_{ch}$ but is still proportional to the charge density in the inversion channel and hence to $(V_{GS} - V_{GS(th)})$. The transconductance becomes constant, as shown by the evenly spaced curves of the saturated drain current in Figure 3.13(c). To a first approximation, the free electron charge in the channel is given by

$$Q_n = \frac{1}{2}C_{ox}zl(V_{GS} - V_{GS(th)}) \quad (10.28)$$

The transit time along the channel is $l/v_{sat}$, so the saturated drain current is

$$I_{D\,sat} = \frac{Q_n v_{sat}}{l} = \frac{1}{2}C_{ox}z(V_{GS} - V_{GS(th)})v_{sat} \quad (10.29)$$

and the transconductance becomes

$$g_{fs} = \frac{1}{2}C_{ox}zv_{sat} \quad (10.30)$$

Before velocity saturation the average drift velocity of the carriers as they travel along the channel is proportional to $V_{ch}$, which at pinch-off equals $(V_{GS} - V_{GS(th)})$.

Since the mobile charge in the channel, $Q_n$, is also proportional to $(V_{GS} - V_{GS(th)})$, the saturated drain current $I_{D(Sat)}$ is proportional to $(V_{GS} - V_{GS(th)})^2$, as given by (10.17).

In looking for an optimum VDMOS configuration, several different cellular designs have been proposed and implemented. Seven are described in detail in the literature [10.3] and three are illustrated in Figure 10.4. Optimisation of the cell pattern with the aim of maximising the transconductance and minimising the on-state resistance has been the subject of a number of theoretical studies. Two-dimensional modelling [10.6] shows that the maximum effective use of the chip area is obtained with hexagonal cells forming a hexagonal lattice (Figure 10.4(b)) and square cells in a square lattice (Figure 10.4(c)). In theory a circular cell geometry offers the best compromise between high $V_{DS(BR)}$, low $R_{DS(on)}$ and high $g_{fs}$ [10.7]. The hexagonal structure which closely approximates this optimum design is shown in more detail in Figure 10.7.

Lateral, double-diffused MOS transistors are usually configured in a comb-like structure similar to power bipolar transistors. An example is shown in Figure 10.8. They are often used in power integrated circuits and are discussed further in Chapter 12.

The dependence of device characteristics on temperature is always important. For power MOSFETs the temperature dependence of the static characteristics is mainly connected with the variation of the carrier mobility with temperature, both in the bulk semiconductor and in the surface layers. However, the threshold voltage $V_{GS(th)}$ decreases with increasing temperature as the Fermi level moves towards the intrinsic level. An example of the temperature dependence of $V_{GS(th)}$ is shown in Figure 10.9.

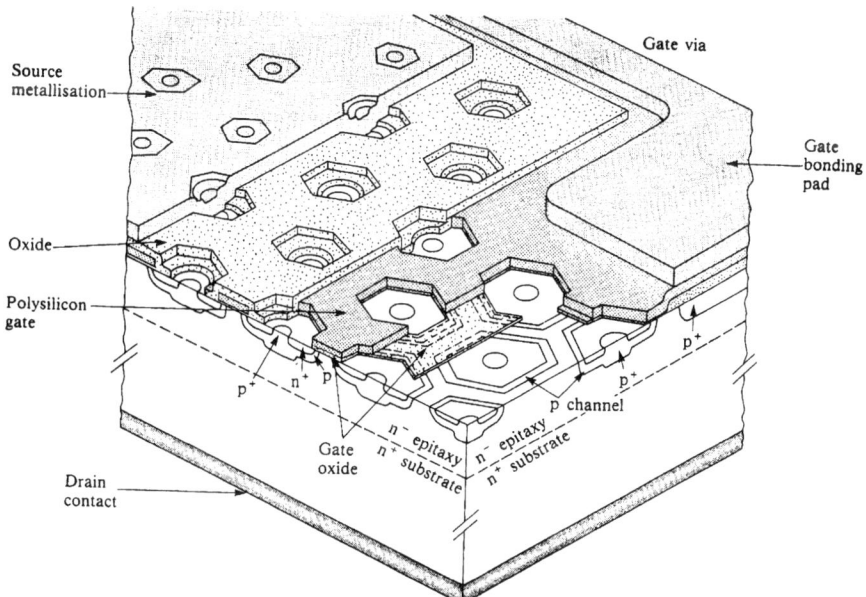

**Figure 10.7** The detailed structure of a hexagonal-cell MOSFET. Reproduced from *Power MOSFETs Theory and Applications* by D. A. Grant & J. Gowar, by permission of John Wiley & Sons, Inc.

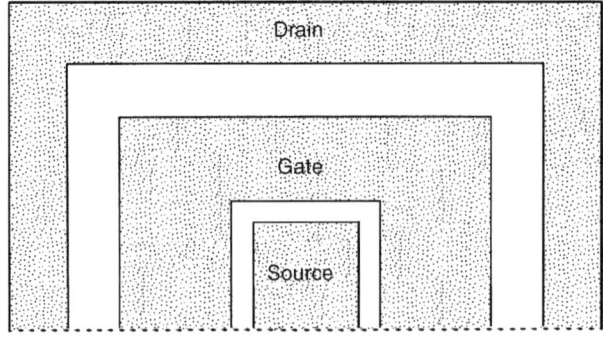

(a)

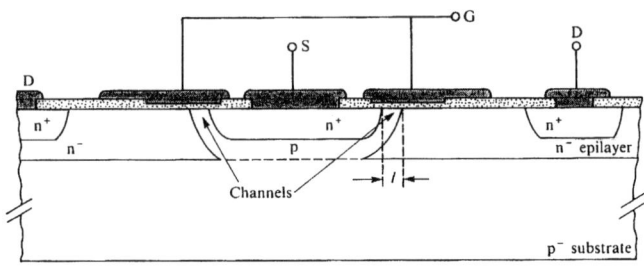

(b)

**Figure 10.8** The typical structure of a LDMOS FET: (a) a plan view of the electrode layout; (b) a section through the active region. Reproduced from *Power MOSFETs Theory and Applications* by D. A. Grant & J. Gowar, by permission of John Wiley & Sons, Inc.

As discussed in Section 1.3.2, carrier mobility decreases with temperature. In lightly doped silicon it varies as $T^{-2.6}$, in surface layers as $T^{-1.5}$. We would therefore expect that

$$R_{ch} + R_a \propto T^{1.5} \qquad (10.31)$$

and

$$R_J + R_D \propto T^{2.6} \qquad (10.32)$$

In low voltage MOS transistors we expect (10.31) to apply and in high voltage transistors, when the on-state resistance is dominated by the resistance of the drain drift region, we expect the variation to be given by (10.32). Being proportional to the carrier mobility, the transconductance also decreases with increasing temperature.

The temperature dependence of the on-state resistance is important in establishing a uniform distribution of the on-state current in power VDMOS transistors. The same voltage, $V_{on}$, is dropped across each individual cell. If the $i$th cell has an on-state resistance $R_{on(i)}$, the power dissipated is $V_{on}^2/R_{on(i)}$. A cell with a lower on-state

## 10.2 VERTICAL POWER MOSFET DESIGNS

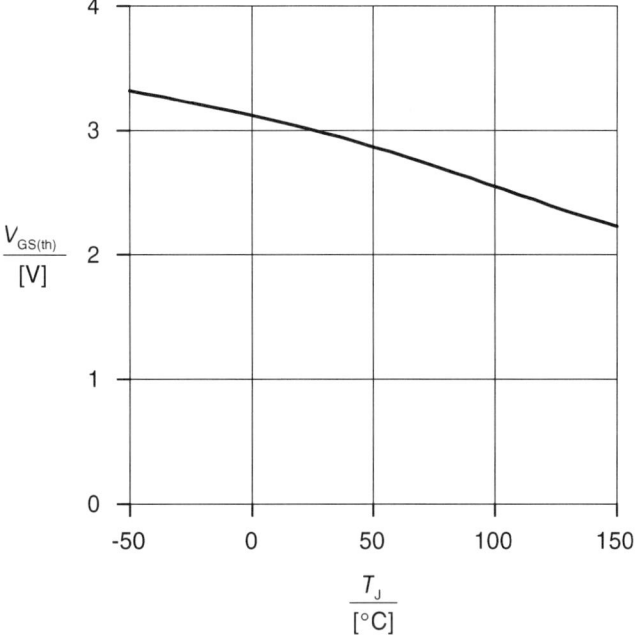

**Figure 10.9** The variation of $V_{GS(th)}$ with temperature

resistance dissipates more power so its temperature increases. Consequently, the on-state resistance increases and the power dissipation falls off. Nevertheless, a high uniformity over the area of the device is important if a homogeneous current distribution is to be obtained during switching.

### 10.2.2 The Frequency Dependence of Parameters

The analysis shows that the power MOS transistor can operate in three stable states controlled by the gate voltage $V_{GS}$:

1. $V_{GS} < V_{GS(th)}$, the off-state, with $I_D = 0$
2. $V_{GS} > V_{GS(th)}$ and $V_{DS} > I_D R_{on}$, the active region, with $I_D = g_{fs}(V_{GS} - V_{GS(th)})$
3. $V_{GS} > V_{GS(th)}$ and $g_{fs}(V_{GS} - V_{GS(th)}) > V_{DS}/R_{on}$, the on-state, with $I_D = V_{DS}/R_{on}$

In MOS transistors, charge transport is by majority carriers (usually electrons) and there is no conductivity modulation by excess carrier injection during transient processes. The speed with which the drain current responds to a change in the gate–source voltage depends on the time needed to establish the inversion layer and the carrier transit time along the channel. The transit time is

$$\tau_T = \frac{l}{v_d} = \frac{l^2}{\mu_{eff} V_{ch}} \qquad (10.33)$$

when $V_{ch}/l \leqslant 10^6$ V/m. It is $l/v_{sat}$ when $V_{ch}/l \geqslant 5 \times 10^6$ V/m. Transit times are typically between 10 and 100 ps. So the dynamic behaviour of MOS transistors is limited by the time needed to charge and discharge the capacitance $C_{ox}$, and the other parasitic capacitances, in order to create or remove the conducting channel. The frequency limitation is determined essentially by the processes of charging and discharging the gate.

An equivalent circuit for a VDMOS transistor, loaded by an impedance Z, is shown in Figure 10.10. This is a simplified version of Figure 10.6, with the parasitic inductances omitted and resistances $R_a$, $R_J$ and $R_D$ included in the load impedance. Transient processes are significantly influenced by the capacitance $C_{GD}$, which by the Miller effect acts as an equivalent input capacitance (the Miller capacitance) given by

$$C_{Mi} = (1 + g_{fs}Z)C_{GD} \tag{10.34}$$

With Z real, the total input capacitance is

$$C_{in} = C_{GS} + C_{Mi} \tag{10.35}$$

The gate–drain capacitance is strongly dependent on the drain–source voltage, whereas the gate–source capacitance is practically independent. This is shown schematically in Figure 10.11. When $V_{DS} < V_{GS}$ the MOS transistor is in the on-state with an accumulation layer of majority carriers in the drain region below the thin gate oxide. The capacitance $C_{DS}$ is then determined by the capacitance of the thin oxide layer only. Increasing the drain–source voltage, so that $V_{DS} > V_{GS}$, causes a depletion layer to form in the drain region below the gate and the thickness of the space charge region increases with increasing $V_{DS}$. Consequently, the capacitance $C_{GD}$ decreases.

The input capacitance $C_{in}$ is charged from a source of voltage $V_{GSS}$ with a series resistance $R_{G1}$. The resistance of the distributed gate contact, usually heavily doped polycrystalline silicon, is $R_{G2}$, so the total resistance of the gate circuit is

$$R_G = R_{G1} + R_{G2} \tag{10.36}$$

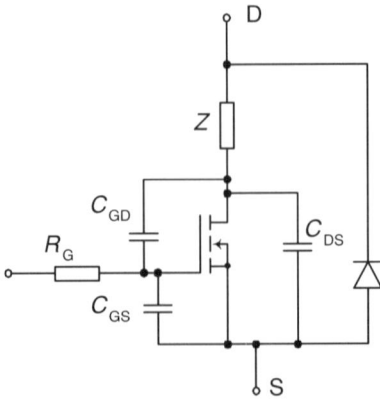

**Figure 10.10** A simplified equivalent circuit for a power MOSFET

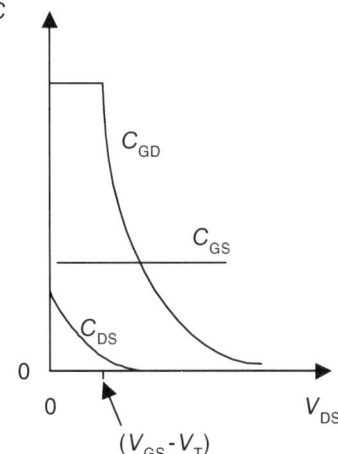

**Figure 10.11** The variation of the parasitic capacitances in a power MOSFET with the drain voltage

In the linear regime, the time constant of the gate circuit limits the cut-off frequency to

$$f_{co} = \frac{1}{2\pi C_{in} R_G} \tag{10.37}$$

This can be shortened by lowering $C_{in}$ or $R_G$. The input capacitance can be reduced, through $C_{GD}$, by minimising the gate overhang or increasing the oxide thickness above the high resistivity n-type region. But this increases $R_{DS(on)}$ and decreases the transconductance. Replacing the polycrystalline silicon gate by metal (e.g. molybdenum), or coating it with a layer of low resistivity metal (e.g. aluminium), lowers $R_{G2}$. Examples of these designs are shown in Figure 10.12. Specially constructed power MOSFETs which combine both techniques have a cut-off frequency in the linear regime of several hundred megahertz [10.8].

The input capacitance is relatively high, so a considerable gate current is needed to operate power MOSFETs at frequencies exceeding 1 MHz and the power dissipated in the gate circuit can be comparable with the power generated by switching within the device. The drain current has to be reduced accordingly. The dynamic processes of turn-on and turn-off are discussed in more detail in the next section.

## 10.3 The Switching of Power MOSFETs

The most common use of power MOSFETs in power electronic circuits is as high frequency switches, when they alternate between the on- and off-states. This enables the control of high load power with minimal dissipation in the device. The type of load influences the voltage and current waveforms and its impedance influences the input capacitance through the Miller effect, as shown by (10.34). General solutions

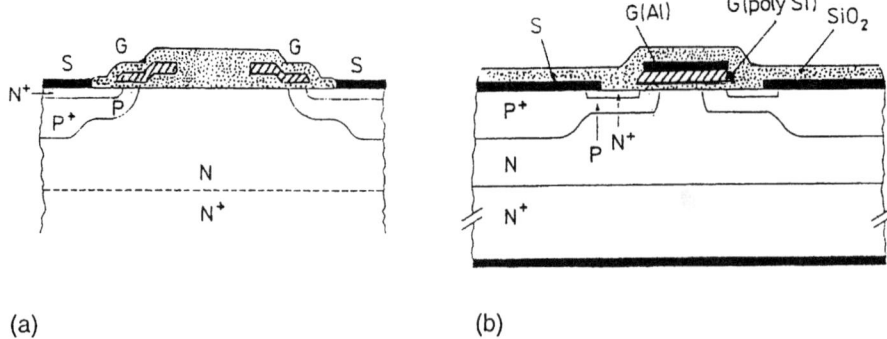

(a)  (b)

**Figure 10.12** Techniques used to improve the high frequency characteristics of a VDMOS transistor: (a) decreasing the gate/drain capacitance; (b) decreasing the gate resistance by the use of a metal gate layer on polycrystalline silicon

are complicated. Very commonly, the load takes the form of a motor winding or a transformer. These are essentially inductive and are usually clamped with a diode, as shown in Figure 10.13. This type of circuit and one with a purely resistive load are analysed in the following sections.

### 10.3.1 The Turn-On Process

With a clamped inductive load, turn-on starts with the gate voltage low ($V_{GS} = V_{GL} < V_{GS(th)}$) and the transistor in the off-state ($I_D = 0$). The constant current $I_{DM}$, flowing in the load inductance, circulates through the freewheeling diode. To

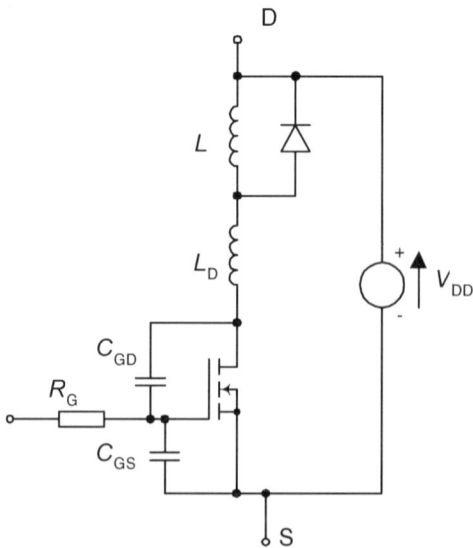

**Figure 10.13** A power MOS transistor switching into a clamped inductive load

## 10.3 THE SWITCHING OF POWER MOSFETS

simplify the equations we assume $V_{GL}=0$. At $t=0$ the gate control voltage is raised abruptly to $V_{GH} > V_{GS(th)}$. In the first phase of the turn-on process, $C_{GS}$ and $C_{GD}$ are charged through $R_{G1}$ and $R_{G2}$ and the voltage across the gate oxide increases initially as

$$V_{GS} = V_{GH}\left\{1 - \exp\left[\frac{-t}{R_G(C_{GS} + C_{GD})}\right]\right\} \tag{10.38}$$

The threshold voltage is reached after the time

$$t_d = R_G(C_{GS} + C_{GD}) \ln\left[\frac{V_{GH}}{V_{GH} - V_{GS(th)}}\right] \tag{10.39}$$

at which point the transistor changes into the active state. The drain current is then given by

$$I_D = g_{fs}(V_{GS} - V_{GS(th)}) \tag{10.40}$$

This part of the turn-on process is relatively complicated and an exact solution needs numerical modelling. If it is possible to omit all parasitic inductance and to assume that $g_{fs}$ is constant, the rise of the drain current can be approximated by

$$I_D(t) = g_{fs}\left[(V_{GH} - V_{GS(th)}) - V_{GH}\exp\left\{\frac{-t}{R_G(C_{GS} + C_{GD})}\right\}\right] \tag{10.41}$$

The current in the freewheeling diode is reduced by $I_D(t)$.

The drain voltage remains practically constant at $V_{DM}$, which is the line voltage plus the voltage across the freewheeling diode. This state of affairs continues from $t_d$ until $t_2 = t_d + t_{ri}$, at which point the drain current reaches $I_{DM}$ and thereafter remains constant. In a simple approximation the current rise time $t_{ri}$ can be expressed as

$$t_{ri} = R_G(C_{GS} + C_{GD}) \ln\left[\frac{g_{fs}V_{GH}}{g_{fs}(V_{GH} - V_{GS(th)}) - I_{DM}}\right] \tag{10.42}$$

After the drain current has reached its maximum value, the drain–source voltage starts to fall but the gate voltage $V_{GS}$ remains constant as long as the drain current is constant. The decreasing drain voltage requires the Miller capacitance $C_{Mi}$ to be charged by the input current. The rate of fall of the drain voltage is constant and is given by

$$\frac{dV_{GD}}{dt} = \frac{J_G}{C_{Mi}} = \frac{V_{GH} - (V_{GS(th)} + I_{DM}/g_{fs})}{R_G C_{GD}} = \frac{dV_{DS}}{dt} \tag{10.43}$$

Thus

$$V_{DS} = V_{DM} - \frac{g_{fs}(V_{GH} - V_{GS(th)}) - I_{DM}}{g_{fs}R_G C_{GD}}(t - t_2) \tag{10.44}$$

The MOSFET is fully turned-on once the drain voltage reaches $V_{on} = R_{DS(on)}I_{DM}$. The voltage fall time $t_{fv}$ is given by

$$t_{fv} = \frac{(V_{DM} - V_{on})R_G C_{GD}}{V_{GH} - (V_{GS(th)} + I_{DM}/g_{fs})} \tag{10.45}$$

The gate capacitance continues to be charged until $V_{GS} = V_{GH}$. The overall turn-on time is given by

$$t_{on} = t_d + t_{ri} + t_{fv} \tag{10.46}$$

Waveforms of $V_{GS}$, $V_{DS}$ and $I_D$ are shown schematically in Figure 10.14.

When some of the load inductance $L_D$ is not clamped by the freewheeling diode, the analysis of the voltage and current waveforms is much more complicated [10.3]. The series inductance decreases the rate of rise of the drain current and reduces the turn-on losses.

In the case of a purely resistive load $R_Z$, the turn-on process takes a slightly different course. After the application of an abrupt step of gate voltage, amplitude $V_{GH} > V_{GS(th)}$, equation (10.38) still represents the rise of $V_{GS}$ up to the time $t_d$. For

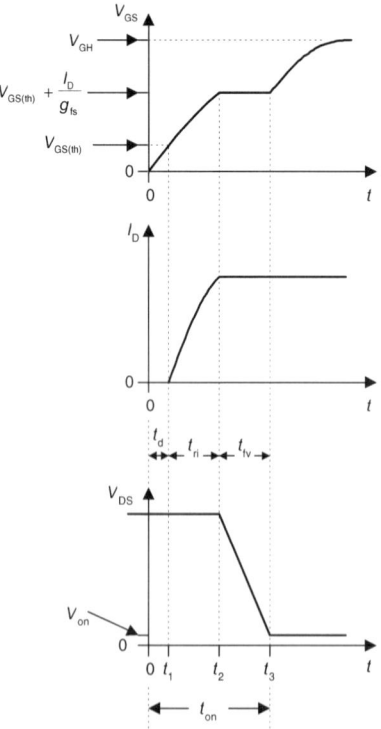

**Figure 10.14** Current and voltage waveforms at turn-on for an MOS-transistor in a circuit with a clamped inductive load, as given by Eqs. (10.41) and (10.44)

$t > t_d$ the drain current increases as described by (10.41) and $V_{DS}$ decreases at the same time:

$$V_{DS}(t) = V_{DM} - R_Z I_D(t) \tag{10.47}$$

Because $C_{GD}$ varies while the drain voltage is decreasing, $I_D(t)$ and $V_{DS}(t)$ differ slightly from (10.41) and (10.47).

### 10.3.2 The Turn-Off Process

In order to illustrate the turn-off process, we again consider a circuit with a clamped inductive load, as shown in Figure 10.13, and carry out an analysis similar to that given for turn-on in the last section.

At the beginning of the turn-off process, the transistor is turned-on, carrying a current $I_{DM}$, and the gate voltage $V_{GS} = V_{GH} > (V_{GS(th)} + I_{DM}/g_{fs})$. Turn-off starts at time $t = 0$ by the abrupt decrease of the gate voltage from $V_{GH}$ to $V_{GL} < V_{GS(th)}$. Usually $V_{GL} = 0$. The gate capacitance $C_G$ is discharged through the resistance $R_G$ until, at the time $t_s$, the gate voltage decreases to the level at which the drain current is saturated:

$$V_{GS}(t_s) = V_{GS(th)} + \frac{I_{DM}}{g_{fs}} \tag{10.48}$$

The delay time $t_s$ is the time needed to discharge the gate:

$$t_s = R_G(C_{GS} + C_{GD}) \ln\left[\frac{g_{fs} V_{GH}}{g_{fs} V_{GS(th)} + I_{DM}}\right] \tag{10.49}$$

For $t > t_s$ the collector current remains at $I_{DM}$ and the drain voltage increases as

$$V_{DS} = V_{on} + \frac{I_D}{C_{GD}}(t - t_s) = V_{on} + \frac{g_{fs} V_T + I_{DM}}{g_{fs} R_G C_{GD}}(t - t_s) \tag{10.50}$$

The drain voltage increases from $V_{DSon}$ to $V_{DM}$ during time interval $t_{rv}$, given by

$$t_{rv} = \frac{(V_{DM} - V_{on}) g_{fs} R_G C_{GD}}{I_{DM} + g_{fs} V_{GS(th)}} \tag{10.51}$$

The freewheeling diode prevents the drain voltage from increasing above $V_{DM} + 0.7\,\text{V}$. Once the drain voltage has recovered to this value, the gate capacitance discharges through the gate resistance, so that

$$V_{GS} = \left(\frac{I_{DM}}{g_{fs}} + V_{GS(th)}\right) \exp\left[\frac{-t}{R_G(C_{GS} + C_{GD})}\right] \tag{10.52}$$

and the drain current is given by (10.40), so it decreases as

$$I_D(t) = (I_{DM} + g_{fs} V_{GS(th)}) \exp\left\{\frac{-t}{R_G(C_{GS} + C_{GD})}\right\} - g_{fs} V_{GS(th)} \tag{10.53}$$

until the gate voltage reaches the threshold voltage. The drain current fall time $t_{fi}$ is therefore given by

$$t_{fi} = R_G(C_{GS} + C_{GD}) \ln \left[ \frac{I_{DM} + g_{fs}V_{GS(th)}}{g_{fs}V_{GS(th)}} \right] \quad (10.54)$$

The total turn-off time is

$$t_{off} = t_s + t_{rv} + t_{fi} \quad (10.55)$$

Typical gate voltage, drain voltage and drain current waveforms are shown in Figure 10.15.

When all of the series inductance in the circuit of Figure 10.13 is not clamped with a freewheeling diode, the drain voltage is likely to overshoot and oscillations occur. A detailed analysis of this situation is given in the literature [10.3]. The series inductance keeps the turn-on losses low but causes the turn-off losses to increase considerably, as shown in Figure 10.16.

In the case of a resistive load, while the drain voltage increases, while the drain current decreases according to (10.40) and the turn-off time is shorter.

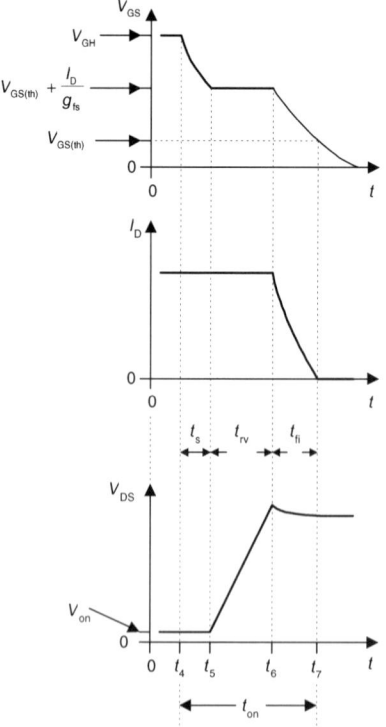

**Figure 10.15** Current and voltage waveforms at turn-off for an MOS-transistor in a circuit with a clamped inductive load

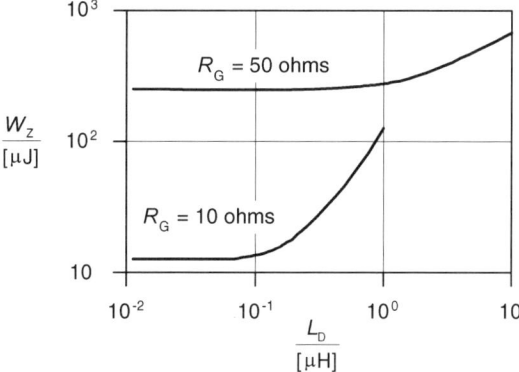

**Figure 10.16** An example of the increase of the sum of the turn-on and turn-off losses in a power MOSFET as a function of the unclamped series inductance

### 10.3.3 The Frequency Dependence of the Switching Parameters

Several effects limit the maximum operating frequency of power MOSFETs. One limitation follows from the turn-on and turn-off times themselves. Because the device must turn on and turn off within one period, the maximum switching frequency is limited by

$$f_{max} \leqslant \frac{1}{\pi(t_{on} + t_{off})} \qquad (10.56)$$

All the intervals that contribute to the switching times depend on the time constant $R_G C_{in}$. Power VDMOS transistors with a polycrystalline silicon gate and $V_{DM} > 200$ V have typical switching times in the range 200–600 ns. Devices with metal gates and reduced input capacitance can have turn-on times and turn-off times an order of magnitude shorter. Generally, the maximum switching frequency, as limited by the turn-on and turn-off times, is in the region of a few megahertz.

Power dissipation may also limit the maximum operating frequency. The instantaneous power dissipation is given by

$$P(t) = I_D(t)V_{DS}(t) + I_G(t)V_{GS}(t) \qquad (10.57)$$

The second term represents losses in the internal gate circuit, which can be considerable at high frequencies. The energy loss during a complete switching cycle (off→on→off) of period $T = 1/f$ is

$$W_z = \int_0^T P(t)\,dt \qquad (10.58)$$

Assuming the power losses in the off-state to be negligible and the duration of the on-state to be $\Delta t$, the energy loss can be expressed as

$$W_z = \int_0^{t_{on}} P(t)\,dt + I_{DM}^2 R_{on}\Delta t + \int_{\Delta t + t_{on}}^{\Delta t + t_{on} + t_{off}} P(t)\,dt \qquad (10.59)$$
$$= \bar{P}_{on} t_{on} + I_{DM}^2 R_{on}\Delta t + \bar{P}_{off} t_{off}$$

and the average power loss at the switching frequency $f$ is now given by

$$P_{AV} = f(\bar{P}_{on} t_{on} + I_{DM}^2 R_{on}\Delta t + \bar{P}_{off} t_{off}) \qquad (10.60)$$

The average power dissipated during turn-on, $\bar{P}_{on}$, and during turn-off, $\bar{P}_{off}$, is much higher than the power dissipated in the on-state, $I_{DM}^2 R_{on}$. However, the relative contributions of the three terms in (10.59) and (10.60) to the total power loss depend on the switching frequency. At low frequency the on-state power loss dominates. With an increase in the operating frequency, $t_{on}$ and $t_{off}$ take up an increasing fraction of the operating cycle and the contribution from the dynamic losses becomes dominant. With a limit on the overall average power loss, the maximum permissible on-state current decreases at higher switching frequencies. Because the switching losses depend on the type of load, the frequency dependence of the maximum drain current also depends on the type of load.

Energy losses connected with turning on and turning off also depend on the load type. An approximation [10.3] using a linear slope of both increase and decrease of drain current and voltage has shown that switching losses in the case of an inductive load clamped with a freewheeling diode (Figure 10.15) are three times higher than those in the case of a purely resistive load.

Another limiting factor is the homogeneity of the integrated structure of the power MOSFET. It is suggested in Section 10.2.1 that connecting VDMOS transistors in parallel is relatively easy because of the temperature dependence of the on-state resistance. However, the transient characteristics depend on the capacitances and the threshold voltage so that, when MOSFETs with different values of $V_{GS(th)}$ and $g_{fs}$ or having an inhomogenous structure, are connected in parallel, oscillations and local overloading can occur. All of these influences result in a lower frequency limitation of the maximum usable on-state current. An example is shown in Figure 10.17.

## 10.4 Safe Operating Area (SOA)

As with bipolar transistors, the safe operating area sets limits for the instantaneous values of the drain voltage and current during device operation that avoid the danger of second breakdown. These limits are determined by the maximum admissible power loss, the maximum admissible current and the maximum admissible voltage, as shown in Figure 3.19.

If power MOSFETs are simultaneously subjected to a high current and a high voltage, second breakdown can occur in theory. The possibility is associated with the parasitic bipolar transistor shown in Figure 10.18. The fact that modern power MOSFETs do not suffer from this phenomenon is the result of the careful design of the features relating to the parasitic bipolar transistor. The resistance $R_B$ corresponds to the lateral resistance of the p-type layer underneath the $n^+$-source, i.e. through the base region of the parasitic bipolar $n^+$–p–n transistor. Part of the drain current flows

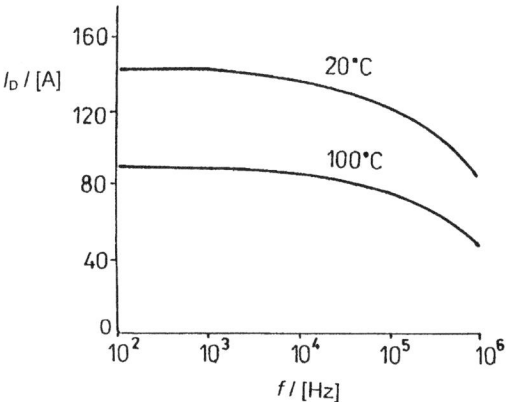

**Figure 10.17** Examples of the frequency dependence of the maximum permitted drain current in a power MOSFET at 20°C and 100°C. Reproduced from *Power MOSFETs Theory and Application* by D.A. Grant and J. Gowar, by permission of John Wiley and Sons, Inc.

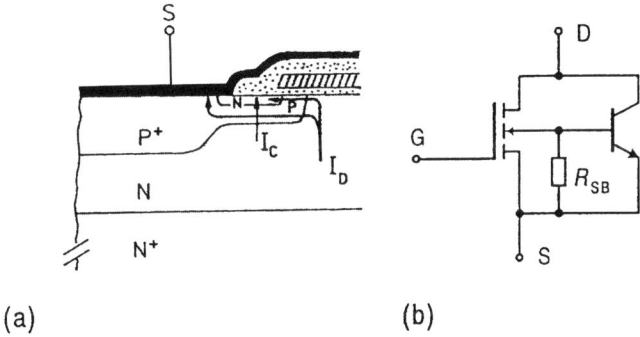

**Figure 10.18** Activation of the parasitic bipolar transistor inherent in the VDMOS structure: (a) the effect of lateral current flow through the base region; (b) a simplified equivalent circuit

through the collector–base junction of the parasitic transistor. If $h_{21B}$ is its common-base current–gain and $I_{ch}$ is the current passing through the MOSFET channel, the currents, $I_D$, $I_S$, $I_E$ and $I_C$ are given by

$$I_D = I_{ch} + I_C \tag{10.61}$$

$$I_S = I_{ch} + I_E + I_B \tag{10.62}$$

$$I_B = I_C - I_E \tag{10.63}$$

$$I_C = h_{21B} M I_E \tag{10.64}$$

where $M$ is the multiplication factor given by (2.44). The base of the transistor is thin (short channel structure) so $h_{21B} \approx 1$ and, following (2.14), the emitter current of the parasitic transistor can be expressed as

$$I_E = I_0 \exp\left(\frac{eV_{EB}}{kT}\right) \tag{10.65}$$

where $V_{EB}$ is the voltage induced by the lateral current in the base, $V_{EB} = R_B I_B$. Combining (10.61) to (10.65), we obtain for small values of $eV_{EB}/kT$

$$I_E = \frac{I_0}{1 - (M-1)I_0(eR_B/kT)} \tag{10.66}$$

Second breakdown occurs when $I_E \to \infty$, i.e. when the denominator of (10.66) approaches zero. Using (2.44) the voltage at which second breakdown occurs is given by

$$V_{DS(SB)} = \frac{V_{CB(BR)}}{(1 + eR_B I_0/kT)^{1/\kappa}} \tag{10.67}$$

which is nearly independent of temperature because both $V_{CB(BR)}$ and the denominator of (10.67) increase with temperature ($R_B \propto T^2$).

It follows from (10.67) that $R_B$ should be made as small as possible by keeping the n$^+$-source region very narrow. If this is effective, the safe operating area is determined by the condition of thermal stability (2.49), i.e. by the maximum admissible power losses. When the on-state current pulses are short, the safe operating area can be extended, as shown in Figure 3.19.

The shorting of the n$^+$–p junction by the source contact creates an integrated antiparallel diode. This diode performs two useful functions. It protects the n$^+$–p junction against breakdown when $V_{DS}$ is negative and it can act as a Zener diode to give very efficient overvoltage protection when $V_{DS}$ is positive. For this to be effective, the structure must be homogeneous enough for the breakdown voltage of all diodes on the chip to lie within a very narrow interval and be lower than the breakdown voltage of the parasitic bipolar transistor. The peak reverse current through this integrated diode can be quite high. To take an example, a 100 V, 28 A MOSFET might be permitted a single 28 A pulse of avalanche current, with a repetitive avalanche energy rating of 15 mJ.

The integrated diode introduces some problems associated with its reverse recovery characteristics. In circuits with an inductive load, the rate of rise of drain voltage can be very high and this represents reverse voltage across the antiparallel diode. It also causes a displacement current to flow laterally through the p-type region to the source contact, like the drain current shown in Figure 10.18. If $dV_{DS}/dt$ is too high, this can activate the parasitic bipolar transistor and result in second breakdown. The critical rate of rise of the drain voltage allowed immediately after diode recovery is typically of the order 4 V/ns. These effects are minimised when the carrier lifetime is reduced. This may be done to reduce the recovery time and recovery charge of the body-drain diode so that it can be used as the freewheel diode in fast-switching inverter circuits, reducing the storage time during the diode reverse recovery process. Power MOSFETs with an integrated fast recovery epitaxial diode are commonly known by the acronym FREDFET (FREDFET is a tradename of Siemens Aktiengesellschaft).

Different techniques have been used to optimise the carrier lifetime without degrading the properties of the gate oxide. Electron irradiation is often used and more recently platinum diffusion from an implanted layer, which has been found to give the best combination of parameters for the MOSFET and the antiparallel diode.

## Summary

The power MOSFET has become the most widely used active power semiconductor device embracing applications up to 10 kW with voltages less than 1 kV. It combines the microscopic cellular structure of MOS integrated circuit technology with the robustness required in a power device. Like the SIT, the high voltage is sustained in the vertical direction, while the control of current is exercised with the economy and simplicity of the MOS gate. The compact cell size of the trench geometry gives a wide gate length per unit area of chip and hence a high transconductance. It is the preferred type for low voltage and low power applications. The double-diffused devices are at less of a disadvantage for higher voltage applications.

The most common type is the normally on, n-channel, enhancement mode device. As a majority carrier device, its switching speed is governed by device and circuit parasitics rather than by internal physical processes. Integral to the design is an antiparallel diode (the body-drain diode). This has the advantage of protecting the MOSFET against negative breakdown and can act as an avalanche diode to limit positive overvoltages. The power MOSFET structure contains an integral parasitic bipolar transistor, which could be activated by an excessive rise rate of the drain–source voltage, particularly immediately after the recovery of the body-drain diode. Good MOSFET design restricts this effect to very high values of $dv/dt$.

## References

[10.1] Hofstein, S. R. and Heimar, F. P. The Silicon Insulated-Gate Field-Effect Transistor, *Proc. IEEE*, **51**, 1190–1202 (1963).

[10.2] Sze, S. M. *Physics of Semiconductor Devices* (John Wiley & Sons, 2nd Edn, 1981).

[10.3] Grant, D. A. and Gowar, J. *Power MOSFETs: Theory and Applications* (John Wiley & Sons, 1989).

[10.4] Chiu, T. G. and Sah, C. T. Correlation of Experiments with a Two-Section-Model Theory of the Saturation Drain Conductance of MOS Transistors, *Solid State Electronics*, **11**, 1149 (1968).

[10.5] Selberherr, S. and Langer, E. Three Dimensional Process and Device Modeling, *Proc. 17th Yugoslav Conference On Microelectronics*, 383–95 (Nis, 1989).

[10.6] Fnoss, D. *IEEE International Devices Meeting (IEDM) Digest*, 250 (1982).

[10.7] Hu, C., Chi, M. H. and Patel, V. M. Optimum Design of Power MOSFETs, *IEEE Trans. on Electron Devices*, **ED-31**, 1693–1700 (1984).

[10.8] Esaki, H. and Ishikawa, K. *IEEE International Electron Devices Meeting (IEDM) Digest*, 447 (1984).

# 11

# POWER BIPOLAR–MOS DEVICES

Figure 3.17 demonstrates the considerable simplification that is possible in the units used to drive the voltage-controlled MOSFET in comparison with those needed for the current-controlled bipolar junction transistor. A bipolar device with a MOS gate structure that would combine the benefits of each device type has long been the goal of device engineers. Intensive research has led to the development of several such devices which do satisfy the requirements of relatively easy voltage control, good current-handling capacity and high frequency operation. Not surprisingly, they are coming to dominate a wide range of applications. The most important of these devices are the insulated gate bipolar transistor (IGBT) and the MOS-controlled thyristor (MCT).

Of these two types of construction, the IGBT has taken a clear lead in acceptance. Early attempts to achieve voltage control, as in MOS devices, along with the benefits of conductivity modulation as obtained in bipolar devices, were centred around the Darlington connection shown in Figure 11.1, initially with discrete devices, later in the form of a hybrid integrated circuit, and finally as a monolithic integrated structure, as shown in Figure 11.2. Unfortunately, the current per unit area of wafer is not improved.

In this chapter we deal in detail with the structure and operating properties of the IGBT and the MCT and other devices that have been derived from them.

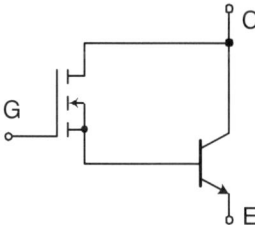

**Figure 11.1**  The Darlington configuration

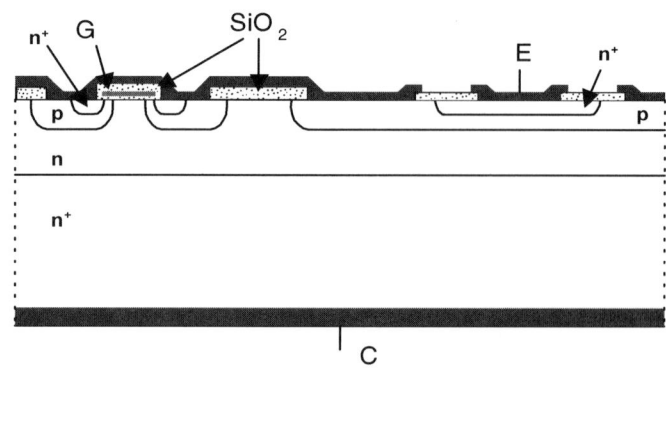

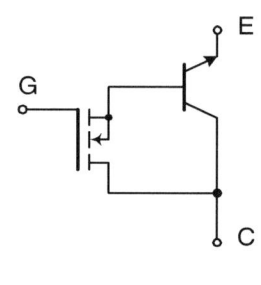

**Figure 11.2** An integrated, Darlington-connected, MOSFET/bipolar combination: (a) section through the active region; (b) equivalent circuit

## 11.1 The Insulated Gate Bipolar Transistor (IGBT)

The structure of the insulated gate bipolar transistor (IGBT) derives from that of the VDMOS FET by the substitution of a $p^+$-substrate for the normal $n^+$-material [11.1]. It has a similar cellular structure. A cross-section through one of the cells is shown in Figure 3.13(a). In the period following its invention, different names for the IGBT were used by different manufacturers: the insulated gate transistor (IGT) [11.2], the conductivity modulated FET (COMFET) [11.3], the gain-enhanced MOSFET (GEMFET) [11.4]. Since 1988 the name IGBT has been generally agreed.

Although the IGBT structure may appear very similar to that of the VDMOS transistor, simply being grown on a $p^+$- rather than an $n^+$-substrate, its function is rather different. Its operating principles combine those of the MOSFET, the bipolar transistor and the power diode. The more detailed section through the active region shown in Figure 11.3 illustrates the origin of the various parasitic elements that make up the device. An equivalent circuit is shown in Figure 11.4.

## 11.1 THE INSULATED GATE BIPOLAR TRANSISTOR (IGBT)

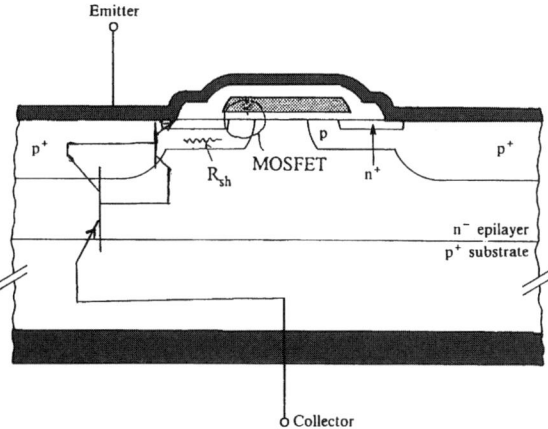

**Figure 11.3** A section through the active region of an IGBT cell, showing the various parasitic components

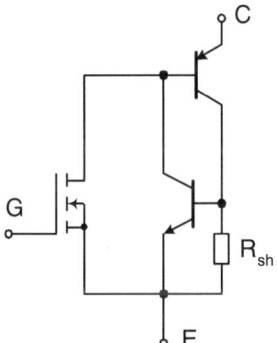

**Figure 11.4** An equivalent circuit for the IGBT

### 11.1.1 Static Parameters

As with a MOSFET, the application to the gate of a positive voltage $V_{GE}$, higher than some threshold voltage, sets up an inversion layer in the semiconductor underneath the gate oxide and creates a conducting channel between the $n^+$- and n-regions. With a sufficiently high gate voltage, the resistance of the channel is low and the application of a positive drain voltage causes an electron current to flow. This forward biases junction $J_1$, and holes are injected from the $p^+$-emitter of the $p^+np$ transistor into the n-base. These holes reduce the series resistance of the structure, enabling a higher on-state current density to be handled in comparison with a MOSFET of similar dimensions.

If the channel and shunt resistances are low, so that the $n^+pn$ transistor is not activated, the equivalent circuit can be simplified to that shown in Figure 11.5. The electron current $I_n$ passing through the channel of the MOS transistor forms the base

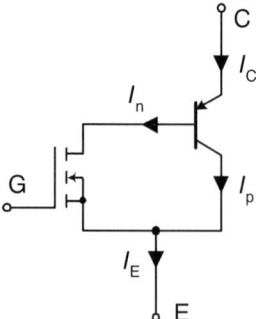

**Figure 11.5** A simplified equivalent circuit for the IGBT used in the analysis of $V_{CE(Sat)}$

current of the p⁺np bipolar transistor. If $\alpha_{pnp}$ is the common-base current gain of the p⁺np transistor, the hole current which crosses junction J₂ is

$$I_p = \left(\frac{\alpha_{pnp}}{1 - \alpha_{pnp}}\right) I_n \tag{11.1}$$

The emitter current of the IGBT is $I_n + I_p$ and under steady-state conditions it equals the collector current. Thus

$$I_E = I_C = \frac{I_n}{1 - \alpha_{pnp}} \tag{11.2}$$

After the application of a gate–emitter voltage, $V_{GE} > V_{GE(th)}$, the electron current flowing through the channel of the MOS transistor depends on $V_{GE}$. If it is high enough, $I_n$ is given by (10.12):

$$I_n = \frac{z}{l} \mu_{eff} C_{ox} (V_{GE} - V_{GE(th)}) V_{ch} \tag{11.3}$$

where $V_{ch}$ is the voltage dropped along the conducting channel, which has length $l$, and width $z$. The capacitance per unit area of the gate oxide is $C_{ox}$ and the average mobility of the electrons in the inversion channel is $\mu_{eff}$. Then, combining (11.2) and (11.3), the collector current of the IGBT can be expressed as

$$I_C = \frac{z}{l} \frac{\mu_{eff} C_{ox}}{(1 - \alpha_{pnp})} (V_{GE} - V_{GE(th)}) V_{ch} \tag{11.4}$$

When the channel voltage exceeds $(V_{GE} - V_{GE(th)})$, the collector current saturates in agreement with (10.17):

$$I_C = \frac{1}{2\alpha} \frac{z}{l} \frac{\mu_{eff} C_{ox}}{(1 - \alpha_{pnp})} (V_{GE} - V_{GE(th)})^2 \tag{11.5}$$

A typical set of gate and collector characteristics is shown in Figure 3.13.
The transconductance of the IGBT in saturation is

$$g_{fs} = \frac{z}{\alpha l} \frac{\mu_{eff} C_{ox}}{(1 - \alpha_{pnp})} (V_{GE} - V_{GE(th)}) \tag{11.6}$$

A typical value of $\alpha_{pnp}$ is about 0.5, so the transconductance of the IGBT is approximately twice that of a MOSFET of similar construction (size, concentration profiles, oxide thickness, etc.).

As with all power switching devices, the on-state voltage–current characteristics determine the on-state power losses and are very important in setting the current-handling capacity. The main conduction path through an IGBT can be modelled as a diode in series with a MOS transistor, as shown in Figure 11.6. The diode current is $I_n$, and from (11.2) the current density flowing vertically between the cells may be expressed as

$$J = \frac{(1 - \alpha_{pnp}) I_C}{l_G z} \tag{11.7}$$

where $l_G$ is the spacing between the cells and $z$ is the gate width.

The forward characteristics of the power diode are discussed in detail in Section 5.1. In (5.18) its forward voltage drop is expressed as

$$V_F = K_0 + K_1 \ln J + K_2 J^m$$

where $m$ is approximately 0.7, $K_1$ is a function of the carrier concentration and the thickness of the $p^+$-emitter, and $K_2$ depends on the ratio of the diode base thickness to the carrier diffusion length $K_2 \propto \cosh(w/L)$.

From (11.4) the voltage drop across the channel of the MOS transistor can be expressed as

$$V_{ch} = \frac{(1 - \alpha_{pnp}) I_C l}{\mu_{eff} C_{ox} z (V_{GE} - V_{GE(th)})} \tag{11.8}$$

The total on-state voltage drop across the IGBT is the sum of the diode and MOS channel voltages, as given by (11.8), (5.18) and (11.7):

$$V_{CE(on)}(I_C) = V_{ch}(I_C) + V_F(I_C) \tag{11.9}$$

As discussed in Section 5.1 in the case of an ordinary power diode, $V_F$ can be represented by a threshold voltage $V_{(TO)}$ and a differential resistance $r_T$. Increasing temperature causes $V_{(TO)}$ to decrease and $r_T$ to increase. The positive temperature coefficient of the differential resistance enables IGBTs to be made from many integrated cells connected in parallel, just like a power MOSFET. The on-state voltage drop of the IGBT increases considerably with the ratio $w/L$ in the same way

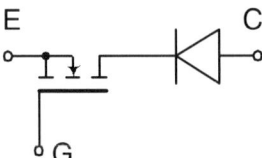

Figure 11.6  An equivalent circuit used for modelling the on-state characteristics of the IGBT

as the forward voltage drop of a power diode. The on-state current–voltage characteristic also depends considerably on the excess carrier lifetime, which influences most parameters, as it does in nearly all types of bipolar device.

In contrast to ordinary power bipolar transistors, the space charge region of the collector–base junction does spread into the base region, whose effective width $w$ decreases with increasing collector voltage. Consequently, $\alpha_{pnp}$ increases with the collector voltage, as described by (6.7), thus increasing the proportion of the current flowing in the bipolar transistor.

The maximum blocking voltage is limited by the breakdown voltage $V_{CEO}$ of the pnp$^+$ transistor. The problem is similar to the one described in detail in Section 7.1.1 for thyristors. To combine a high blocking voltage with a low on-state voltage, a compressed-field structure is often used. This is obtained by forming an n$^+$-layer of about 10 $\mu$m thickness between the p$^+$-collector layer and the n-base, as shown in Figure 11.7. This reduces the injection efficiency of the collector, which alone would cause a slight increase in the on-state voltage drop. In practice, however, this is usually more than offset by the effect of the smaller value of $w/L$. In the context of IGBTs this type of device is often called a punch-through IGBT. The relative merits of punch-through and non-punch-through IGBTs are discussed in Section 11.1.4.

Up to now we have assumed that the emitter–base junction of the parasitic n$^+$pn transistor is effectively fully shunted. Equations (11.6) and (11.8) are derived under the assumption that the shunt resistance $R_{sh}$ is zero. However, part of the current flowing through the p$^+$np transistor has to pass though $R_{sh}$ and, if it is finite, even if it is very low, this develops a voltage that imparts a positive bias on the base of the parasitic n$^+$pn transistor, as shown in Figure 11.8. The two transistors form a thyristor structure, as shown in the equivalent circuit of Figure 11.4. At some critical value of the collector current $I_L$, the turn-on condition (7.9) for the n$^+$pnp$^+$ structure is fulfilled and the parasitic thyristor is turned on. This is known as

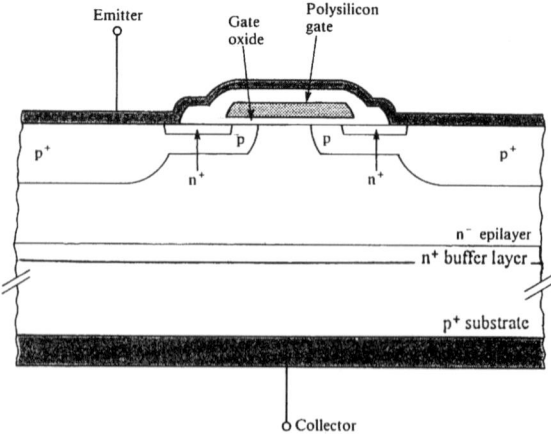

**Figure 11.7** A cross-sectional diagram showing a compressed field IGBT. Reproduced from *Power MOSFETs Theory and Applications* by D. A. Grant & J. Gowar, by permission of John Wiley & Sons, Inc.

## 11.1 THE INSULATED GATE BIPOLAR TRANSISTOR (IGBT)

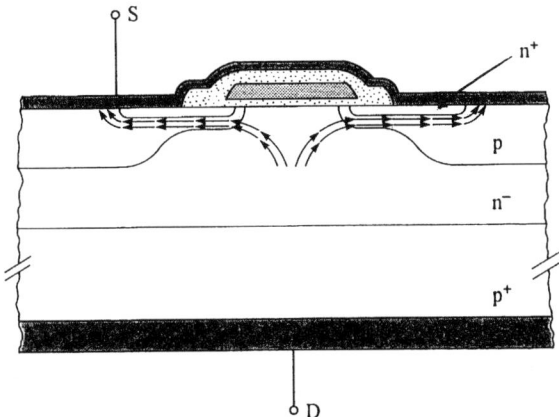

**Figure 11.8** An illustration of the influence of lateral current through the p region in causing latch-up in an IGBT. Reproduced from *Power MOSFETs Theory and Applications* by D. A. Grant & J. Gowar, by permission of John Wiley & Sons, Inc.

*latch-up*. It results in a decrease in the on-state voltage. Once latch-up occurs, the IGBT cannot be controlled by the gate voltage.

Thyristor structures with shunted emitters are discussed in detail in Section 7.2. The conditions for turn-on are given by (7.13) and (7.14). The latch-up effect in IGBTs has a similar origin and a simple approximation for the latching current is

$$I_L \approx \frac{g}{\alpha_{pnp} l_E \rho_p} \quad (11.10)$$

where $l_E$ is the length of the $n^+$-emitter region and $g$ is the thickness of the p-layer, whose resistivity is $\rho_p$.

To increase $I_L$ it is necessary to decrease $l_E$ and $\rho_p$. At the same time it is important that the contact resistance of the interface between the p-base and the emitter contact metal is very low. One-dimensional modelling of the dependence of the latch-up current on the IGBT structure enables the underlying processes to be understood, but an accurate representation of the individual geometry of real IGBT cell structures requires two-dimensional numerical modelling.

Note that a shorter MOS channel, which decreases $V_{CE(on)}$, can decrease $R_{sh}$ and therefore decrease the latch-up current.

### 11.1.2 Switching Characteristics

The beginning of the IGBT turn-on process is closely related to the turn-on process of a power MOSFET. That is to say, after the application of a positive gate voltage greater than the threshold voltage, the input capacitance starts to charge. As soon as the gate voltage reaches the threshold voltage; that is, after the delay time given by (10.39) current starts to flow through the input MOS transistor. This current becomes the base current of the pnp bipolar transistor.

The collector current of the bipolar transistor starts to flow after a delay determined by the carrier transit time through the base. This is given by (6.46). It follows that the total delay time $t_{d(on)}$ from the initial application of a positive gate signal to the rise of the IGBT collector current is the sum of these two delays:

$$t_{d(on)} = \frac{w_{pnp}^2}{2D_p} + R_G(C_{GE} + C_{GC})\ln\left(\frac{V_{GE}}{V_{GE} - V_{GE(th)}}\right) \quad (11.11)$$

where $w_{pnp}$ is the base width of the $p^+np$ transistor. When the IGBT is in the blocking state, $w_{pnp}$ is small, so the first term, the delay in the bipolar transistor, is short. The delay time of an IGBT is then almost the same as that of a power MOSFET of similar structure.

After $t_{d(on)}$ the collector current increases rapidly as excess carriers are injected into the n-base, where the build-up of the excess carrier charge is given by (6.47). The base current $I_B$ is described by (10.41). Simultaneous solution of these equations, taking account of the variation of the capacitance $C_{GE}$ and the current gain of the bipolar transistor, is very complicated. Numerical solution is required.

Because of their lower on-state resistance, IGBTs have a current-handling capacity that is much higher than that of an equivalent area VDMOS transistor. Conversely, the area of an IGBT die is much smaller than that of a similarly rated MOSFET. However, this means that $C_{GE}$ and $C_{GC}$ are lower in the same ratio for devices having the same nominal current rating.

Examples of typical current and voltage waveforms during turn-on are shown in Figure 11.9. As is to be expected, they depend in detail on the design of the device and the parameters of the circuit in which it is operating. However, the turn-on process of the IGBT is very fast and losses are relatively low. This is especially true when the resistance of the gate circuit is small, the transconductance of the MOS section is high and the delay time is short.

More attention has to be given to the IGBT turn-off process. A large charge of excess mobile carriers is stored in the n-base during the on-state. This is the result of the hole injection from the $p^+$-emitter that has caused the conductivity modulation in the n-base. Part of the total on-state electron current that passes through the MOS channel is controlled by the gate charge. For this to be interrupted, the gate capacitance has to be discharged by reversing the gate supply voltage. As soon as $V_{GE}$ decreases below $V_{GE(th)}$, the MOS channel is turned off. As described in Section 10.3.2, this takes time $t_s$, given by (10.49). The collector–emitter voltage then rises according to (10.50), so its rate of rise is given by

$$\frac{dV_{CE}}{dt} = \frac{g_{fs}V_{GE(th)} + I_C}{g_{fs}R_G C_{GC}} \quad (11.12)$$

At the same time the current in the MOS channel decreases abruptly by the amount

$$\Delta I_C = I_n = (1 - \alpha_{pnp})I_C \quad (11.13)$$

After closing the channel, junction $J_2$ is reverse biased and the excess carriers stored in the base of the $p^+np$ transistor are swept out by the expanding space charge region. Compared with a bipolar transistor, there is no negative base current to carry away the excess carriers accumulated in the base. Thus, more charge has to be extracted through the collector junction than in a bipolar transistor of similar base

## 11.1 THE INSULATED GATE BIPOLAR TRANSISTOR (IGBT)

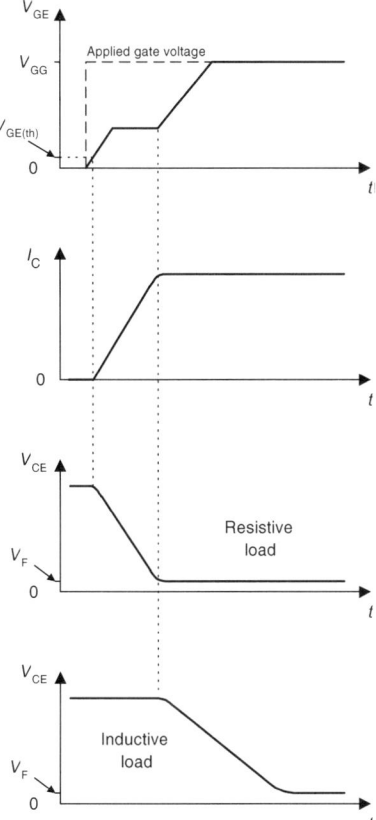

**Figure 11.9** Voltage and current waveforms during IGBT turn-on

thickness and carrier lifetime. An exact solution requires the continuity equation to be solved with moving boundary conditions [11.5,11.6].

Analysing the problem by the stored charge method, equation (5.24) can be used to express the variation of the excess carrier charge that has accumulated in the base during the on-state. By analogy with (5.28), the collector current falls as

$$I_C(t) = \alpha_{pnp} I_C(0) \exp\left(\frac{-t}{\tau_{eff}}\right) \tag{11.14}$$

This is known as the IGBT tail current. The time constant $\tau_{eff}$ depends on the carrier lifetime and the width of the n-base.

A displacement current arising from the rate of change of voltage across the collector junction of the $p^+np$ transistor adds to the current of carriers. Typical current and voltage waveforms during turn-off are shown in Figure 11.10. It is evident that the tail current makes a considerable contribution to the power losses. In the approximation that the collector current falls abruptly by $\Delta I_C$ at $t=0$, and that

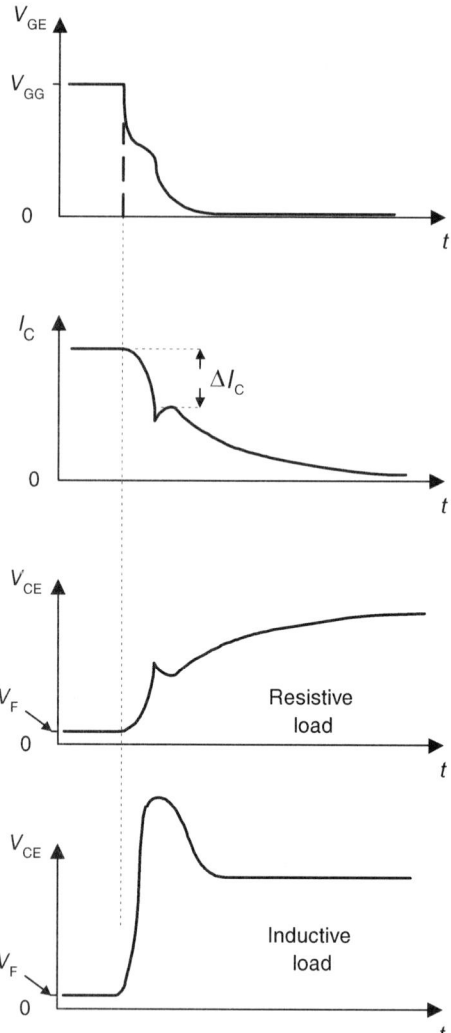

**Figure 11.10** Voltage and current waveforms during IGBT turn-off

the collector voltage rises abruptly at the same time to $V_{CO}$, the energy dissipated during turn-off can be found by integration to be:

$$W_z = V_{C0} I_C \alpha_{pnp} \tau_{eff} \tag{11.15}$$

The detailed analysis of Chapter 10 shows how the rate of rise of the collector voltage and the rate of fall of the MOS channel current depend on the time constant $R_G C_{GC}$. The total gate resistance $R_G$ can be influenced by the external resistance of the gate circuit. These parasitic components are shown in Figure 11.11. In order to minimise energy dissipation and to avoid some unfavourable effects, it

## 11.1 THE INSULATED GATE BIPOLAR TRANSISTOR (IGBT)

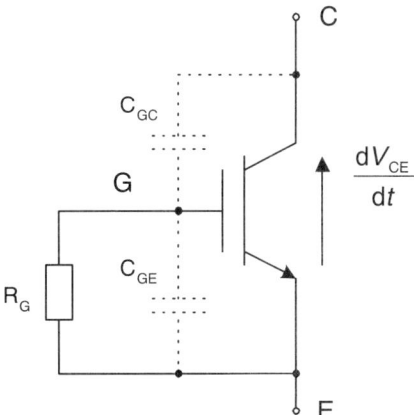

**Figure 11.11** The parasitic capacitances inherent in the IGBT structure

is advantageous if $dV_{CE}/dt$ is not too high. The displacement current induced by a rapid rise of the collector voltage flows laterally through the p-base of the $n^+$pn transistor together with the excess carriers. It forms the gate current of the parasitic thyristor which in theory can be turned on during the IGBT turn-off process if the current exceeds the critical value $I_{CLd}$. The ratio of the dynamic and static critical currents can be expressed as

$$\frac{I_{CLd}}{I_{CL}} = \frac{1 - \alpha_{pnpd}}{1 - \alpha_{pnp}} \tag{11.16}$$

where $\alpha_{pnpd}$ is the transient current gain of the pnp transistor for $V_{CE} > V_{on}$ and $\alpha_{pnp}$ is the static current gain of the pnp transistor when $V_{CE} = V_{on}$. The static current gain $\alpha_{pnp}$ increases with the collector voltage, so the dynamic critical current $I_{CLd}$ is less than the latch-up current $I_{CL}$.

Latch-up limits the maximum permitted turn-on current to a value much less than the critical current. It is an IGBT design aim to minimise $\alpha_{pnpd}$ and so to make the dynamic critical current as high as possible. Two-dimensional modelling [11.7] has shown that this can be achieved by making the elementary IGBT cell in the shape of a strip. However, this tends to increase the influence of the parasitic JFET. Reducing the fabrication design rules (line widths) and shortening the lateral length of the $n^+$-emitter region [11.8] enables the lateral resistance $R_p$ to be decreased. This makes it possible to obtain a high critical current with a cell structure and emitter–gate arrangement similar to that of a power VDMOS FET. Then $I_{CL}$ can be $> 5 I_{CEM}$ while $I_{CLd} > I_{CEM}$.

A higher value of $I_{CLd}$ can also be obtained if $dV_C/dt$ is reduced during turn-off by increasing the external gate resistance through which the input capacitance of the IGBT has to be discharged. In this way it is possible to increase the critical current and at the same time decrease the turn-off losses. A typical variation of $I_{CLd}$ with the external gate resistance is shown in Figure 11.12. A gate circuit like the one shown in

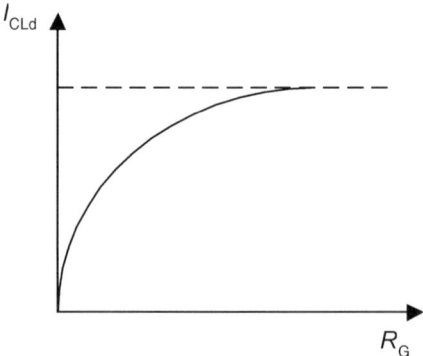

**Figure 11.12** The influence of the gate resistance on the dynamic critical current

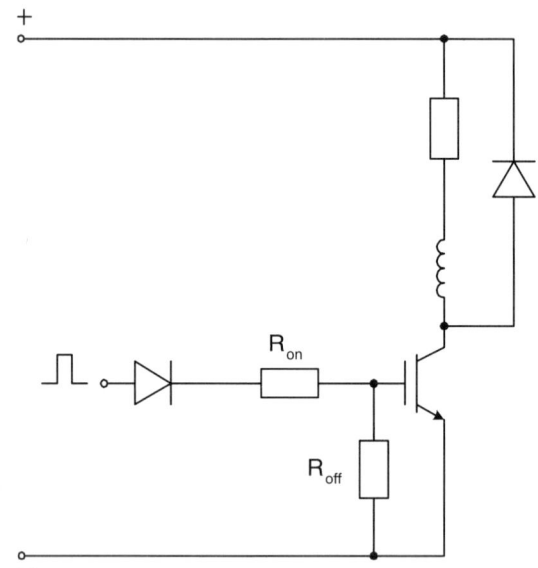

**Figure 11.13** A gate circuit having low $R_{on}$ and high $R_{off}$

Figure 11.13 is able to provide fast turn-on by the use of a small series resistance $R_{on}$, and decreased $dV_C/dt$ during turn-off by making $R_{off}$ much higher.

Another dynamic phenomenon which can cause an IGBT to turn on undesirably is the reapplication of a rapidly rising blocking voltage to a device in the off-state. In this case the IGBT can be modelled by the equivalent circuit shown in Figure 11.14. During the rise of $V_{CE}$, $C_{GE}$ is charged positively through $C_{GC}$ and discharged through $R_{GE}$. If $R_{GE}$ is high, the voltage on $C_{GE}$ can exceed the threshold voltage $V_{GE(th)}$ and turn on the IGBT. Even if it is in the on-state for only a very short time, there is considerable dissipation. It can be prevented by minimising $R_{GE}$, possibly causing the turn-off losses to increase and the critical current to decrease, or by applying a negative gate voltage, typically of about 5 V.

## 11.1 THE INSULATED GATE BIPOLAR TRANSISTOR (IGBT)

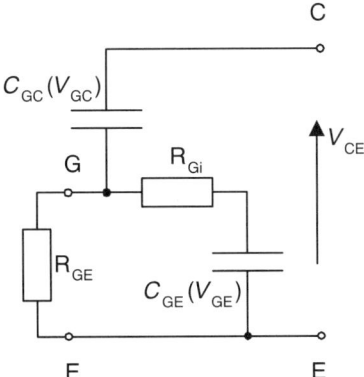

**Figure 11.14** An equivalent circuit used to model the effect of $dV_{CE}/dt$ on IGBT turn-on

### 11.1.3 The Frequency Dependence of Parameters and the SOA

The SOA for an IGBT can vary in nature considerably, depending on the way the device is structured. The IGBT is partly MOSFET and partly BJT. If it is designed so the MOSFET current is large, its safe operating characteristics incline to those of a MOSFET and unrestricted operation up to its peak current capability can be expected. However, as more BJT action is incorporated into the design, latching and hot spot development may restrict the permissible switching trajectories, as described in Section 6.4.

As with bipolar transistors and power MOSFETs, the SOA of IGBTs is set by a maximum power limit and the need to avoid second breakdown. The cellular structure of the IGBT, like the RET-type bipolar transistor, is more resistant to mesoplasma development. In practice the RBSOA of a power IGBT is close to rectangular, being limited by high tail current when high voltage and high current are present together. A typical example is shown in Figure 11.15.

The frequency dependence of IGBT parameters is related to the power losses during turn-on and turn-off. Like the bipolar transistor discussed in Section 6.3 and the power MOSFET discussed in Section 10.3.3, the maximum usable on-state current decreases with increasing operating frequency. With hard switching, where the IGBT is turned on from a high working voltage and turns off to return to the same voltage, the maximum working frequency and the frequency dependence of the on-state current are limited by the high dissipation during the current tail.

It follows from (11.15) that a higher maximum operating frequency can be obtained by reducing the excess carrier lifetime. Either electron irradiation, or platinum diffusion from an implanted layer, can be used. The latter technique is the more effective in shortening the fall time $t_f$ while causing the least increase in the on-state voltage. Note that this is identical with the result for the fast diode integrated into the power VDMOS transistor, as described in Section 10.4. A comparison of the trade-off between $V_{CEon}$ and $t_f$, using the two technologies, is shown in Figure 11.16.

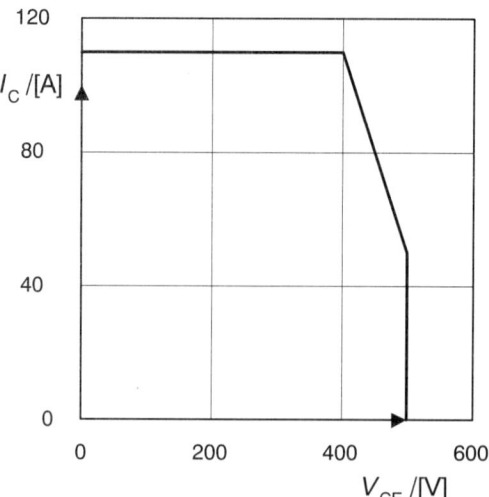

**Figure 11.15** An example of the IGBT safe operating area under reverse bias

In hard-switching conditions, parasitic inductance can cause overvoltage and undesirable oscillations. Hence IGBTs may be protected by a snubber, as described in Section 6.4, for bipolar transistors. The maximum operating frequency in this regime can exceed 100 kHz.

A quite different situation arises in resonant switching circuits of the type shown in Figure 11.17. The current waveform is now sinusoidal. With IGBTs it is advantageous to use zero current switching, where the device is turned off at the instant the decreasing current passes through zero. The power losses are determined by the on-state losses, the turn-on losses and the losses that arise in the gate circuit in charging and discharging the input MOS capacitance. In this switching mode, and with a fast external diode with a soft reverse recovery characteristic connected in antiparallel, it is possible for the operating frequency to be as high as 500 kHz [11.9]. In addition to the considerable reduction in the turn-off losses, there is a much more

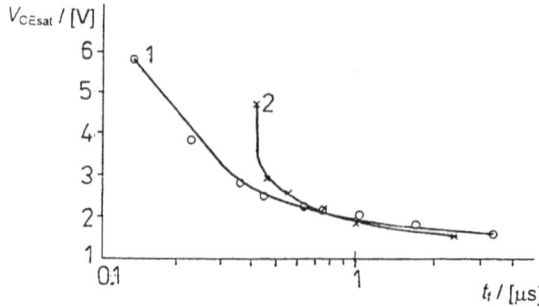

**Figure 11.16** Graphs of $V_{CESat}$ versus fall time for IGBTs with carrier lifetimes reduced (1) by platinum diffusion and (2) by electron irradiation. Reproduced from 'Application of Ion Implantation to the Control of Dynamic Characteristics in Power Devices', by F. Frisina, N. Tavolo, G. Ferla, S. U. Campisano and S. Coffa, in *Proceedings of EPE-MADEP'91*, 53–8, (Firenza, Italy, 1991), by permission of Litografia GEDA, Torino, Italy

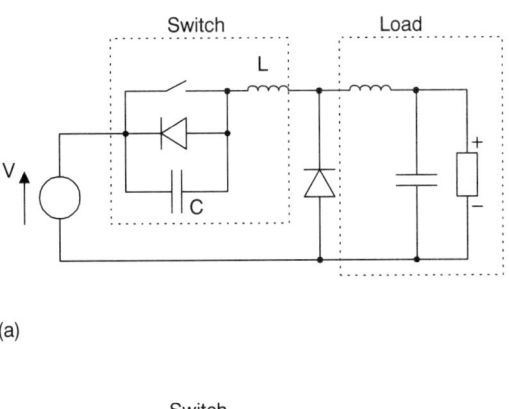

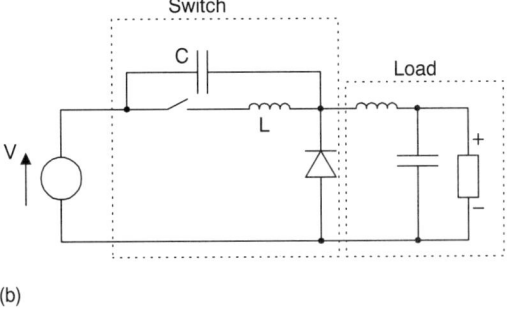

**Figure 11.17** Resonant switching circuits: (a) series; (b) parallel

advantageous working point trajectory within the safe operating area compared to other circuits. This is illustrated schematically in Figure 11.18 [11.10].

### 11.1.4 Punch-Through (PT) and Non-Punch-Through (NPT) IGBTs

As indicated in Section 11.1.1, it is common practice in IGBT construction to use a compressed-field design in which the depletion region is allowed to punch through the wide, lightly doped, $n^-$-region until it is stopped by an $n^+$-buffer layer. The term 'punch-through' is preferred because the $n^-$-region constitutes the base of the pnp transistor inherent in the IGBT structure. The term 'reach-through' is generally applied when the field extends through the collector region. Punch-through devices tend to use epitaxial technology with a lightly doped $n^-$-base region being grown on a substrate already bearing the $n^+$-buffer layer. Reduction of the minority carrier lifetime, perhaps by electron irradiation, enables the switching times to be reduced, subject to the usual consequential trade-off between the forward voltage drop and the turn-off losses.

By comparison, non-punch-through IGBTs are usually constructed in float zone silicon, with the $p^+$-collector layer being diffused in from the back side. The $n^-$-base region has to be wide, to accommodate the depletion region and leave a margin to ensure that punch-through never occurs under test or any conditions of use. The

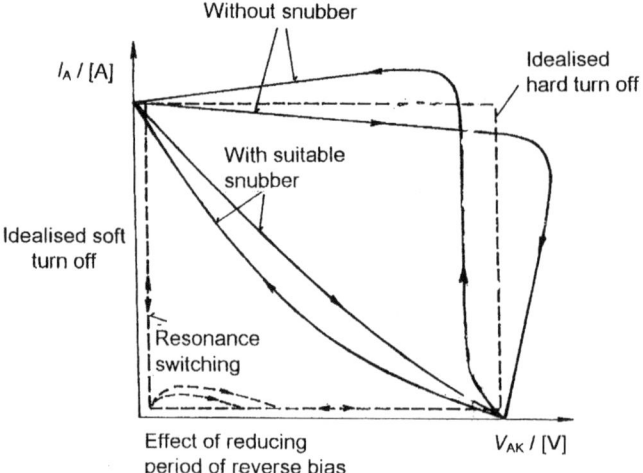

**Figure 11.18** Working point trajectories in resonant and hard-switching circuits

wide base region tends to give the NPT IGBT a higher forward voltage drop than the PT IGBT. However, the construction process is cheaper and NPT devices are less reliant on the reduction of the minority carrier lifetime to control the switching speed. As a result, the amplitude and duration of the current tail on turn-off is less temperature dependent. NPT IGBTs also tend to have a better reverse bias SAO and better short circuit withstand capability.

NPT technology is more appropriate for high voltage devices because the starting wafer for low voltage devices (e.g. 600 V) is very thin. Such devices have been made successfully on starting material with a thickness of only 100 $\mu$m [11.11]. For the most popular voltages, in the region of 1200 V, NPT and PT IGBTs are both well established. In the higher voltage regions (2.2 kV and above), NPT technology tends to be preferred, because high voltage PT devices require a thick epitaxial starting layer which is expensive. Furthermore, the float zone process can provide the very lightly and homogeneously doped starting material that is required.

At very high voltages (around 4.5 kV) the lightly doped base region in NPT IGBTs can be up to 650 $\mu$m wide [11.12], leading to a high forward voltage drop and an unacceptable level of conduction losses. There is thus an incentive at these voltages to return to PT technology with both an $n^+$-buffer layer and a $p^+$-collector layer. Shorts to the buffer layer through the $p^+$-layer improve the switching characteristics [11.13]. They allow the injection efficiency of the $p^+$-emitter to be regulated and a path is provided for the discharge of these regions during turn-off. Interestingly, the use of collector shorts has the effect of creating an integral antiparallel diode, which some circuits (e.g. inverters) require in any case.

As an alternative to the collector shorts, the $p^+$-layer can be made very thin so that minority carriers reach the metallisation without recombination. The layer is then said to be 'transparent' [11.12]. Its transparency can be increased by reducing its thickness or by reducing the doping concentration. Reducing the doping concentration has been shown also to reduce the turn-off losses.

### 11.1.5 IGBT Developments

As described in Section 11.1, the IGBT is essentially a power MOSFET built on a p$^+$-substrate rather than an n$^+$-substrate. It is not surprising therefore that IGBT development has followed MOSFETs with a move to trench gate technology. At first sight, the gains to be achieved in IGBT technology with the adoption of trench gates would not seem to be as great as those achievable in low voltage power MOSFETs. In low voltage MOSFETs a major part of $R_{DS(on)}$ is to be found in the MOSFET channel, and the increase in cell density which trench technology permits therefore has an immediate effect in reducing $R_{DS(on)}$. On the other hand, IGBTs are generally high voltage devices. However, as can be seen in Figure 11.5, the IGBT may be regarded as a Darlington-like combination of a pnp transistor and a MOSFET which links the transistor collector and base. Hence in IGBTs the resistance of the MOSFET channel should be as low as possible, so the required base current for the pnp transistor can be derived with as low a collector voltage as possible.

With trench technology the current flows vertically from emitter to collector, and the JFET resistance encountered in planar MOS technology is eliminated. As the structure becomes finer, the device in the on-state approximates more to a MOSFET in series with a diode (rather than a pnp transistor) and the channel resistance contributes significantly to the forward voltage drop in the conducting state. IGBTs are now produced with trench widths of the order of 1 μm and a forward voltage characteristic close to that of a diode. With trench technology, switching losses are reduced, the short circuit capability is improved and the device is less susceptible to latching.

The voltage ratings of IGBTs now extend up to 4.5 kV [11.14] and a similar increase in the current rating of IGBT-based switches has also been achieved. The IGBT has shown itself amenable to parallel operation, so that modules are made with many dice in parallel. Examples are 3.5 kV, 1200 A [11.15] and 2.5 kV, 1800 A [11.16]. The availability of such devices challenges the GTO in applications such as high power inverters for motor drives and in traction applications such as motor control for trams and trains. At the same time, switching speed is being improved in smaller devices and designers are encouraged to consider using IGBTs in 150 kHz switching applications [11.17].

The IGBT can be described as the workhorse of power electronics at mains supply voltages because of its high current-handling capacity, high breakdown voltage, broad safe operating area and simple voltage control. Its use is expanding to include most applications where high voltage bipolar transistors or GTOs were formerly employed.

## 11.2 The MOS-Controlled Thyristor (MCT)

Of all the different types of device, thyristors have the highest current-handling capacity, i.e. the highest on-state current density. This is made possible by the large concentration of excess carriers that builds up in the inner layers of the thyristor structure where an electron–hole plasma is created by the double injection process. The charge that is stored in the device during the on-state complicates the thyristor turn-off process, since it is necessary to remove these excess carriers, either by a reversal of the anode current or by a large pulse of negative gate current.

It is possible to make a thyristor structure that is turned on by a MOS input stage, using an IGBT-like structure with a high shunt resistance ($R_{sh}$ in Figure 11.4). Quite a low current through the IGBT structure is sufficient to cause latch-up of the $n^+pnp^+$ thyristor structure and so cause it to turn on. Examples are shown in Figure 3.14.

MOS-gated thyristors of this kind were first developed in 1979 [11.18]. They have the advantage that they can be turned on by the use of a simple voltage-controlled gate circuit. As with conventional thyristors, discussed in Section 7.4, turn-off requires the commutation of the anode current or its reduction below the holding current.

The MOS-controlled thyristor (MCT) enables full voltage-controlled switching, both turn-on and turn-off. Its structure can have several variants. An equivalent circuit for the p-channel device shown in Figure 3.15(c) is given in Figure 11.19 and it enables the principle of the MCT to be explained. We start in the forward blocking state, anode positive with respect to the cathode, and consider the effect of an applied positive gate voltage. With this gate polarity, the n-channel MOS transistor is conducting, whereas the p-channel MOS structure is non-conducting. Current flows through the n-channel MOS transistor and the pnp transistor, the $n^+pn$ transistor becomes active and, as a result, the thyristor turns on. In the on-state an electron–hole plasma floods the region between the $n^+$-and $p^+$-layers as a result of bipolar injection from both emitters. When a negative gate voltage is applied, the p-type layer is shorted to the $p^+$-region through the inversion channel. At this moment the injection efficiency of the $n^+$-emitter and the conditions for keeping the thyristor in the on-state are changed.

As discussed in Section 7.4.2, there is a minimum current density $J_H$, corresponding to the holding current of the structure. When the current density decreases below this value, the electron–hole plasma decays and the device turns off. That is, the partial $p^+np$ and $n^+pn$ structures change from saturation to the active state. Thus, the space charge region at junction $J_2$ is restored and the thyristor reverts to the blocking state. The holding current density depends on parameters of the structure such as the thickness of the layers, the concentration profiles and the

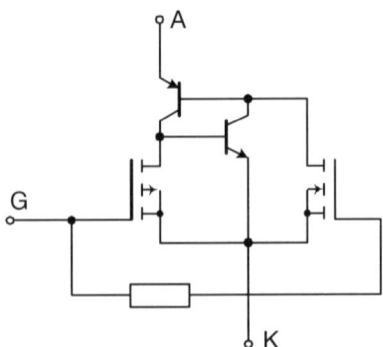

**Figure 11.19** An equivalent circuit for the p-channel MCT shown in Figure 3.15(c)

## 11.2 THE MOS-CONTROLLED THYRISTOR (MCT)

carrier lifetime. One of the most important of these is the density of the emitter shorts [11.19].

When the on-state current density in a structure with an unshorted emitter is less than that corresponding to the holding current with the emitter shorted by the p-type inversion channel, the condition for turn-on (7.31) is no longer met and the thyristor turns off. If the turn-off condition

$$J_T < J_{Hsh} \tag{11.17}$$

is not fulfilled, the thyristor structure stays in the on-state. This MCT turn-off condition is shown schematically in Figure 11.20.

The decay of the electron–hole plasma follows the interruption of electron injection from the n$^+$-emitter. An injection voltage $V_E$ of about 0.5–0.7 V is needed to keep the structure in the on-state. If the resistance shorting the n$^+$-emitter is $R_{sh}$ and with the hole current described by (11.10), condition (11.17) can be expressed as

$$\alpha_{pnp} I_p R_{sh} < V_E \tag{11.18}$$

In this case the resistance $R_{sh}$ is the sum of the lateral resistance of the p-base below the n$^+$-emitter and the channel of the gated MOS transistor. The maximum turn-off current of the MCT is $I_{TMoff}$ and it depends on the applied gate voltage. Because the current gain coefficient $\alpha_{pnp}$ decreases with the width of the n-base of the MCT, proper turn-off capabilities can be achieved even in structures with a high blocking voltage. The maximum turn-off current is given by

$$I_{TMoff} = \frac{V_E}{\alpha_{pnp} R_{sh}} \tag{11.19}$$

and through $\alpha_{pnp}$ it depends on the blocking voltage to which the MCT is turned off.

A considerable increase of the maximum turn-off current $I_{TMoff}$ can be obtained by replacing the p-channel MOS turn-off transistor with an n-channel device. The higher carrier mobility gives a lower channel resistance $R_{sh}$ for the same geometric

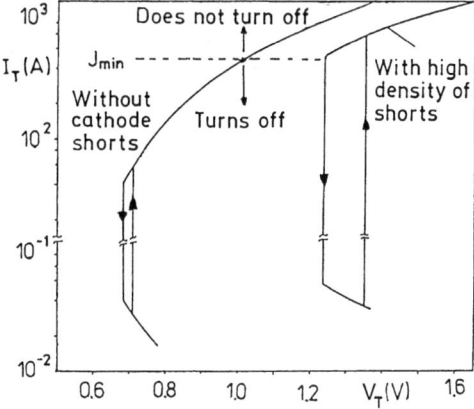

**Figure 11.20** A schematic illustration of the MCT turn-off condition

dimensions. This requires a change of the conductivity type of all the layers of the MCT structure. An example of an n-channel MCT is shown in Figure 3.15(a). The corresponding equivalent circuit is shown in Figure 11.21. Now the maximum turn-off current $I_{\text{TMoff}}$ may be expressed as

$$I_{\text{TMoff}} = \frac{V_E}{\alpha_{\text{npn}} R_{\text{sh}}} \quad (11.20)$$

Comparing n-channel and p-channel MCTs, in which the individual regions have similar dimensions and excess carrier lifetimes, the maximum turn-off current of the n-channel device is some 50% higher than that of the p-channel device. Therefore MCTs are presently constructed as n-channel devices. The disadvantage of MCTs with n-type turn-off channels is the need for the anode side to operate at low voltage.

The temperature dependence of $I_{\text{TMoff}}$ is also unfavourable, because the injection voltage $V_E$ decreases with temperature, while the current gain coefficient, $\alpha_{\text{pnp}}$ or $\alpha_{\text{npn}}$, and the resistance $R_{\text{sh}}$ both increase with temperature.

Typical current and voltage waveforms during the MCT turn-off process are shown in Figure 11.22. These too are influenced by temperature and the duration of the tail current increases considerably above 75 °C as a result of the temperature dependence of the carrier lifetime [11.20].

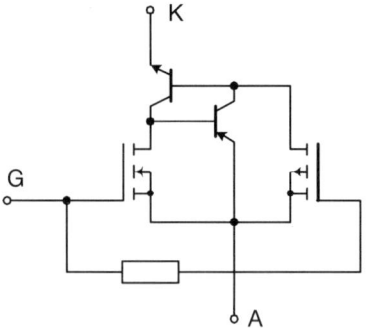

**Figure 11.21** An equivalent circuit for the n-channel MCT shown in Figure 3.15(a)

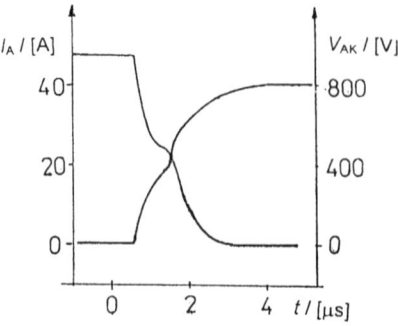

**Figure 11.22** Voltage and current waveforms during MCT turn off. Results taken from Reference [11.9]

MOS-controlled thyristors can also operate in a way that is analogous to the gate-assisted turn-off of conventional thyristors. A gate signal to turn on the turned-off channel is applied at the moment of the reverse recovery of the thyristor structure. It is expected that a p-channel MCT can successfully operate in this regime at above 100 kHz.

High power devices need the parallel connection of many individual MCT structures. In this respect, they are like IGBTs, VDMOSFETs and GTOs. An experimental, 80 mm diameter, 5 kV, 2 kA MCT [11.20] comprises 144 parallel devices, each consisting of 441 cells, turned on by a single n-channel MOS transistor. Each cell contains 62 elementary MCT structures turned off by the p-channel MOS transistor. An illustration of the cell geometry is given in Figure 11.23. For reliable operation, a very high level of homogeneity is necessary and such devices are at the research stage, as are several other MCT variants. These include the base resistance controlled thyristor (BRT), the emitter switched thyristor (EST) and dual-gate MCT structures [11.21]. With the present level of research by all major producers of power devices, future large-scale production of some of these devices is anticipated.

## 11.3 Other Bipolar–MOS Structures

IGBTs and MCTs are constructed as devices able to control high load power, having a high blocking voltage and high current-handling capacity. Some other voltage-controlled bipolar devices have been developed for power electronics applications, especially for use in power integrated circuits, as described in Chapter 12, but also as discrete devices.

### 11.3.1 The Lateral IGBT

Just as power D-MOS transistors can be designed with either vertical (VDMOS) or lateral (LDMOS) geometry, so too can the IGBT. An example of a lateral (LIGBT) structure is shown in Figure 11.24. The disadvantages are the larger area needed, in comparison with the vertical structure, to meet a given voltage rating and, as with LDMOS, the difficulty of connecting many elementary cells in parallel to form a power device. Consequently, the current-handling capacity is also much lower than

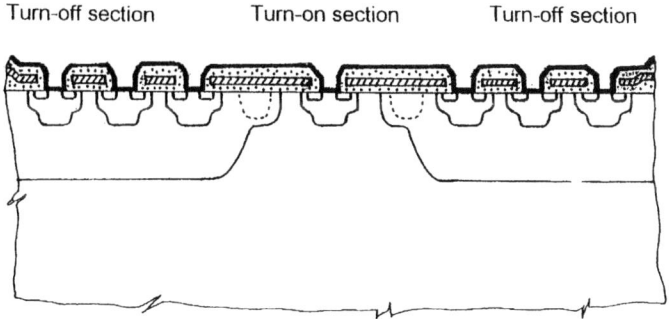

**Figure 11.23** A cross-section through a cellular MCT structure

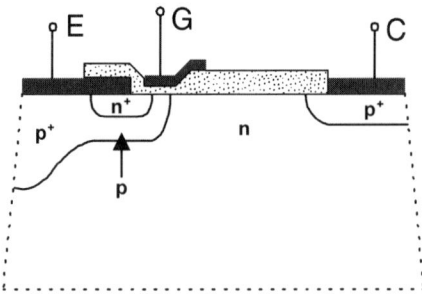

**Figure 11.24** The structure of a lateral IGBT [11.22]

for the vertical structure. Its advantage lies in its use as a component in power integrated circuits. The MOS-controlled thyristor can also be made in a lateral geometry, as shown in Figure 11.25.

### 11.3.2 The IBT

The insulating base transistor (IBT) shown in Figure 11.26 has a structure very similar to the IGBT, but the $p^+$–$n^+$ junction is a tunnelling junction and has no rectifying effect. As in the IGBT, bipolar injection modulates the conductivity of the p-base. Latch-up of the parasitic thyristor structure is eliminated by the effective short circuit of the $p^+$–$n^+$ junction so that, in principle, a higher on-state current density can be obtained. Blocking voltages up to 400 V are claimed.

### 11.3.3 Monolithic Integration of Parallel-Connected Devices

In the literature a number of structures have been described that are effectively a bipolar device and a voltage-controlled device, connected in parallel. Among them the following three may be mentioned, though it has to be said that none has been as widely applied as the IGBT, which remains one of the most important devices in power electronics.

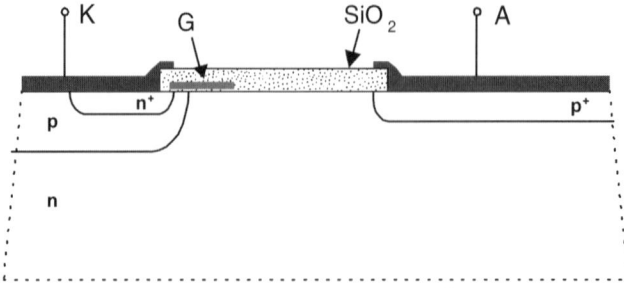

**Figure 11.25** The structure of a lateral-geometry MOS-controlled thyristor

## 11.3 OTHER BIPOLAR–MOS STRUCTURES

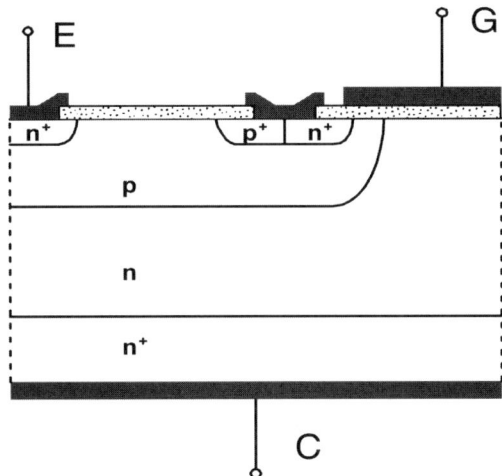

**Figure 11.26** The Insulating Base Transistor (IBT)

1. The structure shown in Figure 11.27 is a parallel connection of a VDMOS and a bipolar junction transistor [11.23]. For turn-on a pulse is applied simultaneously to both devices. The VDMOS transistor has a shorter delay time than the bipolar transistor and lower turn-on losses. The on-state current then flows through the bipolar transistor, which has a lower on-state voltage drop. When the device is turned off, the bipolar transistor is turned off first and then the VDMOS transistor. In practice the MOSFET conducts only during switching, when its turn-on and turn-off losses are lower than those of the bipolar transistor. In this way the device is able to combine the advantages of both the individual devices. Unfortunately, the drive circuits are rather complicated and the device has not been successful in competition with the IGBT.

2. The VDMOS-LIGBT or BIENFET structure [11.24] shown in Figure 11.28 is an integrated structure with a common emitter and gate. The drain of the VDMOS and the collector of the LIGBT are connected externally.

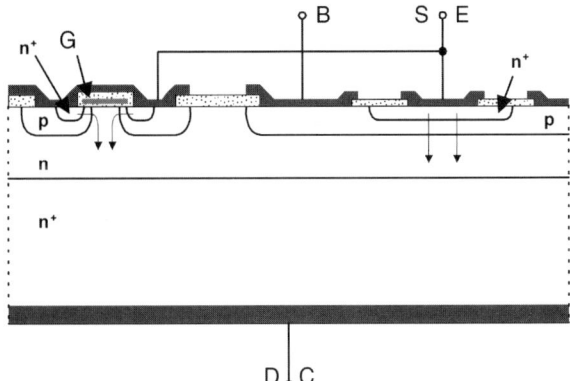

**Figure 11.27** Parallel integration of a bipolar and a VDMOS transistor

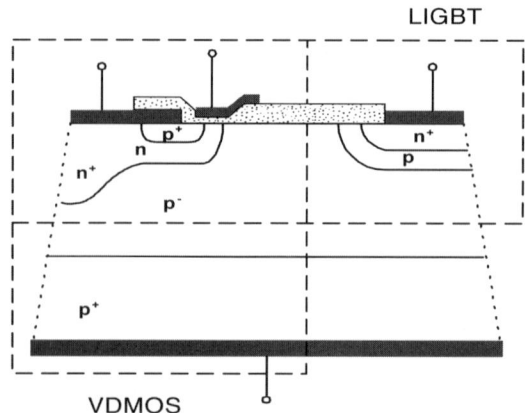

**Figure 11.28** The integration of an LIGBT and a VDMOS transistor (BIENFET)

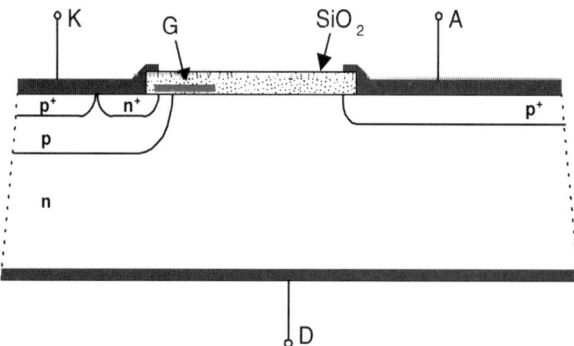

**Figure 11.29** The $T^2$-MOS structure

Integration decreases the total power loss in comparison with the use of individual devices.

3. The transistor-thyristor MOS device ($T^2$-MOS) is an integrated structure of an n-channel VDMOS transistor with a lateral MOS-controlled thyristor [11.25]. The thyristor part of the structure assures a low on-state voltage drop and a high on-state current density; the vertical MOS transistor accelerates the turn-off process of the lateral thyristor. The $T^2$-MOS structure is shown in Figure 11.29.

## Summary

The IGBT combines the high current and voltage ratings of all but the largest thyristors with the simple control circuitry of the power MOSFETs. Not surprisingly, it is the preferred main switching device for most medium power electronics applications and the range of power levels over which this is true is continually expanding. The ideal on-state characteristic approaches that of a diode in series with the additional voltage across the inverted MOS channel. Turn-on is fast and losses are low. The current tail during turn-off can give rise to high levels of dissipation and may cause the parasitic thyristor to latch, although present-day IGBTs are generally considered to be free from latching as long as they are operated within their SOA. High levels of turn-off dissipation reduce the SOA in hard-switching circuits, but operation at frequencies in the region of 100 kHz is possible for appropriately constructed devices.

The MCT is a more complex device that allows full voltage-controlled switching, turn-on and turn-off being controlled by separate MOS transistors of complementary type. The use of an n-channel MOS device as the turn-off transistor increases the maximum anode current that can be turned off, but requires the anode to be on the low voltage side of the circuit. Many variants of these devices are in the experimental stage of development.

# References

[11.1] Becke, H. W. and Wheatley, C. F. U.S. Patent 4,364,073 (1982).
[11.2] Baliga, B. J., Adler, M. S., Love, R. P., Gray, P. V. and Zommer, N. D. The Insulated Gate Transistor: A New Three-Terminal MOS-Controlled Bipolar Power Device, *IEEE Trans. on Electron Devices*, **ED-31**, 821–8 (1984).
[11.3] Russell, J. P., Goodman, A. M., Goodman, L. A. and Nielson, J. M. The COMFET – A New High Conductance MOS-Gated Device, *IEEE Electron Devices Letters*, **EDL-4**, 63–5 (1983).
[11.4] Gauen, K. *Motorola Application Note*, **AN-934** (1985).
[11.5] Benda, V. Physical phenomena during the thyristor turn-off process under conditions of the turn-off time measurements (in Czech), *Acta Polytechnica*, **9 (III-2)**, 79–95 (1980).
[11.6] Hefner, A. R. An Improved Understanding for the Transient Operation of the Power Insulated Gate Bipolar Transistor (IGBT), *PESC'89 Conference Record*, 303–13 (1989).
[11.7] Yilmaz, H. Cell Geometry Effect on IGT Latch-Up, *Electron Device Letters*, **EDL-6**, 419–21 (1985).
[11.8] Mori, M., Nakano, Y. and Tanaka, T. An Insulated Gate Bipolar Transistor with a Self-Aligned DMOS Structure, *IEEE International Electron Devices Meeting (IEDM) Technical Digest*, 813–16 (1988).
[11.9] Huggins, J. L., Menhart, S., Portnoy, W. M. and Sankaran, V. A. Temperature Variation Effects on the Switching Characteristics of MOS-Gate Devices, *Proc. EPE-MADEP Conference*, vol. 0, 262–6 (1991).
[11.10] Grant, D. A. A Comparison of Hard and Soft Switching of IGBTs, *Proc. ISPS'92*, 141–5 (Prague, 1992).
[11.11] Laska, T., Matschitsch, M. and Scholz, W. Ultra Thin Wafer Technology for a new 600 V NPT IGBT, *Proc. ISPSD'97*, 361–4 (1997).
[11.12] Bauer, F., Dettmer, H., Fichtner, W., Lendemann, H., Stockmeier, T. and Thiemann, U. Design Considerations and Characteristics of Rugged Punchthrough (PT) IGBTs with 4.5 kV Blocking Capability, *Proc. ISPSD'96*, 327–30 (1996).
[11.13] Tomomatsu, Y., Suekawa, E., Kondo, H., Enjoji, T., Takeda, M., Hagino, H. and Yamada, T. An Analysis for Transient Characteristics of Shorted Collector IGBTs in Free Wheeling Mode of Operation, *Proc. ISPSD'97*, 209–12 (1997).

[11.14] Minato, T., Thapar, N. and Baliga, B. J. Correlation between the Static and Dynamic Characteristics of the 4.5 kV Self-Aligned Trench IGBT, *Proc. ISPSD'97*, 89–92 (1997).

[11.15] Brunner, H., Hierholzer, M., Laska, T. and Porst, A. Progress in development of the 3.5 kV high voltage IGBT/diode chipset and 1200A module applications, *Proc. ISPSD'97*, 225–8 (1997).

[11.16] Takahashi, Y., Yoshikawa, K., Koga, T., Suotome, M., Takano, T., Kirihata, H. and Seki, Y. Ultra High-Power 2.5 kV – 1800 A Power Pack IGBT, *Proc. ISPSD'97*, 233–6 (1997).

[11.17] Goodenough, F. 150 kHz IGBTs Take on Power MOSFETs, *Electronic Design*, 37–40 (23 June 1997).

[11.18] Baliga, J. B. Enhancement and Depletion Mode Vertical Channel MOS Gated Thyristors, *Electronics Letters*, **15**, 645–7 (1979).

[11.19] Benda, V. Lateral Processes in a Thyristor at Small Forward Currents (in Czech), *Acta Polytechnica*, **10 (III-2)**, 59–67 (1978).

[11.20] Rousisvalle, C., Ferla, G. and Zani, P. E. High Power MOS-Controlled-Thyristor Using the Parallel Contacting Technology for Devices on the Same Wafer, *Proc. EPE-MADEP Conference*, vol. 0, 267–9 (1991).

[11.21] Kurlagunda, R. and Baliga, B. J. 600 V Dual Gate MCT Structures, *Proc. ISPSD'97*, 273–6 (1997).

[11.22] Oppermann, K.-G. and Stoisiek, M. Optimalisation of LIGBTs in a Dielectric Insulated IC-Technolgy Using a Switched Anode *Proc. ISPSD'97*, 239–42 (1997).

[11.23] Morse, J. D. and Navon, D. H. Optimized Design of a Merged Bipolar MOSFET Device, *IEEE Trans. on Electron Devices*, **ED-32**, 2277–81 (1986).

[11.24] Chow, T. P. and Baliga, B. J. Counter-Doping of MOS Channel (CDC) – New Technique of Improving Suppression of Latching in Insulating Gate Bipolar Transistor, *IEEE Electron Device Letters*, **EDL-6**, 29–31 (1988).

[11.25] Behrens, F. H., Charitat, G. and Rossel, P. Analysis of the vertical insulated base transistor, *Proc. EPE-MADEP Conference* vol. 0, 215–19 (1991).

# 12

# POWER MODULES AND INTEGRATED STRUCTURES

Power semiconductor devices are normally used in systems which control the change of electrical energy to other forms such as mechanical or thermal energy or, as discussed in Chapter 4, into a different form of electrical energy: from ac to dc, to ac of a different frequency, or to a different voltage. Such systems can be divided into several blocks, as shown schematically in Figure 12.1: a block of control logic, a block to transform the logic signals into gate signals for the power devices, a block of power semiconductor devices, the equipment being controlled, a block of protection devices (against overvoltage, overcurrent or overheating), and finally, a block designed to sense the correct functioning of the equipment, which is normally connected through a feedback circuit to the logic control block.

When a power electronic circuit is made from discrete power semiconductor devices, it normally occupies quite a large volume determined by the size of the

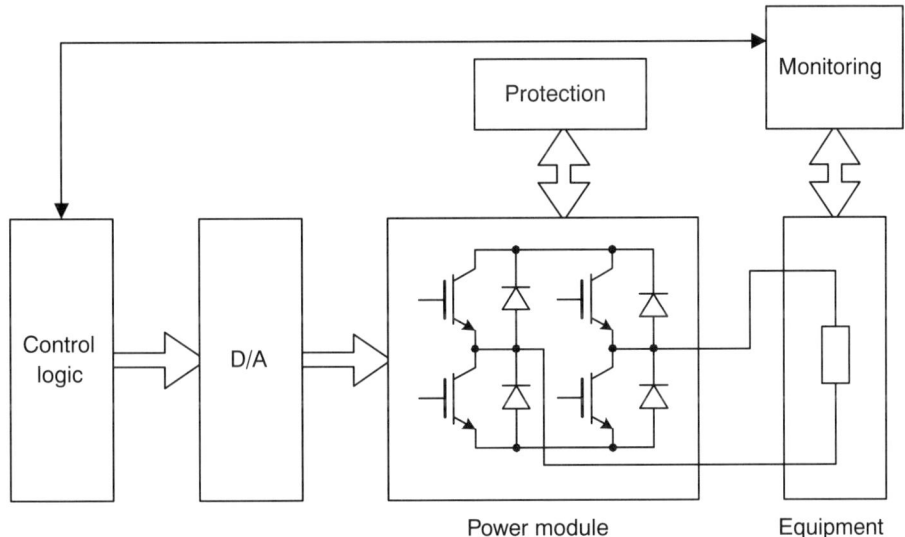

**Figure 12.1** Function blocks of a power electronic system

device housings and heat sinks. At the same time, many devices operate at very different potentials and so must be electrically insulated. This complicates heat transfer. Devices are connected by external conductors which introduce parasitic inductance. It is often possible to make the circuit simpler and its assembly more compact by packaging the individual devices together in functional units in a shared housing. This saves material (housings, heat sinks, insulating materials and interconnecting conductors) and reduces the volume of the equipment. Either monolithic or hybrid integration techniques can be used.

The level of integration of the equipment can be different in different parts. Several examples of integrated structures that are normally treated as a single device in circuit applications are cited in other chapters. These include the power Darlington and the RCT. Devices such as GTO thyristors, VDMOS transistors and IGBTs are also integrated structures consisting of many elementary low power elements connected in parallel on the chip. They also contain parasitic components such as the antiparallel diode of the VDMOS transistor and the thyristor inherent in the IGBT.

Devices can be connected together as functional units in different forms. Examples include

- Connection of several discrete power semiconductor devices in one case (module).
- On-chip integration of part of the logic and driving circuits with the power devices to form a power integrated circuit (PIC).
- Integration of logic circuits, transducers and drive circuits in a special application-specific integrated circuit (ASIC).
- Monolithic integration of a power device and sensors (e.g. temperature), complete with control circuits and, if need be, with overvoltage and overcurrent protection (smart power devices).
- Hybrid integration of power devices, sensors, control circuits and overvoltage and overcurrent protection in one housing known as an intelligent power module.

The basic principles of some of the most important integrated power device structures are discussed in this chapter.

## 12.1 Power Modules

When discrete power semiconductor devices are used in power electronic circuits, devices working at different potentials must be placed separately on electrically insulated heat sinks that are also isolated from the equipment frame.

Mechanical construction can be simplified if devices are placed on electrically insulating material that has a good thermal conductivity. Then the power semiconductor device can be electrically isolated from the radiator and the heat sink by the insulating layer, as shown schematically in Figure 12.2. In this way several devices can be located on a single radiator that does not have to be insulated from the equipment frame, and a function block consisting of two or more devices with appropriate internal connections can be built up in an isolated module. The

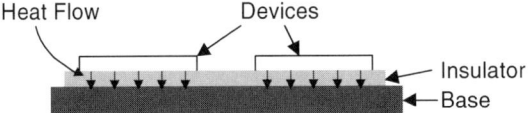

**Figure 12.2** Heat flow through an insulating substrate

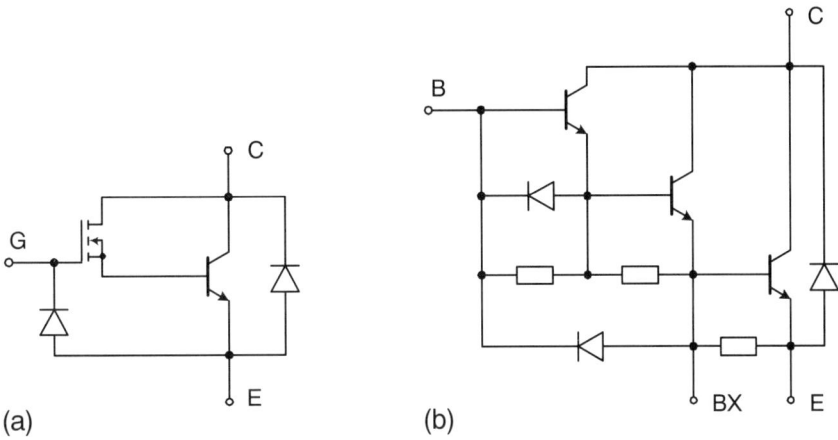

**Figure 12.3** Switching devices made in the form of power modules: (a) a MOS-Bipolar Darlington with anti-parallel diode; (b) a three-stage bipolar transistor Darlington

device parameters can be chosen so that the module is optimised for the particular application, and special devices are sometimes developed for this purpose.

A major design problem is to maximise the heat transfer through the electrically insulating layer. Usually, devices are placed on high thermal conductivity ceramic substrates such as copper-plated alumina wafers. A promising material is an AlN-based ceramic, which has a much higher thermal conductivity than alumina but is still rather expensive. Beryllia-based ceramics also have a high thermal conductivity, but the high toxicity of beryllium has caused producers to reduce their use. Further details of the technology of power modules are presented in Section 13.1.

It is often desirable to design a single switching device, such as a high power MOSFET or IGBT switch, as a power module consisting of the parallel connection of several devices with suitable parameters. Another example of a single device made in the form of a power module is a high power Darlington, possibly consisting of three or more stages. Examples are shown in Figure 12.3.

A different form of power module consists of two or more devices housed together. They form function blocks that can be used to create a range of power

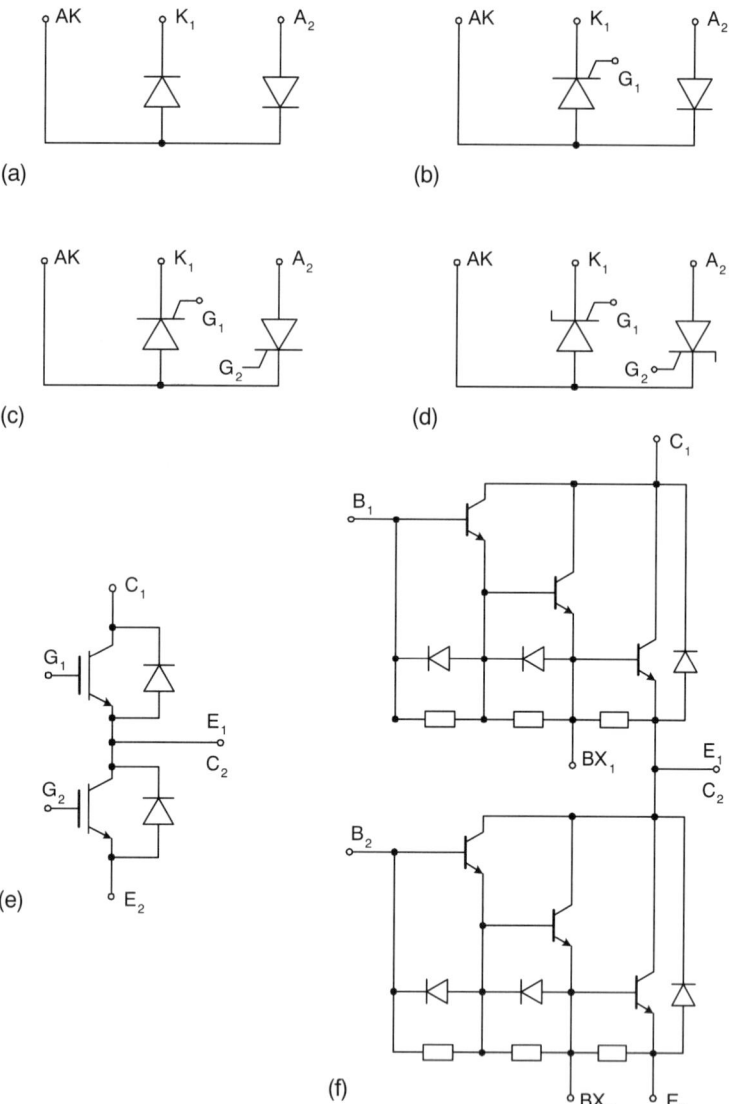

**Figure 12.4** Examples of function blocks made in the form of power modules: (a) diode pair; (b) diode and SCR; (c) pair of SCRs; (d) pair of RCTs; (e) inverter leg; (f) inverter leg using triple Darlingtons, with diodes to aid turn-off

electronic circuits by external connections. A few examples of the internal arrangement of such types of module are shown in Figure 12.4. External connections completing the function unit are usually made when the modules are mounted onto a common radiator.

Another possibility is to house the all the components of a basic power circuit, such as a bridge rectifier (controlled or uncontrolled), an ac switch or an inverter, in a common case. Some examples are shown in Figure 12.5.

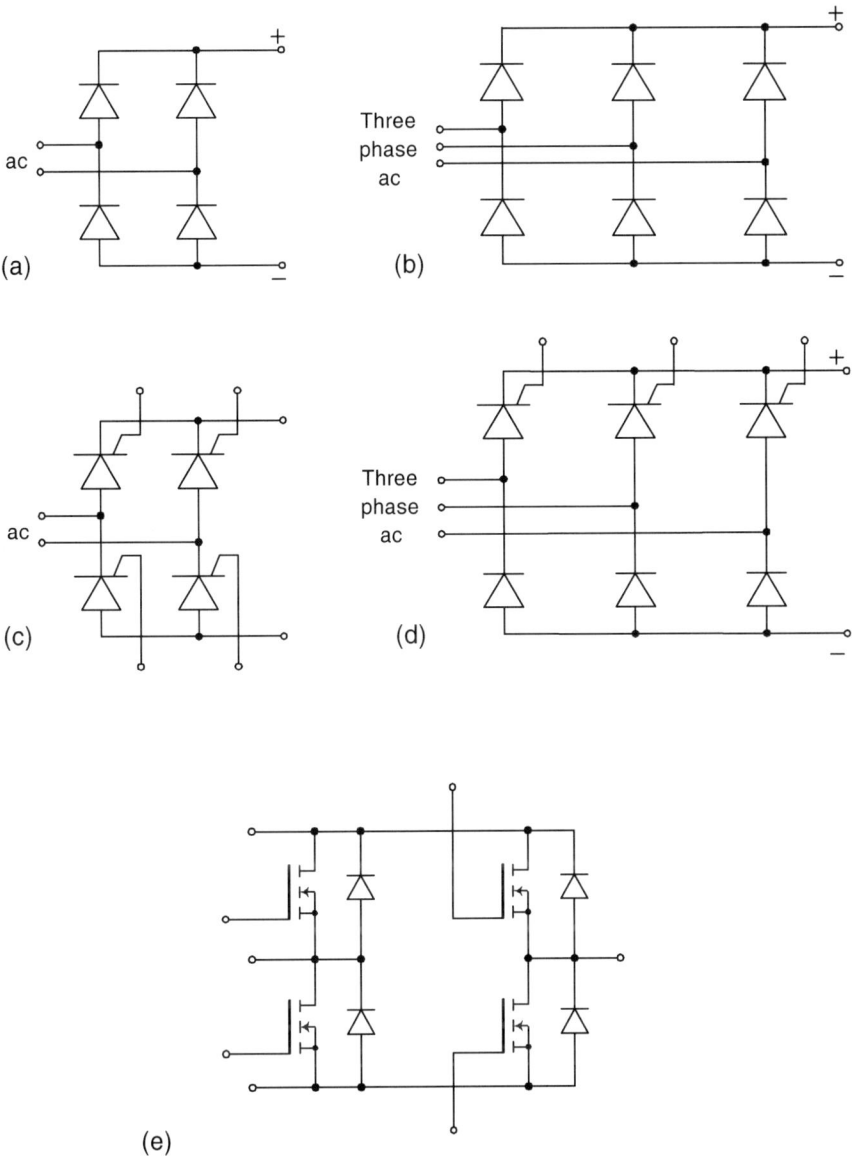

**Figure 12.5** Examples of power electronic circuits made in the form of power modules: (a) single-phase full-wave rectifier; (b) three-phase full-wave rectifier; (c) single-phase controlled rectifier; (d) three-phase half-controlled rectifier; (e) inverter

Hybrid technology can also be applied to combine the basic power circuit and the driving circuits into one common enclosure and so create a higher-level integrated system, as shown in Figure 12.6.

Dielectric insulation of devices placed on a ceramic substrate has been used for relay-like devices, such as ac or dc switches with fully insulated input (gate) and output (current) contacts. These solid-state relays (SSRs) are illustrated schematically

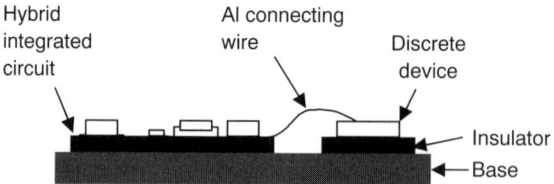

**Figure 12.6** A power hybrid integrated circuit

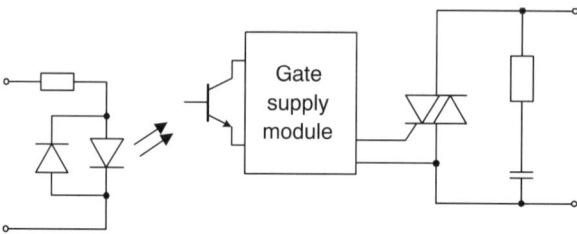

**Figure 12.7** A solid-state relay (SSR)

in Figure 12.7. They are faster and more reliable than conventional relays but suffer the disadvantage that they require cooling because of the relatively high on-state power dissipation. Solid-state relays are produced as switches for dc, single-phase and three-phase ac, with on-state currents in the range 1–10 A or, as power modules, up to 100 A. Control of the SSR is achieved as follows. The input current passes through an LED and the optical power generated is detected by an optically connected phototransistor. This triggers the gate voltage module, which generates a suitable current or voltage to turn on the power semiconductor device in the main circuit. Triacs are often used, as are power bipolar transistors, MOSFETs and IGBTs.

Another useful hybrid interconnection is the series combination of a power MOSFET and a bipolar junction transistor in the cascode circuit shown in Figure 12.8. When the power MOSFET is turned off, the emitter of the bipolar transistor is isolated, and its current gain coefficient is zero. Consequently, the breakdown voltage of the transistor $V_{CE(BR)} = V_{CBO} > V_{CEO}$. When the power MOSFET turns on, the base current $I_B$ starts to flow through the base–emitter circuit and the bipolar transistor turns on. During the turned-off period, the stored charge from the inner layers of the transistor diffuses out only through the base contact, as shown in Figure 12.9. Avalanche injection is minimised and the possibility of second breakdown is greatly reduced. The Zener diode protects the power MOSFET against overvoltage during the turn-off process and during the off-state. The current-handling capacity is limited by the power MOSFET but it is possible to use one with a relatively low breakdown voltage and a very low on-state resistance. The exclusion of the emitter operation during the turn-off process reduces the turn-off time of the power bipolar transistor and allows a maximum working frequency above 100 kHz [12.1].

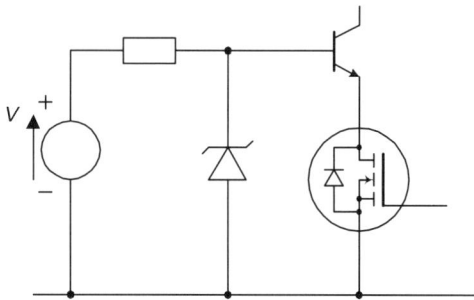

**Figure 12.8** The Cascode connection of a power MOSFET and a power bipolar junction transistor

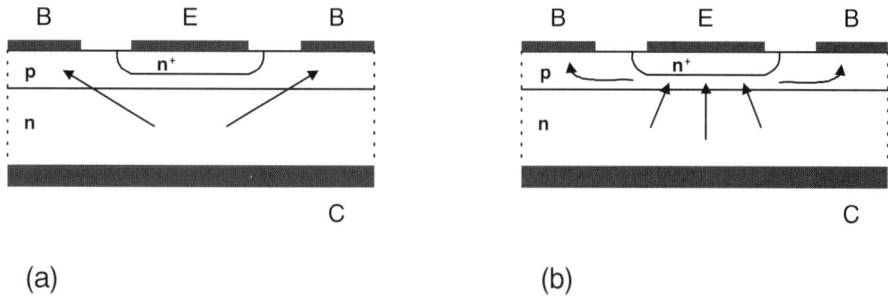

**Figure 12.9** A comparison of the diffusion of excess carriers during transistor turn-off: (a) in Cascode connection; (b) with a normal base drive connection

In the series (cascode) combination of a MOSFET and a bipolar junction transistor, all the collector current of the bipolar transistor is drained through the base contact during turn-off. Unless it is designed to accommodate this, reliability may be compromised. There are also high demands on the parameters of the Zener diode.

As well as the MOS–bipolar transistor cascode circuit, the series connection of a GTO thyristor and a power MOSFET can also be useful [12.2]; it is shown in Figure 12.10. This connection enables the GTO turn-off time to be shortened. Although the frequency dependence of the current-handling capacity can be improved, the circuit becomes very complicated for GTOs with high $I_{TGQM}$. High current MOSFETs suitable for the MOSFET–GTO cascode arrangement are the subject of research and development [12.3].

## 12.2 Power Integrated Circuits

The fabrication of power MOSFETs and IGBTs requires similar technological operations to those used in integrated circuit manufacture. For example, the diameter of a VDMOS cell is of the order of 5–10 $\mu$m, which is comparable in size to

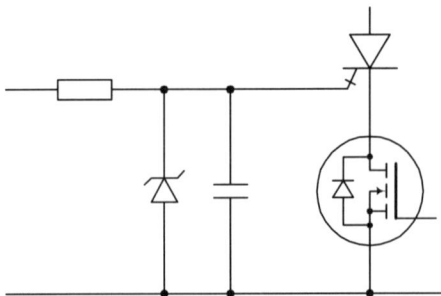

**Figure 12.10** A MOSFET/GTO Cascode arrangement

a CMOS logic cell. The gate voltage needed to turn on the power MOSFET is also similar. Consequently, it is possible to fabricate both integrated logic circuits and power devices on the same chip using the same processes and thus to integrate the driving function with the load power control.

The basic problem of power integrated circuits is electrically screening the low voltage logic circuits from the high voltage power devices. In addition, the power devices require a working temperature of about 150 °C, if the device area is to be exploited effectively. The logic circuits have to work in the same environment at what is for them an unusually high temperature.

In ordinary integrated circuits all contacts to the individual insulated devices are made on the same side of the silicon wafer, and insulated interconnections are formed using multilevel conducting tracks. Lateral power semiconductor structures such as LDMOS or LIGBT are suitable for this monolithic integration technology. As in ordinary integrated circuits, the areas occupied by these devices can be electrically isolated using a reverse-biased p–n junction. This is shown in Figure 12.11(a). An alternative is the self-insulating arrangement [12.4] shown in Figure 12.11(b). When LDMOS is used together with NMOS or $I^2L$ logic circuits, no electrical insulation is necessary but a relatively large area of silicon is needed.

Power integrated circuits with lateral power devices are suitable for applications with relatively low, controlled load currents, say up to about 1 A. For the switching of higher currents, vertical structures such as VDMOSFETs, bipolar transistors or IGBTs must be used. Then the technology of integrating power circuits is very complicated, requiring at least two cycles of epitaxial growth combined with the formation of buried layers. An example of such a structure is shown in Figure 12.12. For these reasons, the silicon-on-insulator (SOI) technique shown in Figure 12.13 is normally used at present for the logic control circuits [12.5] and is the subject of a great deal of research and development.

The configurations described are suitable for power integrated circuits, where only one vertical power device is used. If more are necessary, lateral devices are more suitable. Otherwise, vertical devices must be used, with all the electrodes on one side of the wafer and insulated from one another by means of a dielectric layer or a p–n junction, as shown in Figure 12.14.

## 12.2 POWER INTEGRATED CIRCUITS

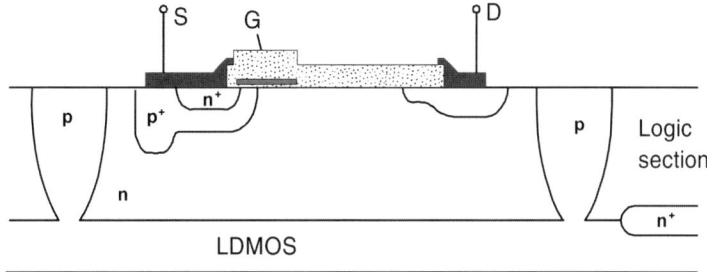

(a)

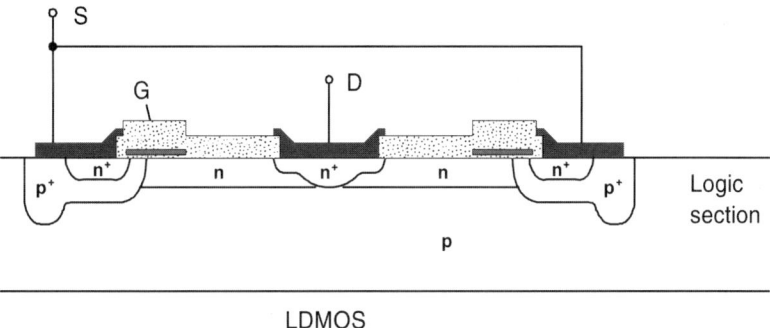

(b)

**Figure 12.11** Power integrated circuits with LDMOS: (a) with junction isolation; (b) a self-isolating arrangement

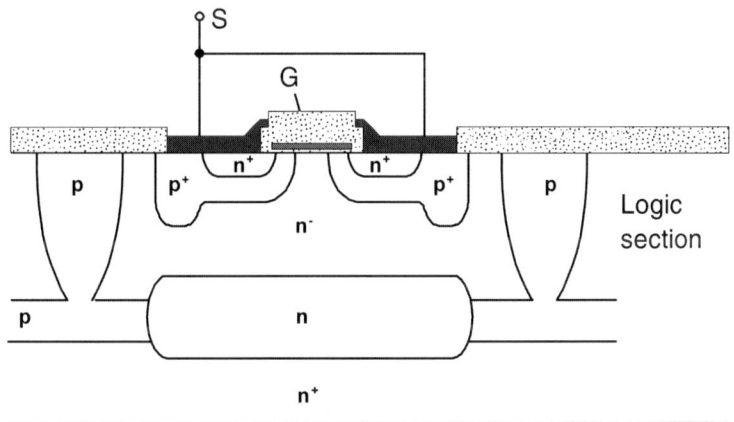

**Figure 12.12** A VDMOS power integrated circuit using a buried layer

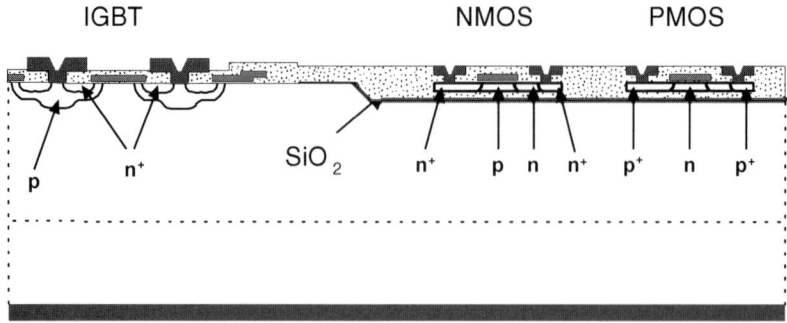

**Figure 12.13** A power integrated circuit with dielectric insulation of the logic circuits

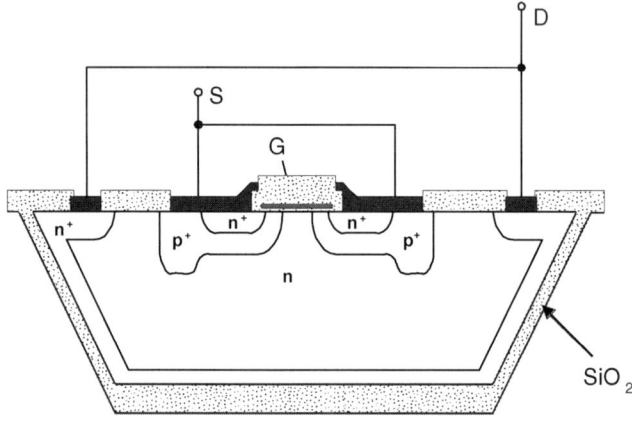

**Figure 12.14** A structure in which two or more power VDMOS transistors can be formed into an integrated circuit

From the viewpoint of economy, monolithic power integrated circuits need to be produced in large volumes. A very important application for power integrated circuits is vehicle electronics, where the automobile industry constitutes a very large market. They also have application in home electronics for the control of washing machines, microwave ovens, etc.

With current technology it is very complicated to make chips of area larger than about 400 mm$^2$. When 50% of the integrated circuit area is occupied by power devices and the rest by control functions, the power-handling capacity in the load is of order 1 kW. When higher power-handling capacity is required, or if very complicated circuits are to be used, the integrated circuit can be realised using hybrid technology, as indicated in Section 12.1. This also enables electrical isolation to be obtained by the use of transformers or optical coupling.

## 12.3 Smart Power: Intelligent Power Devices and Integrated Circuits

Integrated structures known as smart power devices, or intelligent power devices and circuits, differ from ordinary devices and integrated structures in that they integrate sensors of the device state (e.g. sensors of temperature, current or voltage) and incorporate an integrated circuit which actively protects the device or circuit against overvoltage, overcurrent or an excessive increase in temperature. In some cases the device itself can be used to detect unfavourable overloads and initiate protection against them. Then circuit solutions are simpler, heat sinks can be optimised more easily and the reliability of the device and the overall system is increased.

An integrated current sensor can take the form of a resistance of a few milliohms built into one of the contacts; for example, as part of the emitter metallisation in a bipolar transistor. The voltage drop across the resistance is amplified by an operating amplifier which forms part of the logic and control functions of the integrated structure. If the current exceeds a set level, the device is turned off or an external overcurrent protection relay is activated. A disadvantage of this type of current sensor is the need for auxiliary circuits such as a reference voltage source, operational amplifier, etc. There may also be a small power loss in the sensing resistance.

Another current sensor arrangement, which can be applied in power MOSFETs, uses the principle of the current mirror. The equivalent circuit is shown in Figure 12.15. The power VDMOS transistor consists of many parallel-connected VDMOS cells having common gate and source contacts on one surface of the silicon chip. After the gate signal is applied, a conducting channel, through which the drain–source current passes, is induced in each cell below the gate region. If a group of cells has an independent source contact, the current it carries is a proportional part (usually 1/1500) of the drain current. This current is passed through a sensing resistance connected between the sensing terminal (CS) and the Kelvin terminal (K), which is an auxiliary source contact, as shown in Figure 12.15. It generates a voltage proportional to the drain current. A more detailed analysis is given in the literature [12.6].

The ratio of the current in the current sensor to the drain current depends on the working regime of the device [12.7]. In the linear region, when the current is limited

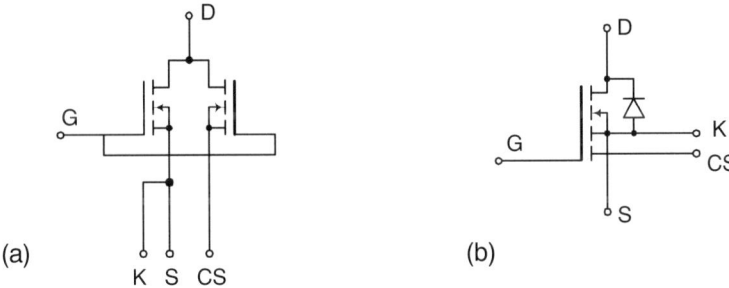

**Figure 12.15** A power MOSFET with an integrated current sensor: (a) device equivalent circuit; (b) device symbol

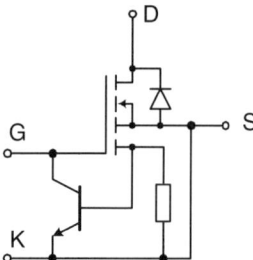

**Figure 12.16** A current-mirror current sensor

by the transconductance, it is divided in the ratio of the channel widths in the main and the sensing parts of the device. In the on-state, the current ratio is determined by the resistance ratio between the main and sensing parts. When monitoring fast rates of change of the drain current, the current ratios can change from the steady state. In addition, an inductive interaction between the main leads and those of the current sensor causes voltage spikes [12.8]. As a result, relatively large errors can occur.

With the current mirror sensor, it is possible to include the sensing resistance in the integrated structure to derive the overcurrent protection for the device [12.9]. The basic equivalent circuit is shown in Figure 12.16. An npn bipolar junction transistor (usually a lateral BJT) is connected between the gate and the Kelvin terminal of the power MOSFET. The sensing resistance is connected between the base and the emitter of the bipolar transistor. If the current is higher than a given limit, the voltage developed turns on the bipolar transistor, the gate voltage of the power MOSFET is reduced and the drain current reduced in consequence. Because the sensing resistance is made from silicon, its resistance increases and the current limit decreases with rising temperature. Similar systems of current monitoring can also be applied to IGBTs, including active overcurrent protection. This ability to monitor current and limit its maximum value decreases the likelihood of a current overload.

To protect against a possible overvoltage of the device, especially to limit voltage spikes connected with device switching in inductive circuits, external resistor/capacitor or resistor/diode/capacitor circuits are the classical methods used. Such snubber circuits occupy quite a large volume and increase the total circuit losses. Another method is to use avalanche diodes, Zener diodes or varistors as voltage limiters. An avalanche diode can be also integrated into the structure of a power semiconductor device. However, current flowing during avalanche breakdown dissipates a considerable amount of power, so there is the possibility that local overheating may disturb the thermal stability of the device. As a result, this type of integrated overvoltage protection can only guard against very short voltage spikes.

A more advanced technique is to create an active integrated overvoltage protection circuit using antiseries diodes connected between the drain (or collector) and the gate (or base) [12.10] of the main switching device, as illustrated by the equivalent circuit of Figure 12.17. When the drain voltage is higher than the breakdown voltage of the diodes, a positive voltage is applied to the gate, the device is turned on and it cannot be destroyed by overvoltage. The antiseries-connected

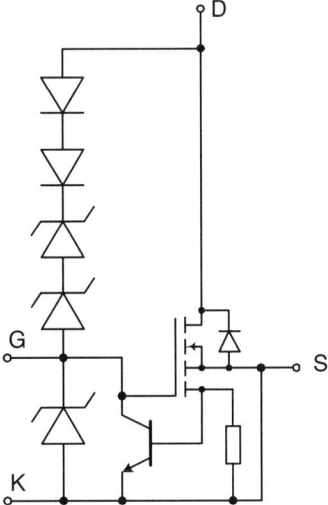

Figure 12.17 A power MOSFET with integrated overcurrent and overvoltage protection

diodes can be formed as a component of the integrated structure, e.g. from polycrystalline silicon [12.9], together with the Zener diode between the source and gate, which protects the gate oxide against breakdown. In this way it is possible to provide overcurrent and overvoltage protection for a power MOSFET or IGBT which also limits the dependence of the maximum on-state current on the device temperature.

The temperature in a power integrated structure can be sensed in several different ways. One is to monitor the temperature dependence of the forward voltage drop of a p–n junction. This decreases linearly with increasing temperature under constant current conditions. Another is to measure the exponential increase with temperature of the reverse current at the junction under constant voltage conditions. A third monitors the temperature-dependent resistance of a piece of single-crystal or polycrystalline silicon. The advantage of the last method is that the resistance sensor can be formed anywhere on top of the oxide layer and does not have to be sited in any special part of the device structure. Neither does it need a p–n junction to isolate it.

Such sensors and self-protection units are the basic properties of 'intelligent' devices. They are used in the design and production of application-oriented power integrated circuits known generically as smart power devices. These structures consist of power devices plus all the required logic and analogue devices, integrated into a single functional unit on a single chip. Reference voltage sources, operational amplifiers, analogue–digital and digital–analogue transducers, can all be formed within the integrated structure. Some of the functions that are fulfilled in smart power integrated structures are illustrated schematically in Figure 12.18.

## 12 POWER MODULES AND INTEGRATED STRUCTURES

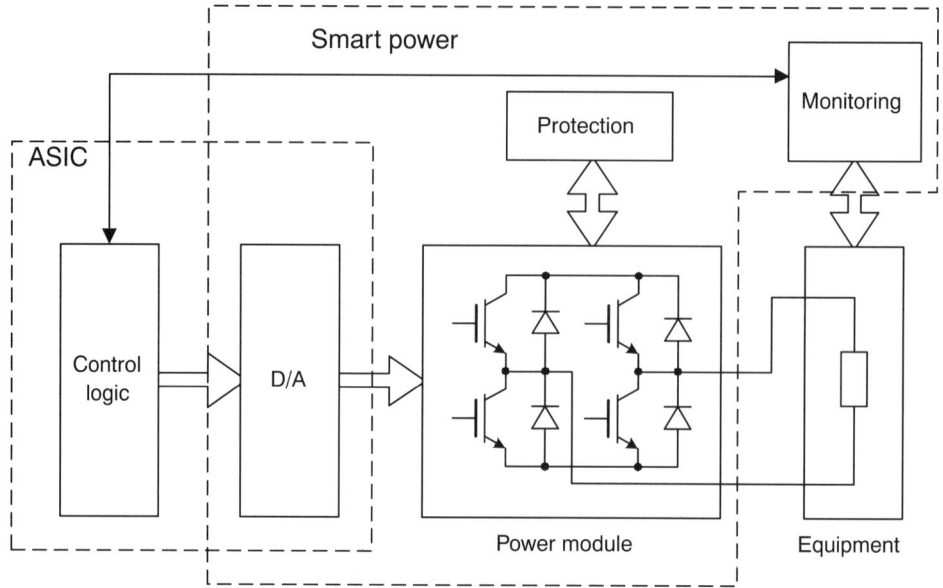

**Figure 12.18** The range of functions of a power integrated structure

The main circuit functions in which smart power devices are applied are

- pulse width modulation of output voltage
- drives of small motors
- controlled voltage or current sources
- special circuits for automobile electronics
- special circuits for home electronics

Monolithic smart power integrated structures have all the limitations of power integrated circuits, and consequently they can be used to control power only up to about 1 kW.

To control higher load power, hybrid technology facilitating the heat sinking and dissipation of much higher losses must be used. Such modules, which can fulfil similar circuit functions to smart devices, but up to levels exceeding 100 kW load power, tend to be known as intelligent power modules.

## Summary

Power semiconductor devices can be integrated together and with other components in functional units for different applications. This can be done in many ways: hybrid integration and the interconnection of multiple power devices in modules; monolithic integration of general-purpose logic and driver circuits in power integrated circuits; further integration of special components in application-specific integrated circuits; integration of built-in or external sensors for protection in smart power devices and intelligent power modules.

## References

[12.1] Penalver, C. M. and Farina, J. An Improved High Frequency Bi-MOS Switch. Application to 300 kHz. Switch Mode Supply, *Proc. EPE Conference*, 553–6 (1987).

[12.2] Clerc, G., Riotte, J. P. and Rolat, G. A New Step Towards the Ideal Switch: The GTO MOS Cascode, *Proc. EPE Conference*, 87–92 (1987).

[12.3] Oetjen, J. and Sittig, R. Hybrid 3000 A MOSFET for GTO Cascode Switches, *Proc. ISPSD'97*, 241–4 (1997).

[12.4] Frank, R. and Janikowski, R. Trends in Power IC development, *Proc. Power Conversion and Intelligent Motion (PCIM)*, **12**, 26–9 (1986).

[12.5] Gabriel, R. An Intrinsic Safe Smart Power IGBT, *Proc. EPE-MADEP Conference*, vol. 0, 179–82 (1991).

[12.6] Grant, D. A. and Gowar, J. *Power MOSFETs: Theory and Applications* (John Wiley & Sons, 1989).

[12.7] Grant, D. A. and Pearce, R. Dynamic Performance of Current-Sensing Power MOSFETs, *Electronics Letters*, **24**, 1129–31 (1988).

[12.8] Grant, D. A. and Williams, R. Current Sensing MOSFETs for Protection and Control, *IEE Colloquium on Power Semiconductors (Digest 1992/179)*, 7/1–5 (Birmingham, UK, 22nd October 1992).

[12.9] Franc, R. and Aloisi, P. Power Devices with Integrated Protection, *Proc. EPE-MADEP Conference*, vol. 0, 110–13 (1991).

[12.10] Castro Simas, M. I. and Simões Piedade, M. *IEEE PESC Record*, 69–75 (1987).

# 13

# CONDITIONS FOR RELIABLE OPERATION

Power semiconductor devices are often used in demanding industrial applications where any breakdown results in serious economic losses. For that reason, everybody who designs, constructs and operates power converters needs to know in detail the characteristics of the devices used and the way their parameters depend on the operating conditions. The proper packaging of devices and the avoidance of dangerous operating regimes is essential to ensure safe and reliable performance. This requires

- A suitable heat-sink system commensurate with the operating load power
- Observing the constraints on the connection of devices in series and parallel
- Protection against current and voltage overloading
- Exclusion or substantial limitation of operating regimes which can cause the device parameters to deteriorate

## 13.1 The Cooling of Power Semiconductor Devices

The power losses that arise in operation are transformed into heat and increase the device temperature. As discussed in Section 1.2.4, when the power losses in a semiconductor device exceed a critical value, thermal instability and thermal breakdown can occur. That aside, many device parameters are adversely influenced by a temperature increase. For example, the reverse current at a p–n junction increases exponentially, the turn-off time of a bipolar device rises and the breakover voltage of a thyristor falls.

The temperature range for satisfactory working is limited. The lower limit is $T_{Jmin}$, which is the lowest working or storage temperature of the device. This is determined by the mechanical construction of the device (e.g. the mechanical stress resulting from the differential expansion of materials) and the temperature dependence of its operating characteristics. The upper limit is $T_{Jmax}$, which is the highest silicon wafer temperature at which the device can operate repeatably or continuously. This may only be exceeded for the short period before overcurrent protection comes into effect, under either steady-state or transient operation. In order to avoid thermal breakdown, $T_{\text{Jmax}}$ is determined by (2.49), the condition for thermal stability. However, it is often set lower because of the temperature dependence of important device parameters such as the breakover voltage $V_{\text{D(BO)}}$, or $V_{\text{CEO}}$, or the turn-off time. By lowering the maximum operating temperature, degradation processes may be retarded and device reliability increased, at the cost of reducing the current-handling capacity of the device.

To keep their operating temperature within the range $T_{\text{Jmin}} < T_{\text{J}} < T_{\text{Jmax}}$, devices have to be cooled. That is, the power dissipated within the device has to be conducted to the surface and dispersed into the ambient. The efficiency with which this can be achieved directly affects the maximum current-handling capacity of the device and is discussed in detail in the following section. The problems associated with this have remained important from the introduction of power semiconductor technology [13.1] to the present [13.2]. It has been the subject of many books and papers [13.3].

### 13.1.1 Thermal Resistance and Transient Thermal Impedance

Power semiconductor devices are typically attached to a heat sink having good thermal contact with the ambient medium. In this section we examine the parameters that determine the temperature difference between the ambient and the interior of the device.

When thermal power is generated at a rate $P_v$ per unit volume in a homogeneous medium of specific heat $c$, density $\rho_m$, and thermal conductivity $\kappa$, the distribution of the temperature $T(x,y,z,t)$ must satisfy the equation

$$P_v(x,y,z,t) = \rho_m c \frac{dT(x,y,z,t)}{dt} - \kappa \, \text{div} \, \textbf{grad}[T(x,y,z,t)] \tag{13.1}$$

In any practical situation, especially where two or more materials are involved and $\rho_m$ and $c$ are functions of position, the solution of (13.1) is very complicated. However, the evolution of the temperature distribution can be found using simplified models.

In a one-dimensional, steady-state analysis, the rate of transfer of thermal energy $P_{\text{th}}$ along a homogeneous rod of uniform cross-section $A$, length $l$, and thermal conductivity $\kappa$, can be expressed as

$$P_{\text{th}} = \kappa A \frac{dT}{dx} = \kappa A \frac{\Delta T}{l} = \frac{\Delta T}{R_{\text{th}}} \tag{13.2}$$

where $\Delta T$ is the overall temperature dropped along the rod and $R_{\text{th}}$ is known as the thermal resistance of the rod. It is assumed that no heat is lost from the sides of the rod. We can make an analogy with Ohm's law by thinking of the temperature difference as being like the potential difference in an electrical circuit and the flux of thermal energy $P_{\text{th}}$ as being analogous to electric current. Likewise, the thermal

## 13.1 COOLING POWER SEMICONDUCTORS

capacity $C_{th}$ of a volume $V$ of material can be deemed analogous to the capacitance in an electrical circuit:

$$C_{th} = \rho_m c V \qquad (13.3)$$

If heat is generated uniformly throughout the volume at the constant rate $P_v$, then

$$P_v V = \rho_m c V \frac{dT}{dt} = C_{Th} \frac{dT}{dt} \qquad (13.4)$$

Note that $P_{th}$ is a power (W), and $P_v$ is a power density (W/m³).

The transient processes involved in the transfer of heat through a body, such as a semiconductor wafer and its heat sink, while simultaneously heating it, can be modelled in terms of the charging of the thermal capacitance through the thermal resistance, as in the equivalent electrical circuit shown in Figure 13.1.

The heat is generated within the silicon chip and it is normally dissipated by means of a heat sink into the ambient surroundings, which are assumed to be at a temperature $T_A$. The thermal system which conveys the heat can be represented as a network of thermal resistances and thermal capacitances. Assume that a power $P_J$ is generated at a device junction, where the temperature is $T_J$. Then

$$T_J = P_j Z_{thJA} + T_A \leqslant T_{Jmax} \qquad (13.5)$$

where $Z_{thJA}$ is known as the transient thermal impedance between the junction and the ambient. Because of the thermal capacitance of the case and the heat sink, $Z_{thJA}$ is a function of time. It increases with time and eventually reaches its maximum value $R_{thJA}$.

Using the analogy of a thermal circuit consisting of partial thermal resistances and capacitances enables solutions to be found for complex systems involving a power semiconductor device and its heat sink. An example is shown in Figure 13.2, where $R_{thSi}$ and $C_{thSi}$ are the thermal resistance and capacitance of the silicon wafer, $R_{thD}$ and $C_{thD}$ are the thermal resistance and capacitance of the expansion plate, $R_{thB}$ and $C_{thB}$ are the thermal resistance and capacitance of the base of the device case, $R_{thBS}$ is the thermal resistance of the contact between the case and the heat sink, $C_{thS}$ is the thermal capacitance of the heat sink and $R_{thSA}$ is the thermal resistance between the heat sink and the surrounding ambient.

Because of the difficulty of measuring the temperature at the various points, the simplified equivalent circuit shown in Figure 13.3 is usually used, where $R_{thJC}$ is the thermal resistance between the junction and the surface of the case in contact with

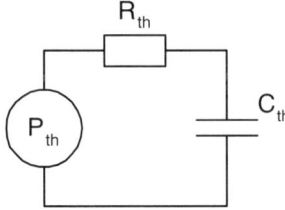

**Figure 13.1** A simple equivalent circuit analogue for representing thermal transfer

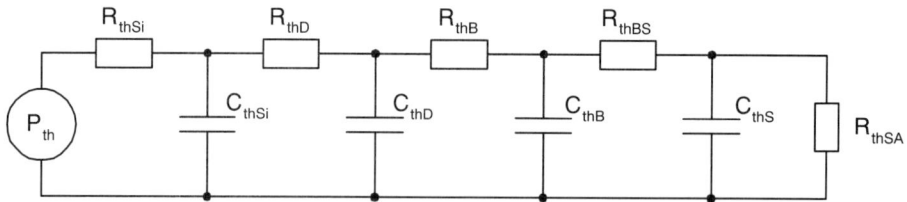

**Figure 13.2** An equivalent electrical circuit to represent the thermal impedance of the device/heat-sink system under transient thermal conditions

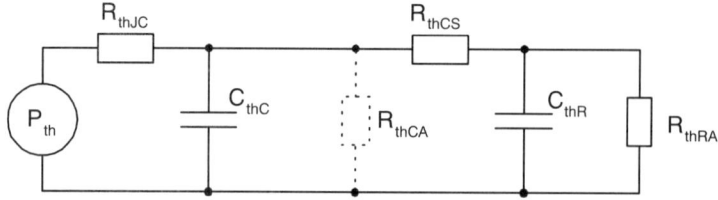

**Figure 13.3** A simplified equivalent circuit for the device/heat-sink system

the heat sink and $C_{thC}$ is the corresponding thermal capacitance. They depend on the case material and its construction. Under steady-state conditions the thermal resistance between the junction, whose temperature is $T_J$, and the surrounding ambient at the temperature $T_A$, is given by

$$R_{thJA} = R_{thJC} + R_{thCS} + R_{thSA} = \frac{T_J - T_A}{P_z} \qquad (13.6)$$

In equation (13.6) $P_z$ is the total power generated in the device; $R_{thJC}=(T_J-T_C)/P_z$ is the contact thermal resistance that depends on the interface between the device and the heat sink, and $T_C$ is the temperature of the device case at its contact with the heat sink; $R_{thCS}=(T_C-T_S)/P_z$ is the thermal resistance of the interface between the case and the heat sink, i.e. between the device case surface and the surface of the heat sink where the temperature is $T_S$; $R_{thSA}=(T_S-T_A)/P_z$ is the thermal resistance of the heat sink, i.e. between the contact with the device where the temperature is highest and the ambient of temperature $T_A$. $R_{thSA}$ depends on the volume, shape and material of the heat sink, on its surface finish and on the cooling medium.

Some heat is dissipated directly into the ambient from the device case, and a thermal resistance $R_{thCA}$ can be defined, as indicated by the dashed line in Figure 13.3. This thermal resistance is usually quite high. Although it is important when the device is operated without a heat sink, it can be neglected when a heat sink is connected.

For a defined combination of device, heat sink and cooling medium (in the case of forced cooling, a given speed of airflow), the maximum steady-state power loss is given by

## 13.1 COOLING POWER SEMICONDUCTORS

$$P_{max} = \frac{T_{Jmax} - T_A}{R_{thJA}} = \frac{T_{Jmax} - T_C}{R_{thJC}} \qquad (13.7)$$

Equation (13.7) relates the current-handling capacity of the device in a given circuit to the ambient temperature and to the method of the cooling as expressed by the thermal resistance $R_{thJA}$.

Equations (13.6) and (13.7) refer to a situation in which a steady-state temperature distribution is established in all parts of the system. If the dissipation varies on a timescale shorter than is needed to establish a steady state, thermal capacitance influences the heat transfer. The transient thermal impedance defined in (13.5) must be used to describe the variation of the junction temperature. The junction–ambient thermal impedance can be defined as

$$Z_{thJA}(t) = \frac{T_J(t) - T_A}{P_z} \qquad (13.8a)$$

the junction–case thermal impedance as

$$Z_{thJC}(t) = \frac{T_J(t) - T_C}{P_z} \qquad (13.8b)$$

and the heat sink–ambient thermal impedance as

$$Z_{thSA}(t) = \frac{T_S(t) - T_A}{P_z} \qquad (13.8c)$$

Because there is negligible thermal capacitance at the case–heat sink contact, the transient thermal impedance is just the thermal resistance $R_{thCS}$. Examples of the time dependence of the transient thermal impedances $Z_{thJC}$ and $Z_{thJA}$ are shown in Figure 13.4.

The transient thermal impedance is always less than the steady-state thermal resistance, so it may be possible, with a given cooling system, to operate devices for

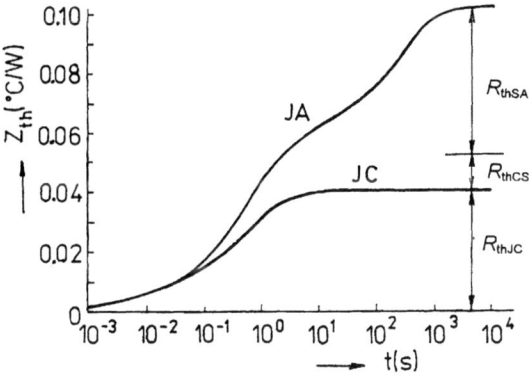

**Figure 13.4** An example of a typical time dependence of the transient thermal impedences $Z_{thJC}$ and $Z_{thJA}$. Reproduced by permission of Polovodiče, a. s., Prague

short periods with higher power losses, and therefore higher on-state currents, than can be maintained indefinitely.

If the current-handling capacity and reliability of power semiconductor devices are to be maximised, all the components which make up the thermal impedance have to be minimised. They are discussed in detail in the following sections.

### 13.1.2 The Encapsulation of Power Semiconductor Devices

The power semiconductor devices and integrated structures described in previous chapters have to be housed in such a way that they can easily be connected into a circuit and, at the same time, have the power losses efficiently conducted away to an appropriate heat sink. Furthermore, it is often necessary to isolate semiconductor structures from the effects of the external environment. This can adversely influence the characteristics and functionality of a device through chemical, physical or mechanical degradation. For all of these reasons, the encapsulation of power semiconductor devices is very important.

The case has to ensure several properties:

- A sufficient mechanical resistance against external pressure, impact and the influence of vibrations

- Connections that resist mechanical tension

- A seal that resists the ingress of moisture, dust and corrosive atmosphere

- Long-term thermal stability

- As appropriate, good electrical and thermal conductivity and good electrical and thermal insulation

- Minimum parasitic inductance and capacitance

- Adequate resistance to thermal cycling

Several types of case are used; the choice depends mainly on the device dissipation.

Lower power discrete devices are normally enclosed in cases cooled from one side. Often the main electrode is connected to a threaded bolt which enables the device to be stud-mounted to the heat-sink body, an easy method of assembly. The inner construction depends on the nominal current. Diodes and thyristors rated at up to a few tens of amperes have an active area less than $50\,\text{mm}^2$ and can be soldered directly to the base of the case with a soft lead solder. Mechanical stress caused by the differential expansion between silicon and the metal of the case during thermal cycling is insufficient to affect the device parameters under normal storage and operating conditions. A section through this type of device case is shown in Figure 13.5. The base of the case is one of the main terminals to be connected in the electric circuit and is the main conduction path to the heat sink. The upper part of the case is hermetically sealed to the base, usually by means of a resistive weld. The terminals are insulated using glass or ceramic.

Two popular cases for low power devices are the TO220 and TO3 types shown in Figure 13.6. In both, the semiconductor is soldered to the base through which the

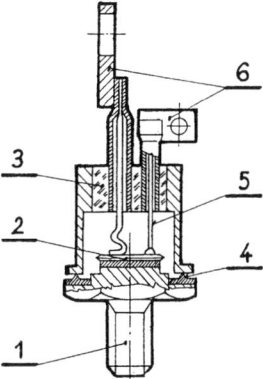

**Figure 13.5** A section through the case of a stud-mounted, soldered semiconductor device: (1) is the threaded stud; (2) is the semiconductor chip; (3) is the upper part of the case with insulated outlets; (4) is the connection between the upper part and the base of the case; (5) is a connection lead to one of the electrodes; (6) are the outlet connection tags. Reproduced by permission of Polovodiče, a. s., Prague

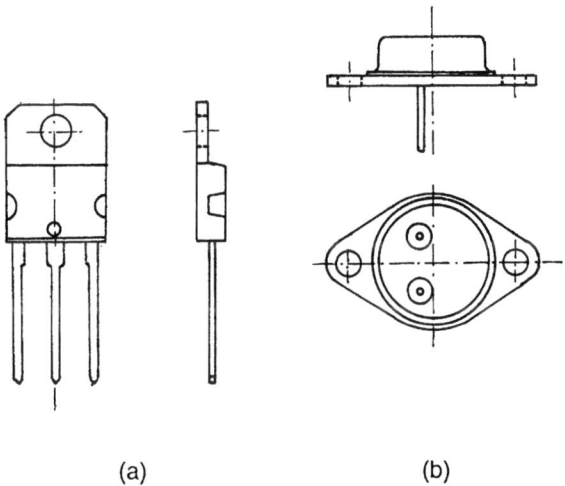

(a)            (b)

**Figure 13.6** Cases for low-power devices: (a) TO220; (b) TO3

device is cooled. In the TO220 and similar cases, the device is encapsulated in plastic. In the TO3 case, it is hermetically sealed in a metal enclosure.

The important physical properties of the materials that may be used for encapsulation are listed in Table 13.1.

The thermal expansion of silicon is less than that of any other material considered. If a silicon device is soldered directly to a copper base, the temperature difference between the solder melting temperature and room temperature causes considerable mechanical stress in the silicon as a result of the large difference in the thermal expansion coefficients of the two materials. Only in small-area devices can the silicon

**Table 13.1** Selected physical parameters of some materials important for power electronic devices

| Material | Thermal expansion coefficient, $10^6 \alpha$ (K$^{-1}$) | Thermal conductivity, $\kappa$ (W m$^{-1}$ K$^{-1}$) | Volume thermal capacity, $C_v$ ($10^6$ J m$^{-3}$ K$^{-1}$) | Electrical resistivity, $\rho$ ($\Omega$ m) |
|---|---|---|---|---|
| Silicon (Si) | 2.6 | 145 | 1.75 | $10^{-6}$ to $10^2$ |
| Gallium arsenide (GaAs) | 5.5 | 50 | 2.0 | $10^{-6}$ to $10^2$ |
| Silicon carbide (SiC) | 3.3 | 490 | 2.35 | $10^{-6}$ to $10^1$ |
| Molybdenum (Mo) | 5.4 | 152 | 2.75 | $5 \times 10^{-8}$ |
| Tungsten (W) | 4.5 | 167 | 2.75 | $5.3 \times 10^{-8}$ |
| Aluminium (Al) | 23.1 | 233 | 2.42 | $2.7 \times 10^{-8}$ |
| Gold (Au) | 14.2 | 294 | 2.50 | $2.35 \times 10^{-8}$ |
| Silver (Ag) | 19.6 | 417 | 2.46 | $1.6 \times 10^{-8}$ |
| Copper (Cu) | 17.3 | 382 | 3.42 | $1.72 \times 10^{-8}$ |
| Kovar | 4.6–5.2 | 16–20 | 3.7–5.4 | $2.1 \times 10^{-7}$ |
| Lead (Pb) | 29.0 | 37.4 | 1.46 | $1.9 \times 10^{-7}$ |
| Al/SiC metal matrix composite (70%/30%) | 6.8 | 180 | 2.37 | |
| Mica | 9–13 | 0.3–0.5 | 0.54–0.66 | $10^{13}$ to $10^{15}$ |
| Alumina (Al$_2$O$_3$) | 7.0 | 25 | 3.31 | $10^{12}$ |
| Beryllia (BeO) | 6.5 | 200 | 3.68 | $10^{11}$ to $10^{13}$ |
| Aluminium nitride (AlN) | 3.9 | 180 | 3.02 | $10^9$ |
| Silicone grease | | 0.67 | | $10^{13}$ |

chips be directly soldered to a copper base. For larger chip areas, say for devices with a nominal current of a few tens of amperes, a better expansion match is required if thermal fatigue is to be avoided. A layer of a material with an intermediate thermal expansion coefficient is placed between the copper base of the case and the silicon wafer. Of the possible materials, tungsten is expensive and the thermal and electrical conductivities of Kovar are too low. At present, molybdenum is normally used. The wafer is either soldered to a molybdenum or tungsten disc, or pressed between two molybdenum discs without soldering. A compressive force of about 10 N/mm² ensures good thermal and electrical connection to the copper case. The difference in the thermal expansion coefficients of the copper base and the molybdenum disc is compensated by allowing the compressed surfaces to slide.

For devices with a nominal current rating up to 400 A, a single-sided, stud-mounted case is often used. The wafer is usually soldered on one side to a molybdenum backing plate using Al–Si eutectic alloy. On the opposite side of the chip, a second molybdenum disc is compressed between the silicon and the copper case using internal springs, as shown in Figure 13.7. Like the TO220 and TO3 cases, the internal thermal resistance $R_{thJC}$ is determined by the materials and construction of the case and cannot be altered by the user, unless the outer surface which contacts the heat sink is mechanically or chemically damaged.

For the dissipation of higher power losses, a different design is needed. The most common solution is the hockey puck, a flat disc cooled on both sides as shown in

## 13.1 COOLING POWER SEMICONDUCTORS

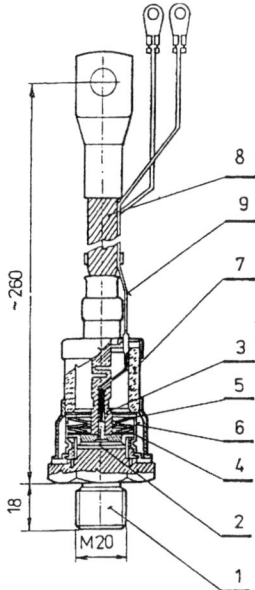

**Figure 13.7** A section through a stud-type case with compression springs: (1) copper base; (2) chip; (3) gate terminal; (4) isolator; (5) spring; (6) collar; (7) ceramic envelope; (8) main cathode lead; (9) gate lead. Reproduced by permission of Polovodiče, a. s., Prague

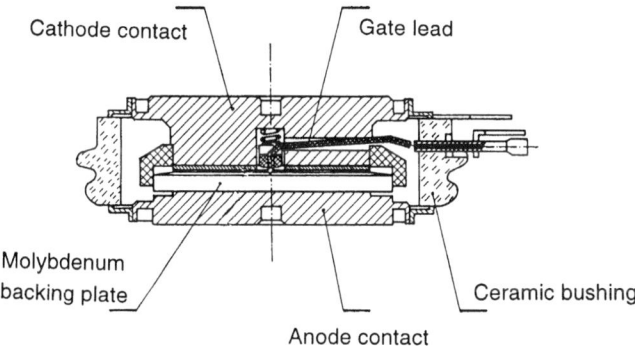

**Figure 13.8** A section through a hockey puck case cooled from both sides. Reproduced by permission of Polovodiče, a. s., Prague

Figure 13.8. The case consists of two ring-shaped copper electrodes which serve as electrical and thermal connections. Both electrodes are flexibly connected to an insulating ring, usually made from alumina ceramic.

To ensure low thermal and electrical resistance, a controlled, external, axial compressive force is applied. As all the thermal parameters, including internal thermal resistances, depend on this compression, the assembly of the device and heat sink must be carried out carefully, with the pressure evenly distributed over the whole area. Figure 13.9 illustrates the effect of the thermal resistance caused by dry

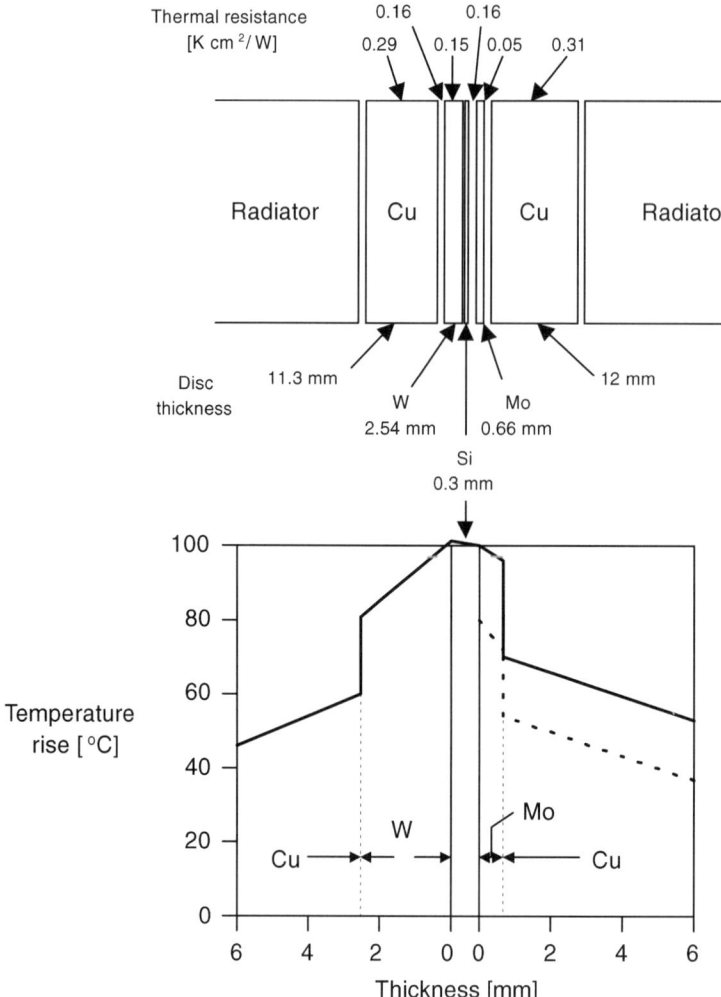

**Figure 13.9** The thermal path and a typical temperature distribution in a compression package cooled from both sides. Example taken from M. S. Adler and V. A. K. Temple, 'Analysis and Design of High Power Rectifiers', in *Semiconductor Devices for Power Conditioning*, Eds. R. Sittig and P. Roggwiller, (Plenum Press, New York, 1982)

interfaces in a package cooled from both sides. The steady-state dissipation in the wafer is adjusted to give a peak temperature rise of 100 °C. The wafer is alloyed to the tungsten backing plate and the solid curve shows the temperature distribution when it is also soldered to the molybdenum strain buffer. Sharp temperature drops occur at the dry interfaces with the copper contacts. When the molybdenum disc is made fully sliding (dashed curve) the dry interface with the wafer introduces additional thermal resistance with the result that the same temperature rise is obtained with 10% less dissipation. Wafers less than 75 mm in diameter are normally soldered. However, the expansion mismatch causes a significant bow on cooling from the soldering temperature (nearly 600 °C) to room temperature and this increases the thermal resistance of the interface. Alloy-free compression is more suitable for larger devices and is used for GTO thyristors up to 150 mm in diameter. An alternative [13.4] technique is to join the wafer to a molybdenum disc with a layer of silver powder sintered at 200 °C at very high pressure.

A hockey-puck encapsulated device can be cooled from the cathode side, from the anode side or from both sides. The internal arrangement shown in Figure 13.8 means that the thermal resistance on the cathode side, $R_{thJCK}$, differs slightly from that on the anode side, $R_{thJCA}$. Usually $R_{thJCK} > R_{thJCA}$. The transient thermal impedances, $Z_{thJCK}(t)$ and $Z_{thJCA}(t)$ are also different.

With double-sided cooling, the total thermal resistance results from the parallel connection of the two thermal circuits (cathode/radiator/ambient and anode/radiator/ambient) and is given by

$$R_{thJA} = \frac{R_{thJAK} R_{thJAA}}{R_{thJAK} + R_{thJAA}} \tag{13.9}$$

A similar expression is valid for the overall thermal impedance.

Hockey-puck encapsulation is used for devices required to dissipate power losses in the range from hundreds to several thousands of watts, depending on the device area.

Up to now it has been assumed that heat is generated homogeneously in the volume of the silicon device. This is not always valid. For example, in a thyristor during the turn-on process, current flows through only a part of the device area. During this period the thermal resistance is higher. As a result, at higher working frequencies when transient processes take up a higher fraction of the operating cycle, the internal junction–case thermal resistance $R_{thJC}$ is increased.

For all types of encapsulation there are conflicting demands on the construction of the case. It needs good mechanical strength for robustness, hence relatively thick layers of materials have to be used. It should present a low thermal resistance, so the materials in the cooling path should be as thin as possible and of high thermal conductivity.

From the viewpoint of device reliability, the case must ensure internal surroundings that cause no corrosion or contamination of the device surface or its contacts. It is necessary to exclude all materials that may be corrosive or easily ionised at higher temperatures or during ageing. To ensure the long-term stability of device parameters, it is necessary to exclude materials which may age at working temperatures and whose physical and mechanical properties, such as elasticity, mechanical strength, thermal conductivity, or electrical strength, may deteriorate with time. It is necessary to make a simple, robust and reliable connection to the case, and both case and connection should have high electrical and thermal conductivities. It may require an external anticorrosion coating and its aesthetic design may be important.

The case of a high power device is normally filled with a dry inert atmosphere and sealed. Some types of medium and low power devices are encapsulated in plastic, e.g. the TO220 case shown in Figure 13.6(a). The silicon chip is then not fully isolated from the ambient, the case is not hermetically sealed, and the surface of the chip must be well passivated, perhaps with glass.

Chapter 12 considers the use of hybrid technology. Several discrete devices or integrated structures are connected together as one function block, placed on an insulating substrate and encapsulated in a single enclosure, usually plastic, to form a power module. The dissipation is normally conducted to a heat sink through a layer of electrically insulating but thermally conducting material. This is normally alumina ($Al_2O_3$), but sometimes substrates made from beryllia (BeO) or aluminium nitride (AlN) ceramics are used. The important electrical and thermal parameters of these materials are included in Table 13.1. Aluminium oxide ceramic is inexpensive but its thermal conductivity is relatively poor. Beryllium oxide ceramic has a high thermal conductivity but its dust is very toxic. Aluminium nitride ceramic has good thermal, electrical and mechanical parameters but is still quite expensive. At present, alumina substrates of 0.5 mm thickness are mostly used for modules where the maximum voltage is less than 2500 V. Above this, aluminium nitride is preferred, because the required thickness of alumina has too high a thermal resistance. It is common for high power MOSFETs and IGBTs to be made into power modules comprising several chips connected in parallel, as discussed in Section 12.1.

Originally, modules were assembled from discrete semiconductor devices soldered to molybdenum expansion discs. These were electrically interconnected and pressed onto an insulating ceramic substrate and a copper base by the mechanical construction of the case, as shown in Figure 13.10. A disadvantage of this construction is the relatively high thermal resistance that results from having several contacts in series.

Thermal expansion coefficients of ceramic materials with high thermal conductivity are relatively close to that of silicon. Therefore, it is possible to cover the ceramic substrate with a conductive layer and to solder silicon devices and other components necessary for a power module or a hybrid power integrated circuit construction to this conductive layer. In modern hybrid devices the basic ceramic substrate is plated on both sides with a copper foil of about 0.3 mm thickness. This uses the direct copper bonding (DCB) technology, in which the copper foil is attached to oxide ceramics such as alumina with a layer of the eutectic alloy $Cu_2O$–Cu.

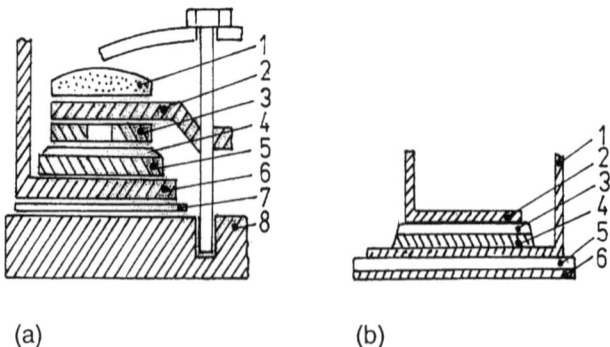

**Figure 13.10** Packages designed to minimise differential expansion stresses: (a) using compression (1 – insulation layer, 2 – copper electrode, 3 – molybdenum expansion disc, 4 – silicon wafer, 5 – molybdenum expansion disc, 6 – copper electrode, 7 – ceramic disc, 8 – copper base); (b) fully soldered (1 & 2 – copper leads, 3 – silicon wafer, 4 – solder, 5 – ceramic disc, 6 – copper base)

The DCB technique can be used with beryllia ceramic substrates, but nitride ceramics need a surface oxidation before copper plating. Plating both sides of a ceramic substrate with copper increases the expansion coefficient slightly, but with lead solder the plasticity during cooling acts as a stress buffer and the process is acceptable.

In a high power module such as a 1200 A/1200 V switch, where the ceramic area may be 45 cm$^2$, the mismatch is considerable, especially with aluminium nitride ceramic. The mechanical stress may cause failure after a number of thermal cycles. In advanced module technology, the copper baseplate is replaced with Al/SiC metal matrix composite, which matches the substrate better. With AlN substrates, the joint to the composite can be made during preparation by infiltrating aluminium into the porous SiC matrix [13.5]. Individual devices can be soldered to mutually insulated islands on the plated ceramic and interconnected using ultrasonic bonding of aluminium wires.

For devices which have the ceramic base plated with copper, it is possible to compress the ceramic base directly onto a heat sink. This variant has considerably lower thermal resistance compared to modules with a metallic (copper) base and an internal mechanical spring construction.

The construction, complete with interconnections, is normally hermetically sealed into a plastic case using silicone gel and epoxy. The use of power modules brings many advantages. More devices can be placed on a single heat sink, and circuit assembly is much easier. As a result, discrete devices are gradually being replaced by modules, function blocks and power integrated circuits.

### 13.1.3 Heat Sinks

The heat sink enhances the transfer of heat from the device package into the surrounding ambient. It is a very important part of the cooling system, and its construction determines the external thermal resistance $R_{thSA}$. This can be expressed approximately as

$$R_{thSA} = \frac{1}{Ah\eta} \qquad (13.10)$$

where $A$ is the surface area of the heat sink, $h$ is the coefficient of heat transfer and $\eta$ is the efficiency of the heat-sink body. The coefficient of heat transfer is the power lost per unit surface area per degree Kelvin. It depends on the mechanism of dissipation into the ambient and on the type of cooling medium. The efficiency term takes account of any non-uniform temperature distribution over the heat-sink surface resulting from the finite thermal conductivity of the material. The product $\eta A$ can be thought of as an effective area of heat-sink surface, at the interface temperature $T_r$.

The cooling medium may be gas (air) or liquid (water or oil). Important parameters of these materials are set out in Table 13.2. We consider each in turn.

**Table 13.2** Physical properties of cooling media

| Cooling medium | Mass thermal capacity, $C_m$ ($J\,kg^{-1}\,K^{-1}$) | Thermal conductivity, $\kappa$ ($W\,m^{-1}\,K^{-1}$) | Density, $\rho$ ($kg\,m^{-3}$) | Kinematic viscosity, $v$ ($kg\,m^{-1}\,s^{-1}$) | Heat transfer coefficient, $h$ ($W\,m^{-2}\,K^{-1}$) |
|---|---|---|---|---|---|
| Air | 1006 | 0.027 | 1.09 | $1.7 \times 10^{-5}$ | 8–20 |
| Oil | 2130 | 0.181 | 850 | 0.98 | 540 |
| Water | 4180 | 0.600 | 995 | $8 \times 10^{-3}$ | 6500 |

#### 13.1.3.1 Air cooling

Air is a natural insulating ambient into which any heat loss from a device or module can be dissipated. It follows from (13.9) that heat sinks should have a large surface area. They are most often made from profiled aluminium, as shown in Figure 13.11. Sheet radiators may sometimes be used for low power devices.

Heat transfer is by convection and radiation, and its overall coefficient is a sum of the coefficients for each of these two basic mechanisms. Thus

$$h = h_{\text{rad}} + h_k \qquad (13.11)$$

where $h_{\text{rad}}$ is the heat transfer coefficient resulting from heat lost by radiation according to the Stefan–Boltzmann law and $h_k$ is the convective heat transfer coefficient for the heat lost to the cooling medium, in this case air.

Air cooling may be by natural or forced convection. In natural convection the flow of the cooling air is stimulated by heating from the heat sink. The air must be

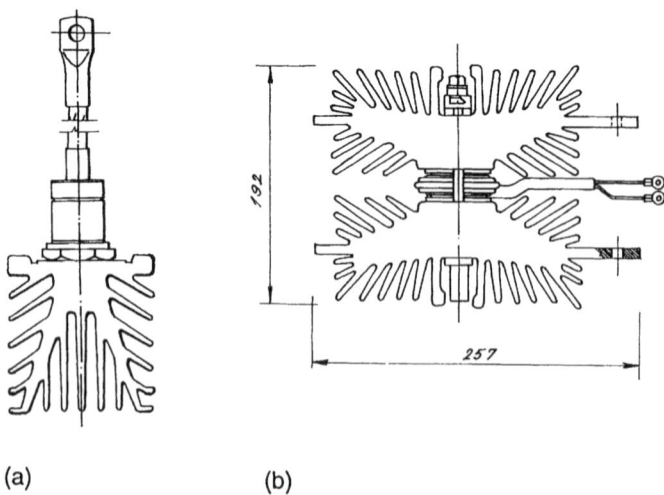

**Figure 13.11** Devices mounted onto finned aluminium heat sinks: (a) a stud-mounted device cooled from one side; (b) a hockey puck package cooled from both sides. Reproduced by permission of Polovodiče, a. s., Prague

allowed to flow along the flanges of the heat sink, which should be arranged vertically. The coefficient of heat transfer $h_k$ depends on the temperature difference between the ambient and the vertically oriented surface area of the heat sink. The coefficient $h_{rad}$ depends on the temperature difference, the heat-sink geometry and the emissivity of the surface. To increase its emissivity the heat-sink surface is often oxidised or lacquered, usually to give a matt black finish. An example of the relationship between the temperature difference $T_S - T_A$ and the dissipation is shown in Figure 13.12.

With forced air cooling, a blower impels the air along the cooling fins. The heat loss is mainly by convection. If $v$ is the air speed and $l$ the length of the heat sink in the direction of the airflow, the heat transfer coefficient is proportional to $\sqrt{v/l}$. Examples of the variation of $R_{thSA}$ and $Z_{thSA}$ with the velocity of the cooling air are shown in Figure 13.13. These important parameters are affected by the design of the air duct, which determines the pressure drop as a function of the air speed, also shown in the figure.

The thermal resistance of a heat sink of optimum shape is determined by its volume and thermal conductivity. An increase in the volume of an aluminium heat sink above $0.006\,\mathrm{m}^3$ does not significantly reduce its thermal resistance. With double-sided air cooling, a thermal resistance of $R_{thSA} \approx 0.04\,°\mathrm{C/W}$ can be obtained and up to 1 kW dissipated. For higher power losses, air-cooled heat sinks become too large and complicated. The finite thermal conductivity of the heat sink material results in a decrease of temperature along the heat-sink flanges and a decrease in the efficiency of heat loss.

These disadvantages can be mitigated by the use of heat pipes, whose basic operation is illustrated in Figure 13.14. A hermetically closed pipe is partially filled with a liquid of suitable boiling temperature. Previously Freon was used but now a mixture of water and methanol is preferred. If the temperature of the wall of the pipe

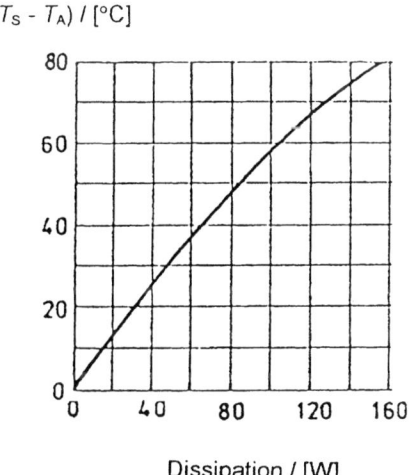

**Figure 13.12** An example of the variation of the heat-loss parameters of a flanged heat sink under natural air cooling. Reproduced by permission of Polovodiče, a. s., Prague

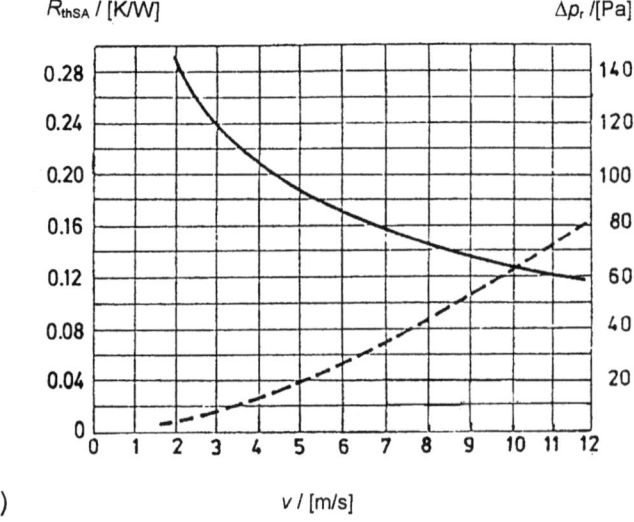

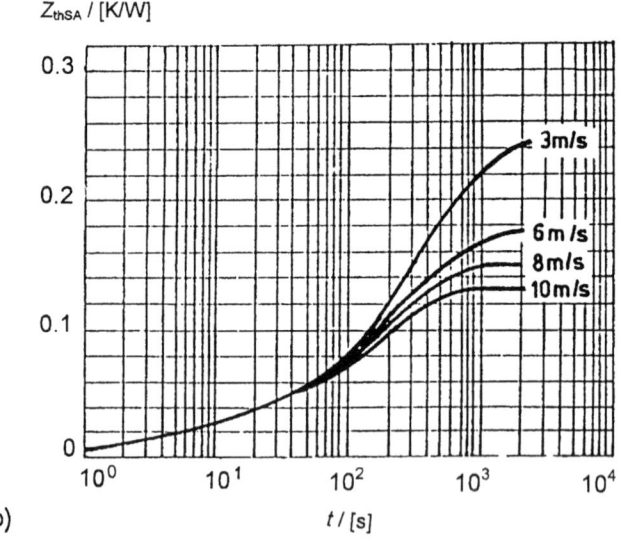

**Figure 13.13** Examples of the variation of the thermal resistance and the transient thermal impedance of a force-cooled flanged heat sink with the air velocity: (a) thermal resistance and air-pressure variation; (b) transient thermal impedance. Reproduced by permission of Polovodiče, a. s., Prague

closest to the heat source approaches the liquid boiling temperature, vaporisation occurs, extracting the latent heat of evaporation. The vapour migrates to the coldest part of the wall and condenses. There the heat of condensation is released and dissipated into the ambient, usually via cooling fins to enhance the heat exchange. The condensed liquid returns to the hot side of the pipe by gravity or capillary

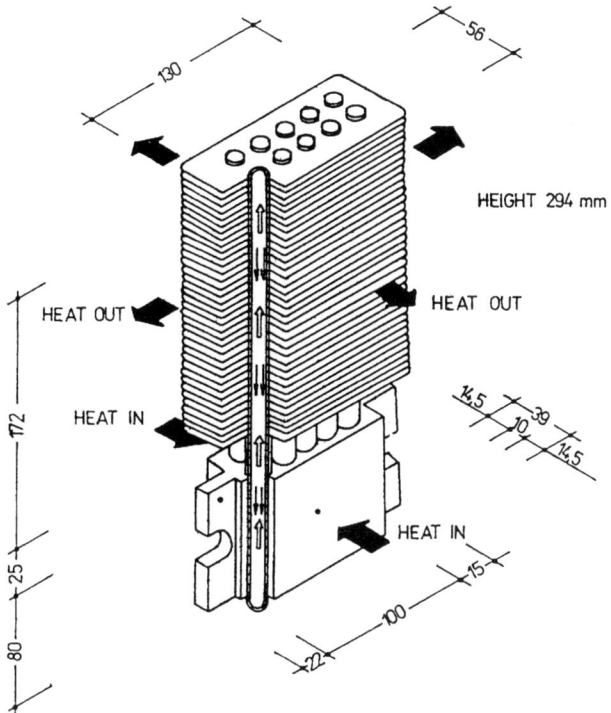

**Figure 13.14** An example of a heat pipe heat-sink. Reproduced by permission of SVÚSS, a. s., Prague

action, depending on the construction. The cooling ribs on the heat pipe have practically the same temperature difference between the surface and the ambient; this makes the heat dispersion more effective. The thermal resistance can be reduced almost to 50% of that of a conventional heat sink and heat can be transferred efficiently to a relatively distant heat sink.

The thermal resistivity of heat pipes depends on the working conditions. It decreases with increasing power dissipation because evaporation is enhanced. However, if the dissipation is so high that a film of boiling liquid covers the contact area, the thermal resistance increases rapidly.

The position of the heat pipe has a considerable influence on the thermal resistivity $R_{thSA}$. The condensed liquid returns to the heated contact area by a combination of gravitational and capillary forces. The inside surface is often grooved to enhance the capillary action. It is important to ensure that the liquid returns to the heated area at a rate sufficient to keep it covered. The variation of the thermal resistance with the dissipated power is shown in Figure 13.15 for heat pipes in vertical and horizontal orientations.

The use of heat pipes enables the cooling of power semiconductor devices with power dissipation up to 2 kW. Another advantage is the smaller volume needed. It is also possible to conduct the dissipated power to a remote heat sink, which can be advantageous in some applications, such as traction.

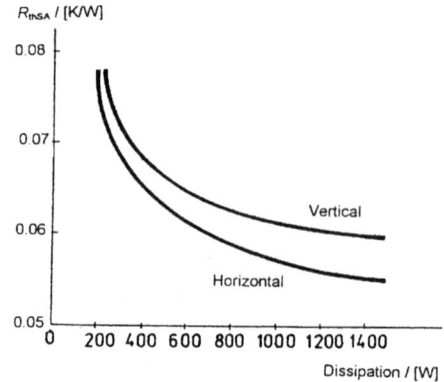

**Figure 13.15** The thermal resistance of heat pipes with vertical and horizontal orientations. Reproduced by permission of SVÚSS, a. s., Prague

### *13.1.3.2 Liquid cooling*

Table 13.2 shows that liquids like oil and water have higher coefficients of heat transfer than air, so they can be used to decrease the thermal resistance of the heat sink.

Liquid-cooled systems are much more complicated than air-cooled systems. They require a heat-sink body, a pump, a heat exchanger and a reservoir of cooling liquid complete with control and auxiliary components. The main part of the liquid cooling system is the heat sink, which effects the heat transfer between the semiconductor device and the cooling medium. The shape of the cooling chamber is important as it determines both the thermal resistance and the hydraulic resistance of the heat sink. An example is shown in Figure 13.16, and the variation of the thermal resistance with the working conditions is illustrated in Figure 13.17.

Its low viscosity and high coefficient of heat transfer make water one of the most suitable of the possible media for liquid cooling. However, because the heat sink is normally at a high potential, only deionised water with a very high resistivity can be used. Deionisation equipment should be included in the water circuit because water can become contaminated during circulation. This also helps to prevent corrosion. A basic water cooling scheme is shown in Figure 13.18. Because of contamination problems, oil is often used instead of water, especially in high voltage applications and when dissipation exceeds 2 kW. A disadvantage of liquid-cooled systems is the number of components needed. This increases the possibility of failure.

With liquid cooling, the heat can be removed by convective flow, by pumping or by evaporation. For evaporation cooling, the devices are clamped between finned heat sinks and immersed in a volatile, insulating liquid in a hermetically sealed vessel. In the past, Freons were used but they have been superseded. The liquid boils at the surface of the heat-sink fins and is evaporated. The vapour condenses in the condensation chamber and the heat is dissipated into the surroundings through a heat exchanger. Up to 20 W/cm$^2$ can be dissipated in this way. The arrangement is shown in Figure 13.19.

## 13.1 COOLING POWER SEMICONDUCTORS

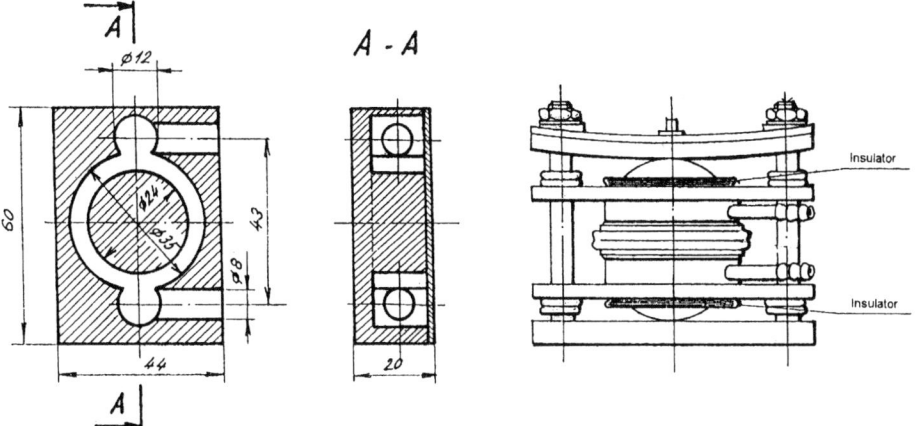

**Figure 13.16** A chamber for liquid-cooling a hockey puck package from both sides. Reproduced by permission of Polovodiče, a. s., Prague

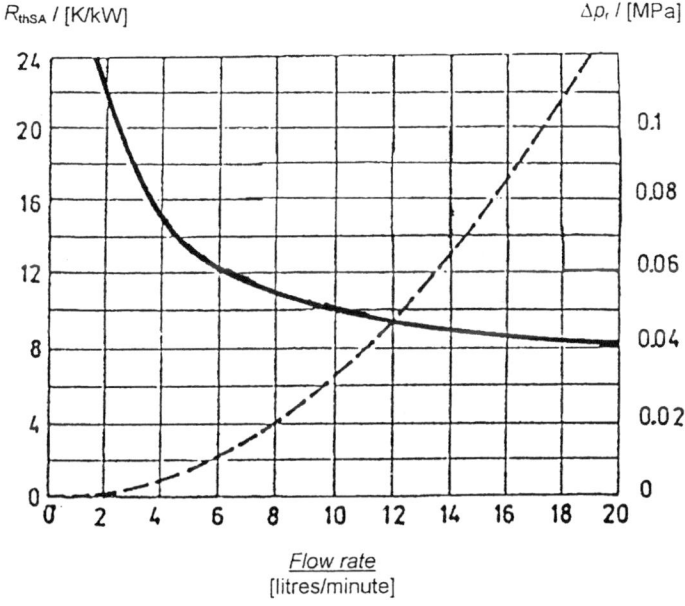

**Figure 13.17** The effect of the liquid flow rate on the thermal resistance (solid line) and the required pressure drop (dashed line). Reproduced by permission of Polovodiče, a. s., Prague

The advantages of this type of cooling system are its compactness and the simultaneous cooling of auxiliary devices such as snubbers and protection circuits. A disadvantage is the complicated technology and construction of the vessel and the other parts of the hermetic system. Such immersion cooling systems are found in traction applications.

# 13 CONDITIONS FOR RELIABLE OPERATION

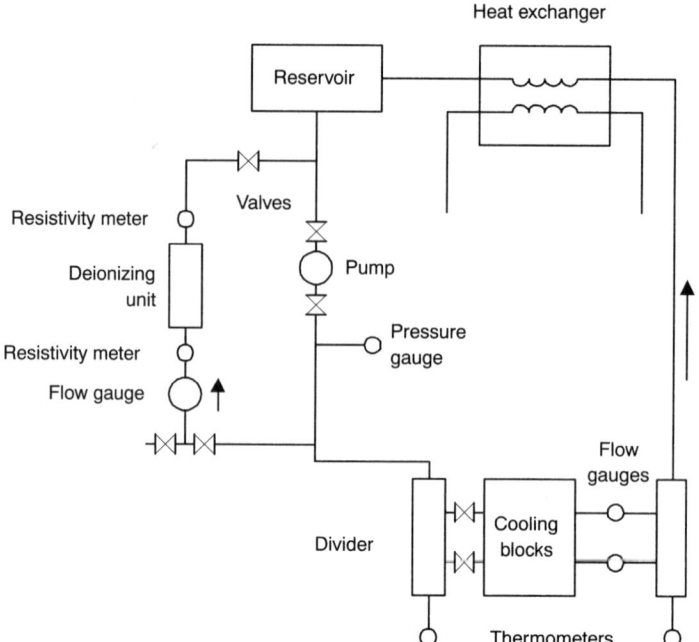

**Figure 13.18** A basic water cooling system

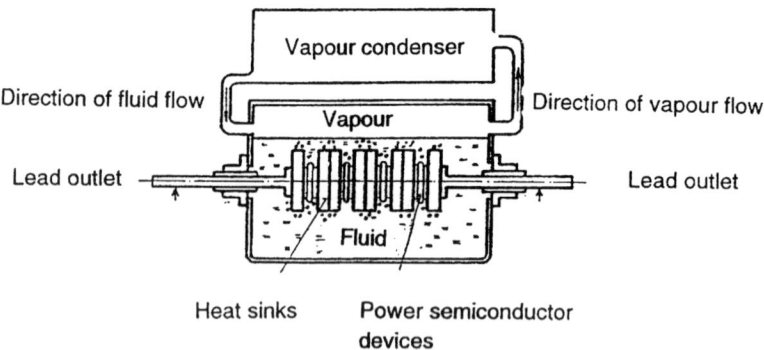

**Figure 13.19** An example of an evaporation cooling system. Taken from *Diody a tyrisroty v průmyslové praxi (Diodes and Thyristors in Industrial Applications)* by J. Zika (SNTL, Prague, 1979)

## 13.1.4 The Thermal Resistance of the Device–Heat Sink Interface

Obtaining a good thermal and electrical contact between the device case and the heat sink that is stable in the long term can be a serious problem. For power modules, a good thermal contact is all that is needed. However, the thermal resistance $R_{thCS}$ is a significant part of the total thermal resistance and can limit the overall efficiency. It depends on the size and quality of the contact area, which must be plane, smooth and clean. A sufficiently high pressure, usually prescribed by the producer, must be applied to the contacting surfaces during fixing, and this pressure must remain constant during the lifetime of the device in the equipment. The contact must remain corrosion proof. A smear of silicone grease is normally applied to fill cavities resulting from surface roughness.

The thermal conductivity of silicone grease is high in comparison with other types of grease, but lower than the other materials listed in Table 13.1. It can be improved by the addition of fine particles of materials such as alumina that have a higher thermal conductivity, but this may increase the electrical contact resistance.

Another way to reduce the thermal contact resistance is to interpose a thin foil of an easily deformable metal between the case and the heat sink. Pure silver or silvered copper foils are often used to improve contacts inside the case, especially when the wafer is pressed rather than soldered to a conducting disc. When a metal with a low melting point (50–60 °C) is used, liquid metal fills all the microscopic cavities in the interface at the working temperature. A eutectic alloy based on Sn, Pb, In and Bi can be spread on the surface of a perforated copper foil.

In the case of thyristors and diodes working at the highest power levels, where liquid cooling is necessary, the contact thermal resistance can be eliminated by integrating the liquid-cooled heat sink into the device package.

## 13.2 The Parallel and Series Connection of Power Semiconductor Devices

In many applications, especially in the fields of traction and power transmission, the required working voltage is higher than the breakdown voltage of any single device and it is necessary to connect devices in series. When the required current is higher than the maximum available from a single device, devices also have to be connected in parallel. Because the characteristics of power semiconductor devices depend on many constructional and technological parameters, device-to-device variations occur and they lead to an uneven loading when they are connected in series or parallel.

### 13.2.1 Devices in Parallel

The on-state voltage drop is the same for all of a parallel-connected group of devices and current is shared according to their static or dynamic $I$–$V$ characteristics. As no two devices are identical, the sharing is unequal. On a larger scale, this is the same problem that concerns individual GTOs, MOSFETs and IGBTs, in which many elementary device cells are integrated in parallel.

A simple way to ensure equal current sharing is to connect a resistance in series with each device. If the resistance is several times higher than the device differential resistance, almost equal current sharing can be achieved. However, there is a substantial increase in power loss.

Another possibility is to put some reactance in series with each device, as shown in Figure 13.20. Each of the diodes shown in Figure 13.20(a) is connected through a balancing transformer winding. If the current $I_{F1}$ passing through diode D1 is higher

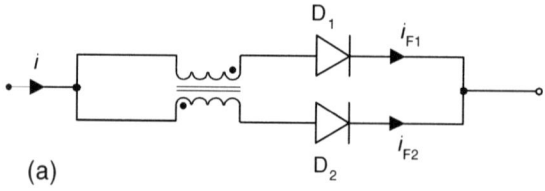

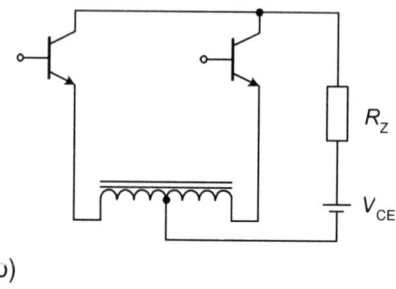

**Figure 13.20** The use of a balance transformer to improve current sharing: (a) between two diodes; (b) between two transistors.

than the current $I_{F2}$ passing through diode D2, the core of the transformer is magnetised with a flux caused by the current difference $(I_{F1} - I_{F2})$. The change of magnetic flux induces a voltage in both transformer arms of such a polarity that it increases the current through D2 and decreases the current through D1:

$$V_{F1} - V_{F2} = 2N \frac{d\Phi}{dt} \qquad (13.12)$$

where $N$ is the number of turns of one winding and $\Phi$ is the magnetic flux induced by the current difference. The magnetic flux in the core is

$$\Phi = \frac{\Delta V_F}{2f} \qquad (13.13)$$

where $f$ is working frequency, and it must not cause the core to saturate.

The balanced transformer can be used with several devices and it can be applied equally to thyristors or bipolar transistors, as shown schematically in Figure 13.20(b). Its disadvantage is the relatively high cost and the increased weight and volume.

To avoid the use of either series resistance or balanced transformers, device on-state characteristics have to be closely matched. Power diodes and thyristors are sorted in groups which are determined from the static characteristics, usually in such a way as to equalise their currents when two devices from the same group are connected in parallel. The method of sorting is indicated in Figure 13.21.

Besides current equalisation under static conditions, devices have to be balanced during switching operations. In the case of thyristors connected in parallel, it is

## 13.2 PARALLEL AND SERIES CONNECTION

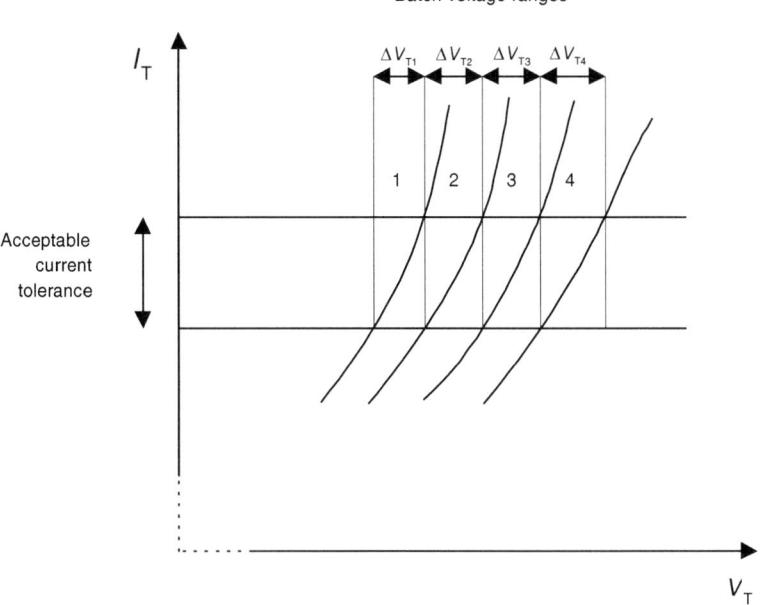

**Figure 13.21** Sorting devices into groups for parallel connection

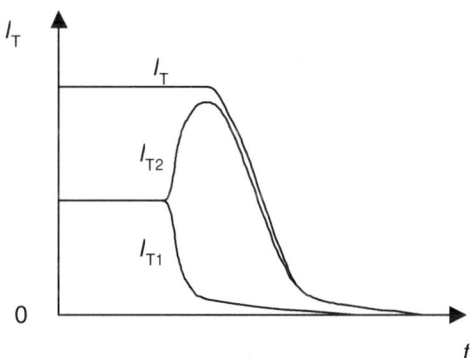

**Figure 13.22** Current waveforms in parallel connected GTOs at turn-off

important for the turn-on delay times to be nearly the same. If the difference is large, the anode voltage on the first thyristor to turn on can fall to such a low level that the second device does not turn on. Steeply rising gate pulses of high amplitude ($dI_G/dt > 1\,A/\mu s$, $I_{GM} > 5I_{GT}$) and sufficient duration are essential. Unequal current distribution is particularly marked with high rates of rise of anode current.

In the case of GTOs very unequal current sharing can occur during turn-off, as illustrated in Figure 13.22. As the thyristor with the shorter turn-off delay time starts to turn off, there is an increase of current in the second thyristor. Turn-off losses increase considerably and there is a possibility of second breakdown. This is minimised by the careful selection of devices with similar turn-off characteristics and the use of very high rise rates for the negative gate current. There is a similar problem when transistors with different storage times are connected in parallel. Antisaturation circuits are then used to shorten storage times and minimise the length of time that any transistor may be overloaded.

### 13.2.2 Devices in Series

Power semiconductor devices (especially thyristors and diodes) are connected in series when the supply voltage in the circuit is greater than the reverse or forward blocking voltage of any individual device. It order to obtain a reasonably equal sharing of the voltage among individual devices under both static and dynamic conditions, a good match of leakage current and switching characteristics is necessary. Series connection is most often used in power thyristor and diode circuits.

Strings of high voltage diodes are used to rectify voltage supplies up to 200 kV at current levels of a few amperes. Up to 80 diodes may be connected in series, normally sealed into an oil-filled alumina tube so that all are effectively cooled. Their static and dynamic characteristics must be closely matched.

We deal first with steady-state voltage sharing by considering two thyristors connected in series. The leakage current through both is the same, and voltage sharing is determined by the static off-state characteristic (Figure 13.23). It can also be influenced by temperature differences caused by on-state power loss differences. To improve on this, resistors can be put in parallel with devices (Figure 13.24). Auxiliary resistances take up a relatively large volume and may cause a significant power loss. They can be avoided if the device reverse characteristics are sufficiently similar, although the resulting uneven voltage distribution may require extra devices to be included in the chain.

During the turn-on process, voltage-sharing problems are associated with differences in the turn-on delay time. The device with the longest delay time turns on last. It is likely to be subjected to an overvoltage for a short time before it turns on, and it may suffer an excessive rise rate of the current when it does turn on. To minimise differences in the delay time, the gate pulses should be steep and they should have a high amplitude.

More significant are differences in the storage time during the turn-off process. The device which turns off first is subject to a voltage overload and can be destroyed

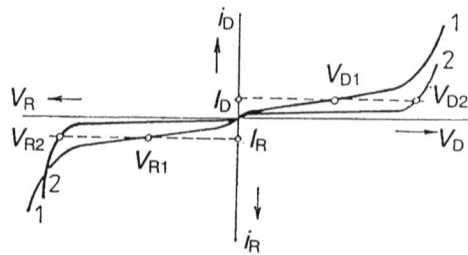

**Figure 13.23** Voltage sharing by series-connected thyristors

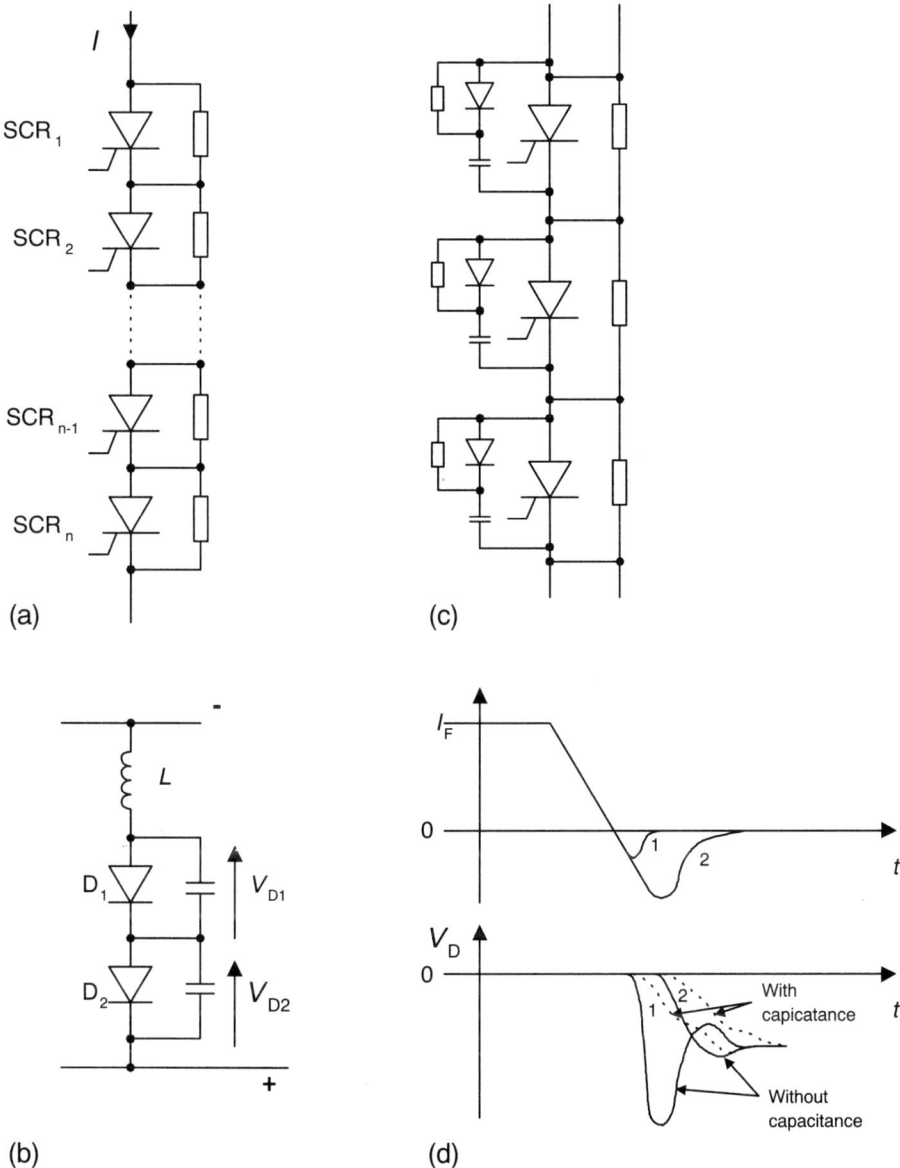

**Figure 13.24** Improving voltage sharing among series-connected devices: (a) steady-state voltage sharing; (b) the use of capacitance for dynamic voltage sharing; (c) a complete divider suitable for thyristors in series; (d) typical waveforms

by the resulting electrical and thermal breakdown. Capacitors tend to resist a change of voltage, so a capacitance divider can maintain reasonable voltage sharing during transients, as shown in Figure 13.24(b). Usually, RC or RDC circuits are used, as shown in Figure 13.24(c).

processes by delaying the gate signals. Thus, the negative gate signal is applied first to the transistor with the longest storage time. A similar system can be used when GTO thyristors are connected in series.

The storage times of power MOSFETs are much smaller and the capacitance divider can be replaced by a semiconductor voltage limiter such as a Zener diode, which protects the device against breakdown as long as there is sufficient voltage sharing between the series-connected devices [13.6].

High power converters need series–parallel connection of devices, which may take either of the forms shown in Figure 13.26. Connection of devices in parallel and series needs either the devices to be carefully chosen or the use of passive components such as resistances, capacitances or balance transformers in auxiliary circuits. The use of auxiliary circuits considerably increases the volume of equipment, and the additional number of devices increases the probability of a failure. It is thus important to minimise the number of devices connected in series or parallel. Note, however, that auxiliary circuits also give protection against a high rate of rise voltage and against overvoltage.

## 13.3 Overcurrent and Overvoltage Protection of Power Semiconductor Devices

### 13.3.1 Overvoltage Protection

In power semiconductor devices, any voltage higher than the repeatable peak values $V_{RRM}$ or $V_{DRM}$ should be treated as an overvoltage. When the voltage across the device exceeds the breakdown voltage, any significant reverse current causes high energy dissipation and a rise in temperature in part of the device. If the power dissipated in a small volume is $P$, the temperature rise in a time $t$ is proportional to $\sqrt{Pt}$. If this is sufficient for the temperature, even locally, to reach the intrinsic temperature $T_i$, the resulting current filament is likely to initiate thermal runaway. Because this can destroy a device, auxiliary circuits are used for protection. These usually take the form of resistor/capacitor (RC) circuits, voltage-dependent resistors (varistors) or avalanche semiconductor devices. It may be necessary to use devices with voltage ratings 30–50% higher than the maximum voltage normally expected in

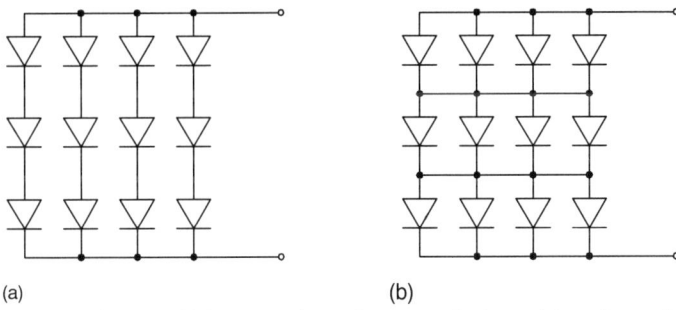

**Figure 13.26** The series–parallel connection of power devices: (a) series only; (b) series/parallel

the application. An overvoltage is usually classified by its place of origin as an *input*, *internal* or *output* overvoltage.

An input overvoltage arises via the input transformer during turn-on and turn-off transients. A simple example is shown in Figure 13.27. When the current in the primary circuit is changed quickly, the magnetic flux in the transformer changes and a voltage appears at the secondary winding which can cause an overvoltage. This can be large if turn-off occurs at the peak of the magnetisation flux in a lightly loaded transformer, or when turn-on occurs at the peak of the input voltage. It can arise when an input fuse is interrupted as a result of atmospheric interference such as lightning.

Internal overvoltage occurs most often when current is interrupted in a circuit with an inductive load. The amplitude of the voltage depends on the rate of fall of the current and on the inductance and capacitance in the circuit. It can also arise when fuses operate. Output overvoltage is caused by the interruption of the load circuit, especially when the load is inductive. The amplitude increases with the rate of fall of the output current.

### 13.3.1.1  The use of RC protection

This type of protection is characterised in terms of the charge absorbed by the capacitor during a given time interval. This is how the energy associated with the overvoltage event is absorbed. It is then dissipated to the ambient via the resistors.

The basic principles of the RC circuit as an overvoltage protection are as follows. When a current $I_k$ passes through an inductance $L$ in series with a semiconductor device, the energy $W_L$ is stored in the magnetic field of the inductance. If the circuit is broken, the capacitance $C$ of the disconnected part is charged to a higher voltage. If the voltage on the capacitance at the time of breaking the circuit is $V_0$, it follows from the conservation of energy that the maximum voltage $V_k$ is given by

$$W_L = \frac{1}{2}LI_k^2 = \frac{1}{2}C(V_k^2 - V_0^2) \tag{13.14}$$

so that

$$V_k = \sqrt{\frac{L}{C}I_k^2 + V_0^2} \tag{13.15}$$

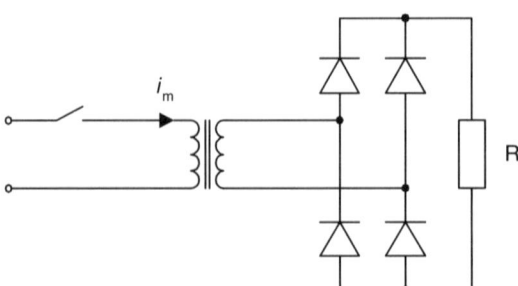

**Figure 13.27**   The origin of an input overvoltage

## 13.3 OVERCURRENT AND OVERVOLTAGE PROTECTION

Increasing $C$ by connecting a capacitor in parallel with a device or a block of devices decreases the voltage peak. The size of the capacitance should be determined from the need to avoid overvoltage under the severest conditions; that is, when the circuit is broken at the full working voltage and current.

The floating protection circuit, shown in Figure 13.28, is typically used to protect against an input overvoltage. An auxiliary rectifier allows the capacitors $C_1$ to be electrolytic, thus minimising the circuit volume. The resistances $R_1$ limit circuit

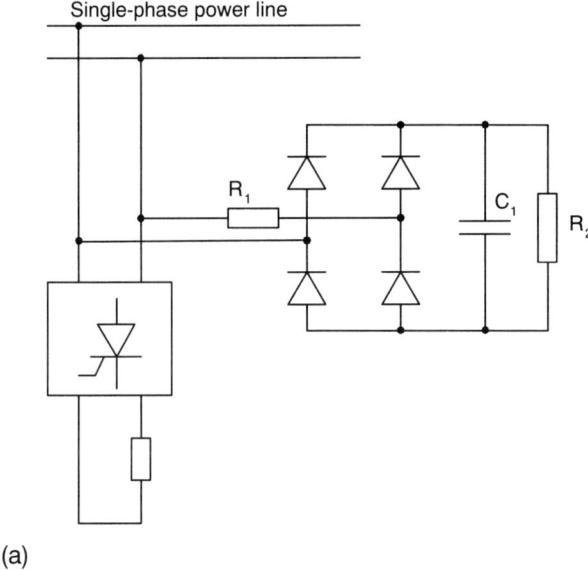

(a)

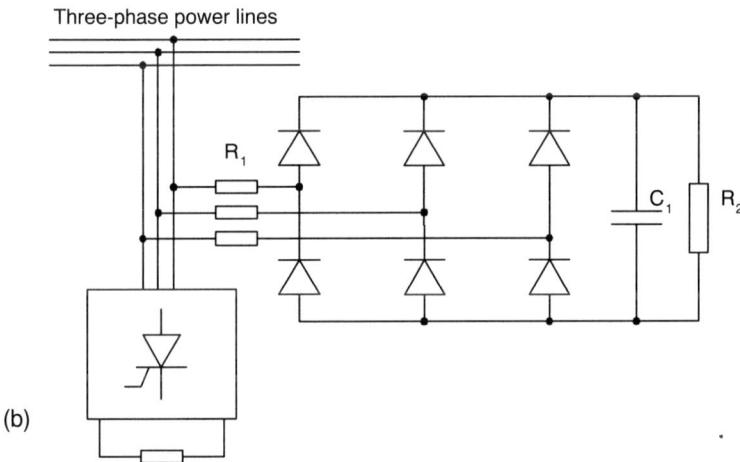

(b)

**Figure 13.28** The floating overvoltage protection circuit: (a) single phase; (b) three phase

oscillations and hence the charging current. The resistances $R_2$ discharge the capacitor. Recommended capacitance and resistance sizes are often published in the device literature.

In circuits where internal overvoltage can occur, RC or RDC snubber circuits are often used to decrease the rate of rise of the blocking voltage during turn-off in bipolar transistors and GTOs, as discussed in Sections 6.4 and 8.3. This is essential with GTOs and some types of power bipolar transistor to ensure that the turn-off current–voltage trajectory remains within the safe operating area, RBSOA. The reverse recovery process of power diodes and thyristors is similarly affected. In all cases the decrease in the rate of rise of the blocking voltage and the reverse voltage produces a decrease in the turn-off losses and hence the total power losses in the device, especially at higher frequencies. The size of the snubber capacitance and resistance depend on the device and the circuit.

With a simple circuit like the one shown in Figure 6.24(a), the collector current can be expected to decrease approximately linearly from $I_C$ to zero during the transistor turn-off period. In order to stay within the safe operating area, the capacitance should satisfy the condition

$$C_s \geq \frac{I_C t_f}{V_{CEOsus}} \tag{13.16}$$

where $t_f$ is the current fall time from 90% to 10% of the on-state value.

The resistance $R_s$ must discharge the capacitance $C_s$ during the time $t_p$ the transistor is turned on. At the same time, it must limit the collector current to something less than $I_{CM}$. If $I_L$ is the load current at the beginning of the turn-on process, $I_{rrM}$ is the peak reverse recovery current of the freewheeling diode and $V_{CW}$ is the off-state collector voltage, the resistance $R_s$, must fulfil the condition

$$\frac{t_p}{C_s} \geq R_s \geq \frac{V_{CW}}{I_{CM} - I_L - I_{rrM}} \tag{13.17}$$

The average power loss in the snubber at a switching frequency $f$ is given by

$$P_Z = \frac{1}{2} C_s V_{CW}^2 f \tag{13.18}$$

Any series inductance in the snubber circuit should be negligibly low. Equations (13.16) to (13.18) show that snubber circuits place quite high demands on the passive components ($R_s$ and $C_s$) and this can be a limitation to converter construction. Many different types of snubber and methods for optimising passive component values are described in the literature.

#### 13.3.1.2 Semiconductor surge protection devices

This group of devices consists of varistors, avalanche surge devices and breakover surge devices.

Varistors are polycrystalline semiconductor devices with a non-linear current–voltage characteristic. Crystals of a wide bandgap semiconductor such as ZnO are sintered into a suitable form such as a cylinder. Individual grains are insulated with a surface potential barrier that decreases with increasing voltage and so results in a considerable increase of current. The current–voltage characteristic can be represented by

$$I = AV^\beta \tag{13.19}$$

where $A$ and $\beta$ are constants that depend on the dimensions and material of the device. When the voltage increases, the resistance of the varistor decreases and the charge associated with the overvoltage is drained off through the varistor in parallel with the protected device. The response time is very short, about 25 ns. Varistors can be produced in a very broad voltage range. A very important parameter is the energy that can be absorbed in a single surge current pulse. This is proportional to the volume of the device and it can reach hundreds of joules. On the other hand, the average power losses are low (a few watts). Therefore, varistors are good surge protection devices against input and output overvoltage.

Another type of surge protection device is the avalanche surge diode. This is often made in the form of a symmetrical pnp structure with a lightly doped n-type region [13.7], as shown in Figure 13.29(a). The current–voltage characteristic is shown in Figure 13.29(b). The abrupt increase of current occurs when the voltage reaches the punch-through voltage given by (7.1). Breakdown takes place over the whole area of the structure, giving a high current-carrying capacity. This type of surge diode is produced for a broad range of voltages up to 4000 V. The average power losses can be 60 W, giving protection briefly even against periodically repeated overvoltage. It may thus protect against internal overvoltage. The peak dissipated power for some types can be 700 kW.

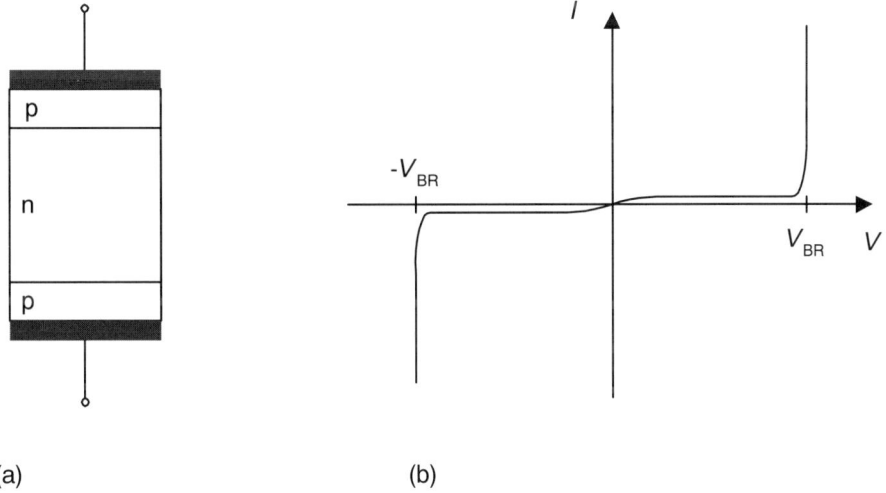

**Figure 13.29** An avalanche surge-protecting device: (a) basic structure; (b) current/voltage characteristics

A different type of semiconductor device used for overvoltage protection of thyristors is the breakover diode (BOD), described in Section 8.6. Essentially, the BOD is a low power, reverse conducting thyristor with no gate terminal; in other words, a diode thyristor. In the circuit of Figure 8.23, when the voltage across the BOD reaches the breakover voltage, the line voltage is connected to the gate of the power thyristor. By turning on the endangered device, the BOD provides active surge protection. A similar type of overvoltage protection is described in Section 12.3 in connection with smart power devices, where a lateral BOD can be used. BODs are produced with breakover voltages up to 1000 V. They can also be used to turn on thyristors at a defined anode voltage.

### 13.3.2 Overcurrent Protection

The current rating of a device is determined by the current that raises its junction temperature to the maximum limit when the device is operating under recommended cooling conditions. Exceeding the current rating can cause an excessive temperature rise and be followed by thermal breakdown. This can happen both under normal working conditions, e.g. in starting up a motor, and under fault conditions. Devices are often able to work for a short time under overcurrent conditions without being degraded. The non-repetitive surge current pulse limit is usually specified as a device parameter.

For thyristors and diodes, the maximum non-repetitive on-state surge current in a half-sine wave of 10 ms duration, $I_{TSM}$, is normally specified and is usually some 10 to 15 times $I_{TAV}$. Graphs are often presented showing the variation of the maximum permissible surge current with the length and shape of the pulse and the device temperature.

Some power semiconductor devices, e.g. thyristors, may be protected against short circuit failure using fuses in the power supply lines. These must be fast acting because of the short thermal time constant of semiconductor devices. A fuse normally has an $I^2 t$ rating, which measures the product of $I^2$ and $t$ up to the point of rupture. This parameter is important for diodes and thyristors in that it specifies the fuse needed to protect them.

The overcurrent capability of power bipolar transistors and IGBTs is more difficult to predict because of second breakdown. It is usually specified through the safe operating area (FBSOA). Exceeding the current rating is permitted for only a very short time, typically tens of microseconds. Passive protective devices like fuses are not easily effective on this timescale and active protection is normally required. The on-state current is monitored. At the moment it exceeds an admissible level, the device is turned off by the driving circuit. This technique can be used with any fully gate-driven devices such as bipolar transistors, MOSFETs, IGBTs, GTOs or MCTs, and also in smart power devices and intelligent power modules.

An alternative arrangement [13.8], known as a crowbar circuit, is shown in Figure 13.30. When the on-state current exceeds a chosen level, a gating signal is generated which turns on the thyristor connected across the supply. The short circuit produced operates conventional overcurrent protection devices such as fuses or circuit-breakers.

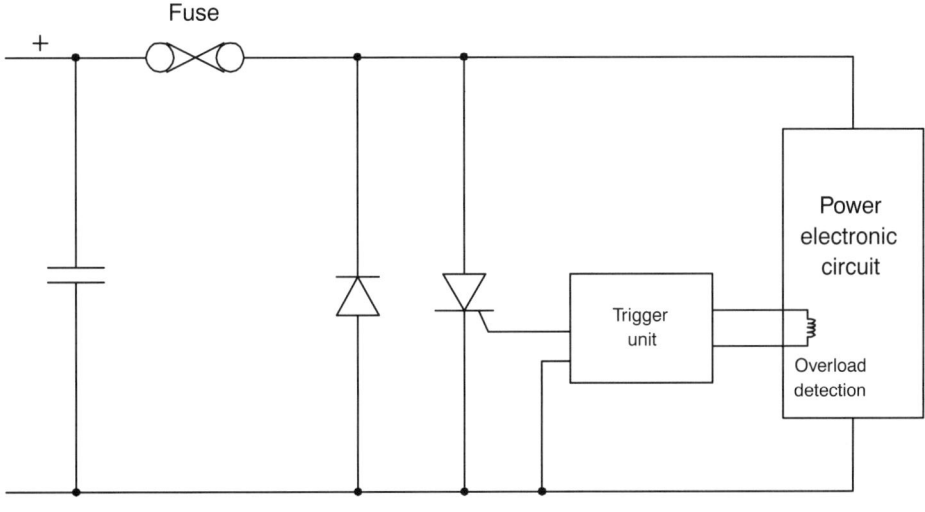

**Figure 13.30** The use of a thyristor to provide overcurrent protection for a load containing vulnerable components

## 13.4 The Operating Reliability of Power Semiconductor Devices

Reliability refers to the capability of devices to perform as specified over an extended operational lifetime; its converse is failure. The following terms are used to describe these properties:

- The failure probability $F(t)$ expresses the probability that the device will fail before time $t$ from the start of its life.
- The reliability $R(t) = 1 - F(t)$.
- The failure probability distribution function $f(t) = dF(t)/dt$.
- The failure rate $\lambda(t) = f(t)/R(t)$.

The failure rate may be written as

$$\lambda(t) = -\frac{1}{R(t)}\frac{dR(t)}{dt} \tag{13.20}$$

There are many possible causes of failure and the parameters that affect them can be classified as follows:

- Internal influences are the construction, material, fabrication and mounting technologies.
- External influences are the device mounting in the equipment, or the operating conditions.

The most important internal influences are

- the homogeneity of the silicon structure
- the high-temperature fabrication operations
- the quality of the termination of the p–n junction at the surface of the wafer
- the metallisation (contacting) technique
- the technique of encapsulation

At each stage of the manufacturing process, measurements are made to detect imperfections so that faulty devices can be removed at once. Nevertheless, devices age during operation and their parameters change. Some aspects of the physical limits and possible destruction mechanisms of a device are connected with its homogeneity. For example, according to (2.36), the breakdown voltage depends on the dopant concentration. An inhomogeneity such as a crystallographic defect in a high resistivity base region can give rise to a locally enhanced field, breakdown, dissipation and a rise in temperature. This can result in electromigration and the aggregation of impurities at the defect. The breakdown voltage may thus decrease over a period of time as the device ages.

In MOS devices the quality of the gate oxide layer is very important. It is often exposed to a high electric field, when the accumulation of oxide defects causes an increase of the gate leakage current. Eventually this may rupture the layer and destroy its control function.

Local inhomogeneities in the surface passivation can lead to the gradual development of instability in some device parameters. In packages that are not hermetically sealed, humidity can cause corrosion as hydroxyl ions react with aluminium contacts and bonding pads. The adhesion of metal contacts may also be affected. All these degradation processes are accelerated at higher temperature.

There is an empirical relation between the device failure rate and its exposure to an elevated temperature over a long period. It can be expressed as

$$\lambda = \exp(A + BT_j) \qquad (13.21)$$

where $A$ and $B$ are constants that depend on the device construction and technology. The increased failure rate following higher operating or storage temperature is indicative of physical or chemical degradation processes whose rates of reaction vary in this way.

When devices are operated at high power with a variable load, it is likely there will be temperature cycling of the silicon wafer and all parts of the case. Differences in the thermal expansion coefficients of the different materials cause mechanical stress which can end in material fatigue and device failure. This is influenced by

- the method of contact between the semiconductor wafer, or chip, and the case
- the type of contact areas and the fabrication method
- the flatness and smoothness of the contact areas

## 13.4 OPERATING RELIABILITY

- the technology of soldering the chip to the expansion disk
- the material and thickness of the expansion substrates

Failure occurs after a certain number of thermal cycles (heating/cooling/heating) that is usually expressed empirically as

$$N = k(\Delta T)^m \tag{13.22}$$

where $\Delta T$ is the difference (in °C) between the maximum and minimum operating temperatures, and $k$ and $m$ are parameters that depend on the device construction and technology. In particular tests [13.9] on a range of power thyristors, it was found that

$$N = 2.5 \times 10^{13}(\Delta T)^{-5} \tag{13.23}$$

As well as thermal cycling of the whole device, local heating can occur during turn-on and turn-off transients. If there is a local increase of temperature $\delta T_{loc}$ in a heavily loaded region of the device compared with adjacent regions, the material expands at the hot spot but is constricted by the surrounding colder material. If the local temperature difference exceeds a critical value, the material at the hot spot is stressed to the limit of its compressive strength, which is reduced under conditions of high injection [13.10]. Cracks occur in the single crystal and the device is ultimately destroyed. In silicon this temperature difference is about 300 °C. With a lower local temperature difference, failure can still occur after a number of cycles $N^*$ found experimentally [13.11] to vary as

$$N^* = \left(\frac{300}{\delta T_{loc}}\right)^9 \tag{13.24}$$

This increases the failure rate of thyristors operating at high $di_T/dt$. There is also risk in applying a positive gate pulse to a reverse-biased thyristor. Electron injection from the $n^+$-emitter causes a large increase of the reverse current density in the surrounding gate region. The resulting effects may increase the failure rate, especially when devices are connected in series. In general, any operating condition that causes local overloading can be expected to increase the failure rate.

Frequent causes of failure are the faults, imperfections and inaccuracies introduced when power semiconductor devices are assembled into equipment. Examples are the damage that can be caused to the case bushing or the fitting areas. The assembly onto a heat sink should not be performed in a dusty atmosphere, because dirt on the fitting area causes an increase in the contact thermal resistance and a decrease in the efficiency of the cooling system. The compressive force used in hockey-puck assemblies must be axial and uniform over all the base area, because any eccentricity can damage the case fitting area or the internal device construction and increase the failure rate.

Power IGBTs and MOSFETs are frequently encapsulated in modules, often together with an antiparallel diode. IGBT chips have the collector soldered onto a substrate made using, for instance, DBC technology while the emitter is connected to another contact point by several aluminium wires of diameter 0.25–0.50 mm. These are ultrasonically bonded to the emitter metallisation. The bonding pressure and temperature are two sensitive factors that influence device reliability. These bonds are considered to be one of the weakest parts of the overall construction, because their severance through electrothermomechanical stress is a frequent cause of failure. For devices encapsulated in this way, the number of thermal cycles $N$ of temperature variation $\delta T$ that caused failure has been found to vary [13.12] as

$$N = 10^7 \exp\left(-\frac{\delta T}{20}\right) \quad (13.25)$$

Results on IGBT failures have also been published [13.13]. These analyses show that operating conditions seriously influence device reliability and must be carefully controlled. For example, if a cyclical temperature variation exceeds 50 °C, modules as presently constructed cannot be expected to survive the $10^6$ cycles desirable in traction applications.

A considerable reduction of thermally induced stress may be achieved by the application of new materials such as the Al/SiC metal matrix composite, and by improvements in jointing and bonding technology. The modelling, simulation and diagnostic tools used in improving power semiconductor device parameters and developing new types of device are very important in this work. Models are used for electrothermal simulation that can help to find local overheating, high local electric fields and other dangerous states which can arise under working conditions. To be effective, these need to be very sophisticated.

Large-area, high power devices also fail as a result of cosmic radiation [13.14]. High energy particles penetrate the case, enter the device and generate electron–hole pairs, creating a conducting channel along their trajectory. This discharges the charge stored at a p–n junction area, which increases with the area and the reverse bias voltage In large-area, high voltage devices, the probability of destruction increases. It particularly limits the length of time that a high power device can sustain a high dc voltage.

Despite the complexity of the potential failure processes and the lack of control over the environment in which components are stored, assembled and operated, there have been a number of attempts to establish general semi-empirical reliability models for silicon devices. This is facilitated by the large numbers of nominally identical samples that are made and are available for test. Where it is possible to establish with confidence the empirical relationships between failure rates and particular stress factors, accelerated life tests can be used to confirm the quality of components and to proof-test samples from a production run. Widely used models for semiconductor devices are MIL-HDBK-217 from the US Department of Defense [13.15] and those published by national public telephone authorities such as BT [13.16] and CNET [13.17]. The relationships quoted here are special cases of these more general models.

## Summary

Ensuring high and predictable reliability of power semiconductor devices requires not only the most stringent control of the materials and processing of the device itself, but also of its mounting, encapsulation and assembly into its operating environment. Thermal expansion coefficients should be well enough matched to avoid fatigue caused by thermal cycling. Only small-diameter chips can be soldered to a copper base. Large-area wafers have to be mechanically clamped, usually between molybdenum discs that give good thermal contact but permit differential expansion. The case should protect the semiconductor against all anticipated mechanical and chemical influences. Devices, say up to a current rating of about 400 A, can be housed in a single-sided, stud-mounted case. Small and medium power devices are often packaged in a flat plastic-encapsulated case, also cooled from one side. For the largest devices, the hockey-puck assembly is preferred.

The internal device temperature must be kept within the prescribed limits. The temperature of the ambient and the thermal impedance between it and the interior of the device together determine the maximum power that may be generated within the device during operation. A heat sink is needed for all but the lowest power operation. Up to about 1 kW, forced air cooling may be used, perhaps 2 kW with the use of a heat pipe. High power assemblies require liquid cooling, either by convection or evaporation.

Parallel and series connection require careful selection of devices to ensure matched characteristics and possibly external voltage and current-sharing components. External snubber circuits also protect against excessive voltage levels and rise rates. Overcurrent protection depends on the surge current rating of the device, matching the characteristics of an external fuse or having an active sensor to trigger a crowbar or to switch the device off. Such sensors are integrated into some power switching devices.

## References

[13.1] Hogarth, C. A., Langridge, A. and Ziman, J. M. Thermal-Resistance Considerations in the Design of Semiconductor Devices, *Proc. IEE, Part B*, **106**, 402–8 (1959).

[13.2] Humpston, G., Jacobson, D. M., Crees, D. E., Newcombe, D. R. and Zambeli, M. Recent Development in Silicon/Heat-Sink Assemblies for High-Power Device Applications, *GEC Review*, **7**, 67–78 (1991).

[13.3] Scott, A. W. *Cooling of Electronic Equipment* (John Wiley & Sons, New York, 1974).

[13.4] Klaga, S. and Sittig, R. Reduction of Thermomechanical Stress by Applying a Low Temperature Joining Technique, *Proc. ISPSD'94*, 259–64 (1994).

[13.5] Anderson, S., Beringer, K. and Romero, G. Advanced Power Module using GaAs Semiconductors, Metal Matrix Composite Packaging Material and Low Inductance Design, *Proc. ISPSD'94*, 21–4 (1994).

[13.6] Grant, D. A. and Gowar, J. *Power MOSFETs: Theory and Applications* (John Wiley & Sons, 1989).

[13.7] Kannam, P. J. Design Concepts of High Energy Punch-Through Structures, *IEEE Trans. on Electron Devices*, **ED-23**, 879–82 (1976).

[13.8] Evans, P. D. and Saied, B. D. Protection Methods for Power-Transistor Circuits, *Proc. IEE*, **B-129**, 359–63 (1982).

[13.9] Zika, J. How to Contribute to Enhancement of Operating Reliability of Thyristors (in Czech), *Elektronik*, **41**, 300–6 (1986).

[13.10] Fulop, W. Influence of High Bipolar Injection Levels of Structural Defects of Single Crystal Semiconductor Devices, Particularly Lasers and Silicon Power Devices, *International Journal of Electronics*, **74**, 209–11 (1993).

[13.11] Somos, I. L., Piccone, D. E., Willinger, L. J. and Tolsin, W. H. Power Semiconductor Empirical Diagrams Expressing Life as a Function of Temperature Excursion, *IEEE Trans. on Magnetics*, **29**, 517–22 (1993).

[13.12] Aloisi, P. A. Failure Diagnostics in Medium Power Semiconductors, *Proc. EPE'91*, vol. 3, 117–20 (1991).

[13.13] Coquery, G., Lallemond, R., Wagner, D. and Gibard, P. Reliability of the 400 A IGBT Modules for Traction Converters – Contribution on the Thermal Fatigue Influence on Life Expectancy, *Proc. EPE'95*, vol. 1, 60–5 (1995).

[13.14] Zeller, H. R. Cosmic Ray Induces Failures in High Power Semiconductor Devices, *Microelectronics & Reliability*, **37**, 1711–18, (1997).

[13.15] *Military Standardization Handbook: Reliability Prediction of Electronic Equipment*, MIL-HDBK-217 (US Department of Defense, Washington, 1982).

[13.16] British Telecom plc, *Handbook of Reliability Data for Electronic Components used in Telecommunications Systems* (Issue 3, 1984).

[13.17] O'Connor, P. D. T. Microelectronic Device Reliability, *IEEE Trans. on Reliability*, **32**, 9–13 (1983).

# 14

# FUTURE MATERIALS AND DEVICES

## 14.1  Materials other than Silicon

In Section 2.1.4 the thermal stability of p–n junction devices is shown to be limited by temperature. When the intrinsic carrier concentration becomes too high, the power dissipated exceeds the possible rate of heat extraction and thermal breakdown occurs. An intrinsic carrier concentration of $\sim 10^{21}\,\text{m}^{-3}$ is a practical limit [14.1]. In silicon this occurs at about 280 °C and the maximum practical operating temperature for silicon devices is between 150 and 200 °C. As the intrinsic concentration depends exponentially on the ratio $W_g/2kT$, materials with a wider bandgap $W_g$ reach this limit at higher temperatures and so have potential advantages [14.2]. However, many of the fabrication processes outlined in Section 3.2 depend on the particular combination of physical and chemical properties exhibited by silicon and its compounds. They give this material a unique advantage over other semiconductors.

The discussion in Chapter 13 shows that for any semiconductor material to be considered for the fabrication of commercial power semiconductor devices, the important parameters include not only the bandgap energy, but also the thermal conductivity, the carrier mobilities and the mechanical strength. The level of material technology is crucial and should be able to provide large and perfect single crystals and support processes capable of realising two- and three-dimensional structures with good reproducibility.

The physical properties of silicon and some other semiconductor materials with a wider bandgap are compared in Table 14.1. The conventional symbols are used:

$W_g$ = band gap energy
$\varepsilon_r$ = relative permittivity
$\mu_n$ = electron mobility
$\mu_p$ = hole mobility
$v_{sat}$ = saturation drift velocity
$E_{BR}$ = breakdown electric field
$\kappa$ = thermal conductivity

$n_i$ = intrinsic carrier concentration
$\alpha$ = thermal expansion coefficient
$H$ = microhardness.

Generally, an increase in the bandgap energy results in an increase of the built-in voltage at a p–n junction and consequently an increase in the forward voltage drop of bipolar devices. This need not be a crucial problem for device application, but it does give an advantage to unipolar devices.

The use of wide bandgap semiconductors can improve the maximum operating junction temperature $T_{Jmax}$, the maximum breakdown voltage and the dynamic behaviour.

The increase of $T_{Jmax}$ with the semiconductor bandgap follows from (2.48) and (2.49). In silicon devices it is normally less than about 150 °C, with GaAs devices it can be in the region of 200 °C, for SiC devices it can reach over 400 °C and diamond devices might operate at 700 °C. As a consequence of (13.7), the increase in the maximum junction temperature results in a considerable increase of the maximum allowed steady-state power loss and hence of the working current density. This is offset by the decrease in the thermal conductivity of semiconductors with increasing temperature.

It can be seen in Table 14.1 that GaAs, GaP and InP are disadvantaged by their relatively low thermal conductivities. They also have a poorer mechanical strength than silicon and so are in greater danger of suffering thermal fatigue. Silicon carbide has excellent mechanical and thermal properties, including an expansion coefficient that better matches the coefficients of encapsulation materials. The expansion coefficient of diamond is very low and can cause serious mismatch problems.

The second advantage of wide bandgap semiconductors is the higher critical breakdown field in comparison with silicon. This allows the same breakdown voltage to be reached at a p–n junction with a higher doping concentration in the more lightly doped material. If the ratio between the critical breakdown field for a wide bandgap semiconductor and for silicon is $\psi$, it follows from (2.36) that the permitted maximum doping concentration is proportional to $\psi^2$. From (2.41) it follows that the space charge region at a p–n junction extends a distance inversely proportional to $\psi$. The thickness of the layers needed to support a given voltage is decreased accordingly. The carrier transit times in bipolar structures are decreased in inverse proportion to $\psi^2$, as can be seen from (6.47). The on-state resistance of unipolar devices such as MOSFETs and Schottky diodes is also decreased by the higher doping levels and reduced layer thickness made possible by the higher dielectric strength of wide bandgap materials.

Various figures of merit for semiconductor materials have been proposed to characterise the potential performance limitations of power semiconductor devices made from them [14.3 to 14.5]. All of these figures demonstrate the intrinsic superiority of GaAs and SiC over silicon and the potential of diamond to outperform either. However, diamond is probably the most difficult of these materials to use in practice. The following three sections give a brief summary of the achieved performance of devices made from these three materials.

**Table 14.1** Physical parameters of undoped semiconductor materials at 300 K

| | Si | GaAs | InP | GaP | GaN | 3C-SiC | 6H-SiC | 4H-SiC | Diamond |
|---|---|---|---|---|---|---|---|---|---|
| $W_g$ (eV) | 1.12 | 1.42 | 1.35 | 2.26 | 3.36 | 2.2 | 2.9 | 3.2 | 5.5 |
| $\varepsilon_r$ | 11.9 | 12.9 | 14 | 11.1 | 10 | 9.7 | 10 | 9.7 | 5.5 |
| $\mu_n$ (m$^2$/V s) | 0.143 | 0.85 | 0.46 | 0.011 | 0.038 | 0.01 | 0.046 | 0.08 | 0.19 |
| $\mu_p$ (m$^2$/V s) | 0.047 | 0.009 | 0.015 | 0.0075 | 0.015 | 0.005 | 0.005 | 0.012 | 0.12 |
| $v_{sat}$ (m/s) | $10^5$ | $2 \times 10^5$ | $2 \times 10^5$ | $2 \times 10^5$ | $2.5 \times 10^5$ | $2.5 \times 10^5$ | $2 \times 10^5$ | – | $2.7 \times 10^5$ |
| $E_{BR}$ (V/m) | $3 \times 10^7$ | $4 \times 10^7$ | $6 \times 10^7$ | $7 \times 10^7$ | $2 \times 10^7$ | $1.5 \times 10^8$ | $2.2 \times 10^8$ | $2.4 \times 10^8$ | $7 \times 10^8$ |
| $\kappa$ (W m$^{-1}$ K$^{-1}$) | 150 | 50 | 70 | 50 | 150 | 500 | 500 | 490 | 2000 |
| $n_i$ (m$^{-3}$) | $1.45 \times 10^{16}$ | $1.8 \times 10^{12}$ | – | $1.3 \times 10^6$ | – | $5 \times 10^6$ | 30 | – | $10^{-16}$ |
| $10^6 \alpha$ (K$^{-1}$) | 2.4 | 5.5 | 4.75 | 5.8 | 5.1 | 3.8 | 3.3 | 3.3 | 0.8 |
| $H$ (GPa) | 10.6 | 6.4 | 4.3 | 9.2 | 12 | 29 | 25 | 25 | 86 |

## 14.2 Gallium Arsenide Devices

Gallium arsenide is the most mature semiconductor material after silicon, and high quality single crystals are available in diameters exceeding 75 mm. As a starting material for electronic devices, gallium arsenide is typically some seven times more expensive than silicon [14.6]. The operations needed for device fabrication are more complicated and thus more expensive compared to silicon. One serious problem is that native oxide layers cannot be grown thermally in the controlled way that is possible with silicon. Indeed, they tend to be conducting. As a result, silicon dioxide or silicon nitride layers, put down by chemical vapour deposition, have to be used for passivation, diffusion masking and insulation. More effective techniques are under development [14.7], but achieving the maximum theoretical breakdown voltage is difficult because of the effects of surface breakdown. Gallium arsenide has a relatively low mechanical strength and its devices are prone to thermal fatigue. They are therefore not suitable for applications in which they are likely to be subjected to a large number of thermal cycles.

All acceptor impurities have high diffusion coefficients in GaAs. This makes p–n junctions inherently unstable and prevents multiple diffusion processes. GaAs devices thus tend to be unipolar, rather than bipolar.

Although some GaAs devices may require less die material than their silicon equivalent, they are likely to be relatively expensive. For them to be used beneficially, some key aspect of their performance must be considerably better than that of a comparable silicon device.

As can be seen in Table 14.1, the breakdown field of GaAs is of the same order as that of silicon but the thermal conductivity is significantly less. The principal advantage of GaAs as the basis for power semiconductor devices is thus its high electron mobility, which is six times that of silicon. The series resistance of a lightly doped epitaxial layer is thus six times less than one of comparable doping level and dimensions in silicon. This results in an improvement in the forward characteristics of GaAs devices and a decrease in their $RC$ time constant. Schottky diodes based on GaAs have been shown to have a much lower $Q_{rr}$ than silicon diodes, particularly if the comparison is made at elevated temperatures [14.8]. They have relatively soft reverse recovery characteristics. Both Schottky diodes and GaAs p–i–n diodes (up to 600 V, 20 A) have come into volume production [14.9].

The lower $Q_{rr}$ associated with GaAs diodes allows a considerable reduction in the switching losses for circuits in which there is forced recovery of a freewheeling diode. The consequent ability to operate at a higher frequency can reduce the size and cost of passive components employed in the circuit, thereby offsetting the relatively high cost of the GaAs rectifier.

Vertical geometry GaAs MESFETs (metal–semiconductor field-effect transistors) [14.10] and also MISFETs (metal–insulator semiconductor field-effect transistors) [14.11] have been reported as being capable of high switching speeds at high power. In either case the benefits sought are high operating frequency and low on-state resistance per unit area of device. These devices offer the prospect of efficient switching at a frequency of 100 MHz and a specific on-state resistance of 0.13 m$\Omega$ cm$^2$.

Notwithstanding the advantages of GaAs devices, more research activity is focused on silicon carbide power devices, perhaps reflecting their potentially greater performance advantages over both silicon and GaAs devices. Nevertheless, it is likely that GaAs Schottky diodes will become well established in power applications.

## 14.3 Silicon Carbide Devices

Silicon carbide has several different stable crystalline structures, known as polytypes. The three listed in Table 14.1 are the most common and, as described in Section 14.1, their properties show great potential for use in future, high temperature, high voltage semiconductor devices. The 3C-form has a cubic lattice of the zincblende structure. The others are hexagonal, differing in the cell height. The number indicates how many basic layers of 0.251 nm spacing form the elementary cell. The hexagonal forms are preferred for electronic devices; 6H-SiC initially received the most attention but, since its commercial availability, 4H-SiC now shows the greater promise. The cubic form is isotropic, whereas the hexagonal forms are anisotropic. This means that parameters such as the carrier mobility and the electrical permittivity take the form of tensors, and the carrier drift velocity and the dielectric polarisation may take up directions that differ from that of the applied electric field. The anisotropy for 6H-SiC is greater than for 4H-SiC.

A major difficulty in producing silicon carbide devices has been the lack of starting material with a sufficiently low level of defect density. Micropipes and dislocations are the principal types of defect. Micropipes are micron-diameter holes that can extend through the full length of a single-crystal boule. Intensive materials research has led to a considerable reduction in such defects [14.12] and three-inch wafers of acceptable quality for the production of power semiconductor devices are a distinct possibility.

The higher breakdown field and thinner layers possible with SiC can give unipolar devices a specific on-resistance 1/300 that of equivalent silicon devices. However, surface passivation remains a major concern. At high fields and high temperatures the reliability and lifetimes of $SiO_2$ layers grown on SiC have proven unsatisfactory. Because SiC devices need to operate at high temperatures in order to realise their full potential, this is a serious problem. However, it has been shown that SiC devices could be much smaller than similarly rated silicon devices with significantly lower power losses, and considerable worldwide research effort is being invested in the development of power semiconductor devices based on SiC.

## 14.4 Diamond Devices

One of the most significant characteristics of diamond as a potential material for power semiconductor devices is its high thermal conductivity, which permits heat to be extracted efficiently from the parts of the device where losses are generated. Its wide bandgap energy of 5.5 eV allows it to operate at temperatures as high as 650 °C. Its breakdown field is many times that of silicon, although only about twice that of silicon carbide. It is therefore particularly suitable for use in high voltage devices.

All of the figures of merit mentioned in Section 14.1 demonstrate the potential of diamond to outperform GaAs and SiC. However, it is probably the most difficult to use in practice. Good adherence of metallisation is difficult to achieve. Doping with boron to make p-type material is relatively easy but n-type doping is not. The obvious candidates are nitrogen and phosphorus, and both present major difficulties. Schottky diodes and field-effect transistors employing hole conduction are therefore expected to be the first practical devices.

Native single-crystal diamond is clearly a very expensive starting material and much research is based on the use of polycrystalline diamond films formed by chemical vapour deposition. The performance of devices based on polycrystalline diamond is generally poorer than the performance of single-crystal devices; this is because of the effects of the grain boundaries in the polycrystalline material. Efforts are being made to grow single-crystal films, which would lead to much improved device characteristics. The high thermal conductivity and low electrical conductivity of intrinsic diamond means that undoped diamond films form excellent insulating layers between active semiconductor devices and their heat sink. They introduce little thermal impedance while providing the required level of electrical isolation.

Research on wider bandgap materials and the development of power semiconductor devices based on them can be expected to continue actively, with the aim of overcoming the operational limitations of silicon devices that have been explored in this book. However, the replacement of silicon in the the vast majority of applications cannot yet be foreseen.

# References

[14.1] Schlangenotto, H. and Niemann, E. Switching Properties of Power Devices on Silicon Carbide and Silicon, *Proc. EPE-MADEP Conference*, vol. 0, 8–13 (1991).

[14.2] Locatelli, M. L., Gamal, S. H. and Chante, J. P. Semiconductor Materials for High Temperature Power Devices, *EPE Journal*, **4**, 43–5 (1994).

[14.3] Johnson, E. O. Physical Limitations on Frequency and Power Parameters of Transistors, *RCA Review*, **26**, 163–77 (1965).

[14.4] Keyes, R. W. Figure of Merit for Semiconductors for High Speed Switches, *Proc. IEEE*, **60**, 225 (1972).

[14.5] Baliga, B. J. Power Device Figure of Merit for High-Frequency Applications, *IEEE Electron Device Letters*, **10**, 455–7 (1989).

[14.6] Anderson, S. GaAs Rectification – An Enabling Technology for High Frequency Operation of Power MOS-Gated Transistors, *Proc. ISPSD'96*, 33–9 (Hawaii, USA, 1996).

[14.7] Wright, N. G., Johnson, C. M. and O'Neill, A. G. Surface Passivation Techniques for GaAs Power Schottky Diodes, *Proc. ISPSD'97*, 141–4 (Weimar, Germany, 1997).

[14.8] Shenai, K., Hodge, D., Feuer, M. D. and Cunningham, J. Novel Ultralow $R_{on}$ MBE GaAs MESFETs for High-Frequency High-Temperature Switched-Mode Power Converter Applications, *Proc. ISPSD'94*, 149–53 (Davos, Switzerland, 1994).

[14.9] *PCIM Europe*, **1/97**, 6 (1997).

[14.10] Wright, N. G., Johnson, C. M., O'Neill, A. G. and Hossin, M. Vertical GaAs MESFET for Smart Power Switching Applications, *Proc. EPE'95*, vol. 1, 620–4 (Seville, Spain, 1995).

[14.11] Thomas, H., Luo, J. K., Morgan, D. V., Westwood, D., Lpika, K., Splingart, E. and Kohn, E. Improvement of the Breakdown Voltage of GaAs-FETs Using Low-Temperature-Grown GaAs Insulator, *Proc. ISPSD'94*, 155–60 (Davos, Switzerland, 1994).

[14.12] Palmour, J. W., Singh, R., Glass, R. C., Kordina, O. and Carter, C. H. Jr., Silicon Carbide for Power Devices, *IEEE Proc. ISPSD'97*, 25–32 (1997).

# Appendix

# THE DIFFUSION EQUATION

## A.1 Basic Concepts

The diffusion equation, which describes the flow and recombination of excess carriers in a semiconductor, is central to the analysis of many aspects of device behaviour. It recurs several times in different chapters. In this appendix we set out some of the methods by which analytical solutions can be obtained when the equation is linear, i.e. when the coefficients $D_a$ and $\tau$ in (1.77) are not functions of $\Delta n$.

Altogether, there are seven different types of mathematical solution [A.1] and, provided the equation is linear, the full solution is the summation of all possible solutions. In practice the initial and boundary conditions imposed by any particular physical problem are satisfied by only one or two of the solution types, causing the others to be eliminated.

Slightly different definitions for the diffusion coefficient and the recombination lifetime in semiconductors are appropriate in different circumstances. For example, under high injection conditions, the ambipolar diffusion coefficient $D_a$ should be used, whereas with low injection, this reduces to $D_n$ or $D_p$, depending on whether the electrons or the holes are the minority carriers. Similarly, the lifetime depends on the dominant recombination process. For generality, in this appendix we use the symbols $D$ and $\tau$ without suffices and assume them to be uniform and unvarying throughout the region under consideration. Equation (1.77) thus becomes

$$\frac{\partial \Delta n}{\partial t} = D \frac{\partial^2 \Delta n}{\partial x^2} - \frac{\Delta n}{\tau} \tag{A.1}$$

The final term in (A.1) represents the net rate of recombination of the excess carriers. It can be eliminated by making the substitution

$$N = \Delta n \exp\left(\frac{t}{\tau}\right) \tag{A.2}$$

Equation (A.1) then becomes

$$\frac{\partial N}{\partial t} = D \frac{\partial^2 N}{\partial x^2} \tag{A.3}$$

and all the standard results from the mathematics of diffusion and thermal conduction [2.6, 2.7] can be applied directly.

The further substitutions $X = x/\sqrt{D\tau} = x/L$, where $L$ is the diffusion length, and $T = t/\tau$ lead to the dimensionless form of (A.3):

$$\frac{\partial N}{\partial T} = \frac{\partial^2 N}{\partial X^2} \tag{A.4}$$

## A.2 The Effect of Recombination

In the absence of recombination, i.e. with $\tau = \infty$, (A.1) and (A.3) are identical. In that situation, let $\Delta n_0(x,t)$ be the solution of (A.1) for a particular set of initial and boundary conditions. That is, $\Delta n_0(x,t)$ is a solution of

$$\frac{\partial \Delta n_0}{\partial t} = D \frac{\partial^2 \Delta n_0}{\partial x^2} \tag{A.5}$$

Under the same initial and boundary conditions, when recombination *is* significant, the solution becomes

$$\Delta n(x,t) = \frac{1}{\tau}\int_0^t \Delta n_0(x,t') \exp\left(\frac{-t'}{\tau}\right) dt' + \Delta n_0(x,t) \exp\left(\frac{-t}{\tau}\right) \tag{A.6}$$

where $t'$ is a dummy time variable.

Note that, when $\tau$ is finite, we cannot simply take a solution of (A.3) and substitute (A.2), because the boundary conditions imposed by $N$ and $\Delta n$ are quite different.

Our approach to any particular problem is to seek solutions $\Delta n_0(x,t)$ to (A.5), under the assumption that there is no recombination and insert the result into (A.6). The use of computational algebra packages, such as Maple and Mathematica, makes such a method practical in many cases. It often gives greater insight and more reliable solutions than a direct numerical analysis.

## A.3 Methods of Solution

### A.3.1 The Laplace Transform

A powerful method for dealing with partial differential equations like the diffusion equation, when they are linear, is to use the Laplace transform to remove the time variable. They are thus converted into ordinary differential equations. We illustrate the method by applying it to solve (A.5) for the particular case of a volume of semiconductor that is unbounded for $x>0$, when, at $t=0$, $\Delta n_0=0$ for $x>0$ and when, for $t>0$, $\Delta n_0=\Delta n_1$ at $x=0$. This is a situation that may arise at an $n^+$–p junction, as discussed in Section 2.1.6. The $x$-coordinate is set at right angles to the boundary between the space charge region and the neutral p-type semiconductor, with its origin at the boundary. If there is no recombination, equation (A.5) follows from the substitution of (2.52) into (2.51). Initially there are no excess carriers. Then, at time $t=0$, the excess electron concentration at $x=0$ is raised to and maintained at the value $\Delta n_1$, by applying a constant forward bias voltage to the junction.

In order to obtain the Laplace transform, each side of (A.5) is multiplied by $\exp(-st)$ and the product is integrated between $t=0$ and $t=\infty$. It is assumed that the new variable $s$ is large enough to ensure that the integral converges. It is also assumed that the functions permit the order of integration and differentiation to be interchanged. The left-hand side may be integrated by parts:

$$\int_0^\infty \exp(-st)\frac{\partial \Delta n_0}{\partial t}\,dt = [\Delta n_0 \exp(-st)]_0^\infty + s\int_0^\infty \Delta n_0 \exp(-st)\,dt = s\mathcal{L}\{\Delta n_0\} \quad (A.7)$$

where $\mathcal{L}\{\Delta n_0\}$ is, by definition, the Laplace transform of $\Delta n_0$. The first term on the right-hand side is zero because $\Delta n_0 = 0$ at $t=0$ and the exponential term is zero at $t=\infty$.

For the right-hand side of (A.5),

$$\int_0^\infty \exp(-st)\frac{\partial^2(\Delta n_0)}{\partial x^2}\,dt = \frac{\partial}{\partial x^2}\int_0^\infty \Delta n_0 \exp(-st)\,dt = \frac{\partial^2 \mathcal{L}\{\Delta n_0\}}{\partial x^2} \quad (A.8)$$

Thus, equation (A.5) transforms to the ordinary differential equation

$$D\frac{d^2 \mathcal{L}\{\Delta n_0\}}{dx^2} = s\mathcal{L}\{\Delta n_0\} \quad (A.9)$$

At $x=0$,

$$\mathcal{L}\{\Delta n_0(0)\} = \int_0^\infty \Delta n_1 \exp(-st)\,dt = \frac{\Delta n_1}{s} \quad (A.10)$$

The solution of (A.9), subject to (A.10) and giving $\mathcal{L}\{\Delta n_0\} = 0$ at infinity, is

$$\mathcal{L}\{\Delta n_0\} = \frac{\Delta n_1}{s}\exp\left(-\frac{x}{\sqrt{D/s}}\right) \quad (A.11)$$

Tables of Laplace transform pairs show that (A.11) transforms back to

$$\Delta n_0(x,t) = \Delta n_1 \operatorname{erfc}\left(\frac{x}{2\sqrt{Dt}}\right) \quad (A.12)$$

where

$$\operatorname{erfc}(y) = 1 - \frac{2}{\sqrt{\pi}}\int_0^y \exp(-\zeta^2)\,d\zeta \quad (A.13)$$

is the well-known complementary error function.

Note that $\Delta n_0$ is a function only of $x/2\sqrt{Dt}$, as mentioned in Section 2.1.6. This means that the carrier concentration at any particular point $x$ increases to a given value in a time that is proportional to $x^2/4D$.

The solution when the carriers have a finite recombination lifetime $\tau$ is obtained by inserting (A.12) into (A.6).

### A.3.2 Separation of the Variables

Solutions to (A.5) that take the form

$$\Delta n_0(x, t) = \xi(x)\eta(t) \tag{A.14}$$

can be found using the method of separation of variables. An example is given at the end of Section 1.6. It can also be applied to the situation discussed in the last section, when a second boundary is introduced at $x=w$, at which $\Delta n_0(w) = l\{\Delta n_0(w)\} = 0$. Such a boundary might take the form of an ohmic contact or another p–n junction. With a reverse-biased p–n junction the minority carrier concentration at the edge of the space charge region may be taken as zero. In the case of an ohmic contact, the equilibrium carrier concentrations are maintained. In either circumstance we can assume that $\Delta n_0(w, t) = 0$.

The solution is found to consist of two parts: the final steady-state solution plus a transient solution that takes the form of a decaying infinite Fourier series:

$$\Delta n_0(x, t) = \Delta n_1 \left(1 - \frac{x}{w}\right) + \frac{2}{\pi}\Delta n_1 \sum_{m=1}^{\infty} \frac{(-1)^m}{m} \exp\left(-\frac{m^2\pi^2 Dt}{w^2}\right) \sin\frac{m\pi x}{w} \tag{A.15}$$

where $m$ is a positive integer. At $t=0$ the Fourier series represents the difference between the initial distribution of carriers and the steady-state solution. Note that the higher harmonics decay the most rapidly. Differentiation of (A.15) shows that the flux of minority carriers at $x=w$ increases as a function of $(Dt/w^2)$ and so reaches any given level after a time that is proportional to $w^2/D$.

It is again necessary to use (A.6) to take account of recombination, which can be neglected only if $w \gg \sqrt{D\tau}$.

As (A.1) is linear, solutions can be combined using the principle of superposition. Thus, the solutions we have given for the consequences of a step change in the carrier concentration at the boundary can be added to any pre-existing steady-state carrier concentration distribution. This is, of necessity, a solution satisfying the same initial and boundary conditions.

## A.4 The Constant Current Condition

As is shown in Section 2.1.6, under normal circuit operating conditions, it is more likely to be the current flow through the p–n junction that is controlled during a switching transient, rather than the voltage across it. The flow of carriers implied by (A.12) and (A.15) can be found from the value of the concentration gradient at $x=0$. For the carrier distribution given by (A.12), the current density is

$$J = eD\left.\frac{\partial n}{\partial x}\right|_{x=0} = \frac{eD\Delta n_1}{\sqrt{\pi Dt}} \tag{A.16}$$

In order to keep the current constant, the excess carrier concentration at the surface must rise as the square root of time. In the absence of recombination, the resulting solution of the diffusion equation is

$$\Delta n_0(x, t) = \frac{2J}{e} \left\{ \sqrt{\frac{t}{\pi D}} \exp\left(\frac{-x^2}{4Dt}\right) - \frac{x}{2D} \operatorname{erfc}\left(\frac{x}{2\sqrt{Dt}}\right) \right\} \quad \text{(A.17)}$$

It can again be seen that there is a strong dependence of the carrier concentration on $x/2\sqrt{Dt}$. To take account of recombination this result, too, has to be put into (A.6). It is illustrated schematically in Figure 2.15.

# Reference

[A.1] Olver, P. J. *Applications of Lie Groups to Differential Equations* (Springer-Verlag, 2nd Edn, 1993). See for example pp. 117–20.

# INDEX

Abrupt junction assumption
  n–n$^+$  57, 58
  n$^+$–i  154
  p–n  33, 42–47, 49, 53–56, 157, 169, 172, 210, 211
  p$^+$–i  154
  p–p$^+$  57, 58
absorption coefficient  16, 268
acceptors  231, 293, 400
acceptor concentration  5–7, 12, 13, 30, 33, 42, 45, 57, 60, 154, 170, 188, 192, 209, 222, 291, 292
acceptor levels  2, 3, 18, 19, 23, 42, 104, 107
accumulation layer  57, 58, 61–64, 289, 294, 297, 298, 304
active region  57, 172, 173, 185–188, 229, 258, 302, 303, 318, 319
air (cooling properties)  371–6
alkali metals (ions)  100, 290
alpha (common-base current gain)  179–182, 216–219
alpha particle irradiation  24, 116–118, 170
Al/SiC metal matrix composite  371, 394
  properties  366
alternator  174
alumina (Al$_2$O$_3$)  345, 367, 370, 379, 382
  properties  366
aluminium (Al),
  as contact  103, 104, 112, 113, 258, 294, 392
  as dopant  2, 3, 106–109
  diffusion coefficient in silicon  107
  heat sinks  372, 373
  properties  366
  solid solubility in silicon  106
aluminium nitride (AlN)  124, 370, 371
  properties  366
aluminium/silicon alloy system  113

ambipolar diffusion (coefficient, length)  28, 148, 153, 154, 241, 405
amplifying gate  134, 228, 229, 232, 233, 248, 250, 256, 267
anisotropic etch  103, 293
annealing  110, 118
anode shorts  238, 239, 244, 271
antimony (Sb)  1, 3, 106, 107, 113
antiparallel diode  79, 136, 137, 206, 207, 251–254, 257, 261, 295, 314, 332, 343, 345, 393
applications  127–145
  BJT  77, 137, 207, 208
  diode  127–129, 140, 143, 174, 175
  GTO  134, 257, 261, 264
  IGBT  132, 134, 136, 141, 143, 333
  MOSFET  136, 139–143
  Schottky diode  140
  thyristor (SCR)  79, 129–134, 247–250, 267
  triac  264
arsenic (As)  1, 3, 106, 107
asymmetric FCD  282
asymmetrical junction  46, 54, 211
asymmetric thyristor  79, 250–253, 261
Auger recombination  18, 24–26, 114, 153, 193, 248
avalanche breakdown  50, 138, 203, 210, 269
avalanche devices (diodes)  295, 354, 385, 388, 389
avalanche injection  202, 203, 260, 348
avalanche ionisation  17, 202
avalanche multiplication coefficient  45, 46, 183
avalanche multiplication factor  48, 216–218

Backporch current  259
balancing transformer  379, 380, 385

band diagram   1–9, 18, 19, 33, 34, 57–66, 178
  impurity energy levels   2, 3, 19
  p–n junction   34, 52
  bipolar transistor   178
  Schottky junction   62–64
band edges   7, 8
bandgap energy   1, 397–399
  function of temperature   4
  narrowing   7, 8
base charge   197–199
base transit time   196–200, 323
base widening (current-induced)   182, 186–192, 200
beryllia (BeO)   345, 366, 371
bevelling   122–124, 212, 213
bipolar diffusion   28, 241
bipolar junction transistor (BJT)   135, 137
  parasitic   211, 295, 312–314, 322
blocking voltage   89
  GTO   259–261
  IBT   338
  IGBT   322
  MCT   335
  p–n junction diode   147
  Schottky diode   74
  SITh   283
  thyristor (SCR)   209–214, 330, 238, 247–253, 269
bonding   371
bonding pad   393
boost regulator   139
boron   2, 3, 11, 12, 97, 106–109, 115, 294, 402
  diffusion coefficient   107
  solid solubility   106
boron nitride (BN)   124
breakdown 38, 45–49, 118–124, 209–213, 282, 293–295, 389
  effect of junction curvature   118, 119
  field   46, 121, 398–401
  gate oxide   294, 299
  in BJT   183–185
  second   50, 194, 202–205, 260–263, 312, 313, 329, 348, 381, 389, 390
  thermal   49–51, 194, 202, 219, 249, 279, 383, 392
  voltage   46–48, 156–159, 171–174, 183–185, 211–213, 238, 252, 257, 314, 322, 348, 392, 398, 400
breakover diode (BOD)   269–271

breakover voltage   281, 360
bridge circuits   127–131, 138
buck-boost regulator   139
buck regulator   138–140
buffer layer   331, 332
built-in field   30, 31, 58, 59, 181
buried gate   80, 273


Capacitance   128, 144, 233, 284, 311, 323, 324, 327, 330, 386–8
  diffusion   29, 39, 45
  effect of   128, 144, 233, 284, 311, 323, 324, 327, 330, 386–8
  Mille   304, 307, 384
  MOSFET   304–312
  oxide   290, 304, 320
  parasitic   263, 304, 305, 327
  p–n junction   45, 221–223, 230, 233, 278
  Schottky   171, 140
  SIT   278, 279
  snubber   166, 263, 388
  thermal   361–363
  thyristor   221
capacitance divider   248, 283
capture cross-section   20, 21
carrier–carrier scattering   13, 14, 28, 153, 175
carrier diffusion length   36, 59, 148, 155, 179–183, 211, 219, 224, 241, 406
carrier generation   16–19, 24, 27, 28, 39, 48, 50–52, 211, 222, 266
carrier lifetime   17–26, 29, 36, 39, 59, 83, 148, 152–158, 168–171, 174, 178, 180, 214, 215, 226, 234, 238, 241, 243, 249, 251, 257, 259, 261, 281, 314, 322, 325, 331, 332, 335
  control   24, 114–118, 170, 234, 244, 248, 257, 261, 315, 329
  effective   29, 40, 167–169, 199, 200, 244, 249, 255
  high level   17, 24, 241
  low level   17, 23
  profile   166, 169, 170
  temperature dependence   36
carrier mobility   10–16, 29
  differential   28
carrier recombination   16–26
carrier saturation drift velocity   15, 16, 300
cathode (emitter) shorts   72, 219, 220, 226, 227, 230–238, 249, 255

cell structure (MOSFET, IGBT) 83, 295, 296, 318
charge
  critical 249, 254
  excess 16–26, 29, 38, 39, 57, 72, 89, 160–167, 197, 199, 225, 233, 234, 239, 241, 242, 244, 262, 266, 324
  inversion layer charge 290–292, 300, 301
  stored 127, 135, 136, 167, 175, 183, 192, 196, 198, 239, 325, 333, 348
charge-control model 28, 29, 166, 220, 239, 325
chemical vapour deposition (CVD) 108, 110–112
chopper circuit 253
clamped inductive load 92, 141, 306, 308–310, 312, 325
clamping diode 143, 306
common-base configuration 177–181, 216, 320
  current gain 179, 183, 184, 216, 313, 320
common-emitter configuration 181–186, 211
  current gain 181, 182, 192–194
commutation 79, 129, 130, 133, 134
  forced 79, 131, 133, 234
  natural 79, 131, 209
compensated material 2, 6, 33
complementary error function 105, 106, 407, 409
compressed-field structure 169, 186, 251
conduction band 1–3
conduction (on-state) losses 83, 114, 136, 137, 241, 330, 332
conductivity 9–11
conductivity modulation 85, 141, 151, 153, 161, 186–189, 192, 200, 273, 284, 303, 324
contact potential 291
contacts 1, 29, 57, 62–66, 175, 228, 284, 350, 392
  formation 112, 113, 293, 294
  ohmic 65, 66, 408
  Schottky (rectifying) 24, 62–65, 147, 171–174, 274
contact potential 291
contact resistance 73, 188, 232, 323, 353
continuity equations 28, 36, 53, 150, 153, 160, 167, 177, 196, 325
converters 138–142
  boost 139, 142, 143
  buck-boost 139, 141, 142

fly-back 139, 141, 142
forward 140, 141
cooling 359–379, 390, 393
copper (Cu) 112, 114, 116, 345, 366, 367, 370, 371, 379
  properties 366
critical charge 249, 254
current
  crowding 183, 193, 194
  holding 79, 215, 218, 233, 236, 259, 264, 266, 334, 335
  latching 218, 220, 224, 323, 33
current filaments 50, 215, 236, 385
current gain
  GTO turn-off 238, 242, 243, 259
  BJT 179–182, 205–207
current-induced base widening 186–192
current tail 244, 257, 262, 325, 329, 336
current–voltage characteristics
  BJT 73, 152, 155, 156
  diode 73, 152, 155, 156
  IGBT 87
  JBS 172
  JFET 81, 276
  MOSFET 84, 86
  Schottky diode 172
  SIT 82, 278
  thyristor 78, 216
  triac 80
cut-off 76, 185
  frequency 278, 279, 305
cycloconverter 130, 132
Czochralski crystal growth 94–98


Darlington configuration 77, 137, 205–208, 317, 318, 333, 343, 345
dc–dc converters 138–142
Debye length 31
deep levels 116
defects 4, 24, 49, 114–118, 174, 212, 392, 401
  density 93, 99, 116
  oxide 100, 392
  surface 123
degenerate semiconductor 8, 9
delay time 196, 222, 224, 225, 307–310, 323, 324, 339, 381, 382
density of states 4, 7, 8

depletion layer  24, 33–36, 40–49, 169, 172, 178–180, 189–192, 198, 200, 234, 236, 239, 241, 257, 273, 283, 288, 293, 300
   capacitance  45, 221, 223, 230
   graded junction  43
   Schottky junction  63, 64
diamond  98, 124, 398, 401, 402
   properties  399
$di/dt$ limitation  226–230, 248–250, 259, 263, 264, 267, 284
dielectric constant (permittivity)  30, 31, 42–46, 60, 61, 65, 124, 157, 158, 190, 210, 211, 274, 275, 278, 290–292, 397, 399, 401
dielectric relaxation time  31
differential capacitance  45, 222, 23
differential expansion  359, 364, 370
differential mobility  28, 149
differential resistance  155, 156, 321, 379
   negative  184, 281
diffused guard ring  118, 120
diffused junction  42–45
diffusion
   capacitance  29, 39, 45
   equation  28, 31
   Fick's laws  26, 27
   of carriers  26–29
      ambipolar  28, 148, 153, 154, 241, 405
      bipolar  28, 241
   of impurities  104–109
diffusion coefficients
   electrons and holes  26–28
   impurities in silicon  42, 107
diffusion length  36, 59, 148, 155, 179–183, 211, 219, 224, 241, 406
diffusion potential  35, 57, 154, 278, 291
direct copper bonding (DCB) technology  370, 371
diode
   anti-parallel  79, 136, 137, 206, 207, 251–254, 257, 261, 295, 314, 332, 343, 345, 393
   applications  174, 175
   double diffused  72–74
   epitaxial  74
   p–i–n  148–152
   rectifier circuits  127–131
   Schottky  62–65, 74, 75, 171–174
direct band-to-band recombination  17, 24
dislocations  401
donor levels  2

doping profile  43, 45, 73–78, 157, 158, 252
double-gate GTO (DGTO)  262
double injection (of carriers)  72, 87, 148, 333
double positive bevel  213
drain resistance  297
drift velocity  14
   saturation  15, 16
dry etching  103, 104
$dV/dt$ limitation  230–232, 248, 260, 266, 267, 283, 314, 328
dynamic characteristics  160, 171, 205, 214, 227, 251, 382

Early effect  182
effective carrier lifetime  29, 40, 167–169, 199, 200, 244, 249, 255
effective mass (electrons, holes)  4, 10
efficiency
   emitter  41, 58, 59, 150, 153, 170, 179–184, 192, 205, 211, 214, 219, 238, 249, 322, 332, 334
   heat transfer  360, 371, 373, 378, 393
   power  135
Einstein relation  27
electromagnetic compatibility (EMC)  132
electron affinity  62
electron irradiation  116, 170, 234
encapsulation  364–371, 398
enhancement mode  83, 84, 287, 288, 293
epitaxial growth  81, 99, 110, 114, 350
etching  74, 98, 99, 103, 104, 123, 124, 212, 213
   anisotropic  103, 293
   dry  103, 104
   isotropic  103
   plasma  104
   reactive ion  104
   wet  103
eutectic alloy
   BiInPbSn  379
   $CuO_2/Cu$  370
   with silicon  51, 109, 113, 366
excess charge  16–26, 29, 38, 39, 57, 72, 89, 160–167, 197, 199, 225, 233, 234, 239, 241, 242, 244, 262, 266, 324
expansion coefficients  366, 399

Failure modes  51, 194, 222, 259, 371, 390–394
failure probability function  391

Fermi–Dirac distribution  3, 4
Fermi level (energy)  3–9, 29, 30
Fick's laws (of diffusion)  26, 27, 105
field-controlled diode (FCD)  81–83, 280–284
field-controlled thyristor  81–83, 280–284
field-effect devices
   JFET/SIT  80–83, 273–280
   MOSFET  83–86, 287–315
field plates  120–122, 169
figure of merit  200, 398
filament (current)  50, 215, 236, 385
float zone silicon  94, 96–98
forced commutation  79, 131, 133, 234
free-wheeling diode  137, 306–310, 312, 388
frequency limitation  311, 312

Gain
   current  179–184, 192–194, 200, 216–219, 237–239, 257, 287, 313, 320–327, 335, 336, 348
   Darlington  205–207
   turn-off  238, 241–243, 257, 259, 263, 282
   voltage  278
gallium (Ga)  2, 3, 107–109, 113
   diffusion coefficient in silicon  107
   solid solubility  106
gallium arsenide (GaAs)  74, 398–402
   properties  366, 399
gallium nitride (GaN) properties  399
gallium phosphide (GaP) properties  399
gamma radiation  24, 116
gate oxide  83, 100, 287, 288, 299, 315, 355, 392
   capacitance  290–3, 320
gate-assisted turn-off thyristor (GATT)  79, 254–257
gate turn-off thyristor (GTO)  79, 134, 236–244, 257–264
gate drive circuit  89, 91, 142, 243, 259, 263
Gaussian profile  105
GCT (gate-commutated turn-off GTO)  263, 264
generation (carrier)  16–19, 24, 27, 28, 39, 48, 50–52, 211, 222, 266
generation–recombination current  38–41, 123, 182, 184
gettering  115, 248
gold (Au)  3, 19, 105, 113–116, 155, 170, 234, 235
   properties  366

Au/Si alloy  113
graded junction  45
guard ring  74, 118–122, 160

Hard recovery  166
H-bridge circuit  138
heat sinks  344, 361–379, 393, 402
heavy doping effects  7–9, 65
heterojunction  24
high level carrier injection  24, 151
high voltage design  118–124, 212, 213
holding current  79, 215, 218, 233, 236, 259, 264, 266, 334, 335
holes  2

IBT (insulated bipolar transistor)  338, 339
IGBT (insulated-gate bipolar transistor)  132, 134–7, 141–143
IGCT (integrated gate-commutated turn-off GTO)  263
impurity energy levels  3, 19
indium (In)  2, 3
indium phosphide (InP) properties  399
inductance parasitic  137, 140, 164, 204, 253, 260–264, 304, 330
inductive load  135, 200–202, 204, 266, 267, 314, 386
   clamped  92, 141, 306, 308–310, 312, 325
interdigitated structures  77, 134, 206, 229, 248, 250, 284
interface charge  290
interface states  65
intrinsic carrier concentration  4–7
inversion layer  61, 62, 287–292, 303, 319
   charge  290–292, 300, 301
inverter circuits  132–138
inverter grade thyristor  134
ion implantation  109, 110
iridium (Ir)  3, 19, 24, 170
iron (Fe)  114


Junction barrier Schottky (JBS) structure  173, 174
junction field-effect transistor (JFET)
   long channel  274–276
   parasitic  297, 327
   short channel  276, 277
junction passivation  74, 111

Kelvin terminal  353, 354
Kirk effect  190–192
Kovar (properties)  366

Lapping  98, 114
latching  72, 329
latching current  218, 220, 224, 323, 333
latch up  334, 338
lead (Pb) (properties)  366
leakage current  38, 116, 179, 181, 185, 217, 218, 235, 382, 392
lifetime of carriers  17–26, 29, 36, 39, 59, 83, 148, 152–158, 168–171, 174, 178, 180, 214, 215, 226, 234, 238, 241, 243, 249, 251, 257, 259, 261, 281, 314, 322, 325, 331, 332, 335
   Auger  24, 25
   control  24, 114–118, 170, 234, 244, 248, 257, 261, 315, 329
   effective  29, 40, 167–169, 199, 200, 244, 249, 255
   high level  24
   profile  166, 169, 170
   temperature dependence  336
light-triggered thyristor  266–269
linearly graded junction  45
long-channel JFET  274–276
low level carrier injection  151

Magnesium (Mg)  64
Mathiesson's law  13
matrix converter  131, 132
merged power Schottky (MPS)  172, 174
mesa  124, 213, 236, 280
mesoplasma  50, 51, 194, 202, 329
metalisation  103, 112, 115, 116, 120, 232, 284, 353, 402
metal–semiconductor junction
   ohmic  65, 66
   rectifying (Schottky)  24, 62–66
mica (properties)  366
microplasma  49, 222
Miller capacitance  304, 307, 384
Miller effect  304, 305, 384
moat etch  124
mobility (of carriers)  10–16, 27
   differential  28, 149
   surface  288, 290, 297
molybdenum (Mo)  112, 113, 305, 368–370
   properties  366

MOSFET  83–86, 135–137, 139–143, 287–315
MOS structure  24, 289
motor control  145, 208, 333
motor drives  174, 175, 333

Natural commutation  79, 131, 209
negative bevel  122, 123, 160, 212, 213
neutron transmutation doping  97
nickel (Ni)  112
nitrogen  402
non-punch-through (NPT) devices  132, 257, 322, 331, 332
non-uniform doping  29–32
non-degenerate material  4
n-type material  3
numerical modelling  124, 125

Ohmic contact  65, 66
oil (cooling properties)  372
on-state (conduction) losses  83, 114, 136, 137, 241, 330, 332
optical triggering  266–269
overcurrent protection  390, 392
overvoltage protection  385–390
oxidation rate  100, 101

Parallel operation  379–382
parasitic components  297, 299, 318, 319, 326, 343
   bipolar junction transistor  211, 295, 312–314, 322
   capacitance  263, 304, 305, 327, 364
   diode  343
   inductance  137, 140, 164, 204, 253, 260–264, 304, 330, 343
   JFET  297, 327
   resistance  304
   thyristor  322, 327, 338, 343
passivation  74, 111, 124, 213, 392, 400, 401
permittivity (dielectric constant)  30, 31, 42–46, 60, 61, 65, 124, 157, 158, 190, 210, 211, 274, 275, 278, 290–292, 397, 399, 401
phase-control thyristor  131
phonon scattering  12
phosphorous (P)  1, 3, 11, 12, 105–109, 114, 294, 402
   diffusion coefficient in silicon  107
   solid solubility  106

photolithography 98, 101–103, 107
plasma spreading velocity 226, 227, 232, 249–251
plasma etching 104
platinum (Pt) 3, 19, 24, 64, 105, 115, 116, 155, 170, 329, 330
p–n junction 24, 33–57
 breakdown 45–49
 capacitance 45, 221–223, 230, 233, 278
 characteristics 33–41
Poisson's equation 30, 31, 42, 49, 53, 60, 62, 122
polarisation 123, 401
polishing 98, 114
polycrystalline silicon 94, 96, 99, 100, 103, 111, 122, 205, 291, 294, 304–306, 311, 355, 389
polyimide 213
positive bevel 123, 213
power factor correction (PFC) 127, 128, 142, 143
power integrated circuit (PIC) 349–352
proton bombardment 24, 116–118, 170
p-type material 3
pulse width modulation (PWM) 133–138
punch through 180, 184, 210, 269, 293, 331, 389
punch-through (PT) devices 257, 322, 331, 332, 389

Quasi Fermi level 17
quasi-resonant circuits 143–145
quasi-saturation 187, 189, 192, 200, 300

Radiative recombination 18
rate of rise
 of current limitation 226–230, 248–250, 259, 263, 264, 267, 284
 of voltage limitation 230–232, 248, 260, 266, 267, 283, 314, 328
reactive ion etching 104
recombination 16–26
 Auger 18, 24–26, 114, 153, 193, 248
 band-to-band 17, 24
 radiative 18
 Shockley–Read–Hall theory 19–24
 surface 26
 velocity 26
 via trapping levels 18–24
recombination centres (trapping levels) 18–24, 114–118, 169, 170

capture cross-section 20, 23, 24
 energy levels 19
rectifier circuits 127–130, 143
regulators 138–140
 boost 139, 142, 143
 buck 138–140
 buck-boost 139, 141, 142
reliability 204, 349, 369, 391–394, 401
resistance (differential) 155, 156, 184, 281, 321, 379
resistance (parasitic) 304
resistivity 12, 31
 silicon 12
resonant converters 143–145
RESURF 122
reverse blocking 78, 79, 116, 132, 175, 209–211, 238, 248–250, 257
reverse conducting thyristor (RCT) 79, 253, 254
reverse recovery charge 114, 115, 136, 139, 140, 142, 166–169, 248


Safe operating area (SOA) 72, 77, 90–93, 136, 143, 202–205, 312–315, 329–333, 388, 390
 forward bias (FBSOA) 390
 reverse bias (RBSOA) 332, 388
saturation drift velocity 15, 16, 300
saturation region
 BJT 76, 185
 JFET 81, 276
 MOSFET 293
scattering 10–13
 carrier–carrier 13
 Coulomb 13
 phonon 12
Schottky barrier 63, 64, 171
Schottky contact 24, 62–66
Schottky diode 74, 75, 140, 171–174, 398, 400–402
Schottky p–i–n structure (SPIN) 174
second breakdown 50, 194, 202–205, 260–263, 312–314, 329, 348, 381, 389, 390
series connection 382–385
series–parallel connection 385
short-base diode 152, 153
short-channel JFET 276, 277
silicon
 properties 366
 single crystal preparation 94–97

silicon carbide (SiC)  99, 398–402
  properties  366, 399
silicon controlled rectifier (SCR)  129–131, 133–135
silicon dioxide
  formation  100, 101, 111
  gate oxide  83, 100, 287, 288, 299, 315, 355, 392
  capacitance  290–3, 320
silicone grease (properties)  366
silicone rubber  123, 124, 213
SIPOS  121, 122, 124, 213
silver (Ag)  366, 379
SIT (static induction transistor)  80–83, 273–280
SITh (static induction thyristor)  81–83, 280–284
smart power  353, 355, 356, 390
snap-off  166
snubber circuits  72, 129, 136–137, 143, 204, 260–264, 330, 388
soft recovery  166
soft soldering  112
soft switching circuits  143–145
solid solubility  106
space charge density  30, 42, 190, 274
space charge neutrality  22, 35, 57, 195
space charge region (depletion layer)  33–36, 40–49, 55, 63, 80, 162–166, 169, 172, 178–180, 189–192, 198, 200, 234, 236, 239, 241, 257, 273, 283, 288, 293, 300
  capacitance  45, 221, 223, 230
  thickness  45, 57, 120, 124, 182, 186, 217, 295, 297, 304
spreading velocity  226, 227, 232, 249–251
static induction thyristor (SITh)  81–83, 280–284
static induction transistor (SIT)  80–83, 273–280
storage time  169, 170, 183, 198–200, 207, 239–242, 259–263, 314, 382–385
stored charge  39, 89, 127, 135, 136, 167, 175, 183, 192, 196, 198, 239, 325, 333, 348
  model  28, 29, 164, 166, 220, 325
surface bevelling  122–124, 212, 213
surface charge  60–62, 290
surface (junction) passivation  74, 111, 124, 392, 400, 401
surge current rating  127
surge voltage rating  128

switched mode power supplies (SMPS)  127, 128, 138
switching losses  136, 137, 141, 144, 145, 312, 333, 400

Tail current  244, 257, 262, 325, 329, 336
temperature distribution  360, 363, 368, 371
temperature limits  5–7, 49–51
thermal
  capacitance  361–363
  conductivity  366, 372, 399
  cycling  364, 393
  fatigue  366, 392, 393, 398
  impedance  360–364
  instability  50, 51, 202, 359
  resistance  360–364
thermal expansion coefficients  365, 366, 370, 371, 398, 399
thermal oxidation  100, 101, 114, 300
threshold voltage  155, 156, 172, 321
  IGBT  319, 323, 328
  MOSFET  288, 301, 303, 312
thyristor (SCR)  77–80, 209–271
  asymmetric (ASCR)  79, 250–253, 261
  blocking voltage  209–214, 330, 238, 247–253, 259–261, 269
  gate-assisted turn-off (GATT)  79, 254–257
  gate triggered  222–226
  gate turn-off (GTO)  79, 134, 236–244, 257–264
  light triggered  266–269
  on-state voltage  214–216
  parasitic  322, 327, 338, 343
  reverse conducting (RCT)  79, 253, 254
  turn-off  232–244
  turn-on  220–232
    losses  227, 228, 232
  two-transistor model  216–220
transconductance  83, 276, 300–302, 305, 320, 321, 324
transient thermal impedance  360–364
transistor (see bipolar junction transistor (BJT), field-effect devices, MOSFET, static induction transistor)
transit time  196–200, 223, 225, 284, 300, 304, 323, 398
trapping levels (recombination centres)  18–24, 114–118, 169, 170
  capture cross-section  20, 23, 24
  energy  19

trench structure  83, 85, 103, 104, 274, 293, 294, 333
triac  79, 209, 264–267, 348
tungsten (W)  112, 113
  properties  366
turn-off  162–171, 198–202, 232–244, 248–251, 253–264, 282, 283, 309–311, 324–328, 335–337
turn-on  160–162, 195–198, 220–232, 254, 283, 284, 306–309, 323, 324
two-transistor model for four-layer device  216–220

U-channel (*see also* trench structure)  83, 85, 103, 104, 274, 293, 294, 133
uninterruptable power supply (UPS)  133

valence band  1–3
velocity (surface recombination)  26
V-groove MOSFET (VVMOS)  83, 85, 293
voltage control  70, 89, 133, 134, 136, 141, 288, 317, 333, 338

Water (cooling properties)  372
wet etching  103
work function  62–64

Zener diode  314, 348, 349, 354, 355, 385
zero turn-off thyristor  256, 257
zero current switching  144, 145
zero voltage switching  144, 145
zone refining  97